Physics

Physics

Physics

Fifth Edition

A. F. Abbott, C. Phys., F.Inst.P.

Formerly Head of Physics Department,
Latymer Upper School, London

With a Foreword by
Sir John Cockcroft, O.M., F.R.S.

HEINEMANN
EDUCATIONAL

Heinemann Educational,
a division of Heinemann Educational Books Ltd,
Halley Court, Jordan Hill, Oxford OX2 8EJ
OXFORD   LONDON   EDINBURGH
MADRID   ATHENS   BOLOGNA   PARIS
MELBOURNE   SYDNEY   AUCKLAND
IBADAN   NAIROBI   HARARE   GABORONE
SINGAPORE   TOKYO   PORTSMOUTH (NH)

First published 1963
Reprinted seven times
Second edition 1969
Reprinted nine times
Third edition 1977
Reprinted nine times
Fourth edition 1984
Reprinted 1984
Reprinted 1985 (twice) 1986
First published as *Physics* 1987
Fifth edition 1989
Reprinted 1990, 1991

ISBN 0-435-67014-X

COVER PHOTOGRAPHS

Front Testing a delta-winged model of the Space Shuttle Orbiter as it goes transonic in a wind tunnel. Photographed using Colour Schlieren technique.

Printed in England by Clays Ltd, St Ives plc

To H. B. K.

Preface to the fifth edition

The primary aim of this book is to cover syllabuses for pupils aged 16+ including GCSE in schools and technical colleges. Also, the requirements of the West African Examinations Council, some of whose questions have been included in this edition, the Hong Kong School Certificate, the East African Examinations Council and the Singapore Ministry of Education have been taken into consideration.

Since the publication of the fourth edition many requests have been received for the addition of new material on semiconductor devices and their applications.

Accordingly, a new chapter has been added which incorporates part of the previous chapter 44 together with much new material provided by Mr B. Johnston. In connection with this, our thanks are due to Mr J. Underwood, Director of Studies at Hampton School, and Mr C. Clare of Dr Challoner's School for reading and constructively advising on the new material.

At the same time we have taken the opportunity of this new edition to amend the text where necessary and to bring it up to date in several places. Our previous policy of emphasis on practical work has been continued.

The whole text has been planned and revised, not only as a firm basis for the studies of future physicists and engineers, but also to present physics to the general reader as a human intellectual discipline with deep roots in the past and largely responsible for the development of our technological culture.

My co-author Mrs U. P. Abbott and I are greatly indebted to the many who have helped with invaluable advice and criticism in the preparation of previous editions, but especially to Dr J. W. Warren of Brunel University, Dr E. J. Wheeler of Surrey University, Mr K. J. B. Constable, M.Sc. of the John Leggott Sixth Form College, Scunthorpe, Mr J. Underwood, B.Sc., M.Ed. of Hampton School, Mr C. F. Tolman, M.A. formerly of Whitgift School, and Mr J. H. Avery, M.A. formerly of Stockport Grammar School.

Finally, we thank the late Sir John Cockcroft for his criticism of the text of the first edition, particularly the section dealing with atomic and nuclear physics.

February 1989

A.F.A.
U.P.A.

Foreword to the first edition
by Sir John Cockcroft, O.M.,F.R.S.

Mr Abbott's book on physics, taken to the Ordinary Level of the General Certificate of Education, should be welcomed by teachers of physics for its combination of theoretical and experimental work, and for its practical outlook, which should make the course interesting also for those students who will not take physics beyond the Ordinary Level. The well-illustrated references to the application of physics in industry should also appeal to the more practically minded student. The author has rightly included an introduction to atomic and nuclear physics.

Contents

Mechanics and General Physics

Electricity and Magnetism

Practical course

Experiments described in the text, which are usually reserved for demonstration, are not included in this list

Mechanics and General Physics

Internal Energy and Heat

(*Measurement of specific heat capacities and specific latent heats will be found under* Electricity)

Optics

Wave Motion

Electricity and Magnetism

Electronics, Atomic and Nuclear Physics

Experiments to study radioactivity begin on page 552

Acknowledgements

Acknowledgement is made to the courtesy of all who have kindly provided prints and given permission for reproduction of the photographs.

By Gracious permission of Her Majesty the Queen, 44.10
Alcan Aluminium Ltd, 37.4
Alcan Industries Ltd, 13.11, 13.12, 13.13
Anglo-American Corporation of South Africa Ltd, 37.2, 37.3
Associated Electrical Industries Ltd, 7.4(b), 41.1(a), 41.2
Associated Press Ltd, 4.11
Australian Information Service, 26.23
Avo Ltd, 39.6
Babcock & Wilcox Ltd, 44.9(a)
Barnaby's Picture Library, 2.1
Birlec-Efco (Melton) Ltd, 41.1(b)
British Aerospace 7.3
British Cast Iron Research Association, 19.3
British Railways, 15.3
Central Electricity Generating Board, 7.4(a), p. 339
Central Office of Information, Crown copyright reserved, 13.2, 22.13, 27.8
CERN (European Organization for Nuclear Research), p. 507, 47.12
Chloride Alcad Ltd, 34.8
Chloride Industrial Batteries Ltd, 34.7
Darwins Ltd, 30.5
Decca Radar Ltd, 26.22
Edwards High Vacuum Ltd, 13.4
Electrical Research Association Ltd, 7.2
The Electricity Council Appliance Testing Laboratories, 44.9(b)
Esso Petroleum Co. Ltd, 2.8
Ferodo Ltd, 2.5, 13.2, 27.9
Fibreglass Ltd, 17.1, 17.2, 27.6
Ford Motor Co. Ltd, 8.17(a)
French Embassy in Great Britain, 2.7
General Electric Co. Ltd, 36.9, 36.13
Griffin & George Ltd, 3.10, 25.4
Colin Grundy, 12.5, 30.1, 32.15
Philip Harris Ltd, 32.16
Hodgson, Northallerton, 15.1
Hot Air Group, 16.6
Italian State Tourist Office, 3.6
Brian Johnston, 45.1, 45.5, 45.6, 45.18(a), (c), 45.22, 45.24, 45.28
Keystone Press Agency, 26.21
Dr L. G. Lajtha, 47.15(d)
J. J. Lloyd Instruments Ltd, 40.2
London Transport Executive, 6.15
Joseph Lucas Ltd, 38.8, 43.9
Eric de Maré and Gordon Fraser Galleries, p. 285

Merryweather & Sons Ltd, 11.4
Microsurgical Administrative Services Ltd, (diamond by Drukker), 24.25(b)
John Moss, 9.3
NASA, 4.11
National Engineering Laboratory, Crown copyright reserved, 2.6
National Physical Laboratory, Crown copyright reserved, 1.1, 1.7, 13.3, 13.14,
  19.8, 27.7, 35.1
National Research Development Corporation, 43.23
Negretti and Zambra Ltd, 10.15
Newport Instruments, 36.8
New Scientist, 9.3
Northern Aluminium Co. Ltd, and South Wales Brattice Cloth and India Rubber
  Co. Ltd, 11.1
Oertling & Co. Ltd, 1.8
Ohaus Scale Corporation, 12.4
Bruce Peebles & Co. Ltd, 43.20
Picture Post, 12.1
Paul Popper Ltd, 4.11, 7.1, p. 157, 29.5
Radio Times Hulton Picture Library, 15.4
Royal Greenwich Observatory, p. 221
Royal Institution of Great Britain, 17.4, 43.7
Tony Sage, 16.6
Science Museum, London, Crown Copyright reserved, 47.5
Science Photo Library, Dr G. Settles (front cover); Fred Espenak (back cover)
Science Research Council, 47.8
Sierex Ltd, 47.14
Standard Telecommunication Laboratories Ltd, 24.25(a)
John Topham, p. 1
Torsion Balance Co. (GB) Ltd, 6.14
United Kingdom Atomic Energy Authority, 7.6, 47.10, 47.7, 47.9, 47.11, 47.14
  (below), 47.15 (a), (b), (c), 47.16
United Press, 21.7
United States Information Service, 4.6, 4.11
Vactric Control Equipment Ltd, 30.12
Vickers Ltd, 8.17(b)
Weir Electrical Instruments Ltd, 39.5
Welcome Trustees, 10.11
Ernest J. Wheeler, 31.5, 31.9, 31.18

Acknowledgement is also made to the following examining bodies for permission to reprint questions from their Ordinary Level G.C.E. physics papers:

Associated Examining Board	(A.E.B.)
University of Cambridge Local Examinations Syndicate	(C.)
Joint Matriculation Board	(J.M.B.)
The Senate of the University of London	(L.)
Oxford and Cambridge Schools Examination Board	(O.C.)
Oxford Delegacy of Local Examinations	(O.)
Southern Universities Joint Board	(S.)
Welsh Joint Education Committee	(W.)
West African Examinations Council	(W.A.E.C.)

Thames Flood Barrier. Piers seven, eight and nine of the anti-flood barrier nearing completion in 1982. Two of the ten gates are seen raised in the flood position. Normally, these gates lie on the bed of the river and are raised only when a combination of high tide and wind are likely to cause floods which threaten the city of London

Mechanics and general physics

Mechanics and general physics

1. Measurements

What is physics?

Until the end of the eighteenth century, the study of material things was treated as a single aspect of human thought and called *natural philosophy*. But, as knowledge increased, it was found necessary to divide the study of nature into two main branches, the physical sciences and the biological sciences. The biological sciences deal with living things, while the physical sciences are concerned with the properties and behaviour of non-living matter.

The two main physical sciences are physics and chemistry. It is difficult to make a clear-cut distinction between the two, but, broadly speaking, chemistry deals with the action of one kind of substance on another while physics is concerned mainly with matter in relation to *energy*. It is the purpose of this book to explain what we understand by energy, to describe the various ways in which it can transfer from one form to another and to show how its laws have been investigated. For elementary purposes, the study of physics may be grouped under such headings as mechanics, general physics, optics, wave motion, magnetism, and electricity, but this list is by no means exhaustive. At a higher level we have *particle* physics which is concerned with the ultimate particles of which matter is composed; *nuclear* physics which deals with atomic nuclei, and *plasma* physics which relates to research on the production of energy by processes similar to those believed to be responsible for the sun's energy.

Since it is the fundamental science, physics underlies physical chemistry and biophysics, which have led to big advances in medicine and surgery. Solid-state physics has led to the development of miniaturization in circuits for calculators, computers, sound and vision equipment as well as a host of other purposes.

The nature of physical knowledge

In physics, certain properties of matter are measured and the results examined to see if there is any mathematical relationship between them. It is important to grasp the true meaning of the equations we find in a physics book. They do not tell us what things are in themselves, but are simply a convenient way of expressing the laws governing their behaviour. This is the main purpose of science, to seek out the laws of the universe and, if possible, to express them in precise mathematical form. Technologists use this information for such purposes as designing electric dynamos and motors, radio, television and radar installations, artificial satellites and space-craft, nuclear power generators, computers and so on, all of which have helped to make our material way of life so different from that of our ancestors.

Measurements in physics

All measurements in physics, even of such things as electric current, are related to the three chosen fundamental quantities of *length*, *mass* and *time*. Until about the year 1800, workers in various countries used different systems of units. Thus, while the English used inches, a continental scientist would measure lengths in centimetres. Fortunately, this unsatisfactory situation has now been changed by the efforts of various international committees of scientists who have met for discussion regularly over many years.

In 1960, the General Conference of Weights and Measures recommended that everyone should use a metric system of measurement called the **International System of Units** (abbreviated **SI** in all languages). The SI units are derived from the earlier MKS system, so called because its first three basic units are the **metre** (m), the **kilogram** (kg), and the **second** (s). These will be explained shortly.

Work of national physical laboratories

Throughout history, people have always realized the need for standard weights and measures, and from time to time these standards have been revised, improved, and established by law.

In most big countries of the world there are laboratories, supported largely by government funds, in which scientists are at work on problems of industrial and national importance. In England we have the National Physical Laboratory at Teddington. These institutions are also responsible for maintaining and improving the standards of all the important physical quantities. Scientists from the various countries compare their results and regularly meet for discussion, and often they are able to announce new definitions of the basic units.

Measurement of length

Fig. 1.1 gives a glimpse of some landmarks in the history of length measurement. The SI unit of length is the metre, originally defined as the distance, at 0 C, between two lines on a platinum-iridium bar kept at the International Office of Weights and Measures at Sèvres near Paris. Copies of this standard were sent to other countries.

Now the trouble with metal standards of this kind is that they are liable to undergo minute changes in length as the years go by. For instance, tests have shown that the imperial standard yard shown in Fig. 1.1 has shrunk by a few parts in a million since it was made in 1845. Small though this error is, the exacting requirements of modern science demand something better. The standard metre is, of course, open to the same objection. **In 1983 the General Conference of Weights and Measures redefined the metre as the length of the path travelled by light in a vacuum during a time interval of 1/299 792 458 of a second.**

For most practical purposes we still have to use metal standards which are checked by an interference comparator (see page 5) and this uses the *wavelength* of light. We cannot go into details of this nor is it necessary to memorise the above definition. However some simple experiments for measuring the wavelength of light will be described in chapter 26.

Various other metric units of length are related to the metre by either multiples or submultiples of 10. Thus,

$$1 \text{ kilometre (km)} = 1000 \text{ metres (m)}$$
$$1 \text{ metre (m)} = 100 \text{ centimetres (cm)}$$
$$1 \text{ centimetre (cm)} = 10 \text{ millimetres (mm)}$$

Very small lengths are measured in micrometres (μm) and nanometres (nm).

$$1 \text{ metre} = 1\,000\,000 \text{ (or } 10^6)\mu\text{m}$$
$$= 1\,000\,000\,000 \text{ (or } 10^9) \text{ nm}$$

For day-to-day work in elementary physics laboratories we use metre and half-metre rules made of boxwood. They are graduated in centimetres and millimetres. Care should be taken to avoid damage to the ends of these rules, as they do not have a short ungraduated portion at the ends to take the wear. However, since a small

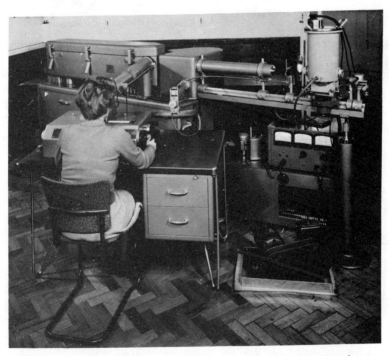

Henry VII yard

Elizabeth I yard

Metre des Archives

Imperial standard yard

International prototype metre

Modern end standard

Fig. 1.1. *Stages in the history of linear measurement.* A 1 metre interference comparator made by Carl Zeiss to a design based on a prototype developed at the Physikalisch-Technische Bundesanstalt (the NPL of Western Germany). It is used at the National Physical Laboratory, Teddington, for highly precise calibrations of all sizes of end standards up to 1 metre length. End standards calibrated in this manner are used as the practical standards of length for reference purposes in precision engineering

amount of wear is almost inevitable, it is best, whenever possible, to measure from the 10 cm graduation and subtract 10 cm from the reading at the other end.

Owing to the thickness of the wood, the eye must always be placed vertically above the mark being read, in order to avoid errors due to *parallax* (see page 228). Fig. 1.2 shows a millimetre scale giving a reading of 28 cm $3\frac{1}{2}$ mm. Since, in science, we invariably use decimals rather than vulgar fractions, we write this as 28.35 cm. The last digit has to be estimated.

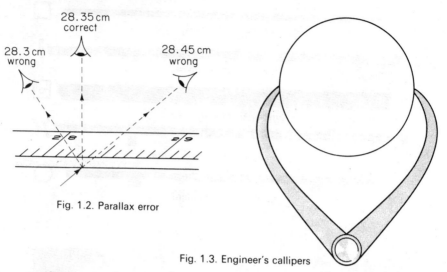

Fig. 1.2. Parallax error

Fig. 1.3. Engineer's callipers

Use of callipers

Fig. 1.3 shows a pair of engineer's *callipers* used for measuring distances on solid objects where an ordinary rule cannot be applied directly. They consist of a pair of hinged steel jaws which are closed until they touch the object in the desired position. The distance between the jaws is afterwards measured on an ordinary scale.

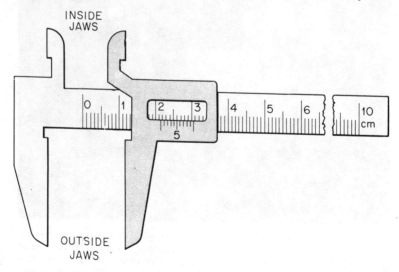

Fig. 1.4. Slide callipers

For some purposes, slide or *vernier callipers* are preferred (Fig. 1.4). These consist of a steel scale with a fixed jaw at one end. The object to be measured is placed between this and a sliding jaw. The callipers shown in the illustration also have inside jaws so that they can be used for such measurements as the internal diameters of tubes.

The short scale of 10 divisions seen on the sliding jaw is called a *vernier*. This is named after its inventor, Pierre Vernier, a French technician who lived in the seventeenth century. The vernier enables us to obtain accurately the second decimal place in centimetre measurements without having to estimate fractions of a division by eye.

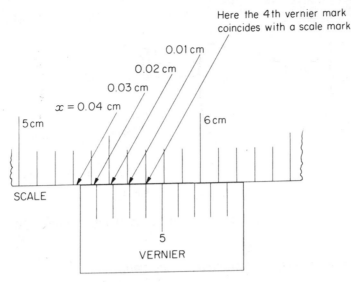

Fig. 1.5. How to read a vernier

How to read a vernier

For use with a scale of millimetres, the vernier is a short scale 9 mm long divided into 10 equal parts, so that the difference in length between a vernier division and a scale division is 0.1 mm or 0.01 cm. The vernier slides along the scale until its zero mark just touches the end of the object being measured (Fig. 1.5).

The diagram shows a scale and vernier giving a reading of 5.34 cm. The fraction of a scale division shown as *x* is given by the second decimal place. Looking along the vernier, we notice that the fourth vernier mark coincides with a scale mark. Counting back from this mark towards the left we see that the differences between successive vernier marks and scale marks increase by 0.1 mm or 0.01 cm each time, finally giving the distance *x* as being equal to 0.04 cm. It follows that *the second decimal place in the measurement made is given by the number of a vernier mark which coincides with a scale mark*.

Besides callipers, many other instruments are fitted with verniers (see page 113).

The micrometer screw gauge

For measuring the diameter of a piece of wire and similar small distances, a micrometer screw gauge is used. The first person to think of using a screw for this purpose was an astronomer named William Gascoigne, who was killed at the battle of Marston Moor in 1644.

The chief features of the instrument are shown in Fig. 1.6. The most important part is a screwed spindle which is fitted with a graduated thimble. The screwed portion is totally enclosed to protect it from damage. The pitch of the screw is 0.5 mm, so that the spindle moves through 0.5 mm or 0.05 cm for each complete turn.

When taking a reading, the thimble is turned until the object is gripped very gently between the anvil and spindle. Some gauges are fitted with a spring ratchet which prevents the user from exerting undue pressure. The sleeve shown in the diagram has a scale of half millimetres, each of which represents one complete turn of the screw. Fractions of a turn are indicated on the thimble, which has a scale of 50 equal divisions. Each division on the thimble, therefore, represents a screw travel of one-fiftieth of half a millimetre or 0.001 cm. Expressing the result in centimetres and remembering that 0.5 mm = 0.05 cm, it follows that the sleeve reading gives the

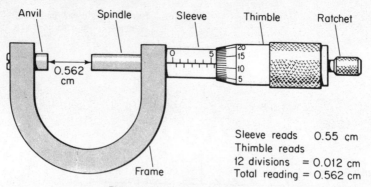

Sleeve reads   0.55 cm
Thimble reads
12 divisions   = 0.012 cm
Total reading = 0.562 cm

Fig. 1.6. Micrometer screw gauge

units and the first two decimal places, while the thimble reading gives the third decimal place.

Precautions when using a micrometer screw gauge

Before use the faces of anvil and spindle should be wiped clean to remove any dirt particles which would cause false readings. Also the instrument may have a zero error. Hence the zero reading must always be checked and recorded and, if necessary, a + or − correction applied to the final answer.

Fig. 1.7. British National copy of the International Kilogram

Mass and weight

The mass of a body is the quantity of matter it contains, and the basic SI unit of mass is the kilogram. (Fig. 1.7.) **The standard kilogram is the mass of a certain cylindrical piece of platinum-iridium alloy kept at Sèvres.** Its various multiples and submultiples are given below:

$$1 \text{ tonne (t)} = 1000 \text{ kg}$$
$$1 \text{ kg} = 1000 \text{ grams (g)}$$
$$1 \text{ g} = 1000 \text{ milligrams (mg)}$$
$$1 \text{ g} = 1\,000\,000 \text{ micrograms (}\mu\text{g)}$$

The kilogram was originally intended to be equal to the mass of 1000 cubic centimetres (cm³) of pure water at the temperature of its maximum density, 4 °C. But a slight error was made at the time and the kilogram is actually equal to the mass of 1000.028 cm³ of water.

The weight of a body is the force it exerts on anything which freely supports it and, normally, it exerts this force owing to the fact that it is itself being attracted towards the earth by the *force of gravity*. This will be discussed in more detail in later chapters.

In everyday conversation the distinction between mass and weight is relatively unimportant: a butcher who had not studied physics would doubtless be surprised if a housewife who had done so, asked him what was the mass of her week-end joint. In science, however, we must be careful to distinguish between them. The mass and weight of a body are different and are measured in different units. As we have already said the unit of mass is the kilogram, but the unit of weight which is a force is the **newton** (N) (see page 37). An important distinction between mass and weight is that the mass of a body does not depend on where the body happens to be, whereas the weight of a body can vary from place to place.

Changes which can occur in the weight of a body are discussed in chapter 2.

Measurement of mass

Fig. 1.8 shows a top-pan balance used for the measurement of mass in the laboratory. Most balances of this type are based on the principle of moments which is dealt with in chapter 6.

Fig. 1.9 shows the general features of a laboratory beam balance. The body whose mass is to be determined is placed in the left-hand pan and standard masses (usually called *weights*) are added to the right-hand pan until the pointer attached to the beam swings equal numbers of divisions on either side of the central mark. The theory of this form of balance, which also depends on the principle of moments, is given on page 68. Balances of this type, being fragile and tedious to use have now been largely replaced by the top-pan balances just mentioned, so we shall not concern ourselves with any further detailed description of their construction and use.

After taking a reading from a balance, the result should be written down immediately and then checked. Never attempt to find the mass of a hot object and always wipe the outside of a bottle or vessel containing liquid before placing it on the balance pan.

Fig. 1.8. Top-pan balance

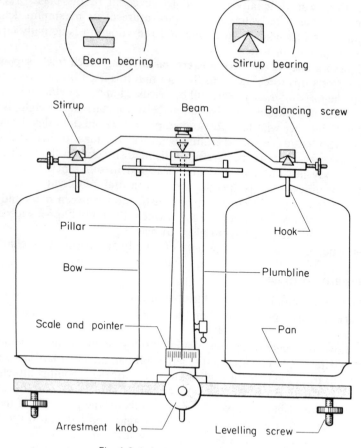

Fig. 1.9. Laboratory beam balance

Volume of liquids

The volume of a liquid is measured in *litres*. The litre has had an unfortunate history.

The litre is 1000 cubic centimetres (cm³) and, when the standard platinum–iridium kilogram was constructed in 1889, it was *intended to be* the mass of 1 litre of pure water at the temperature of its maximum density, 4 °C.

The litre was *then* officially defined as the volume of 1 kg of pure water at 4 °C. This is where the trouble began. In 1907, careful experiments showed that a slight error had been made in constructing the standard kilogram. It was found to be the mass of 1000.028 cm³ of water. Consequently, *as defined above,*

$$1 \text{ litre} = 1000.028 \text{ cm}^3$$

At the time it was decided to leave the matter as it stood and to divide the litre in 1000 equal parts called millilitres (ml) so that,

$$1 \text{ cm}^3 = 0.999\,972 \text{ ml}$$

Scientists thereupon began to use burettes and other liquid-measuring vessels calibrated in ml instead of cm³.

At the present time we are almost back where we started. In 1964, the General Conference of Weights and Measures **redefined the litre as equal to 1000 cm³**. This means that the *litre is, by its new definition, related directly to the metre and not the kilogram.*

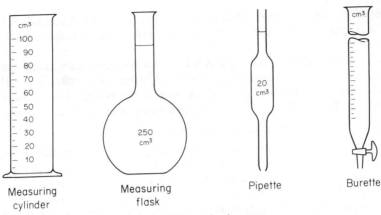

Fig. 1.10. Volumetric glassware

Fig. 1.10 shows a selection of graduated vessels in common use. The **measuring cylinder** is for measuring or pouring out various volumes of liquid; the **measuring flask** and the **pipette** for getting fixed pre-determined volumes. The **burette** delivers any required volume up to its total capacity, usually 50 cm³, and is long and thin to increase its sensitivity. Burette divisions generally represent 0.1 cm³, but measuring cylinders may be graduated at 1, 5 or 10 cm³ intervals according to size.

Readings on all these instruments are always taken at the level of the bottom of the *meniscus* or curved surface of the liquid. Mercury is an exception, as its meniscus curves downwards. Care should be taken to place the eye correctly so as to avoid

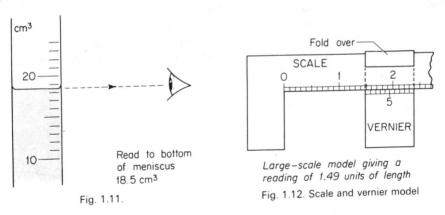

Read to bottom
of meniscus
18.5 cm³

Fig. 1.11.

*Large-scale model giving a
reading of 1.49 units of length*

Fig. 1.12. Scale and vernier model

parallax errors (Fig. 1.11). When taking readings, the pipette and burette must be upright and the cylinder and flask must stand on a horizontal bench, otherwise errors may arise from tilting.

Measurement of time

Everyone knows that a day is divided into 24 hours, each containing 60 minutes of 60 seconds each. At first, water clocks were used and later on mechanical clocks were made to measure time in these units. Until recent years the accurate measurement of time has not been easy. For reasons which we need not discuss, the length of the day varies throughout the year, so that an average value has to be taken.

The Royal Greenwich Observatory uses a number of very accurate quartz-electric clocks which are checked daily against astronomical observations. These, which are controlled by the vibrations of quartz crystals, are themselves checked by the **caesium atomic clock** at the National Physical Laboratory.

The atomic clock is too complex a device to be described in detail, but briefly it is a radio transmitter giving out short waves (about 3 cm long), the frequency of which is controlled by energy changes in gaseous caesium atoms. The great advantage here is that the *frequency* (i.e., the number per second) of the changes is constant and not subject to error.

By using a caesium clock time intervals can now be measured with an error of not more than one second in 3000 years! This is so much better than results obtained by astronomical observations alone, so **in 1967, the second was redefined as the time interval occupied by 9 192 631 770 cycles of a specified energy change in the caesium atom.**

Unit symbols and physical quantity symbols

In this chapter we have mentioned some unit symbols; e.g., m for metre, kg for kilogram, s for second and so on. We shall meet others later on. It is important to note that these symbols are always printed in roman (upright) type.

In contrast, the symbols used to represent physical quantities are printed in *italic* (sloping) type. Thus, the symbol m represents a particular mass in kilograms, while t represents a time in seconds. Now these italic symbols represent not only the numerical value of the quantity but *also* the unit in which it is measured. In order to avoid confusion in certain cases it is therefore advisable to specify the unit being used. Hence, for example, when referring to a radius r it is wise to add (in m) or (in nm) as the case may be.

To sum up, then, we must always remember to write the unit symbol after a numerical value but not after a symbol representing a physical quantity. For example, we might express a particular mass as, say, 5 kg or simply as m, but never as m kg.

To construct a model scale and vernier

Using stiff paper or thin card, construct a model scale and vernier in which the scale divisions are 1 cm long, and the ten vernier divisions are each 9 mm long. This will enable you to measure lengths to an accuracy of 1 mm (Fig. 1.12).

If all the dimensions of your model were reduced to one-tenth of their size you would then have a scale and vernier like the one on this page which measures to within 0.1 mm. The large scale model will, however, enable you to see and understand the principle more clearly.

QUESTIONS: 1

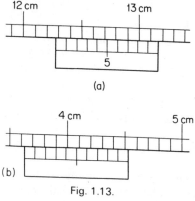

Fig. 1.13.

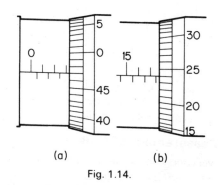

Fig. 1.14.

1. Write down the vernier readings in Fig. 1.13.

2. What are the micrometer readings shown in Fig. 1.14, in which the horizontal scale is in mm above and half mm below?

3. Name the instruments you would use for the following measurements:

    (a) diameter of a piece of wire;

    (b) internal and external diameter of a tube of about 5 cm bore;

    (c) average thickness of a page of this book.

4. State briefly the precautions you would take in finding the external diameter of a glass tube with a micrometer screw gauge. If such a gauge registers 3.218 mm and the *zero reading* of the gauge is −0.013 mm, what is the true value of the distance being measured?

                              (C.)

2. Force of gravity, weight and friction

One thing to which a student must become accustomed is that many words of everyday conversation have a special meaning when used in a scientific sense. Take the words "force" and "power", for example. Most dictionaries describe these words as meaning the same thing. In physics this is not so, as will be explained in due course.

What is force?

The word "force" generally denotes a push or a pull. Now it is not possible to describe a force as we can describe some material object such as an apple. We can only say what force can do. When a body is acted upon by a resultant force it will begin to move. If the body is already moving a force may alter its speed or alter its direction of motion or else bring it to rest. We therefore define force as follows:

Force is that which changes a body's state of rest or of uniform motion in a straight line.

The relation between force and motion will be discussed more fully in a later chapter.

Another way of expressing this would be to say that a force causes a body to accelerate (see page 23). We define our unit of force in terms of the acceleration produced when the force acts on a mass of one kilogram. This unit of force is called the newton (N) and is defined as follows. **A newton is the force required to give a mass of one kilogram an acceleration of one metre per second^2.** The newton received its name in honour of Sir Isaac Newton who laid the foundations of the study of the relation between force and motion in the seventeenth century. We shall see later (page 37) how this definition is related to one of Newton's laws of motion.

It is useful to think of a newton as being roughly equal to the weight of a 100 g mass.

Gravitational force

The force of which we are constantly aware in our daily lives is that which pulls us towards the earth. This is called *gravitational force*. Sir Isaac Newton came to the conclusion that gravitational force exists between all bodies. Thus, two stones are not only attracted towards the earth but also attract each other. Normally, we do not notice this force owing to its smallness, although it can be measured with sensitive instruments. Nevertheless, two 50 000 t ships lying side by side attract each other with a force of about 180 N.

Newton's law of universal gravitation states that any two particles of matter attract one another with a force which is proportional to the product of their masses and inversely proportional to the square of their distance apart. Strictly, this law applies only when the distance is large compared with the dimensions of the particles.

Newton realized that gravitational attraction applied not only to bodies on the earth but was also responsible for holding the moon in its orbit about the earth and also the earth and its fellow planets in their orbits round the sun.

Centripetal force

It is important to grasp the idea that, to keep a body moving in a circle there must be a force on it directed towards the centre. This is called the *centripetal force*. Before Newton's time it was believed that invisible spokes radiated out from the sun and pushed the planets round. Newton's insight into the problem convinced him that a push such as this was not necessary. The planets, carrying their atmospheres with them, go on moving in their orbits because the great vacuum of space offers no opposing force to their motion. Centripetal force is, however, required to produce the continuous *change of direction* which occurs in the orbit and this is provided by gravitational attraction.

We can try a simple experiment to demonstrate centripetal force by securely tying a suitable mass on the end of a string and swinging it round. The pull in the string which is providing the centripetal force can easily be felt and we notice that it varies according to mass, speed, and path radius.

In a laboratory experiment, of course, the circular motion of a mass on a pivoted arm will, if left to itself, rapidly come to rest owing to air resistance and so on. No such resistance is offered to the planets as we have already said; so they continue to move.

Weights of standard masses

In science we frequently use the force of gravity acting on a standard mass to provide a known force. In order to show the relationship between the mass of a body and the gravitational pull on it we have to take another look at the definition of the newton. As we have seen, this is defined as the force which gives a mass of one kilogram an acceleration of one metre per second^2. If a mass of one kilogram is allowed to fall freely in a vacuum it acquires an acceleration which varies slightly from place to place on the earth, but which has an average value of about 9.8 m/s^2. To produce this acceleration the force acting on the mass of one kilogram must therefore be 9.8 N, since a force of one newton produces unit acceleration. Thus we may say that, on the average on the earth, the force of gravity on a mass of one kilogram is 9.8 N. Now, if the kilogram is resting on the earth's surface or is attached to a string it will press down on the surface or pull down on the string with this same force, and this is what we call its weight. So the weight of one kilogram is 9.8 N.

For many purposes it is sufficiently accurate to take the weight of one kilogram as 10 N. On the moon where the acceleration of a freely falling body is only 1.6 m/s^2 the weight of one kilogram would only be 1.6 N (see Fig. 2.1).

Why the weight of a body varies

The force of gravity on a body varies slightly from place to place on the earth for two reasons. First, the shape of the earth, and secondly, its rotation. The earth is not a perfect sphere, but bulges at the equator, so that if a body is taken from a pole to the equator its distance from the centre of the earth will increase. Consequently, in accordance with Newton's law of gravitation, the gravitational pull on it will get less.

Relation between total gravitational force and weight

If we stand on a spring weighing machine, the force or weight we exert on it comes from the earth's gravitational attraction on us. But the weighing machine does not measure the *total* gravitational force.

Fig. 2.1. Astronaut John Young leaps as he salutes the U.S. Flag on the moon during the Apollo-16 mission. The apparent ease with which he jumps conveys a clear sense of the moon's low force of gravity

Owing to the rotation of the earth on its axis we happen to be moving in a circle dependent on our geographical latitude. Consequently, part of the gravitational attraction has to provide centripetal force required to keep us moving in that circle. The remaining part of the gravitational force simply pulls us down on to the earth's surface. This part of the gravitational attraction we call our weight, and this is what the spring weighing machine measures.

Only at the poles, where there is no motion in a circle, would it be true to say that weight is equal to the total gravitational force. We must remember, of course, that the two factors we have mentioned, namely, the bulge effect and the centripetal force are both very small. The reduction in weight of a body as between pole and equator does not amount to more than about 0.5 per cent of the total gravitational force.

To avoid confusion we shall now summarize the terms which will be used in this book in any further discussion relating to bodies at rest on the earth's surface.
(1) By the term **force of gravity** we mean that part of the total gravitational force which *acts on a body* and so enables it, in turn, to exert an equal force on its support. This force *on the support* is called the body's **weight.**
(2) The **centripetal force** is that part of the total gravitational force which is required to constrain the body to move in its circle of latitude.
(3) The sum of the force of gravity and the centripetal force is equal to the **total gravitational force** as given by Newton's law of gravitation.

Action and reaction forces

Sir Isaac Newton pointed out that, whenever a force acts on a body, there must be an equal and opposite force or reaction acting on some other body. This is called the third law of motion and may be expressed in the simple form: **"To every action there is an equal and opposite reaction."**

To take an example, a book presses on a table with a force equal to its weight. The table exerts an equal upward reaction force on the book. We shall meet similar examples in the chapters which follow.

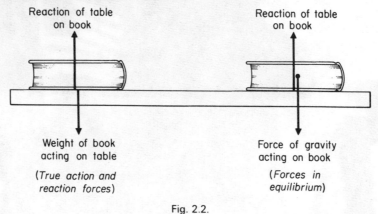

| Reaction of table on book | Reaction of table on book |

Weight of book acting on table

Force of gravity acting on book

(*True action and reaction forces*)

(*Forces in equilibrium*)

Fig. 2.2.

Before leaving the subject, we must warn the reader not to confuse action and reaction forces with the forces which are keeping the book in equilibrium, or at rest, with respect to the table. The book is at rest under the force of gravity on it which is balanced by the upward reaction on it from the table. Both forces are acting on a single body, i.e., the book and hence cannot be called action and reaction forces (Fig. 2.2). For clarity, and to show their points of application, the forces in Fig. 2.2 have been displaced parallel to one another. They do, of course, act in the same straight line.

Weightlessness

We shall now let our imagination run on a "thought experiment". Suppose we were to stand on a spring weighing machine at the equator and that, by some means, the earth's rotation could be speeded up, carrying its atmosphere with it.

With increasing speed, more and more of the earth's gravitational force would be used in providing the extra centripetal force required in the circumstances. Our weight, which is equal to the difference between the total gravitational force and the centripetal force, would therefore be less. Consequently the weighing machine would indicate a smaller weight.

If the earth's rotational speed continued to increase, then at a certain critical speed, the necessary centripetal force would be just equal to the total gravitational force. No resultant force would be left over to provide weight and so the weighing machine would read zero. In other words *we have become weightless although the full gravitational force still continues to act.*

Weightlessness in space vehicles

It is a well-known fact that astronauts experience weightlessness when their spacecraft are in orbit about the earth. In such circumstances, the earth's gravitational pull is just sufficient to provide the centripetal force required for their particular speed and orbital radius. Mechanically speaking they are in a similar position to the person we have just described on a fast rotating earth.

When an astronaut goes outside his cabin, the situation as far as the earth's pull goes is still the same. There is, of course, gravitational attraction between the astronaut and his cabin in accordance with Newton's law but, owing to the comparatively very small masses of both, the attraction between them is exceedingly small. It is far too small, in fact, to bring him back if he jumped off and so he has to use a life line.

Artificial weight in a spacestation

Weightlessness in a space vehicle is highly inconvenient to an astronaut in many ways. For example, he cannot pour liquid into a cup, neither can he drink from it: controlled movement is possible only by the use of handrails and so on.

It has been suggested that the spacestations of the future for use as manned observatories or as staging posts for space exploration might be built in the form of large wheels with hollow rims as in Fig. 2.3. These would be set in rotation so that the outer rim, which acts as the floor, would have to apply a radial centripetal force to the occupants or any objects inside to keep them moving in a circle.

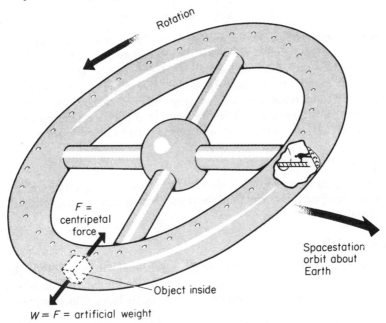

Fig. 2.3. Artificial weight in a spacestation

The equal and opposite reaction to this centripetal force which any person or object exerts on the floor would act as an artificial weight, thus allowing eating, drinking, and working in comparative comfort. The value of this weight would be made equal to or less than the normal earth weight simply by adjustment to the speed of rotation.

As an interesting sidelight, the reader will realize that if a person walked round the station in the direction of its rotation his weight would increase, while walking in the opposite direction would cause it to decrease.

Friction

Friction is the name given to the force which opposes the relative sliding motion of two surfaces in contact with one another. It plays a notable part in our daily lives. For example, we are reminded of the importance of friction every time we slip up on an icy pavement or a polished floor. Walking would be impossible if there were no friction between the ground and the soles of our shoes. If the pavements were perfectly frictionless, then one way we could get about would be with the aid of pegs in the soles of our shoes and holes in the pavements to correspond. Otherwise, some simple form of rocket or jet propulsion might be used (Chapter 4) but this is left to the reader's own imagination. Quite often our lives depend on the force of friction in the brakes of an automobile or a railway train.

There are one or two technical terms used in the study of friction which we shall illustrate with a few simple experiments.

Static friction

A rectangular block of wood placed on a flat surface has a string and spring balance attached to it so that a horizontal force can be applied (Fig. 2.4).

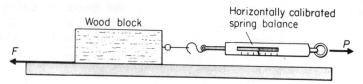

Fig. 2.4. Static and sliding friction

If a gradually increasing force is applied to the block it will, at first, continue to remain at rest since an equally increasing but oppositely directed force of friction F, comes into action at the under surface of the block. At any particular moment, we say that the pull P and the opposing frictional force F are in equilibrium.

If we continue to increase the pull P, a stage will be reached when the block just begins to slip. At this point, the friction brought into play has reached its maximum value for the two surfaces concerned, and this is called the *static friction*.

Kinetic or sliding friction

If the block is pulled along so that it slips at a steady speed we notice that the spring balance gives a reading rather less than the static friction. Kinetic friction is, therefore, less than static.

Coefficient of friction

If the simple experiments we have described are repeated with various weights on the block it is found that both the static and the kinetic friction are increased roughly in simple proportion to the force, perpendicular to the surfaces, which is pressing them together.

The ratios of the static and kinetic friction to the force pressing the surfaces together are called the coefficients of static and kinetic friction respectively. Coefficient of friction is denoted by the Greek letter μ (mu). Thus,

$$\mu = \frac{F}{R} \quad \begin{array}{l} \text{where } F = \text{force of friction} \\ \text{and } \quad R = \text{force pressing surfaces together} \end{array}$$

In the seventeenth century these coefficients were thought to be constants, but we now know that this is far from being the case especially under conditions of fast slipping and high contact pressure.

The nature of friction

We shall now discuss briefly the nature of frictional force. Primarily it depends on the composition of the surfaces, and we are here concerned with the friction between certain smooth surfaces rather than the obvious interlocking of projections which occurs with rough surfaces.

Much research as been done on the subject of friction from which it appears that attractive force between the surface molecules plays a major part. The forces between the atoms and molecules of matter will be discussed in more general terms in chapter 13.

Generally speaking, traces of grease, dust, or other contamination reduce friction considerably. We can illustrate this by a simple experiment in the case of glass. Ordinary glass tubing, even if it looks clean, will usually carry an invisible film of dirt. Also, air will have soaked into the surface of the glass. The absorption of gases by solids in this way is called **occlusion** or **adsorption**. This surface contamination acts as a lubricant so, if one piece of tubing is drawn across another it slips quite easily.

If a piece of glass tubing is now heated and drawn out in a bunsen flame the occluded air and other surface contamination are driven off and a fresh clean surface results. By drawing one such piece of tubing across another, the greatly increased friction can readily be felt, even when the top piece rests on the lower merely by its own weight. Indeed, the mutual attraction of the surface molecules in contact can be as great as that between those in the body of the glass itself. They may be torn away from their neighbours so that the two pieces of glass scratch one another.

A similar effect occurs in the case of clean metals, for example, when a shaft is rotating in a tight bearing, High spots on the metal may bring the surface atoms so closely together that they are held together by their mutual attractions with forces equal to those between atoms in the solid metal. The force required to break these *microwelds*, as they are called, constitutes the frictional force.

Friction and brakes

The brake linings of automobiles and industrial machinery are commonly made from woven or moulded material incorporating the mineral *asbestos*. This has good frictional properties and much research has been carried out in the manufacturer's research laboratories to produce a strong, hard-wearing material with freedom from *fade*. Fade is the name given to the loss of frictional force which occurs when the brakes have been in severe continuous operation. It results from slight chemical changes in the surface caused by the rise in temperature produced. See also Fig. 2.5.

Fig. 2.5. A heavy industrial brake lining under test at the Ferodo Research Laboratories at Chapel-en-le-Frith

Wise motorists do not place full reliance on brakes if the linings get wet. Test trial drivers keep their feet lightly on the brake pedals for a short time after passing through deep water so that the resultant rise in temperature dries the linings and quickly restores their efficiency.

Lubrication

While friction is highly essential in some circumstances, it is a great nuisance in others. In a car engine, for example, oil under pressure is supplied continuously to all bearing surfaces. Failure of the oil supply will allow metal-to-metal contact and the resultant friction often raises the temperature and causes the bearing and pistons to "seize up".

Although liquid molecules attract one another they can interchange partners quite easily. The opposing force which one layer of liquid exerts on another is called *viscosity*. For liquid lubricants this is very much less than the frictional force most solids exert on one another. Some lubricating oils contain certain solid or liquid substances which attach themselves to the bearing surfaces to form a tenacious slippery coating. The possibility of metal-to-metal contact is thus greatly reduced, accompanied by less wear and smoother running.

Air lubrication

As long ago as 1854, the French scientist, Gustav Hirn, suggested the use of air as a lubricant, but nearly a century was to pass before this idea was taken up seriously.

Of recent years, research has been carried out at the National Engineering Laboratory at Kilbride in the development of air lubricated bearings and these are now in highly successful commercial production. They have a special application in machine tools, particularly grinding machines (Fig. 2.6).

 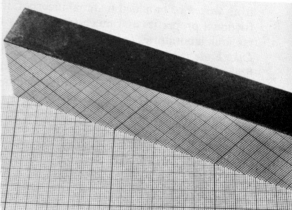

Air bearing wheelhead fitted to a surface grinder at the National Engineering Laboratory, Kilbride. The flexible tube conveys filtered air to the bearing at a pressure of about 40 kN/m²

Typical mirror finish on a workpiece from an air bearing grindwheel

Fig. 2.6. *Air lubrication in machine tools*

When the final accurate finish is required on the component parts of a piece of machinery, it is customary to use a grinding wheel. Obviously, not the slightest wobble can be tolerated in the bearings. In the past oil lubricated bearings were used with very tight clearances, which had to run for some hours before use so that they acquired a steady temperature. The consequent expansion took up any slackness and the bearings were run under conditions in which they just did not seize.

When purified compressed air is employed as a lubricant, very tight clearances are unnecessary. Unlike oil which is practically incompressible, the air forms an elastic cushion which averages out spindle imperfections and gives truer rotational accuracy. The whole problem of friction and heating is eliminated with the further advantage that the outflow of air prevents the entry of grinding dust. The machines can be used immediately they are started up with the net advantage of shorter production time coupled with greater precision of finish on the workpiece.

Further developments

In recent years, other applications of air lubrication have been explored such as the hovercraft which travels over water or land on a cushion of air. Fig. 2.7 shows the French Bertin aerotrain. Powered by an aircraft engine, it sat astride an inverted T-shaped rail and rode on a cushion of compressed air provided by powerful pumps.

Fig. 2.7. One of the first tracked land vehicles without wheels. The first experimental model of this Bertin aerotrain reached a speed of 135 km/h in 1966. A later jet-propelled model reached 400 km/h, and a rocket-propelled one over 1350 km/h. Technical difficulties and expense have, however, stopped further development of this project

In ship holds and other confined spaces, heavy containers may easily be moved for stowage on *hoverpads*. These have a flexible pleated skirt which takes up any unevenness in the ground or floor and so retains the air pumped in. Fig. 2.8 illustrates a novel use of the hoverpad principle made by Esso Petroleum Company when faced with the problem of moving a 60 t oil storage tank from one site to another at its Manchester terminal. During this operation the tank was towed a distance of 320 m on an air cushion, crossing rough ground, a road and a railway line on its way.

QUESTIONS: 2

1. Briefly explain the difference between *mass* and *weight*.

Explain why:

(a) the weight of a body changes if it is taken from the equator towards one of the poles:

(b) an astronaut is often weightless when in an earth-orbiting spacecraft.

2. What is meant by *friction*? Illustrate your answer by reference to the effect of a steadily increasing horizontal force applied to a block of wood at rest on a rough horizontal table.

Discuss the frictional forces between the tyres of a motor-car and the ground when the car:

(a) starts from rest;

(b) skids in the forward direction.   (C.)

3. If the surface of a frozen pond were perfectly smooth so that it would be impossible to walk across it, could you get from one part of it to another by lying down and rolling?   (O.C.)

Fig. 2.8. What are they doing? See the text

3. Speed, velocity and acceleration

Average speed

Unless he is travelling on a motorway, a motorist cannot usually maintain a constant speed for any length of time. In the ordinary way, traffic conditions frequently cause him to change speed or stop. When deciding the time to allow for a particular journey, the motorist must therefore have some idea of the *average speed* at which he will be able to travel.

Speed is defined as the rate of change of distance moved with time. Thus, if a journey from London to Birmingham, 176 km, takes four hours,

$$\text{average speed} = \frac{\text{distance}}{\text{time}} = \frac{176}{4} = 44 \text{ km/h}$$

During such a journey, however, the actual speed of the car at any instant will vary considerably from this figure.

Actual speed

Normally, a driver notes his actual speed at any moment by glancing at the speedometer, but for reasons which we shall not discuss here, speedometer readings are not particularly reliable. In order to obtain an accurate value for the speed of a vehicle at any instant it would be necessary to measure the distance moved in a very short interval of time. This is best done by a roadside observer using special apparatus to time the car over a measured distance. So long as the time interval is short, there is less likelihood that the speed will vary over the measured distance. Otherwise, of course, the value obtained will be an average speed.

Scalar and vector quantities

Most quantities measured in science are classed as either *scalar* or *vector* quantities. A scalar quantity is one which has magnitude (or size) only. A vector quantity is one which has direction as well as magnitude. Thus when we say that a library contains 2000 books or a fuel tank contains 50 litres of petrol, we are dealing with scalar quantities. On the other hand, a good example of a vector quantity is force, since forces always have direction as well as magnitude.

Distance and displacement

We have just explained the difference between *scalar* and *vector* quantities. Now, if we say that a motor-car travels a distance of 100 m, the expression "100 m" is a scalar quantity. But if the car happens to be moving along a straight line and we mention the direction of travel, e.g., 100 m due east, we are now dealing with a vector quantity, and this is called the *displacement of the car*.

Displacement is defined as distance moved in a specified direction.

Velocity

In ordinary conversation the word "velocity" is often used in place of speed. In science, however, it is important to distinguish between these two terms.

Velocity is defined as the rate of change of distance moved with time in a specified direction (or, rate of change of displacement).

Velocity is therefore a vector, whereas speed is a scalar quantity. For example, if a car were travelling at a steady speed of 30 km/h along a perfectly straight road, it would be correct to say that it had a velocity of 30 km/h 30° east of north, or whatever the direction of the road might be. On the other hand, if the car were travelling round a bend at constant speed, its direction of motion would be continuously changing. Hence its velocity would also be continuously changing, although the speed remains constant.

Uniform velocity

A body is said to move with uniform velocity if its rate of change of distance moved with time in a specified direction is constant.

Acceleration

When the velocity of a body is changing, the body is said to be accelerating.

Acceleration is defined as the rate of change of velocity with time.

Acceleration is regarded as positive if the velocity is increasing, and negative if the velocity is decreasing. Ordinarily, however, most people restrict the use of the word acceleration to cases of increasing velocity only, while a decreasing velocity or slowing down is usually called a deceleration or retardation.

Uniform acceleration

A body is said to move with uniform acceleration if its rate of change of velocity with time is constant.

Example. *A motor car is uniformly retarded and brought to rest from a speed of* 108 km/h *in* 15 s. *Find its acceleration.*

$$108 \text{ km/h} = \frac{108 \times 1000}{60 \times 60} = 30 \text{ m/s}$$

therefore   initial velocity $= 30$ m/s
final velocity $= 0$ m/s
change in velocity $=$ final velocity $-$ initial velocity
$= (0 - 30)$ m/s
$= -30$ m/s
acceleration $= \dfrac{\text{change in velocity}}{\text{time}} = \dfrac{-30 \text{ m/s}}{15 \text{ s}}$
$= -2 \text{ m/s}^2$

The minus sign here simply means that the car is accelerating in the opposite direction to its initial velocity. Note the units in which acceleration is measured, namely, m/s^2. Otherwise this may be written as m s^{-2}.

Equations of uniformly accelerated motion

First equation of motion. Suppose a body which is already moving with a velocity of u in m/s begins to accelerate at the rate of a in m/s^2. The velocity will now increase by the numerical value of a in m/s for each second that it moves. The increase in velocity in a time t in s will therefore be equal to at.

Hence the final velocity after a time t is given by

$$v = u + at \quad \ldots \ldots \ldots \ldots \ldots \ldots \quad (1)$$

This is called the first equation of motion.

Second equation of motion. If a body is moving with uniform acceleration its average velocity is equal to half the sum of the initial velocity u, and the final velocity v.

Thus, average velocity $= \dfrac{u + v}{2}$

but $v = u + at$

therefore average velocity $= \dfrac{u + u + at}{2} = u + \frac{1}{2}at$

The distance, x, moved (displacement) = average velocity × time

$$x = (u + \tfrac{1}{2}at) \times t$$

or $\qquad x = ut + \frac{1}{2}at^2 \quad \ldots \ldots \ldots \ldots \ldots \quad (2)$

This is known as the second equation of motion.

Third equation of motion. A useful third equation can be obtained by eliminating t between the first two equations.

Squaring both sides of the equation, $v = u + at$, we obtain

$$v^2 = u^2 + 2uat + a^2t^2$$

Taking out the factor $2a$ from the last two terms of the right-hand side,

$$v^2 = u^2 + 2a(ut + \tfrac{1}{2}at^2)$$

but the bracket term is equal to x

hence, $\qquad v^2 = u^2 + 2ax \quad . \quad \ldots \ldots \ldots \ldots \ldots \quad (3)$

Velocity–time graphs

The distance covered in 5 s by a body moving with a velocity of 6 m/s is given by,

$$\text{distance} = \text{velocity} \times \text{time}$$
$$= 6 \times 5$$
$$= 30 \text{ m}$$

If we plot a graph of velocity against time for this case a straight-line parallel to the time axis is obtained (Fig. 3.1). On this graph, OC represents the time, 5 s, and OA the constant velocity 6 m/s.

Hence, $\quad \text{distance} = \text{velocity} \times \text{time}$
$$= OA \times OC$$
$$= \text{area } OABC \text{ on the scale of the graph}$$

The distance travelled is thus represented by the area between the velocity–time curve and the time axis. This important result applies in all cases, *whatever the shape of the velocity–time curve.*

Uniformly accelerated motion represented graphically

Fig. 3.2 shows the velocity–time graph for a body which starts with a velocity of 3 m/s and moves with an acceleration of 2 m/s² for 4 s.

In this case the velocity increases uniformly, and therefore the average velocity is

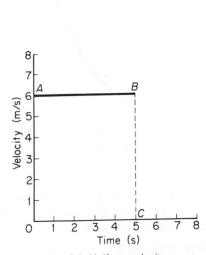

Fig. 3.1. Uniform velocity

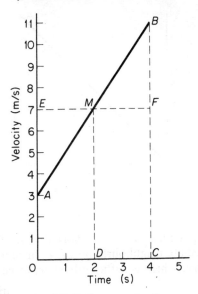

Fig. 3.2. Uniform acceleration

equal to the velocity at half time and is represented by MD on the graph. If we take the right-angled triangle MBF and place it in the position MAE we shall obtain a rectangle $OEFC$ whose area is equal to that of $OABC$.

The height of rectangle $OEFC$ represents the average velocity and its base represents the time.

Therefore,   area of $OEFC$ = average velocity × time = distance
hence also   area of $OABC$ = distance (numerically)

An alternative way of looking at this problem is to take the average velocity as being equal to half the sum of the initial and final velocities, i.e., $\frac{1}{2}(OA + CB) = \frac{1}{2}(3 + 11) = 7$ m/s.

Hence,   distance moved = $\frac{1}{2}(OA + CB) \times OC = 7 \times 4 = 28$ m

But this expression is also equal to the area of $OABC$, since $OABC$ is a trapezium and the area of any trapezium = $\frac{1}{2}$(sum of parallel sides) × perpendicular distance between them.

So, whichever way we look at the problem, the distance moved is seen to be numerically equal to the area between the velocity–time curve and the time axis.

Velocity from distance–time graph

When a body moves with uniform velocity it will travel equal distances in equal intervals of time, and so a graph of distance against time will be a straight line (Fig. 3.3 (*a*)). Now if we take any point A, on the graph and drop a perpendicular AB on to the time axis, it is clear that AB represents the distance moved in the time interval represented by OB.

Hence, $$\text{velocity} = \frac{\text{distance}}{\text{time}} = \frac{AB}{OB}$$

$\frac{AB}{OB}$ is called the gradient or slope of the line OA.

Fig. 3.3 (*b*) is a graph of distance against time for a body moving with a variable velocity. In order to find the velocity at any instant represented by point A on the curve, let us imagine a very small right-angled triangle BCD to be drawn whose

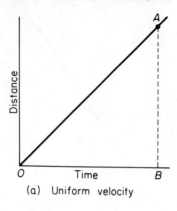

(a) Uniform velocity

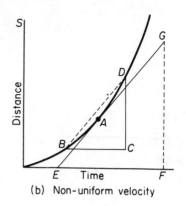

(b) Non-uniform velocity

Fig. 3.3.

hypotenuse *BD* is so short that it effectively coincides with the curve in the immediate neighbourhood of *A*. In other words, we are considering a portion of the curve which is sufficiently short to be regarded as sensibly straight.

Then $$\text{velocity at } A = \frac{\text{distance represented by } DC}{\text{time interval represented by } BC}$$

It would, of course, be of little use attempting to get accurate results from measurements made on so small a triangle. Instead we may find the velocity from a much larger similar triangle obtained by drawing a tangent *EG* to the curve at *A*, and measuring its gradient.

Thus $$\text{velocity at } A = \text{gradient at } A = \frac{\text{distance } GF}{\text{time interval } EF}$$

Acceleration from velocity–time graph

Fig. 3.4 shows velocity–time graphs for a body moving with: (*a*) uniform acceleration, and (*b*) non-uniform acceleration.

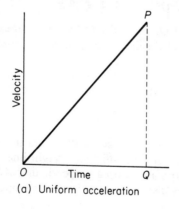

(a) Uniform acceleration

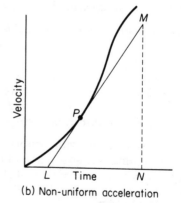

(b) Non-uniform acceleration

Fig. 3.4.

By a process of reasoning similar to that used in the section above relating to the distance–time graphs, it may be shown that,

in Fig. 3.4 (*a*), acceleration = gradient of $OP = \dfrac{PQ}{OQ}$

and in Fig. 3.4 (*b*), acceleration at P = gradient of tangent $LPM = \dfrac{MN}{LN}$

Worked examples

1. *A stone is thrown vertically upwards with an initial velocity of* 14 m/s. *Neglecting air resistance, find:* (a) *the maximum height reached;* (b) *the time taken before it reaches the ground.* (*Acceleration due to gravity* = 9.8 m/s².)

When working problems of this type the reader is recommended to extract the data given in the question and write them down against the appropriate symbols before attempting to substitute in one of the equations of motion.

$$u = 14 \text{ m/s} \qquad v = 0 \text{ m/s} \qquad a = -9.8 \text{ m/s}^2 \text{ (retardation)}$$

To find the height reached, x, we substitute in the equation

$$v^2 = u^2 + 2ax$$
$$0^2 = 14^2 + 2 \times (-9.8) \times x$$

whence
$$x = \frac{-14^2}{2 \times (-9.8)}$$
$$= 10 \text{ m.}$$

The time taken to reach this height is found by substitution in the equation

$$v = u + at$$
$$0 = 14 + (-9.8) \times t$$

Thus,

or
$$t = \frac{14}{9.8} = 1.43 \text{ s}$$

The time taken to fall back to the ground is found by substitution in

$$x = ut + \tfrac{1}{2}at^2$$

in which
$$x = 10 \text{ m}$$
$$u = 0 \text{ m/s}$$
$$a = +9.8 \text{ m/s}^2$$

Thus,
$$10 = 0 \times t + \tfrac{1}{2} \times 9.8 \times t^2$$

or
$$t^2 = \frac{10 \times 2}{9.8}$$

whence
$$t = 1.43 \text{ s}$$

The downward motion is, of course, simply a reversal of the upward motion in every respect.

Answer. Height 10 m
Time 2.9 s

2. *A car starts from rest and is accelerated uniformly at the rate of* 2 m/s² *for* 6 s. *It then maintains a constant speed for half a minute. The brakes are then applied and the vehicle uniformly retarded to rest in* 5 s. *Find the maximum speed reached in* km/h *and the total distance covered in metres.*

First stage $\qquad u = 0 \text{ m/s} \qquad a = 2 \text{ m/s}^2 \qquad t = 6 \text{ s}$

Substituting in the first equation of motion,

$$v = u + at$$
$$v = 0 + 2 \times 6$$
$$v = 12 \text{ m/s}$$
$$= \frac{12}{1000} \times 60 \times 60 \text{ km/h}$$
$$= 43 \text{ km/h}$$

The distance moved in the first stage may be found by substituting in the second equation of motion, thus,

$$x = ut + \tfrac{1}{2}at^2$$
$$= 0 \times 6 + \tfrac{1}{2} \times 2 \times 6^2 = 36 \text{ m}$$

Second stage $u = 12$ m/s (constant) $t = 30$ s
hence   distance moved = speed × time
       $= 12 \times 30 = 360$ m.
Third stage $u = 12$ m/s $v = 0$ m/s $t = 5$ s

$$\text{Acceleration } a = \frac{v - u}{t} = \frac{0 - 12}{5}$$
$$= -2.4 \text{ m/s}^2$$

The distance may be found either by the second or third equations of motion. If we use the latter,

$$v^2 = u^2 + 2\,ax$$

whence $$x = \frac{v^2 - u^2}{2a}$$
$$= \frac{0^2 - 12^2}{2 \times (-2.4)} = 30 \text{ m}$$

Answer. Total distance travelled = 36 + 360 + 30 = 426 m
Maximum speed reached = 43 km/h

Alternative graphical solution

Fig. 3.5 shows the velocity–time graph for the previous problem, in which *OA*, *AB* and *BC* represent the three stages of the motion respectively. The distance moved is numerically equal to the area of the figure *OABC* (a trapezium).

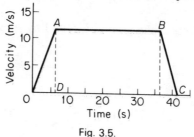

Fig. 3.5.

$$\text{Area } OABC = \tfrac{1}{2}(AB + OC) \times AD$$
$$= \tfrac{1}{2}(30 + 41) \times 12$$
$$= 426 \text{ scale units}$$

thus the total distance = 426 m

Galileo Galilei

The equations of motion explained earlier in this chapter were first worked out by Galileo Galilei, who was born at Pisa in Italy in 1564. Galileo began his university career as a medical student, but later forsook medicine for the study of mathematics and physics. His outstanding ability earned him a lectureship by the time he was twenty-five years of age, and eventually he became professor at the University of Padua. Later in life he was imprisoned for teaching that the sun was at the centre of the solar system. This, being contrary to the teaching of the Church at that time, was regarded as an attempt to undermine its authority.

In the study of mechanics, Galileo is particularly remembered for the work he did on the acceleration of falling bodies. The Greek philosopher Aristotle taught that the speed with which a body fell to the ground depended on its mass. This wrongful assertion had been accepted for centuries, but legend has it that Galileo tested the truth of the matter by a very simple experiment. He ascended the leaning tower of Pisa and, from the top, he simultaneously released three iron balls of different masses. They all reached the ground at the same time. So much for the legend. The truth is that the Flemish engineer and mathematician, Simon Stevin published a book on mechanics in 1586 in which he describes an experiment with two balls of lead, one ten times the mass of the other. These, when dropped simultaneously from a height of ten metres, reached the ground at the same time. This was nearly twenty years before Galileo's supposed experiment at Pisa. See Fig. 3.6.

It has been found that the acceleration due to gravity for a freely falling body has an average value of about 9.8 m/s². The accepted symbol for the acceleration due to gravity is an *italic* letter "g" to distinguish it from the symbol for gram which is a Roman letter "g".

It is very easy to draw a wrong conclusion from a single casual observation. A feather, for example, falls to the ground much more slowly than a stone. The fact is that air resistance as well as force of gravity govern the rate at which a body falls towards the earth. In the case of a light feather of large surface area the air resistance is very large compared with the force of gravity on it. If air resistance is eliminated the feather falls with the same acceleration as the stone. This was first demonstrated by Robert Boyle shortly after Galileo's death. Using his newly invented air pump, Boyle removed the air from a tall glass jar containing a lead bullet and a feather. When the jar was inverted both bullet and feather reached the bottom of the jar simultaneously.

The simple pendulum

A story is told of Galileo, that he was once attending a service in the cathedral at Pisa when his attention was distracted by the swinging of a lamp which was suspended from the roof by a long chain. Using the beats of his pulse as a clock, he noticed that the time of swing of the lamp remained constant even when the oscillations were dying away. The lamp was behaving as a *pendulum*.

In mechanics, **a simple pendulum is defined as a small heavy body suspended by a light inextensible string.**

Galileo was quick to realize the importance of the constant time of swing of the pendulum, and later on it occurred to him that a pendulum might be used to govern a clock more satisfactorily than the horizontal oscillating crossbars which were then in use. Before he died in 1642, Galileo left plans for the construction of a pendulum clock, but it was not until 1657 that the first successful pendulum clock was constructed by the Dutch scientist Christian Huygens.

To study the simple pendulum. Measurement of g

For experimental purposes, a simple pendulum is made by attaching a length of thread to a small brass or lead sphere called the bob. The thread is held firmly between two small pieces of wood (or a split cork) held by a clamp and stand.

One complete to and fro movement of the pendulum is called an *oscillation* or *vibration*. The time taken for one complete oscillation is called the *periodic time*. The *length* of the pendulum is defined as the distance from the point of suspension to the centre of gravity* of the bob.

As the pendulum swings to and fro, the maximum displacement of the bob from its rest position is called the *amplitude*. Alternatively, we speak of the *angular amplitude* of the pendulum or the angle between the extreme and the rest positions of the string.

Provided the amplitude is small, i.e., not more than a few degrees, the periodic time depends only on the length of the pendulum and the acceleration due to gravity. Experiments carried out using bobs of different sizes show that the periodic time does not depend on the mass or material of the bob.

In more advanced books it is shown that the periodic time, *T*, of a simple pendulum is given by,

$$T = 2\pi\sqrt{\frac{l}{g}}$$

where l = length in m;
$\quad g$ = acceleration due to gravity in m/s².

* centre of gravity is explained on page 65, in this case it is the centre of the bob.

In order to verify this equation experimentally it will be found convenient to rearrange it so that all the constant factors are on one side only.

Squaring both sides,

$$T^2 = 4\pi^2 \frac{l}{g}$$

dividing both sides by l,

$$\frac{T^2}{l} = \frac{4\pi^2}{g} = \text{constant}$$

The object of the experiment now to be described is to verify this equation by measuring the periodic time T, for a series of different values of l.

If the equation is true, the results should show that $\frac{T^2}{l} = \text{constant}$.

Fig. 3.7 shows how the length of the pendulum is measured by means of a metre rule and a set-square. If a set-square is not available a small rectangular card may be used instead. The zero end of the rule is held in contact with the under-surface of the blocks and the set-square reading x noted when it is just touching the bottom of the bob. The radius of the bob is now subtracted from x to give the required length, l.

The observer should sit in front of the pendulum and note the rest position of the lower part of the string against some convenient mark. The pendulum is then set swinging with a *small amplitude*.

Fig. 3.6. Scene of a legendary experiment

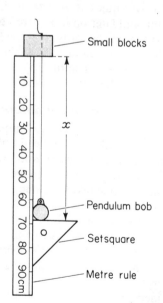

Fig. 3.7. Simple pendulum experiment

Timing of the oscillations is begun by counting "nought" as the bob passes through its rest position and simultaneously a stop-clock is started. Counting is continued, "one, two, three, etc." each time the bob passes through its rest position in the same direction. If a stop-clock is not available the time may be measured by an ordinary clock or watch which has a seconds hand. In this way, the time of 50 oscillations is found for at least half a dozen different lengths varying from about 0.30 to 1.0 m.

In each case the timing should be repeated as a check on the previous reading and the results tabulated as shown.

Length l (m)	Time for 50 oscillations		Periodic time T (s)	T^2	$\dfrac{T^2}{l}$
	1	2			

Constancy of the values of $\dfrac{T^2}{l}$ entered in the last column of the table shows that **the square of the periodic time is proportional to the length of the pendulum.** This may also be demonstrated by plotting a graph of T^2 against l. A straight line through the origin should be obtained.

Calculation of the acceleration due to gravity from the results

Since
$$\frac{T^2}{l} = \frac{4\pi^2}{g}$$

it follows that
$$g = 4\pi^2 \div \frac{T^2}{l}$$

Hence a value for g may be calculated either by dividing $4\pi^2$ by the mean value of $\dfrac{T^2}{l}$ obtained from the last column of the table, or from the gradient of the graph of T^2 against l.

Distance moved by a freely falling body related to time of fall

The distance moved by a body falling freely from rest is given by,
$$x = \tfrac{1}{2} gt^2$$
or, if the acceleration of free fall, g, is constant
$$x = \text{constant} \times t^2$$

The value of the constant here is equal to $\tfrac{1}{2} g$ and we can best illustrate the relationship between distance fallen and time taken by plotting a graph of x against t^2. We should expect to obtain a straight line through the origin, the gradient of which is $\dfrac{g}{2}$ (Fig. 3.8). Hence,
$$g = 2 \times \text{gradient}$$

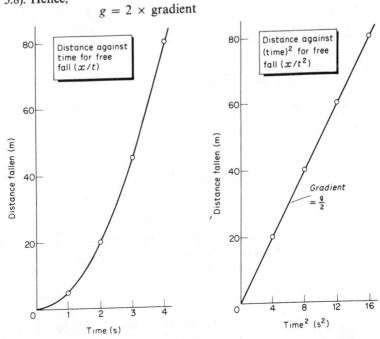

Fig. 3.8. Free fall under gravity

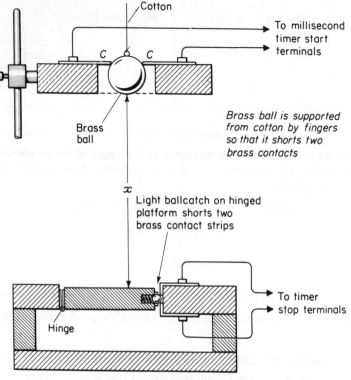

Fig. 3.9. Determination of *g*

Note that this method of illustrating the relationship applies not only to falling bodies but also to any body which is moving with a uniform acceleration.

To measure g by use of a millisecond timer

In this experiment the time taken by a brass ball to fall through a measured height is found by means of an electric clock which measures time intervals in thousandths of a second or *milliseconds* (ms). This works on a different principle from an ordinary electric clock. It generates electric impulses at the rate of 1000 per second, which are recorded on a counter (Fig. 3.10). Sockets are provided to which start and stop switches may be connected and these are operated by a brass ball at the beginning and end of its free fall through a measured height.

Fig. 3.9 shows the method. The brass ball, supported by hand from a piece of cotton, is held so that it bridges two metal contacts, *C*. This acts as a stop switch by shorting out the counter, thus preventing it from receiving impulses from the generator.

The counter is set to zero and the ball is released. At the moment of release, contact is broken, and the counter starts to record milliseconds.

At the end of its fall, the ball strikes a hinged platform, thereby operating a switch which cuts off the impulses from generator to counter.

The time of fall *t* (in seconds) for a series of different heights *x* (in metres) is measured in this way and the results recorded in a table as shown. From these a graph is plotted of *x* against t^2, and its gradient, $\dfrac{x}{t^2}$ is measured.

As we saw in the previous section, a body starting from rest and falling freely under the action of gravity falls a distance *x* in time *t* given by

$$x = \tfrac{1}{2}gt^2$$

whence
$$g = 2 \times \frac{x}{t^2}$$

or
$$g = 2 \times \text{gradient of graph (see Fig. 3.8)}$$

For details of how to measure the gradient of a graph, see page 25.

Height fallen x (in m)							
Time of fall t (in s)							
t^2							

Fig. 3.10. Millisecond timer

QUESTIONS: 3

(*Where necessary, assume g* = 10 m/s²)

1. Define *speed* and explain what is meant by *average speed*. A motorist travels from a town *A* to a town *B*, 145 km distant in 3 h 45 min. Find his average speed:
   (*a*) in km/h;
   (*b*) in m/s.

2. (*a*) Sketch a velocity–time graph for a car moving with uniform acceleration from 5 m/s to 25 m/s in 15 seconds.
   (*b*) Use the sketch graph to find values for
   (i) the acceleration,
   (ii) the total distance travelled during acceleration.
   Show clearly at each stage how you used the graph. (*J.M.B.*)

3. Explain the difference between speed and velocity. Draw a graph of velocity against time for a body which starts with an initial velocity of 4 m/s and continues to move with an acceleration of 1.5 m/s² for 6 s. Show how you would find from the graph:
   (*a*) the average velocity;
   (*b*) the distance moved in the 6 s.

4. A car travels at a uniform velocity of 20 m/s for 5 s. The brakes are then applied and the car comes to rest with uniform retardation in a further 8 s. Draw a sketch graph of velocity against time. How far does the car travel after the brakes are applied?

5. A body starts from rest and accelerates at 3 m/s², for 4 s. Its velocity remains con-stant at the maximum value so reached for 7 s and it finally comes to rest with uniform retardation after another 5 s. Find by a gra-phical method:
   (*a*) the distance moved during each stage of the motion;
   (*b*) the average velocity over the whole per-iod.

6. Define *speed*, *velocity* and *acceleration*.
   (*a*) A stone is released from rest at the top of a tall tower. Draw a distance–time graph of its free fall under gravity during the first 6 s. Show your table of values.
   (*b*) A bullet, fired vertically upwards from a gun held 2 m above the ground, reaches its maximum height in 4 s. Calculate:
   (i) the initial velocity of the bullet,
   (ii) the total distance the bullet travels by the time it hits the ground. [g = 10 m s⁻².]
   (*W.A.E.C.*)

7. A trolley starts from rest on an inclined plane and moves down it with uniform ac-celeration. After having moved a distance of 40 cm its velocity is 20 cm/s. Find its acceler-ation:
   (*a*) in cm/s²;
   (*b*) in m/s².

8. A motorist, travelling at 90 km/h, applies his brakes and comes to rest with uniform retardation in 20 s. Calculate the retardation in m/s².

9. An electric train moving at 20 km/h accelerates to a speed of 30 km/h in 20 s. Find the average acceleration in m/s² and the distance travelled in metres during the period of the acceleration.

10. The speed of a goods truck which has been shunted on to a level siding falls from 10 km/h to 5 km/h in moving a distance of 30 m. If the retardation is constant, how much further will the truck travel before coming to rest?

11. A train, 90 m long, stops in a station with its front buffers in line with a lamp-post on the platform. Later it starts off with an average acceleration of 0.45 m/s². What will be its speed, in km/h, when the tail buffers pass the lamp-post?

12. (*a*) Define the terms "velocity" and "acceleration". Choose one of these terms and explain what is meant when the quantity is said to be "uniform".

(*b*) A car runs at a constant speed of 15 m/s for 300 s and then accelerates uniformly to a speed of 25 m/s over a period of 20 s. This speed is maintained for 300 s before the car is brought to rest with uniform deceleration in 30 seconds. Draw a velocity–time graph to represent the journey described above. From the graph find:

(i) the acceleration while the velocity changes from 15 m/s to 25 m/s;

(ii) the total distance travelled in the time described;

(iii) the average speed over the time described. (*J.M.B.*)

13. (*a*) What is meant by *acceleration*?

(*b*) An object was thrown vertically upwards and its height above the ground was measured at various times. The results obtained are shown in the table.

Time/s	0	1	2	3	5	6	7	8
Height/m	0	35	60	75	75	60	35	0

Plot a graph of height on the *y*-axis against time on the *x*-axis. From your graph find

(i) the maximum height reached;

(ii) the time taken to reach this height.

(*c*) Using either or both of the answers from part (*b*), calculate the initial velocity with which the object was thrown.

(Assume that the acceleration of free fall is 10 m/s².) (*J.M.B.*)

14. (*a*) Describe a laboratory experiment by which you would measure the acceleration of free fall. State clearly the apparatus used, how it is arranged, the observations you would make and their use in obtaining the final result.

(*b*) A stone is thrown from ground level vertically upwards with a velocity of 40 m/s. Find the maximum height to which the stone rises and its total time of flight, neglecting air resistance.

Sketch a graph of the height of the stone against time. (Plot the height on the vertical axis of the graph.) (*L.*)

15. A small solid sphere falls freely from rest, in air, with an acceleration of 10 m/s². How far does it fall in 5.0 s?

Explain why there would be a difference in the distance fallen by a large hollow sphere, of the same mass, in the same time. (*C.*)

16. A body which starts from rest and slides down an inclined air-track covers the following distances *x* in times *t*.

x in m	0	0.128	0.200	0.288	0.392	0.512	0.648
t in s	0	0.8	1.0	1.2	1.4	1.6	1.8

Show by the graphical method that the body moves with uniform acceleration and find its value in m/s².

17. Describe an experiment to measure the acceleration of free fall (due to gravity). Your answer should include

(i) a labelled diagram of the apparatus,

(ii) an account of the measurements you would make,

(iii) an account of how these measurements would be used to calculate the final result. (*J.M.B.*)

18. A body travelling with uniform acceleration covers a distance *s* metres after a time *t* seconds as shown on the table:

s(metres)	0	2	8	12.5	18	24.5	32	40.5	50	60.5	72
t(seconds)	0	1	2	2.5	3	3.5	4	4.5	5	5.5	6

Draw a distance–time graph and determine from this graph the velocities at 2.5 seconds and at 4.5 seconds and from the values obtained calculate the acceleration. (*W.*)

19. A projectile is fired vertically upwards and reaches a height of 125 m. Find the velocity of projection and the time it takes to reach its highest point.

20. A stone is thrown vertically upwards with an initial velocity of 30 m/s from the top of a tower 20 m high. Find:

(*a*) the time taken to reach the maximum height;

(*b*) the total time which elapses before it reaches the ground.

21. A small iron ball is dropped from the top of a vertical cliff and takes 2.5 s to reach the sandy beach below. Find:

(*a*) the velocity with which it strikes the sand;

(*b*) the height of the cliff.

If the ball penetrates the sand to a depth of 12.5 cm, calculate its average retardation.

4. Newton's laws of motion

The study of moving bodies begun by Galileo Galilei in Italy was continued after his death by Sir Isaac Newton in England.

In 1687 Newton published a book written, as was the custom in those days, in Latin and given the title, *Philosophiae naturalis principia mathematica*. Translated, this means, "The mathematical principles of natural philosophy". The Principia, as it is familiarly called, is regarded as one of the greatest scientific works ever written. It is devoted mainly to the study of motion, particularly that of the planets and other heavenly bodies. In the first part of the book Newton sums up the basic principles of motion in three laws. In this chapter we shall discuss these laws and consider some of their applications.

Newton's first law of motion

Every body continues in its state of rest or of uniform motion in a straight line unless compelled by some external force to act otherwise.

It is a matter of common experience that objects at rest do not begin to move of their own accord. If we place an object in a certain place we expect it to remain there unless a force is applied to it. Some simple parlour tricks are based on this principle. For example, if a pile of draughts or coins is placed on a table the bottom one can be removed without disturbing the remainder, simply by flicking it sharply with a piece of thin wood or metal. In this case the force of friction between the bottom disc and the one in contact with it acts for too short a time to cause any appreciable movement of the discs above. Incidentally, the use of a ruler is not recommended for this experiment, as the blows it receives will not improve its straight-edge.

The bottle and coin trick is equally effective in demonstrating the same effect. A small coin is put on a card and placed over the mouth of a bottle (Fig. 4.1). When the card is flicked away with the finger the coin drops neatly into the bottle. For success

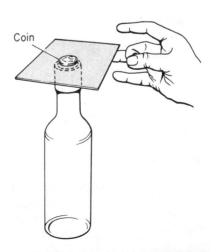
Coin

Fig. 4.1. An old trick

in performing this trick the finger should move in a horizontal plane so that the card is not tilted.

It is not immediately obvious that a body moving with uniform velocity in a straight line tends to go on moving for ever without coming to rest. The fact is that no one has yet found a means of eliminating all the various outside forces which can retard a moving body.

A person riding a bicycle along a level road does not come to rest immediately he stops pedalling. The bicycle continues to move forward, but eventually it comes to rest as a result of the retarding action of air resistance and friction. In a collision between two motor vehicles the passengers are frequently injured when they hit the windscreen. An external force stops the vehicle, but not the passengers who simply continue their straight-line motion in accordance with Newton's first law; hence the advantage of a seat belt.

When a bullet is fired from a gun its motion is opposed both by air resistance and the pull of the earth. Sooner or later it returns to the earth, but it would be reasonable to suppose that, if air resistance and gravitation could be eliminated, the bullet would go on moving in a straight line for ever.

In the early seventeenth century the German astronomer, Johann Kepler, had shown that the planets move in elliptical paths or orbits round the sun, but he was unable to explain why. It was left to Sir Isaac Newton to offer a satisfactory explanation based on the first law of motion and the law of universal gravitation (see page 13). Newton pointed out that the planets move in curved paths because the sun is attracting them. No slowing up occurs, since there is no retarding force. The planets move in the vacuum of space, carrying their atmospheres with them. *If the attraction of the sun suddenly ceased, a planet would continue to move in a straight line making a tangent with its original orbit.*

It is important to realize that, once a body is moving with uniform speed in a straight line, it needs no force to keep it in motion provided there are no external opposing forces.

The tendency of a body to remain at rest or, if moving, to continue its motion in a straight line is described as **inertia**. For this reason, Newton's first law is sometimes called "the law of inertia". We shall now go on to discuss the motion of a body when it is being acted upon by a force.

Momentum

A truck requires a larger force to accelerate it to a given speed in a given time when it is heavily laden than when it is empty. Likewise, far more powerful brakes are needed to stop a heavy goods vehicle than a light car, in the same distance, if both are moving at the same speed when the brakes are applied. The heavier vehicle is said to possess a greater quantity of motion or *momentum* than the lighter one.

The momentum of a body is defined as the product of its mass and its velocity.

The unit of momentum in the SI system is, therefore, 1 kilogram multiplied by 1 metre per second (unit symbol, kg m/s).

When two bodies, a heavy one and a light one, are acted upon by the same force for the same time, the light body builds up a higher velocity than the heavy one. But the momentum they gain is the same in both cases. This important connection between force and momentum was recognized by Sir Isaac Newton and expressed by him in a second law of motion.

Newton's second law of motion

The rate of change of momentum of a body is proportional to the applied force and takes place in the direction in which the force acts.

We shall now show how Newton's second law of motion enables us to define an absolute unit of force which remains constant under all conditions.

Suppose a force F acts on a body of mass m for a time t, and causes its velocity to change from u to v.

The momentum changes uniformly from mu to mv in time t, therefore the rate of change of momentum $= \dfrac{mv - mu}{t}$.

By Newton's second law, the rate of change of momentum is proportional to the applied force and hence,

$$F \propto \frac{mv - mu}{t}$$

Factorizing,

$$F \propto m\frac{(v - u)}{t}$$

but

$$\frac{(v - u)}{t} = \frac{\text{change in velocity}}{\text{time}} = \text{acceleration}$$

$$= a$$

therefore

$$F \propto ma$$

This relationship can be turned into an equation by putting in a constant.

Thus, $F = constant \times ma.$

It is this equation which enables us to define an *absolute unit* of force. If $m = 1$ kg and $a = 1$ m/s², the value of the unit of force is chosen so as to make $F = 1$ when the constant $= 1$.

Using kilograms, metres, and seconds in this way the unit of force so obtained will be the SI unit which is defined as follows:

The SI unit of force is called the newton (N) and is the force which produces an acceleration of 1 m/s² when it acts on a mass of 1 kg.

Thus, when F is in newtons, m in kilograms and a in metres per second squared, we have

$$F = ma$$

To verify experimentally that F ∝ ma

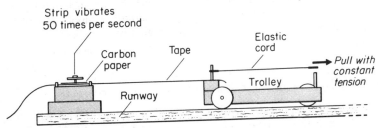

Fig. 4.2. Measurement of acceleration

We may test the truth of Newton's second law by measuring the accelerations produced when various forces are applied to a special trolley running on three ball-bearing wheels (Fig. 4.2).

The equation, $F = ma$, applies to any consistent system of units so we can save time in weighing and calculation if we use a number of identical trolleys which can be fixed, one on top of the other, and taking *the mass of a trolley as a unit of mass.*

Similarly, forces are applied to the trolleys by one or more identical elastic cords, each stretched by exactly the same length, thus using *the tension in a cord as a unit of force.*

Attached to the trolley is a long strip of paper tape which passes under a disc of carbon paper. Immediately above the carbon paper is an iron strip, vibrating 50 times per second, with a stylus which makes a series of dots on the tape as it is pulled along. Constructional details of the *ticker-tape vibrator* are shown in Fig. 4.3 and its action will be better understood after reading chapter 36. The apparatus acts as both clock and distance measurer; distances between the dots representing distances moved by the trolley in successive intervals of $\frac{1}{50}$, or 0.02 s.

How we calculate the acceleration from the dots on the tape will be explained presently.

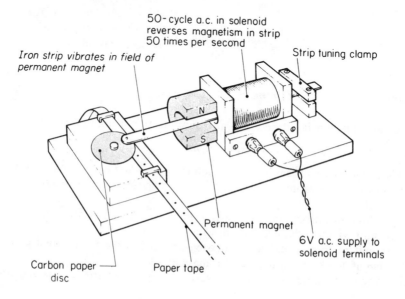

Fig. 4.3. Ticker-tape vibrator

The trolley runs on a smooth flat board about 2 m long, but before each experiment it is necessary to compensate for friction which retards the trolley's motion. This is done by tilting the runway with suitable packing pieces at one end until the trolley moves with uniform velocity after having been given a slight push. Correct adjustment has been obtained if all the dots are equally spaced. The resultant force on the trolley is now zero, so only the force applied by the elastic cord will produce acceleration.

Experiment 1. *To show that a ∝ F when the mass, m, is constant*

Having set up the apparatus as described, a single elastic cord with rings at each end is attached to the vertical rod at the tape end of the trolley and the vibrator switched on. The cord is then pulled by finger and thumb keeping its end ring in line with the two vertical rods at the opposite end. A fairly constant accelerating force can be maintained if one keeps an eye on the end of the cord and rods while walking alongside the runway in time with the trolley. To acquire skill, a few trial runs should be made before attaching the tape.

The experiment is repeated with fresh tapes, but using two and three elastic cords respectively.

In each case the acceleration is found from the tape dots as shown on the next page.

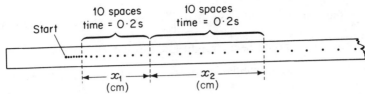

Ignoring the first few dots, the distances x_1 and x_2 occupied by successive 10 dot-spaces are measured. The trolley thus takes $10 \times 0.02 = 0.2$ s to travel each distance x_1 and x_2.
Therefore,

average velocity in cm/s over the distance $x_1 = \dfrac{x_1}{0.2 \text{ s}}$, and

average velocity in cm/s over the distance $x_2 = \dfrac{x_2}{0.2 \text{ s}}$

hence, increase in velocity in 0.2 s $= \dfrac{x_2}{0.2 \text{ s}} - \dfrac{x_1}{0.2 \text{ s}} = \dfrac{x_2 - x_1}{0.2 \text{ s}}$

from which, acceleration $= \dfrac{\text{change in velocity}}{\text{time}} = \dfrac{x_2 - x_1}{(0.2 \text{ s})^2}$

Usually, several pairs of values of x_1 and x_2 may be obtained from the tape and a mean value of $x_2 - x_1$ used in calculating the acceleration. The results may be tabulated as below.

Force F (elastic units)	x_1 (cm)	x_2 (cm)	$x_2 - x_1$ (cm)	Mean $x_2 - x_1$ (cm)	Acceleration $a = \dfrac{x_2 - x_1}{t^2}$ (cm/s²)	$\dfrac{a}{F}$

(Extend table sections as required.)

If $a \propto F$, then the results should show that $\dfrac{a}{F} =$ constant or, alternatively, the graph of a against F should be a straight line through the origin.

Experiment 2. *To show that* $a \propto \dfrac{1}{m}$ *when the accelerating force, F, is constant*

This experiment is carried out in the same manner as the first, but this time we use a constant accelerating force and vary the mass being moved by placing extra trolleys on top of the first. Owing to the fact that friction depends on the weight of the trolley, it will be necessary to make a fresh adjustment to compensate for friction each time the mass of the trolley is altered. The results are recorded in a table as below.

Mass of trolley (m units)	x_1 (cm)	x_2 (cm)	$x_2 - x_1$ (cm)	Mean $x_2 - x_1$ (cm)	Acceleration $a = \dfrac{x_2 - x_1}{t^2}$ (cm/s²)	$m \times a$

(Extend table sections as required.)

If $a \propto \dfrac{1}{m}$ the results should show that $\dfrac{a}{1/m} = ma = $ constant. Alternatively, the graph of a against $\dfrac{1}{m}$ should be a straight line through the origin.

Summary. The two experiments described above show, respectively, that

$$a \propto F$$

and

$$a \propto \frac{1}{m}$$

Combining the two results, $\quad a \propto F \times \dfrac{1}{m}$

whence $\qquad\qquad\qquad F \propto ma$

Note. For simplicity, we have used our own special units of mass and force in these two experiments. We could, of course, have weighed the trolleys and found their mass in kilograms, and also have used a spring balance graduated in newtons to apply the accelerating force. If the distances were then measured in metres, our experiments would have been carried out entirely in SI units.

Weight of a body expressed in newtons

Galileo showed that all bodies, whatever their masses, have the same acceleration when they fall freely under the action of gravity. In equations, the acceleration due to gravity is denoted by the italic letter "*g*" to distinguish it from the Roman letter "g" used as the symbol for mass in grams.

Thus, for a freely falling body of mass m in kilograms, the force of gravity, F in newtons, acting on it is given by

$$F = mg, \text{ where } g = 9.81 \text{ m/s}^2$$

If, however, the body is at rest on the earth's surface this same force will pull it down on to the earth and this is the force we call its *weight* as defined on page 15.

Hence, $\qquad\qquad$ weight of a body in newtons $= mg$

It must, of course, be realized that the acceleration due to gravity varies according to where we happen to be, so 9.81 m/s² must be taken as an average value. In accurate measurements we would use the actual value of g at the place where we were working.

For most purposes, however, it is sufficiently accurate to round off the value of g to 10 m/s², which means that the weight of 1 kilogram is 10 newtons.

To calibrate a spring balance in newtons

Forces or weights are often measured by means of a spring balance. The principle underlying the spring balance was first investigated in the seventeenth century by Robert Hooke. He showed that **when a spring is fixed at one end and a force is applied to the other, the extension of the spring is proportional to the applied force, provided the force is not large enough to stretch the spring permanently.** This is an application of **Hooke's law** (see page 145).

The value of the force for which the extension is such that, on the removal of the force, the spring fails to return to its original length and begins to stretch permanently, is called the *elastic limit*.

In order to verify Hooke's law experimentally, a spiral spring with a scale pan and pointer attached is held vertically by a clamp and stand (Fig. 4.4). Standard masses are then added to the pan, say 10 g (0.01 kg) at a time, and the corresponding extensions to the spring are calculated from the readings of the pointer on a scale of millimetres. A second set of readings is taken as the pan is gradually unloaded. The readings should be entered in a table as shown and the mean extension for each load calculated.

As we have seen in the previous section, the weight of a mass m is given by mg (in newtons) where m is in kilograms and g = acceleration due to gravity in m/s².

For the purpose of this experiment we shall assume $g = 10$ m/s² so that the force readings in column 2 are simply equal to 10 times the numerical value of the mass added to the pan in kg.

Mass in kg	Force in N	Balance readings (mm)		Mean reading (mm)	Extension of spring (mm)	Extension force (mm per N)
		Loading	Unloading			

The results show that, for each reading taken,

$$\frac{\text{extension in mm}}{\text{force in N}} = \text{constant}$$

hence

$$\text{extension} \propto \text{force}$$

This relationship may also be verified by plotting a graph of extension against force, when a straight line through the origin is obtained.

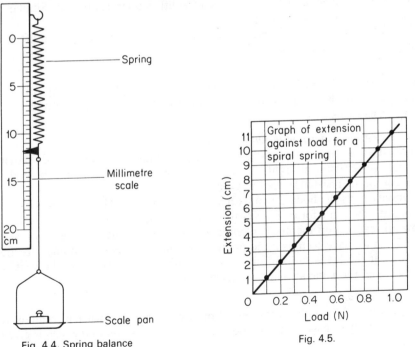

Fig. 4.4. Spring balance

Fig. 4.5.

Fig. 4.5 shows a typical graph obtained in this experiment. Such a graph is called a *calibration curve* for the spring.

Use of a calibrated spring balance

The spring of the previous experiment may be used in conjunction with its calibration graph for the purpose of measuring unknown weights in newtons. Suppose, for example, the extension of the spring is 5 cm when a piece of metal is placed in the pan. From the graph (Fig. 4.5) it will be seen that an extension of 5 cm corresponds to a force of 0.45 N.

Can a spring balance be used to measure mass as well as weight?

A spring balance which has been calibrated to measure force in newtons will measure force or weight anywhere, and is independent of any variations in the force of gravity provided that the effect of gravity on the spring itself can be regarded as negligible.

Such a spring can be used to find the mass of a body hung from it simply by dividing the reading by *g*. Thus suppose a spring balance gives a reading *F* in newtons when stretched by the weight of a body of mass *m* in kilograms, then

$$mg = F$$

or $m = \dfrac{F}{g}$ where *g* = local value of the acceleration due to gravity in m/s^2

Had we plotted extension against mass in the pan instead of force the balance could be used to measure unknown masses directly, *but only at the place where the original calibration was carried out*. Elsewhere the readings will depend on the local value of *g*.

Weight of a body in a lift

There must be few people who have not, at some time or other, travelled in a lift or elevator. So probably everyone is acquainted with the feeling of extra personal weight when the lift accelerates from rest to attain its steady upward velocity, together with a feeling of lightness when the lift decelerates to rest. Similar feelings are experienced in a descending lift.

The following discussion refers to a body of mass *m* in kg inside a lift at a place where the acceleration due to gravity is *g* in m/s^2.

Lift at rest
The case is no different from that when the body is placed on any other support fixed relative to the earth. It therefore presses down on the floor with its rest weight of *mg* in newtons.

Lift moving up or down with uniform velocity
From Newton's first law of motion, no force is required to keep a body moving with uniform velocity in a straight line. The air inside the lift moves with the body, so the question of air resistance does not concern us.

Gravity pulls the body down with a force of *mg* in N. The body presses on the lift floor with the same force. By Newton's third law (page 15) the floor exerts an equal upward reaction on the body. Consequently the resultant force on the body is zero which, as we have seen, is a requirement for uniform velocity in any direction.

Thus, if the lift is moving up or down with uniform velocity, the weight of the body is *mg*, the same as when at rest on earth.

Lift accelerating upwards
Arising out of Newton's second law of motion we saw that when a resultant force, *F*, acts on a body of mass *m*, it will move with an acceleration of *a* which is given by the equation,

$$F = ma$$

If the lift moves upwards with acceleration *a* its floor must push upwards on the body inside to give that the same acceleration also.

By Newton's third law the body will therefore exert an equal and opposite reaction of *ma* downwards on the floor. The force of gravity, however, has not ceased to act and is still causing the body to press down on the floor with a force of *mg*. The resultant force in newtons which the body exerts on the floor is, therefore,

$$mg + ma = m(g + a)$$

Fig. 4.6. Three astronauts take part in weightless manoeuvres. The aircraft in which they are travelling is speeding through a pre-determined arc so that they are effectively under free-fall acceleration. Two crew members with their feet on the floor watch (left to right) Astronauts Edward White, James McDivitt and Neil Armstrong float around the cabin, in preparation for space flights

But the resultant force exerted by a body on its support is defined as its weight (pages 9 and 15).

$$\text{Hence, new weight of body in newtons} = m\,(g + a)$$

Lift moving downwards with acceleration a, less than g

Since the lift floor cannot apply a downward force on the body, what happens in this case is that a portion, ma, of the force of gravity, mg, is being employed to accelerate the body downwards. Only the remainder of the force of gravity, i.e., a resultant force of $(mg - ma)$ in newtons is available to press the body down on to the floor.

Hence, in this case, new weight of body $= m\,(g - a)$.

Lift falling freely

Suppose the worst happened and the lift rope broke. Both lift and body would fall freely. As far as the body is concerned the whole of the force of gravity, mg, would be employed in giving it free-fall acceleration of g. Nothing would be left over to press it on to the floor. Its weight has now become,

$$mg - mg = 0 \text{ newtons}$$

In other words the body is weightless when the lift falls freely. See also Fig. 4.6.

Lift moving downwards with acceleration a' greater than g

In order to make the lift move downwards with an acceleration greater than that of gravity it would have to be driven downwards by some form of propeller or rocket engine. The body inside would have the same acceleration, a', and this would require a force of ma' which is greater than the force of gravity mg on the body. In such circumstances the body would rise to the ceiling and obtain the extra required push from it of

$$(ma' - mg)$$

The body will therefore exert a reaction on the ceiling equal and opposite to this force.

In accordance with our definition, the weight of the body is the resultant force it exerts on its support (in this case the ceiling).

Hence, new weight of body $= ma' - mg = m (a' - g)$.

But this weight acts *upwards*, i.e., in the opposite direction to the body's normal rest weight with respect to the earth. So under these conditions we could say that it had a *negative* weight.

Fortunately, this is not a passenger lift or the unfortunate occupants would either bump their heads on the ceiling or, more comfortably, be able to walk around it upside down!

Newton's third law of motion

Whenever a force acts on one body, an equal and opposite force acts on some other body. This is sometimes stated, **to every action there is an equal and opposite reaction.** It is important to realize that the action and reaction act on different bodies.

This law has already been mentioned with reference to the force which a table exerts on a book placed upon it and also in connection with the weight of a body in an accelerating lift. We shall now discuss another aspect of the law.

When a bullet is fired from a gun equal and opposite forces are exerted on the bullet and gun during the time the bullet is passing down the barrel. Since, therefore, both bullet and gun are acted upon by equal forces for the same time, they will, in accordance with the second law of motion, acquire equal and opposite momenta. The backward momentum of the gun itself is shown by its kick or recoil. Hence,

$$\text{mass of bullet} \times \text{muzzle velocity} = \text{mass of gun} \times \text{recoil velocity}$$

It is to be noted that momentum is a vector quantity, i.e., it has direction as well as magnitude. The momenta of the bullet and gun are equal but in opposite directions. Consequently, the sum total of their momenta is zero. This illustrates an important principle arising out of Newton's second and third laws which is known as the

Law of the conservation of momentum

When two or more bodies act upon one another, their total momentum remains constant, provided no external forces are acting.

To investigate the conservation of momentum for interacting bodies moving in the same straight line

Experiment 1. *Inelastic collision of a moving body with one at rest*

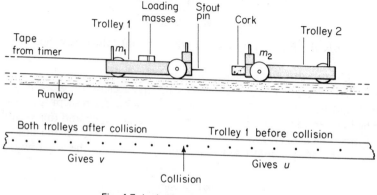

Fig. 4.7. Inelastic collision experiment

The arrangement of the apparatus is shown in Fig. 4.7. Two trolleys 1 and 2 are placed on a runway which is adjusted as far as possible to compensate for friction as explained on page 38.

Trolley 1 is loaded with extra masses to give it greater mass and has a tape attached which passes through a ticker-timer. It is then given a push so that it runs forward with uniform velocity and collides with trolley 2 which is at rest further along the runway.

One trolley is fitted with a stout pin and the other a cork so placed that, on collision, the pin penetrates the cork causing the trolleys to stick together and move on as one. This is described as an *inelastic* collision.

The tape is found to have two consecutive sets of equally spaced dots from which the velocity of trolley 1 before collision and the common velocity of both trolleys after collision may be found in the usual way. Both trolleys are weighed to find their masses, m_1 and m_2 in kg, and the results may be recorded as shown.

Results

Mass of trolley 1 = kg
Mass of trolley 2 = kg

Calculations from tape measurements

Distance x (m)	Time t (s)	Velocity $\frac{x}{t}$ (m/s)	Momentum (kg m/s)
		Trolley 1 before collision	
		$u =$	$m_1 u =$
		Both trolleys after collision	
		$v =$	$(m_1 + m_2)v =$

Total momentum before collision = kg m/s
Total momentum after collision = kg m/s

Experiment 2. *Partially elastic collision of a moving body with one at rest*

This experiment is performed in a manner similar to the previous one except that the pin and cork are replaced by a spring buffer to ensure an *elastic* collision (Fig. 4.8).

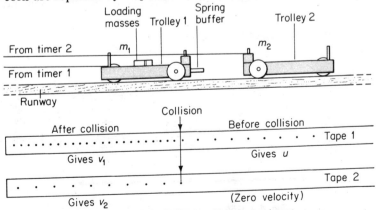

Fig. 4.8. Partially elastic collision experiment

Trolley 1 is loaded to make it more massive than trolley 2 so that, on collision, trolley 1 is slowed up and trolley 2 moves forward with a greater velocity. Two tapes are therefore required, one for each trolley, from which the velocities before and after collision are measured as before.

Suppose the masses of the trolleys are m_1 and m_2 respectively and let the velocity of trolley 1 in m/s be u before, and v_1 after collision. Then, if v_2 is the velocity of trolley 2 after collision, we should expect that,

total momentum before collision = total momentum after collision

or, in SI units (kg m/s),

$$m_1 u = m_1 v_1 + m_2 v_2$$

The various momenta are calculated and recorded as previously.

Results

Mass of trolley 1 =   kg
Mass of trolley 2 =   kg

Calculations from tape measurements

Distance x (m)	Time t (s)	Velocity $\frac{x}{t}$ (m/s)	Momentum (kg m/s)
Trolley 1 before collision			
Trolley 1 after collision			
Trolley 2 after collision			

Total momentum before collision =   kg m/s
Total momentum after collision =   kg m/s

Within the limits of experimental error we should expect to find that the total momentum before collision is equal to the total momentum after collision for both inelastic and partially elastic collisions.

Later on we shall have something to say about the *conservation of energy* involved when two bodies collide as distinct from the conservation of their momentum.

Experiment 3. *"Gun and projectile" experiment*

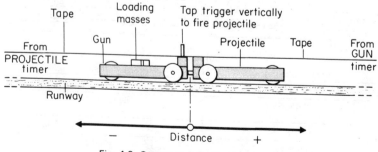

Fig. 4.9. Gun and projectile experiment

A heavy trolley (the "gun") is fitted with a horizontal metal rod which is pushed in against a spring and held in position by a metal plate. This is on a level runway in contact with a lighter trolley (the "projectile"). Both trolleys are attached to ticker-tape vibrators (Fig. 4.9).

The vibrators are switched on and the vertical trigger given a light tap to release the spring. The two trolleys move apart and their velocities v_1 and v_2, are measured from the equally spaced dots on their tapes. The masses m_1 and m_2 of gun and projectile are found by the use of a balance.

Now the total momentum of gun and projectile at the start was zero since both were at rest. How then is the momentum still zero after firing when both are in motion? The answer is simple. Momentum, being a vector quantity, has direction.

From a given starting-point (the origin) we represent distances and velocities to the right by a + sign; those to the left by a − sign. The reader will have already met this convention during the study of graphs in mathematics.

We must, therefore, write the gun's velocity as $-v_1$,

and the projectile's velocity as $+v_2$.

The momentum before firing $= 0$

and momentum after firing $= -m_1v_1 + m_2v_2$.

Errors inevitably occur owing to the impossibility of compensating for friction when two trolleys are moving in opposite directions. Within these limitations we should expect the total momentum after firing to be zero.

Rocket propulsion

Let us now think of a toy rubber balloon which has been blown up with air and its mouth secured with string. If the string is now removed so that the air can escape rapidly, the balloon will, if released, dart round the room until all the air has gone. It behaves as a simple rocket.

The rockets which form a familiar feature of firework displays contain solid chemicals which burn to produce a high-velocity blast of hot gas. Space rockets which are designed to travel large distances have tanks of liquid fuel together with a supply of either liquid oxygen or some other liquid which produces oxygen to enable the fuel to burn.

In either case, a chemical reaction takes place inside the rocket and creates a large force which propels the gaseous products of combustion out through the tail nozzle with tremendous velocity. The reaction to this force propels the rocket forwards. Although the mass of gas emitted per second is comparatively small, it has a very large momentum on account of its high velocity. An equal momentum is imparted to the rocket in the opposite direction, so that, in spite of its large mass, the rocket also builds up a high velocity.

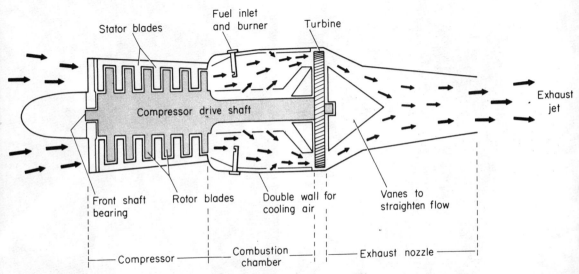

Fig. 4.10. Jet engine or gas turbine (see page 50)

Fig. 4.11. *Beginning the conquest of space.*

Far left: The Atlas rocket, carrying Project Mercury space capsule Friendship 7 with the first U.S. Astronaut John Glenn inside, blasting off the launching pad at Cape Canaveral, Florida, on 20 February 1962 for a tri-orbital space flight around the earth

Left: The second U.S. astronaut Scott Carpenter examining his spacecraft. This honeycomb-like structure dissipates, in the form of heat, the internal energy resulting from the work done against friction when the vehicle re-enters the earth's atmosphere. Scott Carpenter also made a tri-orbital space flight on 24 May 1962

Bottom left: John Glenn's space capsule being hoisted out of the Atlantic after his flights round the earth

Bottom right: *Rehearsal for the moon landing.* Docked Apollo 9 command service module and lunar module with the earth in the background. This picture, taken by Russell Schweickart, shows David Scott standing in the open hatch of the command module. Subsequently the first men on the moon were Neil Armstrong and Edwin Aldrin from Apollo 11 in July 1969. See also Fig. 2.1 on page 15

Below: July 1971: Apollo 15. James Irwin prepares for a trip across the surface of the moon in the lunar rover vehicle. Five kilometres away in the background is the St. George Crater.

Jet engines

An aircraft jet engine works on the same principle as a rocket, the difference between them is concerned only with the method of obtaining the high-velocity gas jet. Fig. 4.10 shows, in simplified form, the construction of an axial-flow gas turbine or jet engine. The fuel used in these engines is kerosene (paraffin). This is sprayed through burners into combustion chambers, where it burns in a blast of compressed air and produces a high-velocity jet of gas which emerges from the exhaust nozzle.

The air supply is drawn in at the front of the engine and compressed by a turbine compressor. A turbine is simply a special kind of fan having alternate sets of fixed and rotating metal blades. The compressor itself is driven by another smaller turbine worked by the exhaust jet.

In order to start the engine it is, of course, necessary to have a powerful starter motor to impart an initial rotation to the compressor and a hot electric spark to ignite the kerosene inside the combustion chamber.

Jet engines of this type are used for aircraft which fly through the atmosphere, but they cannot be used to propel vehicles into outer space where there is no air supply. Space rockets, on the other hand, can travel beyond the earth's atmosphere, since they carry their own oxygen supply (Fig. 4.11).

Worked examples

1. *A car of mass* 1000 kg *travelling at* 36 km/h *is brought to rest over a distance of* 20 m. *Find:* (a) *the average retardation;* (b) *the average braking force in newtons.*

The retardation is found from the formula,

$$v^2 = u^2 + 2ax$$

in which $\quad v = \quad 0 \text{ m/s}$

$$u = 36 \text{ km/h} = \frac{36 \times 1000}{60 \times 60} = 10 \text{ m/s}$$

$$x = 20 \text{ m}$$

By substitution,

$$0^2 = 10^2 + 2 \times a \times 20$$

therefore

$$a = \frac{-10^2}{2 \times 20} = -2.5 \text{ m/s}^2$$

(minus sign means retardation)

Knowing *m* and having found *a*, we now substitute in $F = ma$, to find F. Thus,

$$F = 1000 \times 2.5 \text{ N}$$
$$= 2500 \text{ N}.$$

Answer. Average retardation $\quad = 2.5 \text{ m/s}^2$

Average braking force $= 2500 \text{ N}$

2. *A bullet of mass* 20 g, *travelling with a velocity of* 16 m/s, *penetrates a sandbag and is brought to rest in* 0.05 s. *Find:* (a) *the depth of penetration in metres:* (b) *the average retarding force of the sand in newtons.*

We must first find the average retardation from the formula,

$$v = u + at$$

in which $\quad v = 0 \text{ m/s}$

$$u = 16 \text{ m/s}$$

$$t = \quad 0.05 \text{ s}$$

Substituting,

$$0 = 16 + a \times 0.05$$

from which

$$a = \frac{-16}{0.05} = -320 \text{ m/s}^2$$

(minus sign means retardation)

The depth of penetration may be found using either

$$x = ut + \tfrac{1}{2}at^2$$

or

$$v^2 = u^2 + 2ax$$

Suppose we choose the latter, then

$$0^2 = 16^2 + 2(-320) \times x$$

whence

$$x = \frac{-16 \times 16}{2(-320)} = 0.4 \text{ m}$$

The average retarding force may now be calculated from

$$F = ma$$

in which $m = 20$ g

$= 0.02$ kg   (remember, we must convert to kg)

and $a = -320$ m/s^2

Thus, $F = 0.02 \times (-320)$

$= -6.4$ N   (*minus sign means retarding force*)

Answer. Depth of penetration   = 0.4 m

Average retarding force = 6.4 N

3. *The valve of a cylinder containing 12 kg of compressed gas is opened and the cylinder empties in 1 min 30 s. If the gas issues from the exit nozzle with an average velocity of 25 m/s, find the force exerted on the cylinder.*

The force required to accelerate the gas out of the cylinder is given by,

$$F = ma$$

$$= \text{mass} \times \frac{\text{change in velocity}}{\text{time taken}}$$

$$= \frac{\text{change in momentum}}{\text{time taken}}$$

The velocity of the gas changes from rest to 25 m/s

change in momentum $= 12 \times 25$ kg m/s

time taken $= 90$ s

average force on gas $= \dfrac{12 \times 25}{90}$ newtons

$= 3.3$ N

By Newton's third law, an equal reaction force is exerted on the cylinder.

Answer. Average force on cylinder = 3.3 N

QUESTIONS: 4

(*Where necessary assume g $= 10$ m/s^2.*)

1. A resultant force of 25 newtons acts on a mass of 0.50 kilograms starting from rest. Find:
   (a) the acceleration in m/s^2;
   (b) the final velocity after 20 seconds;
   (c) the distance moved in m.

2. A resultant force of 12 N acts for 5 s on a mass of 2 kg. What is the change in momentum of the mass? What would be the change in momentum of a mass of 10 kg under the same conditions?

3. State Newton's Laws of motion and ex- plain how the second law may be used to define a unit of force, the newton.

A breakdown truck tows a car of mass 1000 kg along a level road, and accelerates at 0.5 m/s^2. What is the tension in the towline? If the towline breaks when the car reaches a speed of 36 km/h, how far will the car travel before coming to rest if a braking force of 5000 N is applied?   (*W.*)

4. A trailer of mass 1000 kg is towed by means of a rope attached to a car moving at a steady speed along a level road. The tension

in the rope is 400 N. Why is it not zero?

The car starts to accelerate steadily. If the tension in the rope is now 1650 N, with what acceleration is the trailer moving?

(*O.C.*)

5. It is required to cause a body of mass 500 g to accelerate uniformly from rest across a smooth horizontal surface so that it will cover a distance of 20 m in 4 s. Determine:

(i) the acceleration which must be imparted to the body;

(ii) the magnitude of the minimum force required to produce this acceleration; and

(iii) the momentum of the body at the end of the 4-s period.

(*L.*)

6. A bullet of mass 0.006 kg travelling at 120 m/s penetrates deeply into a fixed target, and is brought to rest in 0.01 s. Calculate:

(*a*) the distance of penetration of the target,

(*b*) the average retarding force exerted on the bullet.

(*C.*)

7. A wooden trolley of mass 1.5 kg is mounted on wheels on horizontal rails. Neglecting friction and air resistance, what will be the final velocity of the trolley if a bullet of mass 2 g is fired into it with a horizontal velocity of 400 m/s along the direction of the rails?

8. (*a*) Describe, with the aid of a clearly labelled diagram, the experiment which you would perform in order to investigate the relationship between the force acting on a body and the acceleration produced. Show clearly how you would calculate the acceleration from the measurements taken in the experiment and indicate the conclusion reached.

(*b*) An arrow of mass 100 g is shot into a block of wood of mass 400 g lying at rest on the smooth surface of an ice rink. If at the moment of impact the arrow is travelling horizontally at 15 m/s, calculate the common velocity after the impact.

Also calculate the common velocity if the block is then struck by a second similar arrow travelling in the same direction but with horizontal velocity of 12 m/s.

(*A.E.B.*)

9. State the law of conservation of momentum. Explain why the recoil velocity of a gun is much less than the velocity of the bullet. Describe an experiment to illustrate your answer and show how the result is calculated.

10. Two boys of masses 45 kg and 60 kg sit facing one another on light frictionless trolleys holding the ends of a strong taut cord between them. The lighter boy tugs the cord and acquires a velocity of 2 m/s. What is the initial velocity of the other boy? What happens to their motion when they collide? Explain your answers carefully. (*S.*)

11. An inflated balloon contains 2.0 g of air which is allowed to escape from a nozzle at a speed of 4.0 m/s. Assuming that the balloon deflates at a steady rate in 2.5 s, what is the force exerted on the balloon? (*S.*)

12. Explain the terms; velocity and momentum. What is the relation between force and momentum?

A rocket of total mass 5000 kg, of which 4000 kg is propellent fuel is to be launched vertically. If the fuel is consumed at a steady rate of 50 kg/s, what is the least velocity of the exhaust gases if the rocket will just lift off the launching pad immediately after firing?

13. A man whose mass is 70 kg stands on a spring weighing machine inside a lift. When the lift starts to ascend its acceleration is 2.5 m/s^2. What is the reading of the weighing machine? What will it read:

(*a*) when the velocity of the lift is uniform;

(*b*) as it comes to rest with a retardation of 5.0 m/s^2?

14. Describe how you would investigate experimentally the relationship between *extension* of a light spring and the *load* which it supports. State what you understand by *elastic limit* and, assuming that the spring is not loaded beyond this point, sketch the graph which you would expect to obtain from your readings. Explain how you would use the apparatus and graph to find the weight of a stone which is less than the maximum load you put on the spring. (*L.*)

15. A spring with its upper end fixed, hangs vertically alongside a millimetre scale. The lower end of the spring gave the following readings when various masses were hung from it:

Mass (kg)	0	0.02	0.04	0.06	0.08	0.10
Readings (mm)	110	121	129	139	151	161

For each reading calculate:

(*a*) the values of the force applied;

(*b*) the extension in millimetres. Plot a graph of extension in mm against force in N. From the graph find:

(i) the extension for a mass of 0.045 kg;

(ii) the scale reading when a force of 0.15 N is applied.

16. A vertical spring of unstretched length 30 cm is rigidly clamped at its upper end. When an object of mass 100 g is placed in a pan attached to the lower end of the spring its length becomes 36 cm. For an object of mass 200 g in the pan the length becomes 40 cm. Calculate the mass of the pan. Name and state clearly the law you have assumed. (*S.*)

5. Vectors

On page 22 we explained the difference between scalar and vector quantities by saying that vectors have both magnitude *and* direction while scalars have magnitude only.

In this chapter we shall concern ourselves with the addition of vectors by the aid of *vector diagrams*. The sum of two or more vectors is a single vector which is called their *resultant*. Scalars, of course, are simply added by the ordinary rules of arithmetic: 5 kg added to 2 kg makes 7 kg; 25 books added to 10 books makes 35 books and so on.

Vector quantities such as displacement, force and velocity cannot be dealt with so simply unless they all act in the same or directly opposite directions. For example, a force of 5 N added to a force of 2 N can have a resultant of anything between 3 N and 7 N according to the directions in which these forces act.

Let us begin by showing how to add displacements.

Addition of displacements

Suppose a man starts from a point A (Fig. 5.1) and walks a distance of 3 km in a direction due NE to a point B. From B he then walks 4 km NW ending at a point C. His resultant displacement is found as follows.

The first displacement is represented by a line AB drawn 45° east of north, 3 units long and with an arrow on it to show the direction of travel. From the head of this line we draw the second displacement BC as a line 4 units long, 45° west of north, i.e., at right angles to AB.

Clearly the man could have reached his final destination by walking direct from A to C, provided there were no obstructions. Hence AC represents the resultant sum of the two vectors AB and BC both in magnitude and direction. Measurement shows that this resultant displacement is 5 km in a direction 8° west of north.

Polygon of vectors

A number of vectors which are added together or compounded to give a single resultant are called the **components** of the resultant. In the previous example the rule we have used, namely, *to draw each component vector in turn with the head of one starting from the tail of the previous one* can be extended to any number of vectors so that the diagram becomes a polygon instead of a triangle. Also the order in which the vectors are taken does not matter provided they are drawn in the correct direction. Moreover, this method can be applied to any kind of vector; force, velocity, acceleration, and so on. Fig. 5.2 shows how the method could be used to find the resultant of four component forces of 2, 1.5, 3 and 2.5 N respectively acting on a body at a point A in the directions shown.

Resultant force

We shall now apply the above rule to find the resultant force on an ocean liner which is being towed into harbour by two tugs A and B exerting forces of 2.5 kN and 3.5 kN respectively and with their tow-ropes making an angle of 68° (see Fig. 5.3).

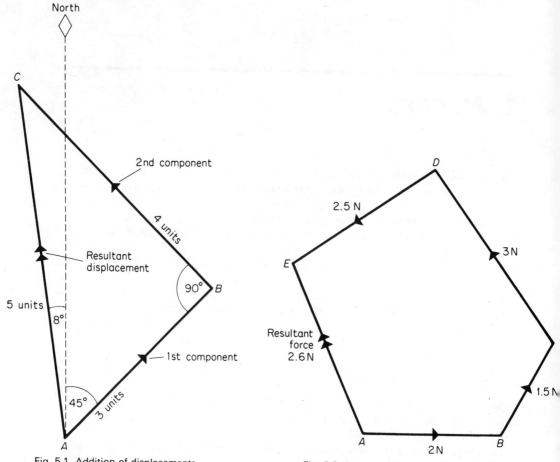

Fig. 5.1. Addition of displacements

Fig. 5.2. Polygon of vectors (forces in this case)

A line *LA* 2.5 units long with an arrow on it is drawn to represent the pull of tug *A*.

From *A* we now draw a line *AB* 3.5 units long parallel to the direction of the second tug (i.e., making an angle of 68° with the direction of *LA*) and also arrowed to represent the pull of tug B.

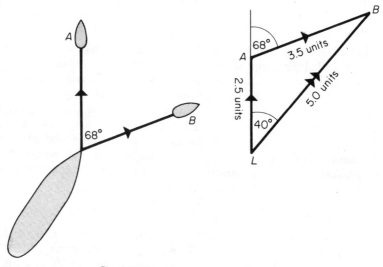

Fig. 5.3. Resultant force on a liner

Measurements made on our vector diagram show that *LB* is 5.0 units long and makes an angle of 40° with *LA*. Hence the resultant pull is 5.0 kN, making an angle of 40° with the tow-rope of tug *A*.

Now the addition of displacements by a vector diagram which we described earlier was fairly obvious, but we merely *stated* that the vector diagram gave the correct resultant when applied to vectors other than displacements. There is no formal proof: rather it is that we *define* the resultant in terms of the rule for constructing a vector diagram. The only real proof is an experimental one: so we shall now devise an experiment to test the truth of our result obtained for the resultant pull on the liner.

An experiment to find the resultant of two forces

To make our experiment manageable we shall reduce all the forces to one ten-thousandth of their actual value so we shall have what is called a small-scale simulation of the real thing. This kind of small-scale reduction is often done both in drawings and models when planning the construction of buildings, bridges, harbours, ships, and so on.

For the forces involved we shall use the weights of standard masses of 0.025, 0.035, and 0.050 kg so that, assuming $g = 10$ m/s², the forces will be 0.25, 0.35, and 0.50 N respectively.

A drawing board with a sheet of paper attached is set up vertically with two freely-running pulleys clamped to the top (Fig. 5.4). A length of thread having a 25 g mass at one end and a 35 g mass at the other is passed over the two pulleys and a second length of thread carrying a 50 g mass is tied to the first at *O*. The threads will take up a position for which the three forces acting at *O* are in equilibrium. Small pencil crosses are now made on the paper, as far apart as possible, to mark the positions of the threads. A fairly accurate method of doing this is to mark the positions of the shadows of the threads formed either by the sun or a *distant* lamp.

The paper is now removed from the board and the crosses joined by pencil lines to represent the force directions. Next, using a scale of 20 cm to represent 1 N, mark off a distance $OP = 5$ cm to represent the force of 0.25 N. Now mark off $PR = 7$ cm, parallel to the direction of the force of 0.35 N. Join *OR*. This line represents the resultant of the two forces of 0.25 N and 0.35 N.

Clearly, these two forces together are balanced by the 0.50 N force. Therefore the resultant of the two forces must be equal and opposite to the 0.50 N force.

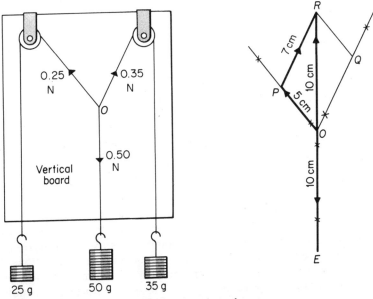

Fig. 5.4. Resultant of two forces

We can measure the length of the line *OR* and show that it corresponds to a force of 0.50 N and that its direction is equal and opposite to this force. Thus we have verified our rule for adding vectors in this particular case.

Equilibrant

In the experiment just described (Fig. 5.4) the three forces acting at *O* are in equilibrium, that is to say they exactly balance one another. Any one of these forces is said to be the **equilibrant** of the other two.

Worked example

Find by means of a vector diagram or otherwise the resultant of two forces of 7 N and 3 N acting at right angles to one another.

The problem may be solved by means of an accurate vector diagram using any convenient unit of length to represent one newton. Thus in Fig. 5.5 *OP* and *PR* are drawn at right angles and are made 3 and 7 units long respectively. *OR* represents the resultant which, when measured, is found to be 7.6 units long and makes an angle 23° with the 7 N force.

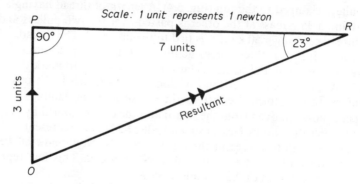

Fig. 5.5. Resultant of two forces at right angles

Otherwise, by Pythagoras's theorem

$$OR^2 = OP^2 + PR^2$$
$$= 3^2 + 7^2 = 9 + 49 = 58$$

hence
$$OR = \sqrt{58}$$
$$= 7.6 \text{ N}$$

also, if θ is the angle *PRO*, $\tan \theta = \frac{3}{7} = 0.4286$
therefore, from tables, $\theta = 23° \, 12'$

It follows that the resultant is a force of 7.6 N in direction making an angle of 23° 12' with the 7 N force.

Resolution of forces

So far we have concerned ourselves with the resultant of two or more vectors. Particularly in the case of forces, it is very often necessary to carry out the reverse process and convert a single force into two components. When this is done the force is said to be **resolved** into two components. Now the number of directions in which a force can be resolved is infinite, since any number of different vector triangles may be drawn, but except in rare cases a force is resolved into two directions at right angles only.

Fig. 5.6 shows the forces involved when a barge is being towed along a canal by a horse. Though rarely seen today, this was an important mode of transport in England during the eighteenth and nineteenth centuries before the advent of rail and road transport.

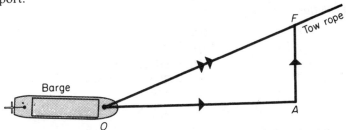

Fig. 5.6. Forces on a towed barge

The actual pull on the barge is represented to scale by the line OF. If a right-angled triangle OAF is constructed on OF as hypotenuse and with its longer side directed along the centre line of the barge, then we have a vector diagram showing that the pull OF is equivalent to two components OA and AF. The component OA is usefully employed in pulling the barge along the canal, while the component AF merely tends to pull it into the bank. This tendency to be pulled into the bank is counteracted by use of the rudder so as to turn the prow of the barge slightly outwards.

Worked examples

1. *A garden roller is pulled with a force of 200 N acting at an angle of 50° with ground level. Find the effective force pulling the roller along the ground.*

As a rule, problems of this type may be solved either by scale diagram or else by calculation, so both methods will be explained here.

(a) *By vector diagram* (Fig. 5.7).

Using any suitable scale, the pull on the roller is represented by a line OF, 200 units long making an angle of 50° with the horizontal, OH.

If a perpendicular is now dropped from F on to OH, the sides OH and HF of the vector diagram OHF represent the horizontal and vertical components of OF

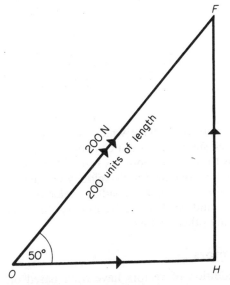

Fig. 5.7. Components of a force

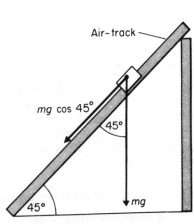

Fig. 5.8. Air-track problem

respectively. Of these the horizontal component OH represents the effective force which pulls the roller along the ground. Measurement shows that $OH = 129$ units long.

Hence the effective force $= 129$ N

(b) *By calculation*

In Fig. 5.7,

$$\frac{OH}{OF} = \cos 50°$$

therefore

$$OH = OF \cos 50°$$
$$= 200 \times 0.6428$$
$$= 129$$

hence

effective force $= \underline{129 \text{ N}}$

Before leaving this problem it is of interest to consider the part played by the vertical force component HF. This acts against the weight of the roller, and therefore reduces the force which the roller exerts on the ground. On the other hand, if the roller is pushed instead of being pulled, the vertical component increases its effective weight.

2. *A 50 g mass is placed on a straight air track sloping at an angle of 45° to the horizontal. Calculate, in m/s², the acceleration of the load as it slides down and also the distance it would move from rest in 0.20 s.*

The force on the load causing it to accelerate is the resolved part of the force of gravity on it along the air track (Fig. 5.8) and is equal to $mg \cos 45°$.

which

$$= 0.05 \times 10 \times \frac{1}{\sqrt{2}} \text{ N}$$

The acceleration is found from the equation $F = ma$

Thus

$$a = \frac{F}{m} = \frac{0.05 \times 10}{\sqrt{2} \times 0.05} = 7.1 \text{ m/s}^2$$

The distance, x, moved from rest is found from

$$x = ut + \tfrac{1}{2}at^2$$

where
$u = 0$ m/s
$a = 7.1$ m/s²
$t = 0.20$ s

substituting,
$$x = 0 \times 0.20 + \frac{7.1 \times 0.20^2}{2}$$
$$= 0.14 \text{ m}$$

Answer. Acceleration $= 7.1$ m/s²
Distance moved $= \underline{0.14 \text{ m}}$

Addition of velocities

A passenger sitting in the corner seat of a railway carriage which is travelling at constant speed along a straight section of track simply has a velocity equal to that of the train. If he now gets up and walks across the carriage floor his resultant velocity with respect to the track is made up of two components: (a) his velocity OA across the carriage, and (b) the velocity AT of the train (Fig. 5.9). His resultant velocity with respect to the railway track is given by OT and is obtained by exactly the same method as we have used previously for adding other vectors.

The parallelogram rule for adding vectors

All of the examples given so far for the addition of vectors have been based on methods involving the drawing of triangles, or where more than two components are

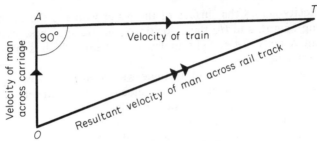

Fig. 5.9. Resultant velocity

concerned, by drawing vector polygons. However, there is another method which the reader may come across which has been used more frequently in the past than the present. This is the parallelogram rule.

In this method the two components are represented by the adjacent sides of a parallelogram, in which case the resultant is given in magnitude and direction by the diagonal of the parallelogram drawn from the point of intersection of the two sides.

Fig. 5.10 shows how our previous problem is solved by this method. Note that the parallelogram, in this case a rectangle, consists of two congruent triangles, either of which alone would have served equally well for finding the resultant velocity, since it does not matter in which order the component vectors are drawn.

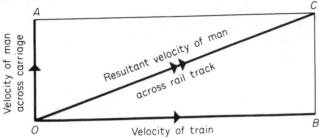

Fig. 5.10. Resultant velocity using parallelogram rule

Note that the parallelogram rule can also be applied to forces. There is a parallelogram of forces in Fig. 5.4.

The ferryman's problem

Those who have ever crossed a flowing river in a small ferry boat will have noticed that the boatman has to point the nose of the boat upstream in order to make a course straight across the river.

Suppose the boat starts from point *O* (Fig. 5.11). It has two component velocities:

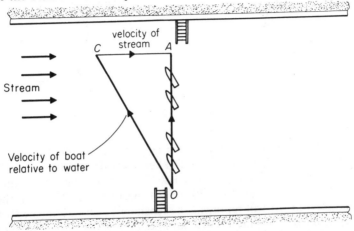

Fig. 5.11. The Ferryman's problem

(*a*) its own velocity across the surface of the water which is the same whether the water is moving or not, and (*b*) the velocity of the stream. These two components have a resultant velocity *OA* at right angles to the bank, which is given by completing the vector triangle *OCA*.

In order to construct this vector triangle we require to know *two only* of the following data: the stream velocity *CA*, the velocity of the boat relative to the water *OC*, the angle *COA* or the required resultant velocity *OA*.

Having constructed the vector triangle, the remaining quantities may be found from it. As in previous cases where we have used the rule for adding vectors, we can do the problem either by an accurate graphical construction or else by calculation from a rough diagram using Pythagoras's theorem.

Notice that the prow of the boat is pointed in the direction *OC*.

Three forces in equilibrium

We shall conclude our study of vectors by considering the vector diagram for three forces which are in equilibrium.

If we know the magnitudes and directions of the forces we should expect the vector diagram to form a closed triangle with the force arrows all following one another round in the same direction. The vector triangle will be closed, since forces in equilibrium have zero resultant, so there can be no open side to represent it.

If, however, we have three forces known to be in equilibrium but we know the magnitude of only one of them together with the directions it makes with the other two we can find the value of these forces from a vector diagram.

Such a case is shown in Fig. 5.12. Here we have a 15 kg mass suspended from a hook in the ceiling which is pulled aside by a horizontal string until the supporting string makes an angle of 30° with the vertical. The three forces in equilibrium are the tensions T_1 and T_2 in the strings and the weight of the 15 kg mass which, assuming $g = 10$ m/s², is $15 \times 10 = 150$ N.

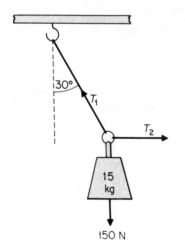
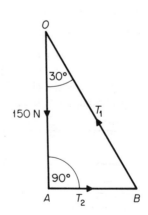

Fig. 5.12.

Let us construct a scale vector diagram (Fig. 5.12). Using a scale of 1 cm to represent a force of 10 N we draw a line *OA* vertically downwards, and 15 cm long, to represent the weight of the 15 kg mass. The vector triangle is completed by drawing *AB* at 90° to *OA* and *OB* at 30° to *OA* meeting at *B*. Then, on the scale we have chosen, *BO* and *AB* will represent the tensions T_1 and T_2 in the strings. If we have drawn the diagram accurately we can find T_1 and T_2 by measurement. Otherwise we can calculate the values by trigonometry:

$$\frac{150}{T_1} = \cos 30°$$

therefore $\qquad T_1 = \dfrac{150}{\cos 30°} = \dfrac{150}{0.866} = \underline{173 \text{ N}}$

Also, $\qquad \dfrac{T_2}{15} = \tan 30°$

therefore $\qquad T_2 = 15 \tan 30° = 150 \times 0.577 = \underline{87 \text{ N}}$

QUESTIONS: 5

(*Where necessary assume g = 10 m/s²*)

1. Which of the following are vector and which are scalar quantities; velocity, mass, speed, force, displacement, acceleration, weight?

2. What is meant by:
 (a) the resultant;
 (b) the equilibrant of two or more forces?

3. Find by drawing or calculation the resultant of two forces each 5.0 N acting at a point at an angle of 60° with each other. (C.)

4. Three strings are attached to a small metal ring, Two of the strings make an angle of 70° and each is pulled with a force of 7 N. What force must be applied to the third string to keep the ring stationary?

5. Two forces acting at a point make angles of 25° and 65° respectively with their resultant which is of magnitude 15 N. Find the magnitudes of the two component forces.

6. A body is in equilibrium under the action of three forces. One force is 6.0 N acting due East and one is 3.0 N in a direction 60° North of East. What is the magnitude and direction of the third force? (O.C.)

7. The ends of a light string are tied to two hooks A and B in a ceiling which are 100 cm apart horizontally, so that the length of string between hooks is 140 cm. A 650 g mass is then attached by a second length of string to a point C on the first, 80 cm from A, and hangs freely. Find, by drawing or calculation, the tensions in the portions AC and BC of the string.

8. A motor vessel tows a small dinghy by a rope which makes an angle of 20° with the horizontal. If the tension in the rope is 150 N, find:
 (a) the force which effectively pulls the dinghy forward;
 (b) the force which lifts its bows out of the water.

9. A man, using a 70 kg garden roller on a level surface, exerts a force of 200 N at 45° to the ground. Find the vertical force of the roller on the ground:
 (a) if he pulls;
 (b) if he pushes the roller.

10. (a) Describe and explain an experiment to show how the resultant of two non-parallel forces acting at a point may be determined.
 (b) Three forces of magnitude 6 N, 2 N, and 3 N act on a small object in directions North, South, and West respectively. Find the direction and magnitude of the resultant force.
 If the object is free to move and its mass is 0.2 kg calculate the initial acceleration. (S.)

11. A microphone of mass 500 g hangs from the end of a long wire fixed to the ceiling. A horizontal string attached to the microphone exerts a pull which keeps the wire at an angle of 20° to the vertical. Find the tensions in both string and wire.

12. A smooth board placed at 30° to the horizontal supports a mass of 10 kg which is prevented from slipping by attaching it to a spring lying along the board and fixed at its upper end. Find the tension in the spring. If the spring is removed, what horizontal force would have to be applied to the mass to keep it at rest?

13. A man can row a boat in still water at 6 km/h. He wishes to row due north across a river 3 km wide which is flowing due East at 2 km/h.
 Either by scale drawing or by calculation, find:
 (i) the direction in which he must head the boat;
 (ii) the time taken to reach the other bank;
 (iii) how far from his destination he would land if he mistakenly steered due North. (A.E.B.)

6. Turning forces

Every time we open a door, turn on a tap or tighten up a nut with a spanner, we exert a turning force.

Two factors are involved here, first, the magnitude of the force applied, and secondly, the distance of its line of action from the axis or *fulcrum* about which turning takes place. A very large turning effect can be produced with a comparatively small force provided the distance from the fulcrum is large. For this reason, it is easier to loosen a tight nut with a long spanner than with a short one. Experienced mechanics are aware that long spanners must be used with care, as it is easy to strip screw threads by too large a turning force.

The combined effect of the force and distance which determines the magnitude of the turning force is called the *moment* of the force and is defined as follows.

The moment of a force about a point is the product of the force and the perpendicular distance of its line of action from the point.

Two experiments to study moments

Experiment 1. A small hole is drilled in a metre rule on the 50 cm mark but slightly offset from centre. The rule is then pivoted on a knitting needle so that it balances horizontally (Fig. 6.1). If one side of the rule happens to be slightly heavier it will tilt, but this can be corrected by means of a U-shaped rider made of thick copper wire. The rider is placed on the lighter side of the rule and its position adjusted until the rule becomes horizontal.

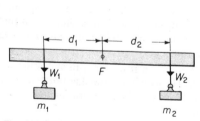

Fig. 6.1. Experiment to study moments (1)

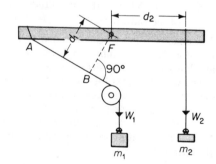

Fig. 6.2. Experiment to study moments (2)

Unequal masses, m_1 and m_2 are now hung from cotton loops on either side of the rule and their distances, d_1 and d_2, from the pivot are adjusted until the rule once more comes to rest horizontally. The weights, W_1 and W_2, of the masses are now exerting equal and opposite movements about the pivot.

The experiment is repeated a number of times using different pairs of weights and distances and the results tabulated.

Making allowance for experimental error, it will be seen that in every case the product, *force (or weight) × distance* on the left-hand side is equal to the product, *force (or weight) × distance* on the right-hand side. We therefore conclude that the product, *force × distance from the fulcrum* measures the turning effect of the force.

In each case the weight is given by mg (in N) where m = mass in kg and g is assumed equal to 10 m/s^2.

m_1 (kg)	$W_1 =$ $(m_1 \times 10)$ (N)	d_1 (m)	m_2 (kg)	$W_2 =$ $(m_2 \times 10)$ (N)	d_2 (m)	$W_1 \times d_1$ (N m)	$W_2 \times d_2$ (N m)

Experiment 2. In the previous experiment the strings supporting the weights were at right angles to the rule, and therefore the distance along the rule was also the perpendicular distance from the fulcrum.

In the second experiment the same apparatus is used, except that one of the strings is passed over a pulley as in Fig. 6.2, so that the line of action of the force W_1 is not now at right angles to the rule.

The rule is balanced horizontally as before by suitably adjusting the positions of the strings along it. The perpendicular distance of the line of action of W_1 from F is now equal to FB. This distance, d_1, is measured with a half-metre rule and is entered in a table together with W_1, W_2 and d_2.

When the products $W_1 d_1$ and $W_2 d_2$ are worked out they are once more found to be equal. This shows that the turning effect of W_1 in this experiment is given by $W_1 \times FB$ and not $W_1 \times FA$.

These two experiments verify the *principle of moments*.

Principle of moments

When a body is in equilibrium, the sum of the anticlockwise moments about any point is equal to the sum of the clockwise moments.

Resultant moment

When dealing with problems involving a number of moments acting on a body which is not in equilibrium, the first step is to draw a sketch indicating the forces and their distances from a fulcrum.

It is customary to give a positive sign to anticlockwise moments and a negative sign to clockwise moments. The various moments are written down with appropriate signs and are added algebraically. The sign of the answer will then give the direction of the resultant moment.

Example. A rod AE of negligible weight, 40 cm long, is pivoted at a point D. Weights of 1, 2, 3, and 4 N act on the rod as shown in Fig. 6.3.

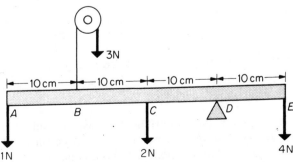

Fig. 6.3. Resultant moment

Taking the forces acting from left to right, we have, measuring in metres

sum of moments = (1 × 0.30) − (3 × 0.20) + (2 × 0.10) − (4 × 0.10) = −0.50.

Hence the resultant moment is 50 N m units acting in a clockwise direction.

Parallel forces. Couples

Parallel forces which act in the same direction are called *like* forces, and it is always possible to find their resultant or the single force which exactly replaces them.

The same usually applies in the case of parallel forces which are *unlike* or act in opposite directions. There is, however, one special case in which a single resultant force cannot be found, namely, where the forces can be shown to be equivalent to two equal and opposite parallel forces. **Equal and opposite parallel forces form what is called a couple** and the reader should be able to show quite easily that the moment of a couple is equal to the product of one of the forces times the perpendicular distance between them. It is not possible to find a single force to replace a couple. A couple simply produces rotation and can only be balanced by an equal and opposite couple.

To study parallel forces in equilibrium

A metre rule is weighed and then suspended by vertical threads attached to two spring balances held by clamp and stand (Fig. 6.4). Two weights are attached as shown so that one exerts an upward and the other a downward pull on the rule. Lastly, the spring balances are raised or lowered as necessary to bring the rule into a horizontal position.

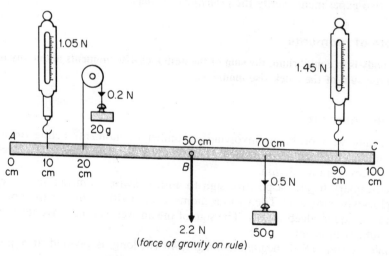

Fig. 6.4. Parallel forces in equilibrium

We shall assume, for the time being, that the force of gravity on the rule itself acts downwards through its centre. In this case the centre of the rule is also its centre of gravity, and this will be explained more fully later in the chapter.

The results of a typical experiment are indicated in Fig. 6.4. It will be noted that

the sum of the upward forces = 1.05 + 0.20 + 1.45 = 2.7 N, and

the sum of the downward forces = 2.2 + 0.5 = 2.7 N.

Furthermore, as the rule is in equilibrium, there is no resultant moment acting which would cause it to turn about *any* point. This may be verified by working out the moments of the forces about any point we choose and showing that the sum of the anticlockwise moments is equal to the sum of the clockwise moments.

It is convenient to do this by entering the various forces and distances in a table. In accordance with convention, anticlockwise moments are given a positive sign and clockwise moments a negative sign. Moments have been taken about the end *A*.

Force (N)	Distance from A (m)	Moment about A (N m)
1.05	0.10	+0.105
0.20	0.20	+0.040
2.20	0.50	−1.100
0.50	0.70	−0.350
1.45	0.90	+1.305

Sum of anticlockwise moments (+) 1.45 N m
Sum of clockwise moments   (−) 1.45 N m

The experiment shows that, when a number of parallel forces are in equilibrium:

(i) the sum of the forces in one direction is equal to the sum of the forces in the opposite direction;

(ii) the sum of the anticlockwise moments about any point is equal to the sum of the clockwise moments about the point.

When performing this experiment, the reader should use different weights and distances from those used here and draw an appropriate diagram. The results should be entered in a table and a different point chosen about which moments are taken.

Centre of gravity

We have already seen a body is attracted to the earth by the force of gravity. This statement, however, says nothing about the point of application of the force, so we shall now discuss this.

Any particular body, a stone, for example, may be regarded as being made up of a very large number of tiny equal particles of mass *m*, each of which is pulled towards the earth with a force *mg* (Fig. 6.5). The earth's pull on the stone thus consists of a very large number of equal almost parallel forces. These will have a resultant which is equal to the total force of gravity *Mg* on the stone, and it will act through a point *G* called the centre of gravity.

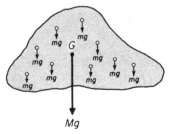

Fig. 6.5. Centre of gravity

The centre of gravity of a body is defined as the point of application of the resultant force due to the earth's attraction on it.

The centre of gravity of a body also coincides with its **centre of mass.**

A ruler can be made to balance on a finger-tip if the finger is placed immediately below the centre of gravity. Under these conditions the ruler is in equilibrium under the action of two forces; the force of gravity acting vertically downwards and the equal and opposite reaction of the finger to the weight of the ruler acting upwards. Should the centre of gravity not be exactly above the finger, the force of gravity will have a turning moment about the finger and so cause the ruler to topple over.

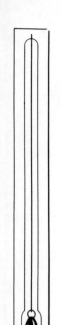

Note carefully the distinction we have made above between the *force of gravity which acts on the ruler* and the *weight of the ruler which acts on the finger.*

The essential point to grasp is that *three forces all equal in magnitude* are involved in the above discussion, namely:

(a) the force of gravity acting *on the ruler*;
(b) the weight of the ruler acting *on the finger*;
(c) the reaction of the finger acting *on the ruler*.

In this particular case, as indeed for all bodies at rest, or on a support moving with constant velocity, the force of gravity on the body is equal to its weight on its support.

We saw, on pages 42–3 that this is not so for a body resting on an accelerating support. Here, the force of gravity on the body is effectively constant, while the weight varies according to the acceleration of the support.

The plumbline

When a body is freely suspended by a string and is at rest the force of gravity on it, acting vertically downwards, is balanced by an equal and opposite force or tension in the string. Being flexible, the string therefore sets in a vertical direction. This is the principle of the plumbline, which consists of a small leaden bob supported by a thin cord. From the earliest times, plumblines and plumbrules (Fig. 6.6) have been used by builders for the purpose of testing the uprightness of walls, pillars, and so on during the course of construction.

Fig. 6.6.
The plumbrule

To locate centre of gravity by a balancing method

The centre of gravity of a long thin object such as a ruler or billiard cue may be found approximately by simply balancing it on a straight-edge. The same method could also be used for a thin sheet or *lamina* of cardboard or metal, except that in this case it is necessary to balance in *two* positions. Fig. 6.7 shows this method being

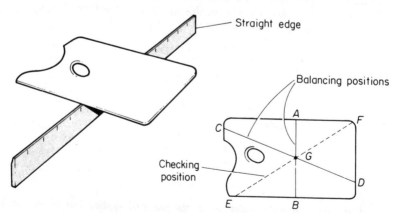

Fig. 6.7. Locating centre of gravity by balancing

used to find the centre of gravity of an artist's palette. The palette is balanced on the edge of a straight ruler in two directions *AB* and *CD*, and the lines of balance are marked with pencil lines. Since the centre of gravity, *G*, is situated on both lines, it must actually lie at their point of intersection. This may be checked by balancing once more in a third position *EF*. The edge of the ruler should now pass through the intersection of the two previous lines.

To locate centre of gravity by means of a plumbline

One of the best ways of finding the centre of gravity of a body is by use of a plumbline. Suppose, for example, we wish to find the centre of gravity of a flat irregularly shaped piece of cardboard.

First of all, three small holes are made at well-spaced intervals round the edge of the card. A stout pin is then put through one of the holes and held firmly by a clamp and stand so that the card can swing freely on it (Fig. 6.8). The card will come to rest with its centre of gravity, G, vertically below the point of support. The vertical line through the support can now be located by means of a plumbline. A suitable plumbline is made from a length of cotton with a loop at one end and a weight tied at the other. This is hung from the pin and the position of the cotton marked on the card by two small pencil crosses. These crosses are joined by a pencil line.

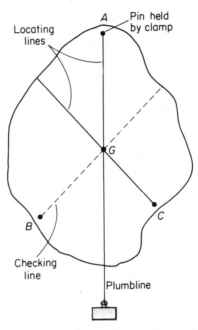

Fig. 6.8. Locating centre of gravity by plumbline

The experiment is repeated with the card suspended by one of the other holes. Since the centre of gravity lies on both of the lines drawn on the card, it must be situated at their point of intersection.

As a check, the card is suspended by the third hole. It should be found that, within the limits of experimental error, the plumbline should pass through the intersection of the two lines.

The above method can be used to show that the *centre of gravity of a triangular lamina lies at the point of intersection of its medians.* (A median of a triangle is a line through one of the corners which bisects the opposite side.)

Centre of gravity of a stool or tripod

Sometimes the centre of gravity of a body is not situated in the actual material of the body, but may be at a point in the air near by. Examples are an iron tripod or a laboratory stool. The plumbline method can still be employed, but requires more skill and ingenuity to carry out than in the case of a lamina. It is necessary to suspend the stool or tripod in such a way that the plumbline positions can be located by lengths of cotton fixed with soft wax.

To find the mass of an object by means of a metre rule

A metre rule with a small hole drilled near its centre as described on page 62 can be used in conjunction with a single known mass to measure the mass of an object.

Having balanced the rule on a knitting needle as described on page 62, the known mass and the object whose mass is to be found are suspended from the rule by cotton loops, one on either side of the pivot. Their distances from the pivot are then adjusted until the rule once more balances horizontally (Fig. 6.9).

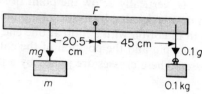

Fig. 6.9. To measure mass by means of a metre rule

Maximum accuracy will be obtained if these distances are made as large as possible. Since it is inevitable that a small error will be made when the distances of the weights from the pivot are measured, the percentage error in the final result will be less if large distances are used rather than small ones.

Suppose that, in a particular experiment, a 0.1 kg mass placed 45 cm from the pivot balances an object of unknown mass m placed 20.5 cm from the pivot on the opposite side of the rule.

Applying the principle of moments

$$\text{(weight of mass } m) \times 20.5 = \text{(weight of mass 0.1 kg)} \times 45$$

or

$$mg \times 20.5 = 0.1\ g \times 45$$

therefore

$$m = \frac{0.1 \times 45}{20.5}$$
$$= 0.22 \text{ kg}$$

Note that g disappears from the final calculation.

As an exercise, this method should be used to find the masses of a number of different objects and the results recorded in a table as below.

Object	Known mass m_1	Distances from pivot		Mass of object =
		Object d (cm)	Known mass d_1 (cm)	$m = m_1 \times \dfrac{d_1}{d}$ (kg)

Principle of the beam balance

We are now able to explain the principle of the beam balance, the construction of which was discussed in chapter 1. The unknown mass m_1, is placed in the left-hand pan and standard masses, m_2, are placed in the right-hand pan to obtain equilibrium (Fig. 6.10).

The balance has been constructed so that the distances, d, of the pans from the centre beam bearing are equal.

Thus equating moments

$$m_1 g \times d = m_2 g \times d$$

or

$$m_1 = m_2$$

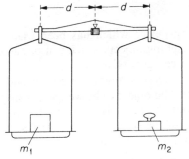

Fig. 6.10. Principle of the beam balance

From this we see that the balance measures mass, not weight, and it is quite independent of the value of g. Furthermore, it will correctly measure mass anywhere so long as the masses concerned possess weight.

Should we want to know the weight of the body we could calculate it by multiplying the mass by the local value of g.

What has been said above about the beam balance should be compared with our discussion on the spring balance in chapter 4.

To measure the mass of a metre rule by using a single known mass

A metre rule is supported in a loop of thread hung from a clamp and stand and adjusted until it balances horizontally. The loop will then be at the centre of gravity, G, of the rule. If the rule is not of uniform thickness (few rules are) the centre of gravity will not be at the 50 cm mark. However, this will not affect the accuracy of the experiment so long as the correct position is noted and later used in the calculation.

A 0.1 kg mass is now hung from the rule by thread at a point near one end. The position of the rule in the supporting loop is then adjusted until it again balances horizontally (Fig. 6.11). Let the mass of the rule be m kg. The rule is now in equilibrium under the action of two moments about the supporting loop, namely, that of its own force of gravity mg acting through the centre of gravity G and the weight of the 0.1 kg mass on the opposite side.

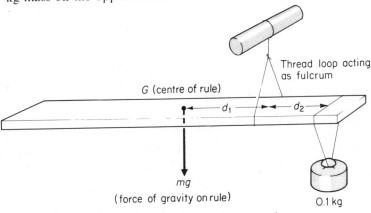

Thread loop acting as fulcrum

G (centre of rule)

d_1

d_2

mg
(force of gravity on rule)

0.1 kg

Fig. 6.11. To find the mass of a metre rule

If d_1 and d_2 respectively are the distances of G and the 0.1 kg mass from the loop, then equating moments we have

$$mg \times d_1 = 0.1\,g \times d_2$$

$$m = 0.1 \times \frac{d_2}{d_1}$$

Note that g disappears from the final calculation

The steelyard

The steelyard is a weighing device which has been used since Roman times. It is particularly suitable for weighing objects which can be suspended from a hook, e.g., sacks of farm produce and carcases of meat.

It consists of a graduated steel rod pivoted near one end and balanced by a heavy piece of metal *B* (Fig. 6.12). The object to be weighed is suspended from a hook. The moment of the object about the pivot is balanced by that of a small rider weight which slides along the arm of the steelyard. This arm is calibrated to read the mass being measured directly in appropriate units.

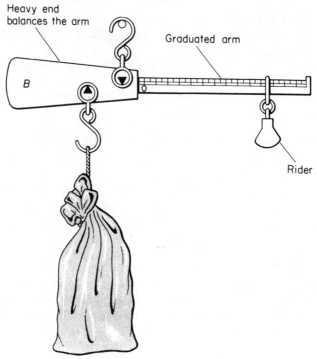

Fig. 6.12. The steelyard

Some types of platform weighing machines employ the steelyard principle. In these, the object is placed on a metal platform and its weight is transmitted to a balancing arm by a system of levers (see also Fig. 6.13 opposite).

Stable, unstable and neutral equilibrium

A child playing with a pencil soon learns that it is scarcely possible to make it balance on its point. On the other hand, it is comparatively easy to make the pencil stand upright on a flat end.

In order to understand the difference between these two cases let us consider a wooden cone placed on a horizontal table. A cone cannot be made to stand on its tip. Theoretically, this feat might be possible if the cone could be placed with its centre of gravity exactly in a vertical line through the tip. The cone would then be in equilibrium under the action of the force of gravity, *mg*, on it acting downwards and an equal and opposite reaction to its weight exerted on it by the table. But even if this condition could be achieved momentarily, the slightest vibration or draught would inevitably cause the cone to tilt. (The force of gravity, *mg*, would then exert a turning force about the tip, and this would cause the cone to topple over (Fig. 6.14 (*a*)). A cone placed on its tip is said to be in *unstable equilibrium*.

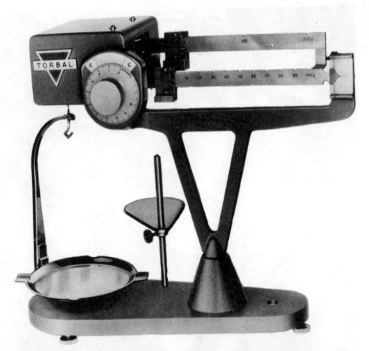

Fig. 6.13. A precision balance incorporating torsion bearings together with the steelyard principle. The V-shaped platform is clamped to support a vessel of water during density determinations (chapter 12)

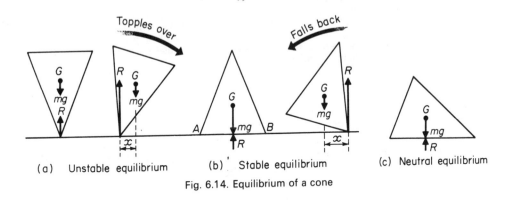

Fig. 6.14. Equilibrium of a cone

Fig. 6.14 (*b*) shows the cone standing on its base. It tilted from this position, even through a fairly large angle, the vertical line through the centre of gravity, *G*, will still fall inside the base *AB*. Consequently, the force of gravity on the cone will have a moment *mg* × *x* about an edge of the base which will pull the cone back into its original position. Under these conditions, it is not easy to knock the cone over, and it is said to be in *stable equilibrium*.

In Fig. 6.14 (*c*) the cone is lying on its side. The base is now simply a straight line, and if the cone is rolled into a new position the vertical line through the centre of gravity still continues to pass through exactly the same point in the base.

Whatever the position of the cone, the reaction from the table will act in the same straight line as the force of gravity through *G*, and so the cone will be in equilibrium. The force of gravity exerts no moment about the base as axis and, if displaced, the cone will therefore remain at rest in its new position. This condition is described as *neutral equilibrium*.

Fig. 6.15. This stability test used to be a legal requirement for London's buses. A double-decked vehicle had to be stable with the chassis tilted to 28° with the vertical whilst loaded with weights to represent a full complement of passengers on the top deck only, plus the driver and conductor, at 64.5 kg per person. The two indicators gave the tilt of the chassis and body respectively. The difference in readings is attributable to the vehicle springing.

Summary. It should be clear from the above explanation that the stability of a body depends on the direction of the turning moment exerted by the force of gravity on the body about the edge of the base, when the body is given a small displacement.

If a small displacement brings the vertical through the centre of gravity outside the base the body will be unstable. If, however, the vertical remains within the base the body will be stable.

When a displacement causes no change in the position of the vertical through the centre of gravity with respect to the base the body is in neutral equilibrium.

Fig. 6.15 illustrates the method which was used at one time for testing the stability of a bus.

Worked example

A uniform wooden lath AB, 120 cm long and weighing 1.20 N rests on two sharp-edged supports C and D placed 10 cm from each end of the lath respectively. A 0.20 N weight hangs from a loop of thread 30 cm from A and a 0.90 N weight hangs similarly 40 cm from B. Find the reactions at the supports.

Fig. 6.16 shows the lath *AB* resting on the supports *C* and *D*, and the force arrows represent the various loads and reactions in magnitude, direction, and position.

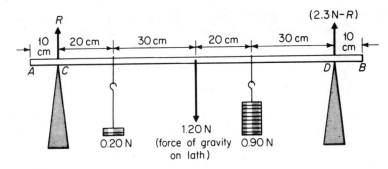

Fig. 6.16. Forces on a supported lath

The total downward force $= 0.20 + 1.20 + 0.90 = 2.30$ N
hence the sum of the upward reactions at the supports also $= 2.30$ N

Let the reaction at $C = R$
then the reaction at $D = (2.30$ N $- R)$

Moments may be taken about *any* point, but, in this problem, the easiest equation will be obtained if **moments are taken about** D. This will eliminate the moment due to the force $(2.30$ N $- R)$ which acts through D
Hence, taking moments about D,

Clockwise   *Anticlockwise*
$R \times 100 = (0.20 \times 80) + (1.20 \times 50) + (0.90 \times 30)$
$R \times 100 = 16 + 60 + 27 = 103$ N cm units

Therefore, $R = 1.03$ N
and $(2.30$ N $- R) = 1.27$ N

Answer. Reaction at $C = 1.03$ N
Reaction at $D = 1.27$ N

QUESTIONS: 6

(*Where necessary, assume g* $= 10$ m/s²)

1. What do you understand by the *moment of force* about a point? How would you use a metre rule, a 100 g mass and some thread in order to find the mass of an apple?

2. A uniform half metre rule is freely pivoted at the 15 cm mark and it balances horizontally when a body of mass 40 g is hung from the 2 cm mark. Draw a clear force-diagram of the arrangement and calculate the mass of the rule. (S.)

3. A pole AB of length 10.0 m and weight 800 N has its centre of gravity 4.0 m from the end A, and lies on horizontal ground. The end B is to be lifted by a vertical force applied at B. Calculate the least force required to do this.
Why would this force, applied at the end A, not be sufficient to lift the end A? (C.)

4. Given a spring balance limited to loads of less than 25 N and simple additional apparatus, how would you arrange to measure loads of between 50 and 100 N? (O.C.)

5. It is found that a uniform wooden lath 100 cm long and of mass 95 g can be balanced on a knife-edge when a 5 g mass is hung 10 cm from one end. How far is the knife-edge from the centre of the lath?

6. State the conditions of equilibrium when a body is acted upon by a number of parallel forces.
A uniform metal tube of length 5 m and mass 9 kg is suspended horizontally by two vertical wires attached at 50 cm and 150 cm respectively from the ends of the tube. Find the tension in each wire.

7. A uniform metre rule of weight 0.9 N is suspended horizontally by two vertical loops of thread A and B placed at 20 cm and 30 cm from its ends respectively. Find the distances from the centre of the rule at which a 2 N weight must be suspended:
(a) to make loop A become slack;
(b) to make loop B slack.

8. Explain the principle of a simple beam balance with equal arms when it is being used to find the mass of a body.

How is the weight of a body calculated from a knowledge of its mass?

9. A cylindrical steel shaft of radius 2 cm and 50 cm long is turned down on a lathe to one-half its radius for a distance of 20 cm from one end. Find the distance of its centre of gravity from the thicker end.

10. (*a*) Explain the meaning of the term *centre of gravity*. How would you determine, by experiment, the centre of gravity of a thin triangular sheet of metal? Show how you could check the accuracy of your result by:

(i) a geometrical construction and measurement, and

(ii) by a second simple experiment. Where should the centre of gravity be located?

(*b*) Explain how the position of the centre of gravity of a body determines whether it is in stable or unstable equilibrium. Illustrate your answer by diagrams.   (*L.*)

11. Explain what is meant by *stable, unstable,* and *neutral* equilibrium. Give one example of each.

12. Explain what is meant by

(i) *moment of a force about a point*;

(ii) *centre of gravity* of a body.

Describe how you would determine, by experiment, the centre of gravity of a piece of uniform cardboard of irregular shape.

The diagram (Fig. 6.17) shows a wheel of mass 15 kg and radius 1 m being pulled by a horizontal force *F* against a step 0.4 m high. What initial force is just sufficient to turn the wheel so that it will rise over the step?

What happens to the size of this horizontal force as the wheel rises?   (*W.*)

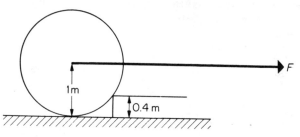

Fig. 6.17.

7. Work, energy and power

Work

In ordinary conversation the word "work" refers to almost any kind of physical or mental activity, but in science and mathematics it has one meaning only. Work is done when a force produces motion. A locomotive pulling a train does work; so does a crane when it raises a load against the pull of the earth. Similarly, a workman who is employed to carry bricks up a ladder and on to a scaffold platform also performs work. On the other hand, the Greek god Atlas, who spent his time supporting the world on his shoulders, must have become very tired, but, technically speaking, he did no work on the world itself, since he merely exerted an upward force on it without moving it.

Nevertheless, when a person holds a heavy load some internal work is done by continuous expansion and contraction of his muscular fibres.

Work is said to be done when the point of application of a force moves and is measured by the product of the force and the distance moved in the direction of the force.

Work = force × distance moved in direction of force.

The SI unit of work is called the joule (J) and is the work done when the point of application of a force of 1 newton (N) moves through 1 metre (m) in the direction of the force.

Larger units used are the kilojoule (kJ) and the megajoule (MJ).

$$1 \ kJ = 1000 \ J \ (\text{or } 10^3 \ J)$$
$$1 \ MJ = 1 \ 000 \ 000 \ J \ (\text{or } 10^6 \ J)$$

It follows that a locomotive which exerts a force of 9000 N over a distance of 6 m will do $9000 \times 6 = 54\ 000$ J, or 54 kJ.

If we wish to calculate the work done by a man of mass 65 kg in climbing a ladder 4 m high, we multiply his weight, mg, by the vertical height raised.

Thus the work done, assuming $g = 10$ m/s², is

$$(65 \times 10) \times 4 = 2600 \ J$$

Energy

Anything which is able to do work, as defined above, is said to possess **energy**, and therefore

Energy is the capacity to perform work.

Work and energy are, of course, both measured in the same units, namely, joules.

The world we live in provides energy in many different forms, of which the most important has been chemical energy. The utilization of the latent chemical energy in coal, oil, and gas, released in the form of heat to drive steam turbines and internal-combustion engines, has been a major factor in the development of modern civilization.

Many of the material comforts which we enjoy today come from the use of electric energy. The first electricity generating plants were powered by coal-fired steam engines, but by the middle of the twentieth century large hydro-electric power installations had been built in countries all over the world. "Hydro-electric" means the production of electricity by generators driven by water turbines. The rapid flow of water required for this purpose comes from big reservoirs formed by building dams across valleys and large rivers (Fig. 7.1).

Windmills which transfer the energy in wind to mechanical energy in machinery have long been in use for working water pumps as well as for milling grain or sawing timber. See also Fig. 7.2. In some parts of the world where the sun shines uninterruptedly for long periods, large concave mirrors have been set up to collect energy directly from the sun by focusing its rays on to special boilers which provide power for running electric generators. Fig. 7.3 shows a space telescope solar cell array.

The utilization of atomic energy, which began after the middle of the twentieth century, has made available a new source of heat as a link in the production of electricity.

Mechanical energy

In mechanics, energy is divided into two kinds called potential and kinetic energy (abbreviated p.e. and k.e. respectively).

Kinetic energy is the energy which a body has by reason of its motion.

Potential energy is the energy something has by reason of its position in a field of force or by its state.

Obvious examples of kinetic energy are moving bullets or hammer heads. These are able to do work by overcoming forces when they strike something. A heavy flywheel stores energy in the form of rotational kinetic energy and so keeps an engine running smoothly in between the working strokes of its pistons.

One of the commonest forms of potential energy is that possessed by a body when it is above the level of the earth's surface. When something is lifted vertically, work is done against its weight and this work becomes stored up in the body as *gravitational* potential energy.

An example of the potential energy a body has by reason of its state is the *elastic* potential energy stored up in a wound clock-spring.

Interchange of energy between potential energy and kinetic energy

A swinging pendulum bob is an example of a body whose energy can be either kinetic or potential or a mixture of both. It is all potential at the extreme end of the swing and all kinetic when passing through the rest position. At intermediate points it is partly kinetic and partly potential.

If we turn our attention to a falling stone, it is obvious that at any particular moment, it possesses both potential and kinetic energy. As it falls its speed increases so that it gains in k.e. at the expense of its p.e. If we ignore the energy the stone gives to the air molecules as it pushes them out of its way, then the loss in p.e. of the stone is exactly equal to its gain in k.e. This is an example of the *law of conservation of energy* which states that energy cannot be destroyed; it only becomes transferred to a different form of energy.

Internal energy

We remarked on the obvious nature of the two kinds of energy possessed by a falling stone. *What is not quite so obvious is the internal energy of the molecules inside the stone.* The molecules of a substance are in continual motion. According to the state of the substance, they possess kinetic energy of motion or vibration together with potential energy resulting from the attractions and repulsions they exert on one another. Internal energy increases as the temperature of a body increases.

Some natural sources of energy

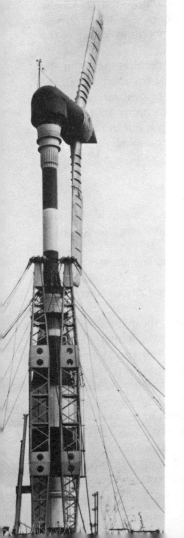

Fig. 7.1. **Above:** The Kariba dam wall as it appeared when the Zambesi hydroelectric project was officially opened in May 1960. This dam is over 120 metres high. The potential energy of the water which it stores is transferred first to kinetic energy and then to electric energy by two power stations, one in each bank, with total power output of about 1500 megawatts

Fig. 7.2. **Left:** This windmill provides power to drive a 100 kW electric generator in N. Africa

Fig. 7.3. **Below:** This shows the silicon solar cell array for a Space Telescope, in course of construction by British Aerospace at Bristol. The array carries 48 760 solar cells which transfer energy in the sun's radiation to energy in an electric current. This provides power for all the telescope's experimental, control and computer systems. Note the operator's special clothing: it is important to prevent dust or other contamination from entering the moving parts

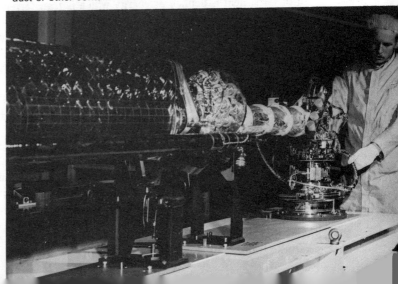

Scientists sometimes refer to the k.e. and p.e. of a body taken as a whole as being *macroscopic* (= visible to the naked eye). Internal energy may be described as *microscopic* (= invisible to the naked eye). However, it is possible to demonstrate the reality of molecular motion as we shall see when we discuss the Brownian motion (page 137).

Transfer of energy from one kind to another

The law of conservation of energy which was mentioned earlier states that energy is never destroyed but is only transferred from one form to another. As an example we shall now discuss the way in which the law applies to a grandfather clock.

The energy we spend in winding up the weights is derived from chemical changes in our muscles and this is provided from the food we have eaten.

If the weights have a total mass of 5 kg and are raised through 2 m then, assuming $g = 10$ m/s^2, a force of 5×10 N has to be exerted through a distance of 2 m. The work done is transferred to potential energy in the weights, so that

$$\text{potential energy} = \text{work done} = \text{force} \times \text{distance}$$
$$= 5 \times 10 \times 2 \text{ J}$$
$$= 100 \text{ J}$$

As the weights slowly descend, their potential energy is transferred to kinetic energy and potential energy in the moving parts of the clock.

At this stage we may ask what has become of the original potential energy of the weights when the clock has completely run down.

The answer is that it has been transferred to internal molecular energy in the various parts of the clock resulting from the work done against friction between the wheel spindles and their bearings, motion of the pendulum against air resistance, and so on. This increase in internal energy raises the temperature of the clock slightly, and as it cools down, the excess internal energy is given out in the form of heat by conduction, convection, and radiation to the surroundings where it is absorbed and becomes internal molecular energy once more.

Heat energy

At this point it will be advantageous to say a word about the meaning of the term *heat*. In everyday life we sometimes loosely refer to the "heat energy *in* a body", instead of using the term: "internal energy".

In physics, **heat is defined as energy which is transferred from one place to another owing to a temperature difference between them.**

There are three main processes of heat transfer, namely, conduction, convection, and radiation, which we shall study in detail in chapter 17.

In *thermodynamics* (the branch of physics concerned with the relation between heat and work), we have to deal with such things as engines in which internal energy changes take place in a substance both as a result of heat flow and also as a result of work being done by or on the substance. To avoid confusion, special symbols U (internal energy), Q (heat), and W (work) are used.

Thus, the internal energy of some steam, for example, is represented by U in joules. This internal energy can be changed by adding or subtracting Q in joules of heat. The internal energy of steam can be increased by compressing it, i.e., doing W in joules of work on it, or the steam can be allowed to expand and transfer some of its internal energy into W in joules of useful work in, say, driving a turbine.

In a case such as this it will be realized that confusion will certainly follow if we talk of the "heat energy" in a substance being transferred to work when actually we mean the transfer of internal energy to work.

The sun as a source of energy

It is interesting to trace back to their origin the series of energy transfers which lead to the release of light energy by pressing an electric switch.

The light is given out from a thin incandescent tungsten filament inside a glass bulb. The internal energy required to raise the temperature of the tungsten wire is derived from the work done when the electric current moves against the resistance offered to its passage through the wire.

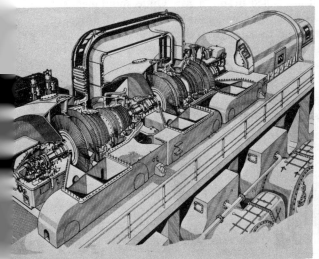

a) Cut-away drawing of a turbine generating set in which the turbine blades are clearly visible

(*b*) The turbine room at Portishead, Somerset

Fig. 7.4. *Making electricity*

The electricity itself is generated by a dynamo run by a steam turbine (Fig. 7.4). The turbine, in turn, is driven by expanding high-pressure steam, which we shall suppose has been obtained by heating water in a coal-fired boiler.

Now coal consists mainly of carbon and is formed from the remains of giant forests which flourished millions of years ago. Changes in the earth's crust led to these forests becoming submerged beneath layers of sediment and subjected to pressure. The plants from which the coal is formed derived their growth from the action of the sunlight in which they were bathed.

Thus we see that the electric light and power from a coal-fired generating station comes to us by a series of transfers of energy which was poured out by the sun, millions· of years ago, and stored as chemical potential energy in coal.

The reader may trace for himself the energy changes back to the sun in the case where electric generators are run by water power.

In the examples of energy transfer considered above the last link in the energy chain is the transfer to internal energy. This is found to be true in all cases. Even the light from the lamp finally turns back into internal energy when it is absorbed by bodies on which it falls. This fact is referred to as the "degradation" of energy into internal energy and has led some physicists to suppose that the end of life, as we know it, will come when all the energy in the universe is uniformly distributed as internal molecular energy at the same temperature.

Nuclear energy

We have just discussed the energy changes which lead to the generation of electricity in a coal-burning electric power station. In a nuclear electricity generating installation the heat required for raising steam is provided by a *nuclear reactor* instead of a coal furnace.

Fig. 7.5 is a simplified diagram to show how a reactor works. It consists of a strong steel pressure vessel enclosing a core made of graphite bricks (Fig. 7.6). This graphite core has a number of vertical channels which are filled with rods of a very heavy metal called uranium. Interspersed among the uranium rods are a set of boron steel rods which may be raised or lowered in similar channels in the graphite. These are the *control rods*, and their function will be explained in due course.

The uranium used in the reactor consists of a mixture of two different kinds of atoms, of which the most important are called uranium-235. Quite spontaneously, some of these uranium-235 atoms explode or disintegrate to form other atoms of smaller mass. When this happens, energy is radiated from the central core or nucleus of the atom together with small high-speed particles called *neutrons*. If one of the

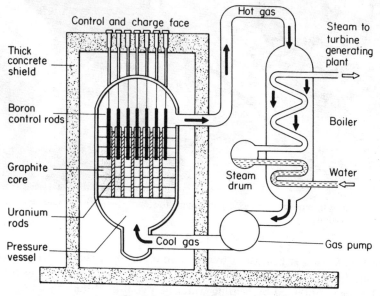

Fig. 7.5. Gas-cooled reactor

Fig. 7.6. The graphite core of the nuclear reactor at Berkeley power station. Note the channels for reception of uranium rods and for the circulation of hot gas. (See also page 82.)

neutrons happens to strike the nucleus of a neighbouring atom this may also disinte-grate, with a further evolution of energy and the production of more neutrons. This splitting up of the nucleus is called *fission*.

The graphite of which the core is composed is called a *moderator* (Fig. 7.6). Its function is to slow down the speed of the neutrons, as it is found that fission of uranium-235 is more likely to occur with slow neutrons than with fast ones.

In a small piece of uranium mixed with moderator most of the neutrons escape through the surface. If, however, the amount of material is increased the chances that a neutron will collide with an atomic nucleus will also increase, since there are more atoms present. Each nuclear fission which occurs produces two or three fresh neutrons which are, in turn, capable of promoting the fission of further nuclei. When the lump of uranium and moderator is above a certain *critical size* the fission process proceeds cumulatively in what is called a *chain reaction* (Fig. 7.7). This is where the above-mentioned boron steel rods play their part. Before the uranium rods are

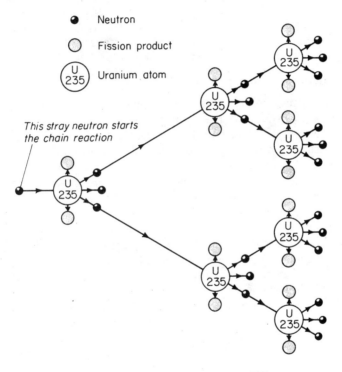

Fig. 7.7. Chain reaction in uranium-235

loaded into the graphite core the boron rods are already in position, and these have the property of being able to absorb neutrons which are shot out from the uranium, and so prevent the chain reaction from starting. When sufficient uranium rods have been added to effect critical conditions the pressure vessel is sealed and the boron rods raised out of the core. The uranium rods are now freely bombarded by one another's neutrons and the chain reaction begins. The rate at which fission occurs can, of course, be controlled by raising or lowering the boron rods. If these are fully inserted into the graphite core the reaction shuts down completely, and only the normal spontaneous nuclear fission takes place.

The heat energy released by the fission process is carried away as internal energy in a stream of high-pressure carbon dioxide gas which is continuously pumped through the pressure vessel. This hot gas circulates through a special steam boiler, and the steam so raised is used to drive an electric turbo-generator in the usual way.

Nuclear power installations working on the principle just described are called Magnox reactors, and the first of these was built at Calder Hall, Cumberland in Britain in 1956. Magnox is the name of the aluminium alloy in which the fuel rods are encased.

In recent years, reactors have been designed which use other types of nuclear fuel with different moderators. Owing to the higher temperatures and pressures involved compared with Magnox reactors, some of these have presented safety design problems.

The future of nuclear power installations

Nuclear reactors like most other industrial plants do not last for ever and, at the time of writing (1988) plans are in operation for decommissioning the early Magnox reactors. The Berkeley one, mentioned in Fig. 7.6 on page 80 is one of the first for demolition. The first advanced gas-cooled reactor (AGR) which was built at Windscale as a model for the second generation of reactors was closed down in 1981. A new generation of reactors is now coming into operation. One of the main problems in closing down a reactor is the safe disposal of radioactive waste. Also some parts of the plant will have to be enclosed in thick concrete for more than a century to allow for a safe natural decay of their natural activity.

The conservation of energy and mass

Early in the twentieth century Albert Einstein put forward new ideas regarding the relationship between space, time, mass, and energy which have come to be known as the theory of relativity. It had long been accepted that matter could not be destroyed. This assumption was expressed in the *law of conservation of matter*, which states that the total quantity of matter in the universe is fixed and cannot be increased or decreased by human agency. Similarly, another law, called the *law of conservation of energy*, states that the total quantity of energy in the universe is also constant and can be neither created nor destroyed.

While we cannot here enter into a discussion of the meaning of relativity, it may be pointed out that Einstein has simplified our picture of the universe by showing that *the mass of a body is a measure of the quantity of energy contained in it*. We find that, in a nuclear reactor, the sum total of the masses of the atoms produced as a result of fission is slightly less than the mass of the original uranium nuclei. The difference represents the mass of the energy liberated as heat, radiation, and kinetic energy of fission products. Thus, in the light of modern physics, we have to consider the laws of conservation of energy and mass as separate aspects of a single principle. We now take the view that the sum total of mass plus energy in the universe is fixed.

Thermonuclear energy

At the present time research is being carried out to investigate the possibility of obtaining energy by the *fusion* of hydrogen nuclei to form heavier ones. This is called a *thermonuclear reaction* and is believed to be the source of the sun's energy. Under the extremely high-temperature conditions in the interior of the sun, hydrogen nuclei fuse together to form helium nuclei, and the resulting loss in mass is emitted in the form of radiation.

Power and its units

Machines may be classified by the speed with which they transfer energy or do work; thus there are motor-car engines of small "power", as they are rated, or large power.

Power is defined as the rate of transfer of energy,

or
$$\text{average power} = \frac{\text{energy transferred}}{\text{time taken}}$$

The SI unit of power is called the watt (W) and is a rate of transfer of energy of 1 joule per second

Thus, $1 \text{ W} = 1 \text{ J/s}$

Larger units used are the kilowatt (kW) and the megawatt (MW)

$$1 \text{ kW} = 1000 \text{ W (or } 10^3 \text{ W)}$$
$$1 \text{ MW} = 1\,000\,000 \text{ W (or } 10^6 \text{ W)}$$

Example. Calculate the power of a pump which can lift 200 kg *of water through a vertical height of* 6 m *in* 10 s. (Assume $g = 10$ m/s².)

Force overcome $= 200 \times 10 \text{ N}$
Distance $= 6 \text{ m}$
Work done (energy transferred) $= 200 \times 10 \times 6 \text{ J}$
Time taken $= 10 \text{ s}$

$$\text{Power} = \frac{\text{energy transferred}}{\text{time taken}} = \frac{200 \times 10 \times 6}{10} = 1200 \text{ J/s}$$
$$= 1200 \text{ W}$$
$$= 1.20 \text{ kW}$$

To measure personal power

Anyone may find the power he or she is able to develop when running upstairs by measuring the total vertical height of a stairway and using a stop-watch to find the time taken to run up it.

Example. A boy whose mass is 40 kg finds that he can run up a flight of 45 steps, each 16 cm high, in 5.2 s. (Assume $g = 10$ m/s².)

Force overcome $= 40 \times 10 \text{ N}$
Distance $= 45 \times 16 = 720 \text{ cm} = 7.2 \text{ m}$

Energy transferred = Work done = force × distance = $40 \times 10 \times 7.2$ J

$$\text{Power} = \frac{\text{work done}}{\text{time taken}} = \frac{40 \times 10 \times 7.2}{5.2} \text{ J/s (or watts)}$$
$$= 554 \text{ W}$$
$$= 0.55 \text{ kW (to 2 significant figures)}$$

The result is creditable to the boy, but it must be remembered that he can maintain this high power only for a comparatively short time.

Experiment shows that the average power of a man walking upstairs at an ordinary pace is only about 0.33 kW.

Kinetic energy

We saw on page 36 that, where there are no opposing forces, a moving body needs no force to keep it moving with a steady velocity. If, however, a resultant force does act on a moving body in the direction of its motion, then it will *accelerate* and the work done by the force will become transferred to increased kinetic energy in the body.

In order to calculate the kinetic energy of a body of mass m moving with velocity v, we begin by supposing that the body starts from rest and is acted upon by a force F (no friction or other forces acting).

This force will give the body a uniform acceleration a, and it will acquire a final velocity v, after travelling a distance x. These quantities, a, v, and x will be related by the equation $v^2 = u^2 + 2ax$ (page 24).

In accordance with the law of conservation of energy, the work done by the force F in pushing the body through distance x will become transferred to kinetic energy of motion in the body.

Thus,
$$\text{work done} = \text{force} \times \text{displacement}$$
$$= F \times x$$
but
$$F = ma$$
therefore, substituting for F,

$$\text{work done} = ma \times x \quad \ldots \ldots \ldots \ldots \quad (1)$$

Applying the equation $v^2 = u^2 + 2ax$ and remembering that $u = 0$
$$v^2 = 0 + 2ax$$

whence
$$a = \frac{v^2}{2x}$$

Substituting this value of a in equation (1), we obtain,

$$\text{work done} = m \times \frac{v^2}{2x} \times x = \text{kinetic energy}$$

or **kinetic energy** (k.e.) $= \frac{1}{2}mv^2$ (in joules)
where
$$m = \text{mass in kg}$$
and
$$v = \text{velocity in m/s}$$

Worked examples

1. *A stone of mass 500 g is thrown vertically upwards with a velocity of 15 m/s. Find: (a) the potential energy at greatest height; (b) the kinetic energy on reaching the ground. (Assume $g = 10$ m/s^2 and neglect air resistance.)*

To solve this problem we use the equation of motion, $v^2 = u^2 + 2ax$, replacing a by g since we are dealing with gravitational acceleration. Thus,

$$v^2 = u^2 + 2gx$$

in which $u = \quad 15$ m/s
$v = \quad 0$ m/s
$g = -10$ m/s^2

hence, by substitution,

$$0^2 = 15^2 + 2(-10) \times x$$
whence
$$x = \frac{-15^2}{2 \times (-10)} = 11.25 \text{ m}$$

$$\text{potential energy} = \text{weight in newtons} \times \text{height raised in metres}$$
$$= mg \times x \ (m \text{ in kg, of course})$$
$$= 0.5 \times 10 \times 11.25 = 56.25 \text{ J}$$

In accordance with the principle of conservation of energy, the whole of this potential energy becomes transferred to kinetic energy when the stone reaches the ground again. Hence kinetic energy on reaching the ground $= 56.25$ J.

Answer. P.E. at greatest height $= 56$ J
K.E. on reaching ground $= 56$ J

2. *During a shunting operation, a truck of total mass* 15 *metric tonnes* (t) *moving at* 1 m/s, *collides with a stationary truck of mass* 10 t. *If the two trucks are automatically connected so that they move off together, find their velocity. Also calculate the kinetic energy of the trucks: (a) before; (b) after collision. Explain why these are not equal.* (1 t = 1000 kg.)

By the principle of conservation of momentum,

$$\text{momentum before collision} = \text{momentum after collision}$$

Let v = common velocity after collision, then using t m/s units of momentum,

$$(15 \times 1) + (10 \times 0) = (15 + 10) \times v$$

or
$$v = \frac{15}{25} = 0.6 \text{ m/s}$$

Using the formula K.E. $= \frac{1}{2}mv^2$ (m in kg; v in m/s)

K.E. before collision $= \frac{1}{2} \times 15\,000 \times 1^2 = 7500$ J

K.E. after collision $= \frac{1}{2} \times 25\,000 \times 0.6^2 = 4500$ J

Answer. Velocity before collision $= 0.6$ m/s

K.E. before collision $= 7.5$ kJ

K.E. after collision $= 4.5$ kJ

In accordance with the principle of conservation of energy, the total energy after collision is the same as that before.

Before collision the whole of the energy is kinetic in the moving truck, but when collision occurs part of this becomes transferred to *internal energy* in both trucks (k.e. and p.e. of molecules, see page 76) and part into *sound energy* (k.e. and p.e. of air molecules). The remainder is left as mechanical kinetic energy in both trucks. Consequently, mechanical kinetic energy after collision is less than mechanical kinetic energy before collision.

QUESTIONS: 7

(*Where necessary assume* g = 10 m/s²)

1. When does a force do work? How is the work it does measured? What is meant by the term "power"? (*C.*)

2. Define the *watt* and the *kilowatt*.

A man whose mass is 75 kg walks up a flight of 12 steps each 20 cm high in 5 s. Find the power he develops in watts.

3. A body of mass 50 kg is raised to a height of 2 m above the ground. What is its potential energy? If the body is allowed to fall, find its kinetic energy:

(*a*) when half-way down;

(*b*) just before impact with the ground. What has become of the original energy when the body has come to rest? ·

4. A cable car is pulled up a slope by a constant force of 5000 newtons at a uniform speed of 6 metres per second. It takes the car 4 minutes to complete the journey.

(i) How much work is done in getting the car to the top of the slope?

(ii) How much work would be done if the speed were 12 metres per second (the force remaining the same)?

(iii) How does the power developed compare in (i) and (ii)? (*L.*)

5. Define *potential energy* and *kinetic energy*.

State the principle of the conservation of energy and illustrate it by discussing the energy changes which occur when a pendulum bob is drawn to one side and allowed to oscillate. Why does the bob eventually come to rest and what has become of its energy?

6. State four of the transfers of energy which occur at a power station which uses coal as its fuel. (*S.*)

7. A jet aircraft climbs at an increased speed, using a large amount of fuel to provide energy. Write brief notes on the energy transformations which occur. (*C.*)

8. Define the *newton* and the *joule*.

A mass of 8 kg is pulled by a force of 20 N along a smooth floor. Find:

(*a*) the acceleration;

(*b*) the velocity after 4 s;

(c) the distance moved in 4 s;

(d) the work done by the force.

9. A ball of mass 1 kg is dropped from a height of 7 m and rebounds to a height of 4.5 m. Calculate:

(a) its kinetic energy just before impact;

(b) its initial rebound velocity and kinetic energy. Account for the loss of kinetic energy on impact.

10. A motor car of mass 1000 kg travelling at 90 km per hour is brought to rest by the brakes in 100 m. Calculate:

(i) the car's initial momentum;

(ii) its initial kinetic energy; and

(iii) the average braking force required.

(L.)

11. Define *momentum* and *kinetic energy*.

A car is moving at 36 km/h. Express this velocity in m/s. What velocity will:

(a) double its momentum;

(b) double its kinetic energy?

12. A bullet of mass 12 g strikes a solid surface at a speed of 400 m/s. If the bullet penetrates to a depth of 3 cm, calculate the average net force acting on the bullet while it is being brought to rest.   (J.M.B.)

13. State the energy changes which occur when a moving car is brought to rest by its brakes, and the car is then driven to the top of a hill.   (C.)

14. What is meant by *power*? Explain the meaning of *kilowatt*. A car of mass 1500 kg is driven from rest with uniform acceleration and reaches a speed of 50 km/h in 30 s. Find:

(a) the useful force exerted by the engine in newtons;

(b) the power developed in kilowatts at 50 km/h. (Assume all friction forces are constant.)

15. A man has to raise a box of mass 40 kg on to a platform 160 cm vertically above the ground. He decides to do this by pushing the box up a smooth straight plank (with negligible friction) set at an angle to the ground.

Calculate in joules the potential energy gained by the box when on the platform. Explain as fully as possible how the work done and effort force needed to raise the box depend on the angle between the plank and the horizontal. Discuss briefly the practical value of using the plank.

If the box falls from the platform calculate its kinetic energy and velocity just before it hits the ground.   (S.)

16. 400 kg of air, moving at 20.0 m/s, impinge on the vanes of a windmill every second. At what rate in kilowatts is the energy arriving at the windmill? What is the maximum mass of water that could be pumped each second through a vertical height of 5.0 m? Mention two reasons why this calculated amount would not be achieved in practice.   (O.C.)

17. A motor van tows a trailer of mass 1000 kg at a steady speed of 48 km/h along a level road. If the tension in the coupling is 800 N find the useful power expended by the engine on the trailer.

If the van ascends an incline of 1 in 10 measured along the road, and maintains the same speed, find:

(a) the new tension in the coupling;

(b) the increased power output required.

18. A body of mass 5 kg is projected up a board inclined at 30° to the horizontal with an initial velocity of 6 m/s. If the frictional force opposing its motion is 4.5 N, find the distance it travels before coming to rest and its increase in potential energy at the end of the run.

19. A railway truck of mass 2.4 t is shunted on to a stationary truck on a level track and collides with it at 4.7 m/s. After collision the two trucks move on together with a common speed of 1.2 m/s. Find:

(a) the mass of the stationary truck;

(b) the original kinetic energy of the first truck;

(c) the total kinetic energy of both trucks after collision. Account for the apparent loss in kinetic energy.

8. Machines

Most people think of a machine as being a more or less complicated piece of mechanism which includes gear wheels, levers, screws, and so on. But however complex a machine may appear to be, its various parts can always be shown to be applications of a limited number of basic mechanical principles.

Essentially, *a machine is any device by means of which a force applied at one point can be used to overcome a force at some other point.*

The lever

The simplest form of lever in common use is a steel rod known as a crowbar, but the term lever may be applied to any rigid body which is pivoted about a point called the *fulcrum*. Levers are based on the principle of moments, which was discussed in chapter 6. A force called the *effort* is applied at one point on the lever, and this overcomes a force called the *load* at some other point. Incidentally, the terms effort and load are not restricted to levers but apply to all types of machine.

Fig. 8.1 illustrates some simple machines based on the lever principle.

Mechanical advantage (M.A.)

If a lever can be used to overcome a load of 50 N by applying an effort of 10 N, the lever is said to have a mechanical advantage of $\frac{50}{10}$ or 5.

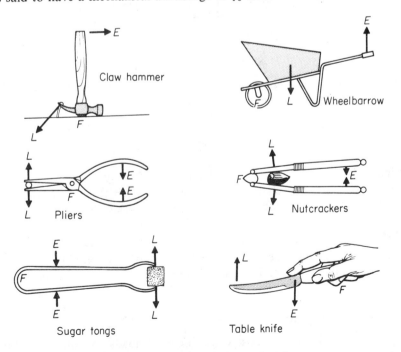

Fig. 8.1. Applications of the lever principle

The mechanical advantage of a machine is defined as the ratio of the load to the effort, or

$$\text{mechanical advantage} = \frac{\text{load}}{\text{effort}}$$

Some machines are designed to overcome a load much greater than the effort used, for example, a spanner used to undo a tight bolt or a screw jack to lift a motor-car. In such cases the mechanical advantage is greater than 1.

In certain other machines the mechanical advantage is less than 1, and in these the effort is greater than the load. It is not generally realized that a bicycle is a machine with a mechanical advantage of less than 1. Under ordinary conditions the resistance to the motion of a bicycle along a level road is comparatively small, and therefore a large mechanical advantage is unnecessary. Thus, although a cyclist works at a "mechanical disadvantage", he nevertheless gains in the speed with which he can travel.

On the other hand, the fact that the mechanical advantage of a bicycle is less than 1 becomes painfully obvious when we begin to ascend a hill. Whereas previously only a small amount of work had to be done against friction and air resistance, we now have to do a vastly increased amount against the force of gravity. Under these conditions it is usually easier to dismount and walk, unless the mechanical advantage of the bicycle can be increased by using a low gear (see gears in Fig. 8.14).

Mechanical advantage of a lever

If we neglect friction at the fulcrum and the weight of the lever itself (both being comparatively small in most cases) the mechanical advantage in any particular case may be obtained by writing down the equation of moments for the load and effort

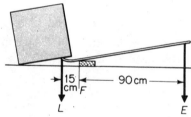

Fig. 8.2. The crowbar

about the fulcrum. Remembering that *moment = force × perpendicular distance from fulcrum*, we have, for the crowbar in Fig. 8.2,

$$L \times 15 = E \times 90$$

therefore

$$\text{mechanical advantage} = \frac{L}{E} = \frac{90}{15} = 6$$

A similar equation of moments may be obtained for any of the other lever devices shown in Fig. 8.1. The mechanical advantage in any particular case will depend, of course, on the position of the fulcrum in relation to the effort and load.

Pulleys

A pulley is a wheel with a grooved rim, and there may be several of these mounted in a framework called a block. The effort is applied to a rope which passes over the pulleys.

The single fixed pulley

This is often used for the purpose of raising small loads contained in a bucket or basket to the top of a building during construction or repair work (Fig. 8.3). The tension is the same throughout the rope, so that, neglecting the weight of the rope itself and any friction in the pulley bearings, we have,

$$\text{load} = \text{effort}$$

and

$$\text{mechanical advantage} = \frac{\text{load}}{\text{effort}} = 1$$

In this case, although the effort applied is equal to the load raised, we obtain the greater convenience and ease of being able to stand on the ground and pull downwards, instead of having to haul the load upwards from the top of the building.

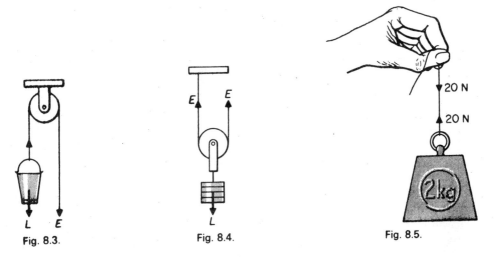

Fig. 8.3.　　　　Fig. 8.4.　　　　Fig. 8.5.

The single moving pulley

This is shown in Fig. 8.4. Here the tension in the string or rope is equal to the effort applied, so that the total upward pull on the pulley is twice the effort E.

Suppose a load of 4 N is supported by the pulley and that the weight of the pulley block and string is negligible. Then, since the load is supported by the tension in two sections of string, the effort applied need only be 2 N. Thus,

$$\text{mechanical advantage} = \frac{\text{load}}{\text{effort}} = \frac{4\text{ N}}{2\text{ N}} = 2$$

Direction of a tension in a string

In all pulley diagrams some consideration must be given to the direction in which force arrows are drawn on the strings.

Take the case of a 2 kg mass supported by a string held in the hand (Fig. 8.5). Assuming $g = 10$ m/s^2 the tension in the string is 20 N and acts equally in both directions. The arrow pointing upwards represents the force which the string exerts on the load, while the arrow pointing downwards shows the force exerted by the string on the hand. This is another example of Newton's third law, "action and reaction are equal and opposite" (page 15).

In general, we do not put in both arrows, but only the one which gives the direction of the force in which we are interested. Thus in Fig. 8.4 both arrows are drawn upwards as we wish to indicate the force exerted by the string on the load. We are not here concerned with the downward pull of the string on the support.

The block and tackle

This is by far the most important pulley system of all, being commonly used for lifts and cranes.

Two blocks are employed containing from two to eight pulleys in each, according to the mechanical advantage required. To illustrate the principle, Fig. 8.6 has been drawn to show two pulleys in each block. For simplicity, the pulleys are shown on separate axles placed one above the other. In practice, however, the pulleys in each block are mounted side by side and run independently on a common axle (Fig. 8.7). A single string is used which passes round each pulley in turn.

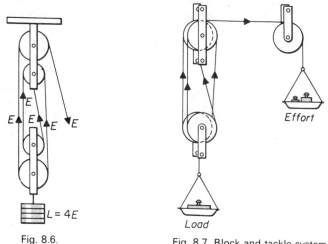

Fig. 8.6.

Fig. 8.7. Block and tackle system

It will be seen from Fig. 8.6 that the lower block is supported by four sections of string. Incidentally, the number of sections of string supporting the lower block is always equal to the total number of pulleys in the two blocks together. It follows that, if an effort E is applied to the free end of the string, then the total upward force on the load will be $4E$.

If we neglect friction and the weight of the moving parts of the system, then,

$$\text{mechanical advantage} = \frac{4E}{E} = 4$$

In practice, however, the practical mechanical advantage in a case such as this is always *less than* 4, since extra effort must be applied to overcome friction and the weight of the moving pulley block and string.

An experiment to measure the practical mechanical advantage of a pulley system is described later.

Velocity ratio (or speed ratio) (V.R.)

In the pulley systems we have already considered where the mechanical advantage is greater than 1, it might appear at first sight that we are getting more out of the machine than we are putting into it. But while in such cases the load is greater than the effort, it must be remembered that the effort moves through a much greater distance than that of the load. Consequently, the work obtained from the machine is equal to the work put into it, less any work wasted in the machine, but this is discussed more fully in the next section.

From Fig. 8.6 it will be clear that, in order to raise the load by 1 m, each string supporting the load must be shortened by 1 m. The effort must therefore be applied through a total distance of 4 m.

The ratio of the distance moved by the effort to the distance moved by the load in the same time is called the velocity ratio of the machine, i.e.,

$$\text{velocity (or speed) ratio} = \frac{\text{distance moved by the effort}}{\text{distance moved by load in same time}}$$

Work done by a machine. Efficiency

If the pulley system shown in Fig. 8.7 were a "perfect machine", i.e., composed of weightless and frictionless strings and pulleys, then a load of 40 N would be raised through a distance of 1 m by an effort of 10 N moving a distance of 4 m.

The work done by the machine on the load is then 40 N × 1m = 40 J, while the work done by the effort is 10 N × 4 m = 40 J also. These are equal, as we should expect for a perfect machine. In practice, however, some work is always wasted in overcoming friction and raising moving parts, and therefore the useful work done by a machine is always less than the work done by the effort.

The ratio of the useful work done by the machine to the total work put into the machine is called the efficiency of the machine.

Usually, this ratio is expressed as a percentage, so we may write

$$\text{efficiency} = \frac{\text{work output}}{\text{work input}} \times 100\%$$

Relation between mechanical advantage, velocity ratio, and efficiency

Since,

$$\text{work} = \text{force} \times \text{distance}$$

it follows that

$$\text{efficiency} = \frac{\text{load} \times \text{distance load moves}}{\text{effort} \times \text{distance effort moves}}$$

$$= M.A. \times \frac{1}{\text{velocity ratio}}$$

$$= \frac{M.A.}{V.R.} \times 100\%$$

This equation will be found very useful for working out problems, but it is not a fundamental definition of efficiency and should not be used as such.

To study the variation of the mechanical advantage of a pulley system with load

We shall use a block and tackle with two pulleys in each block for this experiment, but the same principle may equally well be applied to other types of machine.

The pulleys are set up as in Fig. 8.7, scale-pans being provided for the addition of weights to represent load and effort. Since the scale-pans are more or less an essential convenience, they are best treated as part of the machine itself. Some people, however prefer to include the weights of the pans as part of the load and effort respectively. Whichever course is adopted should be mentioned in the account of the experiment.

An initial mass of, say, 0.05 kg is added to the load pan, and further masses are then added to the effort pan until the load just rises slowly with a steady velocity. The load and effort, in newtons, are given by *mg*, where *m* is the mass in kg and *g* = 10 m/s². The experiment is repeated for a series of increasing loads and the results recorded in a table.

The velocity ratio of this machine may be found by measuring a pair of corresponding distances moved by effort and load.

For each pair of readings of effort and load obtained, the mechanical advantage and the efficiency should be calculated from the appropriate formulae and entered in the table.

Load (N)	Effort (N)	$M.A. = \dfrac{Load}{Effort}$	$Efficiency = \dfrac{M.A.}{V.R.} \times 100\%$

In connection with this experiment, the following points should be noticed:

(1) The useless load consists of the weight of the lower pulley block and the string and friction in the string and bearings. The weight of string lifted depends on the distance between the pulley blocks, but the weight of the lower block is constant. The friction varies with the load, but is small in most cases. Thus, although the useless load varies somewhat, it becomes a smaller proportion of the total load as the total load increases. Consequently, the mechanical advantage increases with load.

(2) The efficiency also increases with load for the same reasons.

(3) Owing to the work wasted in overcoming friction and raising moving parts, the efficiency is less than 100 per cent. Also, since there are only four pulleys altogether, the mechanical advantage cannot exceed 4.

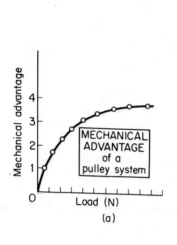

(a)

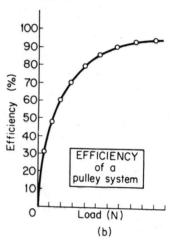

(b)

Fig. 8.8.

Graphs should be plotted of mechanical advantage against load, and efficiency against load. The shape of the curves obtained will illustrate the above remarks. Typical graphs are shown in Fig. 8.8.

The inclined plane

A heavy load may be raised more easily by pulling it along a sloping surface than by lifting it vertically.

Heavy packing cases are often loaded into vans by hauling them up an incline formed by two stout planks held apart by iron stays (Fig. 8.9).

It is believed that the large blocks of stone used in the construction of the Egyptian pyramids were raised into position by dragging them, on rollers, up a long ramp of earth. On completion of the building, the earth was taken away.

In Fig. 8.10 a load is being pulled up an inclined plane AB. In order to raise the load through a vertical height h, the effort has to be exerted through a longer distance equal to the length of the plane l. It must be clearly understood that, because the weight of the load acts vertically downwards, the distance through which the load is overcome is h and not l.

Fig. 8.9. Inclined plane

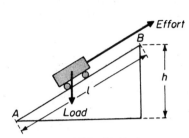

Fig. 8.10. Inclined plane

The velocity ratio is given by,

$$\text{velocity ratio} = \frac{\text{distance moved by effort}}{\text{distance moved by load}} = \frac{\text{length of plane}}{\text{height of plane}} = \frac{l}{h}$$

The mechanical advantage may be obtained by applying the law of conservation of energy (or principle of work). We neglect the work done against friction and assume that the work done on the load is equal to the work done by the effort.

Thus, for a "perfect" inclined plane,

$$\text{load} \times \text{distance load moves} = \text{effort} \times \text{distance effort moves}$$

therefore

$$\text{mechanical advantage} = \frac{\text{load}}{\text{effort}} = \frac{\text{distance effort moves}}{\text{distance load moves}} = \frac{l}{h}$$

The screw

Millions of screws and bolts are used daily for the purpose of holding things together. We have already mentioned the importance of the screw in connection with micrometers (page 7). In addition, the screw is an essential feature of machines such as vices and screw jacks (Fig. 8.11, 8.12).

The distance between successive threads on a screw is called its pitch. For one

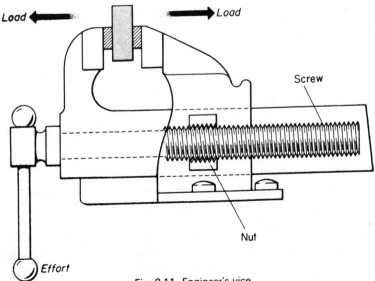

Fig. 8.11. Engineer's vice

complete turn, therefore, a screw moves through a distance equal to its pitch. As an example of the screw as a machine, we shall consider the working of a car jack. Fig. 8.12 shows one type of jack which consists of a long screw carrying a nut. This nut is hinged to a short steel bar which fits into a slot in the car chassis. To raise the car, the screw is turned by hand using a short steel rod known as a tommy bar.

Approximate data for this type of jack are,

Length of tommy bar	24 cm
Pitch of screw	2 mm
Load lifted	5000 N

The effort required to lift the car may be found by applying the principle of work. Thus, ignoring friction,

$$\text{work done by effort} = \text{work done on load}$$

or, effort × circumference of circle traced out by effort = load × screw pitch

Therefore, taking $\pi = 3.14$,

$$\text{effort} \times 2 \times 3.14 \times 0.24 = 5000 \times 0.002$$

or

$$\text{effort} = \frac{5000 \times 0.002}{2 \times 3.14 \times 0.24} = 6.6 \text{ N}$$

In practice, of course, the effort must be much greater than this in order to overcome friction. Even so, the total effort required will still be very small compared with the load being lifted.

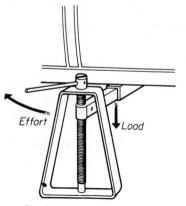

Fig. 8.12. Car lifting jack

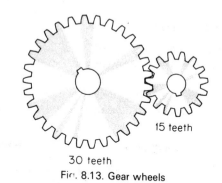

15 teeth

30 teeth

Fig. 8.13. Gear wheels

Wheel and axle principle. Gears

Fig. 8.14 opposite shows some examples of devices using the *wheel* and *axle* principle. The steering-wheel of a car is another obvious example, but one of the main applications of the principle in modern engineering is found in gear-boxes where toothed wheels of different diameters engage to give turning forces at low speed (large mechanical advantage), or high speed (small mechanical advantage), according to which gear is the "driver" and which the "driven" (Fig. 8.13).

In the laboratory, the velocity ratio and mechanical advantage of the wheel and axle may be investigated by using two wheels of different diameters rigidly fixed on the same axle.

Fig. 8.15 shows how the effort is applied by a string attached to the rim of the larger wheel while the load is raised by a string wound round the axle or smaller wheel. For one complete turn, the load and effort move through distances equal to the circumferences of the wheel and axle respectively. The velocity ratio is therefore given by,

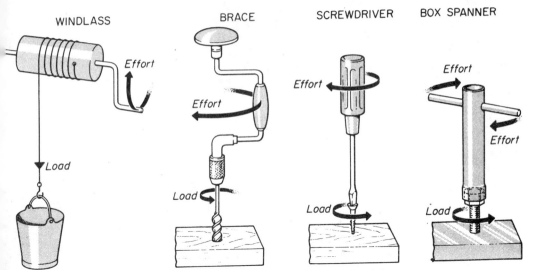

Fig. 8.14. Applications of the wheel and axle principle

$$\text{velocity ratio} = \frac{\text{distance moved by effort}}{\text{distance moved by load}} = \frac{2\pi \times \text{radius of wheel}}{2\pi \times \text{radius of axle}} = \frac{R}{r}$$

The mechanical advantage for a "perfect" wheel and axle may be found, as in previous cases, by applying the principle of work. Otherwise it may be found by taking moments of the load and effort about the axis of rotation. Using the latter method (see Fig. 8.15),

$$\text{load} \times \text{radius of axle} = \text{effort} \times \text{radius of wheel}$$

therefore $$\text{mechanical advantage} = \frac{\text{load}}{\text{effort}} = \frac{\text{radius of wheel}}{\text{radius of axle}} = \frac{R}{r}$$

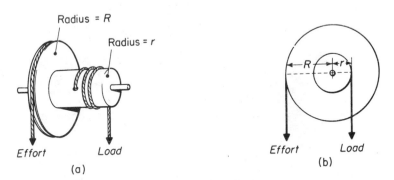

Fig. 8.15. Wheel and axle

Note. For **gear wheels**, remembering that the effort and load are applied to the *shafts* of the gears, the reader should not find it difficult to show that,

$$\text{velocity ratio} = \frac{\text{no. of teeth in } \textit{driven} \text{ wheel}}{\text{no. of teeth in } \textit{driving} \text{ wheel}}$$

The hydraulic press

When a fluid completely fills a vessel, and a pressure is applied to it at any part of the surface, that pressure is transmitted equally throughout the whole of the enclosed fluid. This is known as *Pascal's principle* (see also page 122).

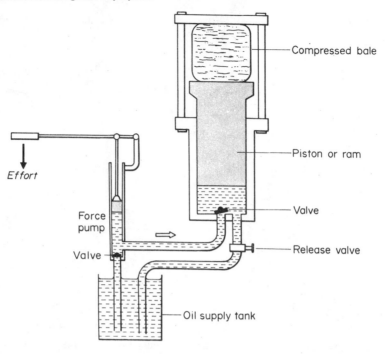

Fig. 8.16. Hydraulic press

This principle finds an important industrial application in the hydraulic press (Fig. 8.16). This type of machine has numerous uses, from the compression of soft materials such as waste paper and cotton into compact bales to the shaping of motor-car bodies and the forging of steel armour plate and light alloys (Fig. 8.17, 13.11).

Fig. 8.17. Pascal's principle in operation

(a) Hydraulic presses at the Ford Motor Works

(b) One of the largest hydraulic presses in the world built f- John Thomson Motor Pressings Ltd.

In its simplest form the hydraulic press consists of a cylinder and piston of large diameter, connected by a pipe to a force pump of much smaller diameter. Oil from a supply tank is pumped into the cylinder and the piston (or ram) moves out, exerting considerable force. A valve is provided to release the pressure and allow the oil to return to the tank, after the press has done its work.

In order to understand how very large forces may be so easily produced by this press, look at Fig. 8.18.

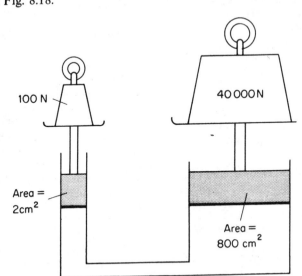

Fig. 8.18. Transmission of pressure

Suppose the pump barrel has an area of 2 cm² and that a force of 100 N is applied to its plunger,

$$\text{the pressure produced} = \frac{\text{force}}{\text{area}} = \frac{100}{2} = 50 \text{ N/cm}^2$$

This pressure is transmitted equally throughout the whole of the liquid and so also to the piston in the large cylinder. If the area of the large piston is 800 cm², then the total force or thrust exerted is given by,

$$\text{thrust} = \text{pressure} \times \text{area} = 50 \times 800 = 40\,000 \text{ N}$$

A force of 40 000 N is therefore obtained simply by exerting a force of only 100 N.

The velocity ratio between the two cylinders of a hydraulic system may be found by using the fact that the volume of liquid which leaves the pump cylinder is equal to that which enters the ram cylinder.

If x is the distance moved by the pump piston and y the distance moved by the ram piston, then equating volumes,

$$x \times \text{area of pump piston} = y \times \text{area of ram piston}$$

or $\quad$ velocity ratio $= \dfrac{x}{y} = \dfrac{\text{area of ram piston}}{\text{area of pump piston}} = \dfrac{\pi R^2}{\pi r^2} = \dfrac{R^2}{r^2}$

where R = radius of ram piston, and
$\qquad r$ = radius of pump piston.

The above expression gives the velocity ratio between the two pistons only. If the *total* velocity ratio of the whole press is required we must multiply $\dfrac{R^2}{r^2}$ by the velocity ratio of the pump handle treated as a lever.

QUESTIONS: 8

(*Where necessary assume* $g = 10$ m/s^2)

1. Define velocity ratio and mechanical advantage. A common windlass is used to raise a 480 N load of earth from an excavation by the application of an effort of 200 N at right angles to the handle and crank. If the handle is 33 cm from the axis and the radius of the axle is 11 cm, find:
(a) the velocity ratio;
(b) the mechanical advantage.

2. What do you understand by the *efficiency* of a machine?

A system of levers with a velocity ratio of 25 overcomes a resistance of 3300 N when an effort of 165 N is applied to it, calculate:
(a) the mechanical advantage of the system;
(b) its efficiency.

3. Define *work* and *power*. Name and define the SI unit of power. By using a block and tackle a man can raise a load of 720 N by an effort of 200 N. Find:
(a) the mechanical advantage of the method;
(b) the man's useful power output if he raises the load through 10 m in 90 s.

4. Draw a diagram of a single-string pulley system with a velocity ratio of 4.

5. An electric pump raises 9.1 m^3 of water from a reservoir whose water-level is 4 m below ground level to a storage tank above ground. If the discharge pipe outlet is 32 m above ground level and the operation takes 1 hr, find the minimum power rating of the pump if its efficiency is 70 per cent. (1 m^3 of water has a mass of 1000 kg.)

6. (a) For a single-string pulley system (block-and-tackle), a man has an upper block of 5 pulleys and the choice for the lower block of *either* 4 pulleys of total mass 20 kg *or* 5 pulleys of total mass 30 kg. State, giving your reasons and showing any necessary calculations, which block you would recommend if the system is to lift a mass of 50 kg and if friction is negligible.

(b) An experiment was carried out to investigate the performance of a single-string pulley system with a velocity ratio of *five*. The following results were obtained:

Load (N)	50	100	200	300	400	500	600
Effort (N)	30	45	65	85	105	125	145

Plot a graph of Effort against Load.
Hence find:
(i) the effort;
(ii) the mechanical advantage, and
(iii) the efficiency corresponding to a load of 450 N.

(c) A man uses the system in (b) to lift a body of mass 50 kg which rises with a velocity of 0.1 m/s. Determine the *power* developed by the man. (A.E.B.)

7. Give a diagram of a single-string pulley system with a velocity ratio of 6. Calculate the efficiency of it if an effort of 1000 N is required to raise a load of 4500 N. Find the energy wasted when a mass of 500 kg is lifted through 2 m.

8. Draw a diagram of a pulley system having a velocity ratio of 5. Draw also two other mechanical devices having the same velocity ratio. Explain, in each of the three examples, why you assert that the velocity ratio is indeed 5.

Give one reason common to them all, why the mechanical efficiency is less than 100 per cent, and one way of increasing the mechanical efficiency of any of the machines you have described. (S.)

9. A box of mass 12 kg is pulled up a straight smooth incline, at 30° to the horizontal, for a distance of 5 m (metres). Calculate the work done. (S.)

10. A man uses a rope to haul a packing case of weight 750 N up an inclined wooden plank of effective length 4.50 m and on to a platform 1.50 m high. The frictional force between case and plank is 200 N. Find:
(a) the effort he must exert on the rope;
(b) the velocity ratio;
(c) the mechanical advantage;
(d) the useful work done on the packing case in joules.

11. A lift of mass 500 kg containing a load of mass 700 kg rises through 25 m in 20 s. In the absence of friction, calculate the average power output of the motor driving the lift. Explain why, in practice, the power output will not be constant during this time. (J.M.B.)

12. What properties of liquids make them suitable for use in hydraulic machines such as car brakes?

Make a simple cross-sectional diagram of a hydraulic press, showing the two cylinders, the oil supply reservoir, the pump, and any necessary valves.

If the larger cylinder is 20 times the diameter of the smaller one, and if the pump is operated by a lever of velocity ratio 6, what is the velocity ratio of the whole machine? If the efficiency is 90 per cent, what force can the machine exert when a force of 100 N is applied to the handle?

13. What is meant by the expression *efficiency of a simple machine*? Derive the relation between efficiency, mechanical advantage, and velocity ratio.

A nut, threaded on a bolt fixed to a bicycle frame, is tightened by means of a spanner. The perpendicular distance between the axis of the nut and the line of action of the force F applied to the free end of the spanner is 0.15 m. F is also perpendicular to the axis of the nut.

What magnitude of F produces a couple of 10.5 N m about the axis of the nut? Where does the second force making up the couple act? Draw a diagram to illustrate your answer.

Regarding the spanner, nut, and bolt as a simple machine, what is its velocity ratio if the pitch of the thread is 1 mm?

Assuming that the efficiency is 10 per cent, what tension is applied to the bolt (i.e., what force is applied parallel to the axis) when the couple of 10.5 N m acts? (*O.C.*)

14. A bicycle has wheels 66 cm in diameter. Its crank wheel has 44 teeth and the rear sprocket 16 teeth. If the crank radius is 16.5 cm, calculate the velocity ratio. Find the efficiency of the bicycle if its mechanical advantage is 0.14.

15. A driving gear wheel having 25 teeth engages with a second wheel with 100 teeth. A third wheel with 30 teeth on the same shaft as the second, engages with a fourth having 60 teeth. Find:
(*a*) the total velocity ratio;

(*b*) the mechanical advantage of the gear system if its efficiency is 85 per cent.

16. (*a*) A block is released and allowed to slide down a rough slope.
(i) Describe the motion of the block by referring to the forces acting on it and the energy changes which take place during the slide.
(ii) If the slope were well-lubricated, describe what differences in the motion there would be compared with those described in (*a*) (i); account for the differences mentioned.

(*b*) The slope mentioned in (*a*) has a vertical height of 0.5 m and measures 1.5 m along the slope. The block has a mass of 2 kg and takes 2.5 s to slide the whole distance down the slope. Calculate
(i) the total energy lost by the block and
(ii) the average power developed.

(*c*) The same block as in (*a*) and (*b*) is allowed to fall freely from rest from a vertical height of 5.0 m and penetrates a distance of 0.02 m into soft sand at the end of its fall. Calculate
(i) The velocity of the block as it strikes the sand,
(ii) the kinetic energy of the block as it strikes the sand,
(iii) the average force exerted on the block by the sand in bringing the block to rest.

(*d*) Name two practical devices which depend on the change of potential energy into kinetic energy. [Assume that the acceleration of free fall (g) = 10 m/s^2 (N/kg)].
(*J.M.B.*)

9. Density and relative density

One often hears the expressions, "as light as a feather" and "as heavy as lead". Equal volumes of different substances vary considerably in mass. Aircraft are made chiefly from aluminium alloys, which provide a structure as strong as steel but which, volume for volume, weigh less than half as much. In physics we refer to the lightness or heaviness of different materials by the use of the word *density*.

The density of a substance is defined as its mass per unit volume.

One way of finding the density of a substance is to take a sample and measure its mass and volume. The density may then be calculated by dividing the mass by the volume. The symbol used for density is the Greek letter ρ (rho).

Thus,
$$\text{density} = \frac{\text{mass}}{\text{volume}} \text{ (in appropriate units: see below)}$$

or in symbols
$$\rho = \frac{m}{V}$$

hence also
$$m = V \times \rho$$

and
$$V = \frac{m}{\rho}$$

The densities of all common substances, solids, liquids and gases, and all chemical elements have been determined and are to be found listed in books of physical and chemical constants.

The SI unit of density is the kg/m^3, but when densities are being measured in the laboratory it is generally most convenient to work in grams and cubic centimetres, thus obtaining a result expressed in g/cm^3. There is no difficulty in converting this to SI units (kg/m^3).

Water happens to have a density of about $1\ g/cm^3$ or $1000\ kg/m^3$ owing to the fact that the kilogram was originally intended to have the same mass as $1000\ cm^3$ of water at 4 °C. Mercury is a metal which is a liquid at ordinary temperatures and it has the very high density of $13.6\ g/cm^3$. It is a very useful substance in scientific laboratories and plays a part in many experiments. When one lifts up a bottle of mercury for the first time one is surprised by its weight.

Importance of density measurements

Architects and engineers refer to tables giving the densities of various building materials when engaged in the design of bridges, flyovers, and other structures. From the plans drawn up, they can calculate the volume of any part of the structure, which, multiplied by the density of the material, gives the mass and hence the weight. Such information is essential for calculating the strength required in foundations and supporting pillars.

Chemists often make a density determination as a test of the purity of a substance. One such test made many years ago led to the discovery of a new gas which has since become very useful industrially. It had been known for a long time that nitrogen obtained from air was slightly denser than that obtained from other sources. Otherwise the two kinds of nitrogen appeared to be identical. The difference in density led Lord Rayleigh and Sir William Ramsay to suspect that atmospheric nitrogen contained small quantities of a heavier gas. Following this clue they began a series of experiments and eventually succeeded in isolating a new gas to which the name of *argon* was given.

Argon at low pressure is used in gas-filled electric lamps. It enables the filament to be run at a higher temperature than in a vacuum, and so to give more light per unit of electric energy.

Simple measurements of density

Liquids. A convenient volume of the liquid is run off into a clean, dry, previously weighed beaker, using either a pipette or burette. The beaker and liquid are then weighed and the mass of the liquid found by subtraction.

Solids. The volume of a substance of regular shape, e.g., a rectangular bar, cylinder, or sphere may be calculated from measurements made by vernier callipers or a micrometer screw gauge.

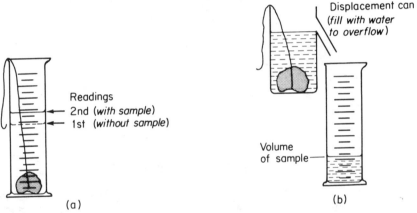

Fig. 9.1. Volume measurement

The volume of an irregular solid, e.g., a piece of coal may be found by either of the methods shown in Fig. 9.1, which is self-explanatory. For solids soluble in water, e.g., certain crystals, some liquid such as white spirit would be used in the measuring cylinder. The mass of the solid is found by weighing.

In each case the density is calculated from,

$$\text{density} = \frac{\text{mass}}{\text{volume}}$$

The above descriptions have purposely been kept short as we shall be describing better methods later on.

Relative density

In the last experiments described we had to make two measurements to find the density, namely, a mass and a volume. Now we can always measure mass more accurately than volume, and so, in the accurate determination of density, scientists have overcome the necessity to measure volume (hence eliminating one source of error) by using the idea of *relative density*.

The relative density of a substance is the ratio of the mass of any volume of it to the mass of an equal volume of water, or

$$\text{relative density} = \frac{\text{mass of any volume of the substance}}{\text{mass of an equal volume of water}}$$

In normal weighing operations in a laboratory, the mass of a body is proportional to its weight, so it is also true to say,

$$\text{relative density} = \frac{\text{weight of any volume of the substance}}{\text{weight of an equal volume of water}}$$

Note that relative density has no units: it is simply a number or ratio. On the other hand, density is expressed in kg/m³ (SI unit) or g/cm³ (sub-multiple SI unit).

To measure the relative density of a liquid

The easiest way to ensure getting identically equal volumes of a liquid and water is. to use a *density bottle* (Fig. 9.2). This bottle has a ground glass stopper with a fine hole through it, so that, when it is filled and the stopper inserted, the excess liquid rises through the hole and runs down the outside. So long as the bottle is used with the same liquid level at the top of the hole, it will always contain the same volume of whatever liquid is put in provided the temperature remains constant.

Fig. 9.2. Density bottle.

The bottle is weighed empty and then when full of the given liquid. The liquid is then returned to the stock bottle. Having well rinsed or cleaned the bottle, it is filled with water and weighed again.

Two precautions are necessary. The outside of the bottle must be wiped dry before weighing. Secondly, the bottle should not be held in a warm hand or some of the liquid may be lost through expansion.

The results are set out as below. For neatness, decimal points, equals signs and so on should be arranged uniformly underneath each other.

Mass of empty bottle	=	g
Mass of bottle full of liquid	=	g
Mass of bottle full of water	=	g
Mass of liquid	=	g
Mass of water	=	g

$$\text{R.D. of liquid} = \frac{\text{mass of liquid}}{\text{mass of water}}$$

$$= \frac{g}{g}$$

$$= \underline{\hspace{3em}}$$

Calculation of density from relative density

Suppose we have measured the relative density of a liquid as described above and found it to be 0.8. If we now assume that 1 cm³ of water has a mass of 1 g, it follows that the mass of any volume of water is numerically the same as its volume in cm³.

Consequently, we can say straight away that the density of our liquid is 0.8 g/cm³. However, a word of warning is necessary. To say that 1 cm³ of water has a mass of 1 g, is only a very close approximation. Owing to expansion, the density of water depends on its temperature.

In work of very high accuracy, scientists make due allowance for this when calculating density from relative density. Other corrections also have to be made concerned with the weighing process, but any further discussion would take us into the realm of more advanced studies.

Finally, it is worth noting that the figures expressing the density of a substance depend on the units used. For example, the density of lead is 11 400 kg/m³ or 11.4 g/cm³, but *its density relative to water is a number or ratio, namely, 11.4 and this is the same whatever system of units is used.*

Advantage of the density bottle

We have already pointed out the advantage of the idea of relative density as a step towards the accurate measurement of density. There are no volume measurements to worry about. Weighings only are required and these can be carried out with a top-pan or beam-balance to a fairly high degree of accuracy. Thus, a good value for the relative density and hence the density of the liquid can be obtained by this method.

Contrasted with the simple measurements of density described earlier a measuring cylinder of the size ordinarily used can be read only to within about 0.5 cm³; even a burette to within 0.1 cm³ only. Consequently, unless a very large volume of sub-

Fig. 9.3. *Density measurement at the NPL*
These two cubes were made at the National Physical Laboratory, and used to determine the density of mercury to one part in a million.

Measurements were made of the mass of mercury which filled the hollow cube and of the mass displaced by the solid one. A mean value from the double experiment eliminated errors due to surface contamination which caused a positive error in one case and a negative error in the other. The volumes of the cubes were calculated from their dimensions measured in terms of light wavelengths. It is of interest to note that the fused quartz plates of the hollow cube were made so accurately flat that they stuck together by molecular cohesion without need for cement. The solid cube was made from a sintered mixture of tungsten carbide and cobalt

stance is used the percentage error in the volume measurement will be large. It follows that density determinations of liquids which depend on the direct measurement of volume will be less accurate than those obtained by the use of a density bottle.

Relative density of a solid

An accurate method for finding the relative density of a solid, based on Archimedes' principle, is described in chapter 12.

Table of densities in kg/m³

Aluminium	2 700	Mercury	13 600
Brass (varies)	8 500	Methylated spirit	800
Copper	8 900	Platinum	21 500
Glass (varies)	2 600	Sand (varies)	2 600
Gold	19 300	Steel (varies)	7 800
Ice (at 0 °C)	920	White spirit	850
Lead	11 300	Zinc	7 100

QUESTIONS: 9

1. A piece of anthracite has a volume of 15 cm^3 and a mass of 27 g. What is its density:
   (a) in g/cm³;
   (b) in kg/m³?

2. Calculate the mass of air in a room of floor dimensions 10 m × 12 m and height 4 m. (Density of air = 1.26 kg/m³.)

3. (a) What is meant by the *density* of a substance?
   (b) A spring balance has a maximum reading of 10 N and the length of the calibrated scale is 20 cm. A rectangular metal block measuring 10 cm by 3 cm by 2 cm is hung on the balance and stretches the spring by 15 cm. Calculate
   (i) the weight of the block;
   (ii) the mass of the block;
   (iii) the density of the metal from which the block is made. (Assume that the acceleration of free fall is 10 m/s².)   (*J.M.B.*)

4. What is the volume occupied by a tonne of sand of density 2600 kg/m³?

5. A tin containing 5000 cm³ of paint has a mass of 7.0 kg.
   (i) If the mass of the empty tin, including the lid, is 0.5 kg calculate the density of the paint.
   (ii) If the tin is made of a metal which has a density of 7800 kg/m³ calculate the volume of metal used to make the tin and the lid.
                                    (*J.M.B.*)

6. Define density and relative density.
   An empty 60 litre petrol tank has a mass of 10 kg. What will be its mass when full of fuel of relative density 0.72?

7. A bottle full of water has a mass of 45 g; when full of mercury its mass is 360g. If the mass of the empty bottle is 20 g calculate the density of mercury. Assuming that you were carrying out this experiment state the order in which the readings would be taken.

8. The mass of a density bottle is 18.00 g when empty, 44.00 g when full of water, and 39.84 g when full of a second liquid. Calculate the density of the liquid.

10. Pressure in liquids and gases

"Pressure" is a word we use from day to day without worrying too much as to its exact meaning. Before setting out on a car journey we make sure that the air pressure in the tyres has been correctly adjusted, not only to keep within the law, but for our own safety and that of others. Steam boilers have gauges fitted to measure the steam pressure. At home, in the kitchen, there may be a "pressure cooker" and so on.

In physics, however, little progress is made until we are able to measure the things we talk about. So we must always define or state the exact meaning of the words we use.

Pressure is defined as the force acting normally per unit area. (Here, the word "normally" means perpendicularly.)

The SI unit of pressure is **1 newton per metre² (N/m²)**. This unit is called the pascal (Pa), in honour of Blaise Pascal, the French mathematician and scientist who did important work on fluid pressure in the seventeenth century. For high pressures the pressure is expressed in kilopascals (1 kPa = 1000 Pa).

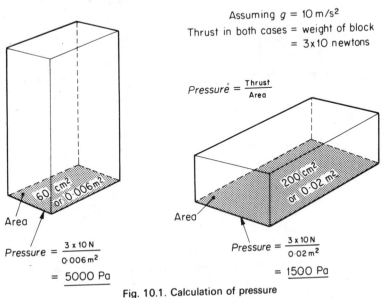

Assuming $g = 10 \text{ m/s}^2$
Thrust in both cases = weight of block
$= 3 \times 10$ newtons

$$Pressure = \frac{Thrust}{Area}$$

Area 60 cm² or 0·006 m²

$$Pressure = \frac{3 \times 10 \text{ N}}{0·006 \text{ m}^2}$$
$$= \underline{5000 \text{ Pa}}$$

Area 200 cm² or 0·02 m²

$$Pressure = \frac{3 \times 10 \text{ N}}{0·02 \text{ m}^2}$$
$$= \underline{1500 \text{ Pa}}$$

Fig. 10.1. Calculation of pressure

Fig. 10.1 shows how to calculate the average pressure exerted by a rectangular block of mass 3 kg: (a) when standing on end; (b) when lying flat. The pressure has been worked out in N/m² or Pa. The total force or *thrust* exerted by the block is the same in both cases.

Thus,
$$\text{pressure} = \frac{\text{thrust}}{\text{area}}$$

or
$$\text{thrust} = \text{pressure} \times \text{area}$$

One does not have to use great force when using a very sharp knife. Owing to the very small area of the cutting edge, even a moderate force will create a very high pressure. The same principle applies to the piercing action of a needle or other sharp point.

Atmospheric pressure

Here on earth we are living at the bottom of a sea of air. Air has weight, and later in this chapter we shall show how its density can be measured. Owing to its weight the atmosphere exerts a pressure at the surface of the earth of approximately 100 000 N/m² (or 100 kPa).

Now the atmospheric pressure acts not only on the earth's surface but all over the surface of objects on the earth, including ourselves. An average-sized man having a surface area of 2 m² will therefore have a total thrust acting over his body of something in the region of 200 kN! He is not of course, conscious of this enormous load, since his blood exerts a pressure slightly greater than the atmospheric pressure, and so a balance is more than effected. At high altitudes, where the pressure of the air is less, nose-bleeding may occur owing to the greater excess pressure of the blood.

Until the seventeenth century it was not realized that the atmosphere did exert a pressure, and people explained its effects by saying that "nature abhorred a vacuum". Thus the common experience that when air is sucked out of a bottle it immediately tries to rush back in again was explained by the theory that nature could not tolerate a vacuum. Nowadays, we say that the excess atmospheric pressure outside the bottle causes a flow of air into the bottle until the pressures inside and outside are equalized.

Crushing can experiment

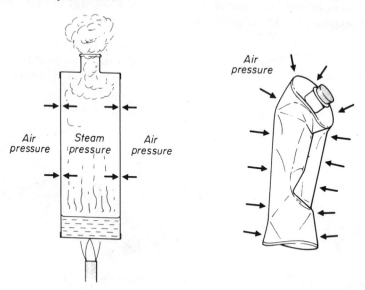

Fig. 10.2. Crushing can experiment

The large forces which can be produced by atmospheric pressure may be demonstrated by means of a metal can fitted with an airtight stopper. The stopper having been removed, a small quantity of water is boiled in the can for a few minutes until the steam has driven out the air. The cork is then tightly replaced and simultaneously the flame beneath the can is turned out.

Cold water is then poured over the can. This causes the steam inside to condense, producing water and water vapour at very low pressure. Consequently, the excess atmospheric pressure outside the can causes it to collapse inwards (Fig. 10.2).

The Magdeburg hemispheres

In 1654 a German scientist named Otto von Guericke made two hollow bronze hemispheres, one of which had a stop-cock. After the rims of these had been placed together with a greased leather ring in between to form an airtight joint the air was pumped out and the stop-cock closed (Fig. 10.3).

Before this was done the hemispheres could be pulled apart quite easily, since the pressure outside was balanced by an equal pressure inside. On removal of the air from inside, only the external atmospheric pressure acted and pressed the hemispheres tightly together. This experiment was first performed before the Imperial Court assembled at Ratisbon, on which occasion two teams of eight horses each were harnessed to the hemispheres and driven in opposite directions. They proved unable to separate the hemispheres until air had been readmitted through the stop-cock.

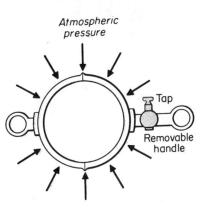

Fig. 10.3. Magdeburg hemispheres

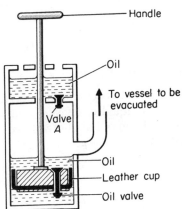

Fig. 10.4. Vacuum pump

The vacuum pump

The invention of the air pump by Otto von Guericke led to big advances in the study of air pressure. Fig. 10.4 shows the principle of a modern type of pump. Each time the piston is at the bottom of the cylinder some air from the vessel to be evacuated expands into the space above the piston. On each upstroke the air above the piston is carried out through the valve *A*. The oil on top of the piston acts not only as a lubricant and air seal but also fills the dead space between piston and valve at the top of the stroke and ensures removal of all air trapped there.

Pressure in a liquid

Liquids exert pressure in the same way that air does.

The pressure in a liquid increases with depth.

This may be shown by means of a tall vessel full of water with side tubes fitted at different heights (Fig. 10.5). The speed with which water spurts out is greatest for the lowest jet, showing that pressure increases with depth.

The pressure at any point in a liquid acts in all directions.

This may be demonstrated by the apparatus shown in Fig. 10.6. Several thistle funnels bent at different angles have thin rubber securely tied over their mouths. These are connected one at a time by rubber tubing to a U-tube containing water. If pressure is exerted on the rubber by hand the air inside is compressed and pushes up the water in the U-tube. So we may regard the whole device as an arrangement for indicating (though *not measuring*) pressure.

When the funnels are lowered into water the U-tube indicates a pressure, whichever funnel is in use, thereby showing that pressure in the water acts in all directions.

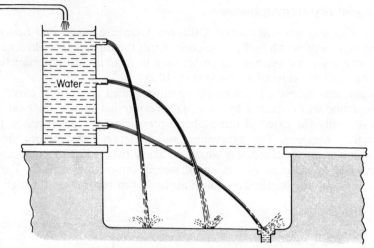

Fig. 10.5. Pressure increases with depth

Calculation of pressure in a liquid

Suppose we consider a horizontal area A, in m², at a depth of h, in m, below the surface of a liquid of density ρ, in kg/m³ (see Fig. 10.7). Standing on this area is a vertical column of liquid of volume hA, in m³, the mass of which (volume × density) is given by $hA\rho$, in kg.

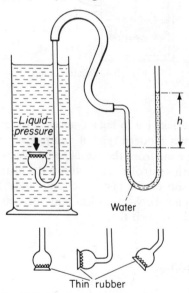

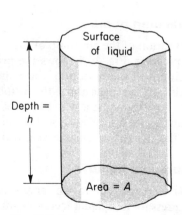

Fig. 10.6. Pressure acts in all directions

Fig. 10.7. Pressure of a liquid column

Now we saw on page 40 that the weight or thrust, in newtons, of a body on its support is given by mg, (m in kg, $g = 10$ m/s²). Hence,

thrust (in N) on area $= hA\rho g$

Therefore,   pressure in N/m² (or Pa) $= \dfrac{\text{thrust}}{\text{area}} = \dfrac{hA\rho g}{A}$

$$= h\rho g$$

It is important to notice that the area A does not appear in the final expression for the pressure. *The pressure at any point in a liquid at rest depends only on the depth and density.* This result can be verified by the following experiments.

To study the variation of pressure with depth in a liquid

A glass tube, 2 or 3 cm in diameter, with its lower end ground perfectly flat is clamped vertically over a large beaker. A flat metal plate attached by thread to a spring balance is held against the lower end of the tube (Fig. 10.8).

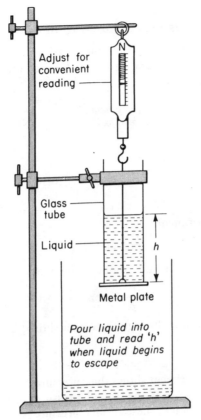

Fig. 10.8. Liquid pressure experiment

The spring balance is clamped so as to give a convenient reading, and water is then slowly poured into the tube. The water will start to leak past the plate when the downthrust of the water becomes equal to F = (spring balance reading − weight of metal plate). The depth, h, of the water is noted when the water just begins to escape.

The experiment is repeated for several spring balance readings and the results entered in a table. Note that, for convenience, we have used N/cm^2 as units of pressure in this case.

Weight of metal plate =   N

Depth h (cm)	Spring-balance reading (N)	F (N)	$\dfrac{F}{h}$ (N/cm^2)

If the area of cross-section of the tube at its base is A in cm^2, then

$$\text{pressure in N/cm}^2 = \frac{F}{A}$$

or, since A is constant, pressure $\propto F$.

Hence, if pressure $\propto h$, the results should show that

$$\frac{F}{h} = \text{constant}$$

Alternatively, a straight line through the origin should be obtained if F is plotted against h on graph paper.

To verify that pressure is proportional to density

Starting with a large spring reading each time, various liquids of known density, ρ, are poured in to the *same depth*. The balance is lowered and its reading noted when the liquid just begins to escape.

Density of liquid ρ (g/cm^3)	Spring-balance reading (N)	F (N)	$\dfrac{F}{\rho}$

As in the previous experiment,

$$\text{pressure} \propto F$$

Hence, if pressure $\propto \rho$, the results should show that

$$\frac{F}{\rho} = \text{constant}$$

Alternatively, a straight line through the origin should be obtained if F is plotted against ρ on graph paper.

Conclusion

The first experiment shows that, pressure $\propto h$
and the second, pressure $\propto \rho$
Combining these results, **pressure** $\propto \boldsymbol{h} \times \rho$

A liquid finds its own level

When water or any other liquid is poured into the communicating tubes shown in Fig. 10.9 it stands at the same level in each tube. This illustrates the popular saying that, "water finds its own level".

When the liquid is at rest in the vessel the pressure must be the same at all points along the same horizontal level, otherwise the liquid would move until the pressures were equalized. The fact that the liquid stands at the *same vertical height* in all the tubes whatever their shape confirms that, for a given liquid, the pressure at a point within it varies only with the vertical depth of the point below the surface of the liquid.

Measurement of gas pressure by the manometer

Earlier in this chapter it was stated that the pressure of the atmosphere was about 100 kN/m^2 (or kPa). Before explaining how this is measured let us make a study of the *manometer*, an instrument for measuring the pressure of gas.

The manometer consists of a U-tube containing water. When both arms are open to the atmosphere the same atmospheric pressure is exerted on the water surfaces A and B, and these are at the same horizontal level (Fig. 10.10 (*a*)).

In order to measure the pressure of the gas supply in the laboratory, the side A is connected to a gas-tap by a length of rubber tubing (Fig. 10.10 (*b*)). When the tap is

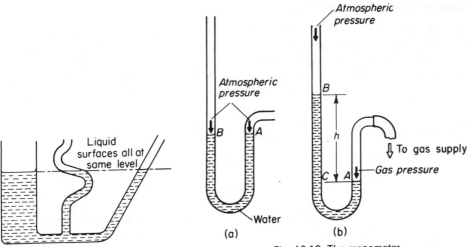

Fig. 10.9. Communicating tubes

Fig. 10.10. The manometer

turned on the gas exerts pressure on the surface *A*, with the result that the level *B* rises until the pressure at *C* on the same horizontal level as *A* becomes equal to the gas pressure. Thus,

pressure of gas = atmospheric pressure + pressure due to water column *BC*

It follows that the excess pressure, in N/m^2, of the gas above that of the atmosphere is given by the pressure of the water column *BC*, and is therefore equal to $h\rho g$ as explained on page 108.

The height, *h*, is called the *head of water* in the manometer and it is often convenient to express the excess pressure simply in terms of *h* only. In this case the units generally used are millimetres of water (mmH_2O).

For measuring higher pressures than in the example above, mercury (density 13.6 g/cm^3) is used in the manometer, in which case the units are mmHg.

Fig. 10.11. This *sphygmomano-meter*, used for measuring blood pressure employs a mercury mano-meter

H_2O and Hg are the chemical symbols for water and mercury respectively. Doctors use an instrument called a sphygmomanometer (Fig. 10.11) for measuring a patient's blood pressure in mmHg which they find adequate for their purpose. As the readings are not used in calculations, the strict use of SI units (pascals) is unnecessary.

Care in the use of mercury

The usefulness of mercury for scientific purposes has already been mentioned on page 100. It does, however, give off a small quantity of odourless vapour which is poisonous if breathed for any length of time in a poorly ventilated room. Great care should be taken not to spill it anywhere it is difficult to recover, for example, in cracks in the floor. All experiments with mercury should therefore be performed over a wooden or plastic tray.

Torricelli's experiment. Simple barometer

About the middle of the seventeenth century an Italian scientist named Torricelli, living at Pisa, suggested an experiment to discount the theory that nature abhorred a vacuum. Torricelli believed that nature's supposed horror of a vacuum was caused simply by atmospheric pressure. In a famous experiment, first performed in 1643, he set up the first *barometer*, an instrument for measuring the pressure of the air.

In the laboratory a simple barometer can be made by taking a stout-walled glass tube about a metre long and closed at one end, and filling it almost to the top with clean mercury. This is done with the aid of a small glass funnel and short length of rubber tubing. Small air bubbles will generally be noticed clinging to the walls of the tube, and these must be removed. With the finger placed securely over its open end, the tube is inverted several times so that the large air bubble left at the top of the tube travels up and down, collecting the small bubbles on its way. More mercury is then added so that the tube is completely full. The finger is again placed over the open end of the tube, which is now inverted and placed vertically with its end well below the surface of some mercury in a dish (Fig. 10.12).

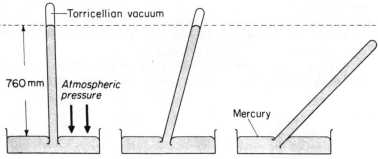

Fig. 10.12. Simple barometer

The finger is then removed and the column of mercury in the tube falls until the vertical difference in level between the surfaces of the mercury in tube and dish is about 760 mm. The vertical height of the mercury column remains constant even when the tube is tilted, unless the top of the tube is less than 760 mm above the level in the dish, in which case the mercury completely fills the tube.

Torricelli explained that the column of mercury was supported in the tube by the atmospheric pressure acting on the surface of the mercury in the dish, and pointed out that small changes in the height of the column, which are noticed from day to day, are due to variations in the atmospheric pressure. The space above the mercury in the tube is called a *Toricellian vacuum*; it contains a little mercury vapour, and in this respect differs from a true vacuum.

Pascal's experiments with barometers

Torricelli died a few years after the barometer experiment had been performed, and did not live to see his explanation of it, in terms of atmospheric pressure, generally accepted among scientists.

After Torricelli's death Pascal repeated the experiment in France and set up two barometers. The first of these was placed at the foot of a mountain in Auvergne called the Puy-de-Dôme, while the other was carried up the mountainside and the height of the column read at intervals on the way up. Owing to the decreasing height of the atmosphere above this barometer, its mercury column showed a progressive fall due to the reduced atmospheric pressure. The barometer at the foot of the mountain showed practically no change.

It was this final experiment which brought about the downfall of the theory that nature abhors a vacuum and established the principle that the atmosphere exerts a pressure.

The Fortin barometer

Accurate measurements of the barometric height, often required in the laboratory, are made with a Fortin barometer (Fig. 10.13). This precision barometer was designed in the late eighteenth century by Nicolas Fortin, a French instrument maker.

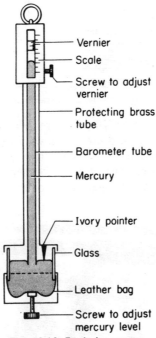

— Vernier
— Scale
— Screw to adjust vernier
— Protecting brass tube
— Barometer tube
— Mercury
— Ivory pointer
— Glass
— Leather bag
— Screw to adjust mercury level

Fig. 10.13. Fortin barometer

The tube containing the mercury is protected by enclosing it in a brass tube, the upper part of which is made of glass so that the mercury surface may be seen. Readings are taken by a vernier moving over a millimetre scale of sufficient length to cover the full range of variation in barometric height.

The zero of the scale is at the tip of an ivory pointer fixed to the lower end of the brass tube. The mercury reservoir is a leather bag which can be raised or lowered by a screw. Before taking a reading of the barometric height, the screw is adjusted until the tip of the pointer just touches its image in the surface of the mercury. If the surface of the mercury is dusty, this adjustment can still be made with reasonable accuracy. A Fortin barometer thus measures the height of a column of mercury supported by atmospheric pressure. On page 108 we saw that the pressure due to a column of liquid of height h and density ρ is given by $p = h\rho g$, where g = acceleration of free fall. If we ignore the variation in the density of mercury with temperature and the variation in g with position on the earth, we see that the pressure is proportional to the height of the mercury column. For many purposes therefore it is sufficiently accurate to quote the pressure in terms of the height of the column, i.e., in millimetres of mercury (mmHg).

The aneroid barometer and altimeter

Barometers of the aneroid (without liquid) type are commonly used as weather glasses, the idea being that low pressure, or a sudden fall in pressure, generally indicates unsettled weather while a rising barometer or high pressure is associated with fine weather. The essential part of an aneroid barometer is a flat cylindrical metal box or capsule, corrugated for strength, and hermetically sealed after having

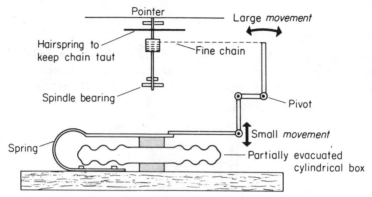

Fig. 10.14. Aneroid barometer movement

been partially exhausted of air (Fig. 10.14). Increase in atmospheric pressure causes the box to cave in slightly, while a decrease allows it to expand. The movements of the box are magnified by a system of levers and transmitted to a fine chain wrapped round the spindle of a pointer. The chain is kept taut by means of a hairspring attached to the spindle, while the pointer moves over a suitably calibrated scale. (See also Fig. 10.15.)

Fig. 10.15. This aneroid *barograph* has a sensitive partially evacuated bellows, the movement of which is transmitted by a system of levers to a pen. It plots a graph of atmospheric pressure against time on paper wrapped round a clockwork-driven drum which makes one revolution per week

Aneroid barometer movements are also used in the construction of altimeters for aircraft. In these the scale is calibrated in metres of ascent. Roughly speaking the pressure falls by 10 mmHg per 120 m of ascent in the lower atmosphere.

The water barometer

Since the density of mercury is 13.6 g/cm³, it follows that if water were used as the liquid in a simple barometer the water column would have to be 76×13.6 cm $= 10.3$ m long.

Such a barometer was constructed in the seventeenth century by von Guericke and fixed on the outside wall of his house. The upper level of the column was indicated by a small wooden float inside the tube on the surface of the water. With the aid of this barometer von Guericke made the first recorded scientific weather forecast. Having noted a sudden fall in the height of the water column, he correctly predicted the imminence of a severe storm.

Pressure in relation to diving and aviation

It should be clear from the previous paragraph that for every 10.3 m (approx. 10 m in sea-water) a diver descends the pressure on his body increases by one *atmosphere*.

The aqualung diving-suit incorporates a rubber helmet fitted with a circular window and supplied with air from compressed-air cylinders carried on the wearer's back. Using this apparatus, experienced divers can descend for very short periods to a maximum depth of about 60 m, where the total pressure is seven atmospheres. At depths in the neighbourhood of 45 m they can work for periods of about 15 minutes. It is dangerous to stay longer at these depths, since, as a result of the high pressure, an excess of nitrogen dissolves in the blood, and on return to the surface nitrogen bubbles form in the blood in the same way that bubbles form in a bottle of soda-water when the cap is removed. Such a condition causes severe pain or even death. and in cases of emergency the diver is immediately placed in a decompression chamber. This is a steel tank full of compressed air and, by slowly reducing the pressure òver a long period, the nitrogen becomes gradually eliminated from the blood without forming bubbles.

The danger to health from the painful *diver's bends*, as this condition is called, is greatly reduced if a mixture of 8 per cent oxygen and 92 per cent helium is used in the gas cylinders. An American diving team claim to have achieved, in 1981, a safe working depth of 686 metres using *Trimix*, a mixture of oxygen, nitrogen, and helium.

In contrast with the problems encountered by the diver, the crew and passengers in aircraft flying at high altitude would experience difficulty in breathing and consequent danger owing to the low atmospheric pressure. The problem is overcome by "pressurizing" the aircraft. All openings are sealed, and a normal atmospheric pressure is maintained inside by the use of air pumps.

Calculation of atmospheric pressure in various units

For ordinary laboratory purposes, atmospheric pressure is measured in mmHg by a Fortin barometer (page 113). The average value of the barometeric height at sea-level over a long period is 760 mmHg, and this has been taken as a unit of pressure and called **standard atmospheric pressure** or **one atmosphere**.

The unit of pressure used for meteorological purposes is called the **bar** and is defined as a pressure of 10^5 N/m².

The more commonly used sub-unit is the **millibar** or one-thousandth of a bar, thus

$$1 \text{ millibar (mbar)} = \frac{10^5}{10^3} = 100 \text{ N/m}^2$$

Example. Express standard atmospheric pressure in: (*a*) newtons/metre²; (*b*) bars; (*c*) millibars. (Assume $g = 9.81$ m/s².)

We know that,

$$\text{standard atmosperic pressure} = 760 \text{ mmHg}$$
$$\text{density of mercury} = 13.6 \text{ g/cm}^3$$
$$= \frac{13.6 \times 100^3}{1000} \text{ kg/m}^3$$
$$\text{acceleration due to gravity} = 9.81 \text{ m/s}^2$$

Hence, using the expression, $h\rho g$, derived on page 108 we have,

$$\text{standard atmospheric pressure (in N/m}^2\text{)} = h\rho g \text{ (in m, kg and s units)}$$
$$= \frac{760 \times 13.6 \times 100^3 \times 9.81}{1000 \times 1000} \text{ N/m}^2$$
$$= 101\,400 \text{ N/m}^2$$
$$= \frac{101\,400}{10^5} \text{ bar}$$
$$= 1.014 \text{ bar}$$
$$= 1014 \text{ mbar}$$

Answer: (*a*) 101 400 N/m² (or Pa)

(*b*) 1.014 bar

(*c*) 1014 mbar

Note. As a matter of interest, atmospheric pressures over the British Isles rarely go below 960 mbar or above 1040 mbar.

To measure the density of dry air

A clean, dry round-bottomed flask is fitted with rubber bung through which passes a short glass tube carrying a short length of rubber pressure tubing fitted with a screw clip.

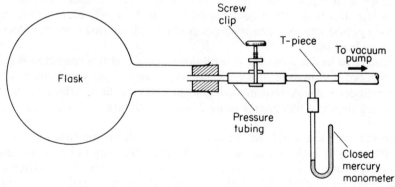

Fig. 10.16. Density of air experiment

With the clip open, the air is evacuated from the flask by connecting it to a good vacuum pump (Fig. 10.16). All the air will have been removed, when the levels are the same on both sides of the *closed* mercury manometer. The clip is then closed and the mass of the flask measured with a balance.

The flask is now connected to a calcium chloride drying tube and the clip opened slowly to admit dry air into the flask. When no more air enters, the clip is closed and the mass of the flask redetermined. The atmospheric pressure is read from a barometer and the temperature is also noted.

The volume of the air is found by filling the flask with water and inserting the bung with the clip open. The bung is removed and the volume of the water found by using a suitable measuring cylinder. Record the results as below, in actual values instead of symbols.

Temperature	$= \theta$ (°C)
Barometric pressure	$= H$ (mmHg)
Volume of air $= v$ in cm³	$= V$ (m³)
Mass of evacuated flask	$= m_1$ (g)
Mass of flask full of dry air	$= m_2$ (g)
Mass of air $= (m_2 - m_1)$ in g	$= m$ (kg)

$$\text{Density of air} = \frac{\text{mass}}{\text{volume}} = \frac{m}{V} \ (\text{kg/m}^3)$$

Density of air at temp. θ and pressure H = _____ kg/m³

Why do we record temperature and pressure?

Densities of gases vary to a much greater extent with changes in temperature and pressure than do those of solids and liquids; hence it is very necessary to record temperature and pressure in the above experiment. For fair comparison, therefore, the densities of gases are calculated at the **standard temperature, 0°C, and standard pressure, 760 mmHg (s.t.p.)**. How this is done is explained on page 180.

Note that we have used SI units in our experiment. Various workers use other density units for gases, e.g., micrograms/cm³ (μg/cm³) or g/m³, to suit their own particular convenience.

QUESTIONS: 10

(*Where required, assume* $g = 10$ m/s²)

1. Define *pressure*. How is it measured in the SI system of units?

2. The base of a rectangular vessel measures 10 cm × 18 cm. Water is poured in to a depth of 4 cm. What is the pressure on the base? What is the thrust on the base?
(*J.M.B.*)

3. A television tube has a flat rectangular end of size 0.40 m by 0.30 m. Calculate the thrust exerted on this end by the atmosphere, if the atmospheric pressure is 1.01×10^5 N/m².
(*O.C.*)

4. Describe any two experiments which show that air exerts a considerable pressure.
(*L.*)

5. Calculate the pressure exerted at the point of a drawing-pin if pushed against a board with a force of 20 N, assuming the area of the point to be 0.1 mm². Give the answer in pascals.

6. Describe, with the aid of a labelled diagram, how you would measure the excess pressure, in Pa, of the laboratory gas supply, using a manometer.
(*A.E.B., 1982*)

7. An open U-tube pressure gauge containing water shows a difference in level of 15 cm when connected to a gas supply. Find, in N/m², the excess pressure of the gas above atmospheric pressure.

8. State how the pressure in a liquid depends on:
(*a*) the depth of the liquid;
(*b*) the density of the liquid.

Describe an experiment to verify one of your statements.

Draw a diagram to show how a water manometer is used to measure the pressure of gas at a gas-tap. Such a manometer connected to a gas main reads 12 cm. Calculate the pressure of the gas, in excess of atmospheric pressure, stating clearly the units in which your answer is expressed. If xylene were used as the manometric liquid, what would be the reading of the instrument? (Density of xylene = 880 kg/m³.)
(*A.E.B.*)

9. Explain how to calculate the thrust on a horizontal surface at a depth h in cm below the surface of a fluid of density ρ in g/cm³.

A solid right circular cylinder of length 10 cm and radius 2 cm has its axis vertical, and its top end is 15 cm below the surface of a fluid of relative density 1.3. Calculate the thrust on:
(*a*) the upper;
(*b*) the lower end of the cylinder due to the fluid.
From your results deduce the loss of weight of the cylinder on immersion in the fluid.
(*O.C.*)

10. Describe how you would set up a simple barometer, pointing out the precautions necessary to obtain a reasonably accurate form of instrument. How would you test whether you had a vacuum above the mercury?

11. A strong dry glass tube of uniform bore and about 1 m long is open at both ends. The tube is supported vertically with its lower end fixed well below the surface of mercury in a large dish. The upper end is connected by suitable tubing to a high-vacuum pump. Describe with the aid of a diagram, and account for, what would you expect to observe when the pump is switched on.

What differences would you expect to observe if the tube was:

(i) not of uniform bore;
(ii) not vertical;
(iii) of very small bore?* Give reasons in each case.   (S.)

12. (*a*) What is meant by *pressure*? Explain simply how the pressure exerted on a surface by a rectangular block depends on which face of the block rests on the surface.

(*b*) Describe **three** experiments, **one** for each case, to show that the pressure exerted by a liquid

(i) varies with the depth of the liquid;
(ii) varies with the density of the liquid;
(iii) is the same in all directions at a given depth.

(*c*) A tank with a base area of 4 m² is connected at the bottom to a vertical tube of cross sectional area 0.01 m² by a horizontal tube. A liquid of density 1000 kg/m³ is poured into the tank until the depth of liquid in the tank is 0.5 m.

Sketch the arrangement of the tank and tubes (not to scale) showing clearly the depth of liquid in the tank and in the vertical tube.

Calculate

(i) the pressure due to the liquid on the base of the tank;
(ii) the pressure due to the liquid at the base of the vertical tube.

If the atmospheric pressure at the time were 120 000 Pa (N/m²), what would be the total pressure on the base of the tank?

(Assume that the acceleration of free fall is 10 m/s² (N/kg).)   (*J.M.B.*)

* See page 144.

13. Describe an experiment:

(*a*) to find the mass of 1 m³ of air, and
(*b*) to demonstrate air pressure. Give details of each experiment.

Calculate the mass of air in a room 5 m × 4 m × 2 m given that the density of air is 1.3 kg/m³.

14. How would you show experimentally that the pressure at a given point in a liquid is the same in all directions?

Fig. 10.17.

Fig. 10.17 shows a section through a mug which has the shape of a truncated cone and which is full of liquid. Is the total thrust exerted by the liquid on the base equal to the weight of the liquid, or more, or less?

The area of the base is 0.003 m² (30 cm²) and the depth of the liquid is 0.15 m (15 cm). The density of the liquid is 1100 kg/m³ (1.1 g/cm³). What is the thrust in newtons exerted by the liquid on the base of the mug?
                                                   (*O.C.*)

15. The air pressure at the base of a mountain is 75.0 cm of mercury and at the top is 60.0 cm of mercury. Given that the average density of air is 1.25 kg/m³ and the density of mercury is 13 600 kg/m³, calculate the height of the mountain.   (*W.*)

16. Draw a labelled diagram of an aneroid barometer.

17. State **two** factors on which the pressure exerted by a liquid depends.

The atmospheric pressure on a particular day was measured as 750 mm of mercury. What is this pressure when it is measured in pascals (Pa) (N/m²)? [Assume that the density of mercury is 13 600 kg/m³ and that the acceleration of free fall (due to gravity) is 10 m/s² (N/kg).]   (*J.M.B.*)

11. Application of atmospheric and liquid pressure

The rubber sucker

A very useful application of atmospheric pressure is found in a circular shallow rubber cup known as a *sucker*. These are used for attaching notices to shop windows and for similar purposes. When the rubber is moistened to obtain a good air seal and pressed on a smooth flat surface the cup is flattened and air squeezed out from beneath it. Atmospheric pressure then holds the sucker firmly on to the surface.

Fig. 11.1. *Lifting an aluminium sheet with giant rubber suckers.* The weight of the beam expels air from the suckers as they are lowered on to the sheet, and a firm hold is then obtained by atmospheric pressure. To release the sheet, air is admitted between suckers and sheet by means of a lever-operated valve.

The discs of thin plastic material used for attaching licences to motor-car windscreens are based on the same principle.

Large rubber suckers are used in industry. Fig. 11.1 shows a lifting beam with two suckers each capable of supporting 500 kg each. The weight of the beam expels the air from the suckers as they are lowered on to the aluminium sheet, and when the sheet has been lifted the suction is released by admitting air through a lever-operated valve.

The common pump (lift pump)

Pumps were used successfully to raise water from wells long before their action was properly understood, and are still to be seen in country villages, carefully preserved as relics of the past. They consist of a cylindrical metal barrel with a side tube near the top to act as a spout (Fig. 11.2). At the bottom of the barrel, where it joins a pipe leading to the well, there is a clack valve *B*. The latter is a hinged leather flap weighted by a brass disc so that it normally falls shut. A plunger carrying a leather cup and fitted with a second clack valve *A* is moved up and down inside the barrel by a handle *H*.

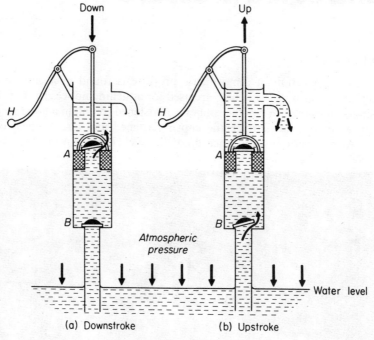

Fig. 11.2. The common pump

To start the pump working it is first primed by pouring some water on to the top of the plunger. This makes a good air seal and prevents leakage of air past the plunger during the first few strokes which are needed to fill the pump with water. Once the pump is filled the action is as follows:

The downstroke (Fig. 11.2 (*a*))
When the plunger moves downwards the valve *B* closes owing to the force of gravity on it and the weight of water above it. At the same time water inside the pump passes upwards through the valve *A* into the space above the plunger.

The upstroke (Fig. 11.2 (*b*)).
On the upstroke the valve *A* closes owing to the force of gravity on it and the weight of water above it. Also as the plunger rises, water is pushed up the pipe through the valve *B* by atmospheric pressure acting on the surface of the water in the well. At the same time, the water above the plunger is raised and flows out of the spout.

Limitations of the common pump
Owing to the fact that the atmospheric pressure cannot support a column of water more than about 10 m long, it follows that 10 m is the theoretical maximum height to which water can be raised by a common pump. An imperfect vacuum, however, is usually obtained owing to bubbles from dissolved air forming near the top of the

water column. For this reason the practical working height of a pump is rather less than 10 m.

Occasionally one finds a pump which can lift water to a height greater than the theoretical maximum. This will occur if air can leak into the pipe near the bottom. Air bubbles then rise in the pipe and break up the water column into a series of shorter columns. Thus, although the total length of water in the pipe is not more than about 10 m, the total length of water plus air is greater than 10 m. Consequently, water can enter the pump. A similar thing can happen when one is using a pipette. If the pipette is inadvertently lifted out of the liquid while it is being filled air will enter and the bubbles will carry the liquid up into the mouth almost immediately.

The force pump

For raising water to a height of more than 10 m, the force pump is used (Fig. 11.3). It consists of a pump with a solid plunger and foot valve B, connected by a pipe to a chamber C through a valve A.

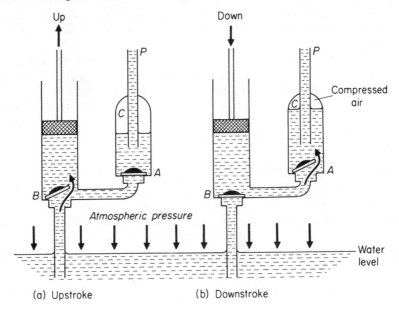

Fig. 11.3. The Force pump

The upstroke (Fig. 11.3 (*a*))
On the upstroke valve A closes and the atmospheric pressure pushes water up into the pump through valve B.

The downstroke (Fig. 11.3 (*b*))
On the downstroke, valve B closes and water is forced into the chamber C through valve A by the pressure due to the mechanical force exerted on the plunger.

The exit pipe P projects into the chamber C so that some air becomes trapped at the top of the chamber. This is compressed and acts as a cushion, thus preventing a sudden jolt to the pump when the water column in P falls slightly and sharply closes valve A at the beginning of the upstroke. C also helps to expel water on the upstroke.

The maximum height to which water may be raised by this means depends on:

(*a*) The force which is exerted on the plunger during the downstroke.
(*b*) The ability of the pump and its working parts to withstand the pressure of the long column of water in the exit pipe P.

Early fire pumps

Towards the end of the eighteenth century manual fire pumps came into use which consisted of a pair of force pumps connected to a long handle and worked by a team of four men. Both pumps fed alternately into a chamber with a compressed air space similar to that described above. At the moment of change-over of feed from one pump to the other the compressed air expanded, and so maintained a steady flow of water to the hoses (Fig. 11.4).

The preserving jar

Atmospheric pressure is used in the kitchen in connection with the preservation of fruit which is unsuitable for deepfreezing.

The preserving jar, is a glass jar covered with a metal cap seated on a flat rubber ring (Fig. 11.5). Clean fruit and water are placed in the jar, leaving a small air space at the top. Several of these jars are placed in a large vessel of cold water, which is then slowly brought to the boil. During this process the metal caps with their rubber rings are loosely held in position by a metal screw cap. About 10 minutes' boiling is generally sufficient to sterilize the fruit and to cause air to be driven from the jars by steam from the water inside. The screw caps are then tightened and the jars removed from the water.

After cooling, the space at the top of the jars contains only water vapour at low pressure. As a result, the metal cap is then firmly pressed down by atmospheric pressure. No bacteria-laden air can afterwards enter, and so the contents remain in good condition for a long period. It is important to notice that, when the jars have cooled the presence of the metal screw cap is not strictly necessary, as the seal is now maintained by atmospheric pressure.

The transmission of pressure in fluids. Hydraulic brake

When the fluid is completely enclosed in a vessel and a pressure is applied to it at any part of its surface, as for example, by means of a cylinder and piston connected to the vessel, then the pressure is transmitted equally throughout the whole of the enclosed fluid. This fact, first recognized by the French scientist and philosopher, Pascal, in 1650, is called the *principle of transmission of pressure in fluids*.

Fig. 11.4. *Two centuries of progress*

A late eighteenth-century fire engine (specification not available, but see text)

A twentieth-century Merryweather diesel pumping set as installed in London's fireboats and capable of delivering over 9 000 litres per minute at a pressure of 700 kN/m²

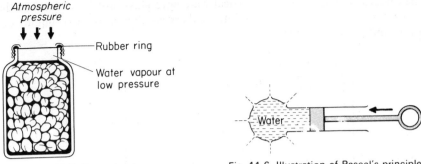

Fig. 11.5. Preserving jar

Fig. 11.6. Illustration of Pascal's principle

Fig. 11.6 shows a piece of apparatus to demonstrate this principle. It consists of a glass barrel fitted with a plunger and ending in a bulb pierced with holes of uniform size. It is filled with water by dipping the bulb in water and slowly raising the plunger. When the plunger is pushed in the water squirts equally from all the holes. This shows that the pressure applied to the plunger has been transmitted uniformly throughout the water.

The principle of transmission of pressure has a number of practical applications. Indeed, our lives may often be said to depend on it whenever we ride in a motor vehicle, since the brakes of the majority of road vehicles are worked by hydraulic pressure.

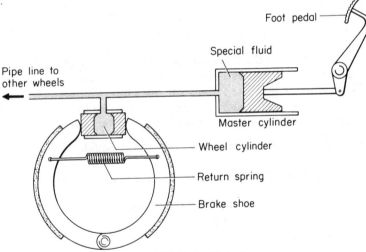

Fig. 11.7. Hydraulic brake

This system of braking is shown diagrammatically in Fig. 11.7. The brake-shoes are expanded by a cylinder having two opposed pistons. These are forced outwards by liquid under pressure conveyed by a pipe from the master cylinder. The piston of the master cylinder is worked by the brake pedal. When pressure on the pedal is released the brake-shoe pull-off springs force the wheel pistons back into the cylinders, and the liquid is returned to the master cylinder.

A very important advantage of this system is that the pressure set up in the master cylinder is transmitted equally to all four wheel cylinders so that the braking effort is equal on all wheels. (See also page 96 for a description of the hydraulic press.)

To compare the densities of two liquids by means of Hare's apparatus

In chapter 9 we saw how the relative density of a liquid can be found by the use of a density bottle. The following method for measuring the relative density of a liquid is based on the pressure exerted by a liquid column.

The apparatus consists of two vertical wide-bore glass tubes connected at the top by a glass T-piece. These tubes dip into beakers containing the two liquids of densities ρ_1 and ρ_2 (Fig. 11.8 (a)).

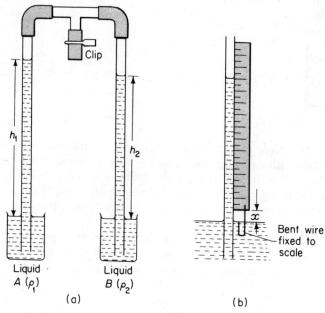

Fig. 11.8. Hare's apparatus

Some air is sucked out of the tubes through the centre limb of the T-piece and the clip closed. Removal of air causes a reduction of pressure inside, with the result that atmospheric pressure pushes the liquids up the tubes. The liquids rise until the pressures exerted at the base of each column are each equal to atmospheric pressure.

The pressure at the base of a column is made up of two parts:

(i) the pressure, P, of the air in the tube above the liquid,
and
(ii) the pressure, $h\rho g$, of the liquid column itself (see page 108),

hence
thus

$$P + h_1\rho_1 g = P + h_2\rho_2 g$$
$$h_1\rho_1 = h_2\rho_2$$

or

$$\frac{\rho_1}{\rho_2} = \frac{h_2}{h_1}$$

If liquid B is water, then $\dfrac{\rho_1}{\rho_2}\left(\text{or } \dfrac{h_2}{h_1}\right)$ will be equal to the relative density of liquid A.

A certain amount of difficulty may arise when measuring the height of the liquid columns, owing to the meniscus which forms when a boxwood scale touches the surface of the liquid. This may be overcome by the use of a bent wire attached to the lower end of the scale, as shown in Fig. 11.8 (b). The scale is adjusted until the tip of the wire is just level with the liquid surface. The scale reading of the liquid level in the tube is then taken, and added to the distance x between the tip of the wire and the zero of the scale.

Several pairs of values of h_1 and h_2 are taken, entered in a suitable table and the mean value of the ratio of the densities calculated.

Alternatively, we may plot a graph of h_2 against h_1 and obtain the ratio from its gradient.

The siphon

Most people are familiar with the use of a siphon for removing water from fish aquaria or other receptacles which cannot otherwise be emptied conveniently.

The siphon is a bent tube made of glass, rubber, or plastic tubing with its short arm dipping into the tank of liquid and its longer arm outside (Fig. 11.9).

To start the siphon it must first be filled with liquid. After this the liquid will continue to run out so long as the end E is below the level of the liquid in the tank.

At one time it was generally accepted that a siphon worked by atmospheric pressure. But there is now strong evidence to support the view that cohesion between the liquid molecules plays an important part.

Older explanation

The pressures at A and D in the two limbs of the tube are both equal to atmospheric pressure, since they are at the same horizontal level as the surface of the liquid in the tank. The pressure at the outlet E is equal to atmospheric pressure plus the pressure $h\rho g$ due to the column DE. The excess pressure, $h\rho g$, therefore causes the liquid to flow out of the tube at E.

Also, since the liquid has to rise a distance AB up the tube, it follows that the siphon will fail to work if AB is greater than the barometric height in terms of the liquid being used.

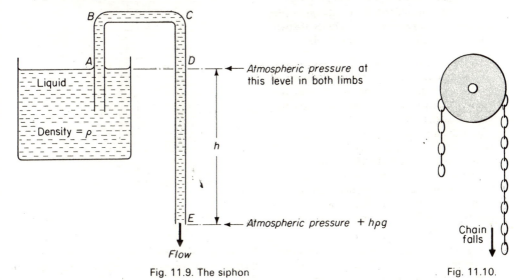

Fig. 11.9. The siphon   Fig. 11.10.

Newer explanation

Experiment shows that *siphons can be made to work in a vacuum* and that, in certain cases, the flow will continue even if AB is somewhat greater than the appropriate barometric height.

The presence of atmospheric pressure, therefore, does not appear to be essential. It would seem that the flow of liquid occurs owing to the greater force of gravity on the column CE, which pulls the shorter column AB through the cohesion (attractive force) between the liquid molecules. The action can be likened to that of a chain passing over a freely running pulley which will run off in the direction of the longer side (Fig. 11.10).

Critical comment

Only *pure* liquids will siphon satisfactorily in a vacuum. If dissolved gases are present the cohesive force between the molecules is greatly lowered and bubbles readily form under reduced pressure. Hence, atmospheric pressure is a necessary condition in the case of impure liquids as it compresses the liquids in the tubes and prevents breakage of the liquid columns through the formation of bubbles from dissolved gases.

A more detailed discussion of the action of the siphon will be found in an article written by M. C. Nokes in the *School Science Review*, Vol. 29, p. 233.

Anyone who is really interested in the subject may read *Hydraulic design of syphons**, A. G. Kelly B.Eng. (Proc. Inst. Mech. Engrs., 1965–6) vol. 180, p. 981. They may come to the conclusion that the last word on the siphon has yet to be written.

* Alternative spelling.

QUESTIONS: 11

(*Where required assume* $g = 10$ m/s^2)

1. With the aid of two labelled diagrams, describe the construction and action of the common pump.

2. Describe a *force* pump and describe its action.

3. Distinguish between *force* and *pressure*.

What do you understand by the *principle of transmission of pressure* in fluids. State two practical applications of this principle.

4. Define relative density.

Describe with a sketch, how you would set up Hare's apparatus for measuring the relative density of a liquid, and show how the result is calculated.

In such an experiment using methanol and water, the lengths of the methanol and water columns were found to be 16 cm and 12.8 cm respectively. Find the relative density of methanol. In a second experiment the length of the methanol column was altered to 21.5 cm. What would be the new height of the water column?

5. Fig. 11.11, which is drawn to scale, shows the heights to which the liquids *A* and *B* have risen in the two sides of an inverted U-tube

when some air has been pumped out at *P*. Obtain the required readings from Fig. 11.11 and deduce the density of liquid *B* if the density of liquid *A* is 800 kg/m^3.

(*O.C.*)

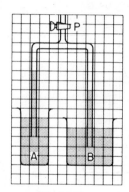

Fig. 11.11.

6. Describe and explain the action of a siphon.

7. Explain why water in the bottom of a floating boat cannot be siphoned over the side.

12. The principle of Archimedes

When it was first proposed to build ships made of iron many people laughed at the suggestion. Owing to the fact that a piece of iron sinks when placed in water they held the view that iron ships would be a failure.

When anything is placed in a liquid it receives an upward force or **upthrust.** In this chapter we shall describe a number of experiments to investigate the forces exerted on a body which is immersed or is floating in a liquid.

Apparent loss in weight

A simple but striking experiment to illustrate the upthrust exerted by a liquid can be shown by tying a length of cotton to a brick. Any attempt to lift the brick by the cotton fails through breakage of the cotton, but if the brick is immersed in water it may be lifted quite easily. The water exerts an upthrust on the brick, and so it appears to weigh less in water than in air.

Stone boulders immersed in water have an upthrust on them equal to about four-tenths of their weight. This explains why boulders can be moved so easily by flood water. In July 1952, for example, a disastrous flood carried debris containing boulders weighing up to 20 t each into the streets of Lynmouth in Devon. Totally immersed they had an apparent weight of only about 12 t, and thus were more easily transported by the force of moving water (Fig. 12.1).

Heavy shingle piles up on a sea-beach for the same reason.

Fig. 12.1. The loss in weight on immersion in water allowed these massive boulders to be easily moved by a flooding river

Archimedes' principle

Experiments to measure the upthrust of a liquid were first carried out by the Greek scientist Archimedes, who lived in the third century B.C. The result of his work was a most important discovery which is now called **Archimedes' principle.** In its most general form, this states:

When a body is wholly or partially immersed in a fluid it experiences an upthrust equal to the weight of the fluid displaced.

It should be noticed that the word "fluid" is used in the above statement. This word means either a liquid or a gas. The application of Archimedes' principle to gases will be discussed later.

To investigate Archimedes' principle for a body in liquid

A displacement can is placed on the bench with a beaker under its spout (Fig. 12.2). Water is poured in until it runs from the spout. When the water has ceased dripping the beaker is removed and replaced by another beaker which has been previously dried and weighed.

Any suitable solid body, e.g., a piece of metal or stone, is suspended by thin thread from the hook of a spring-balance calibrated in newtons and the weight of the body in air is measured. The body, still attached to the balance, is then carefully lowered into the displacement can. When it is completely immersed its new weight is noted.

The displaced water is caught in the weighed beaker. When no more water drips from the spout the beaker and water are weighed.

Owing to the difficulty in weighing a beaker containing water with an extension spring balance, we shall probably use an ordinary balance which gives *mass* readings

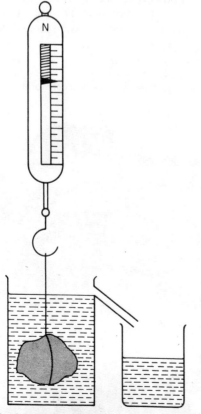

Fig. 12.2. Verification of Archimedes' principle

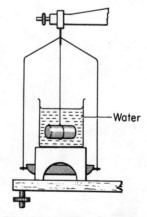

Fig. 12.3. Weighing in water

and hence find the weights of the beaker and displaced water by multiplying the masses by g, assumed equal to 10 m/s² (page 40).

The results should be set down as follows:

Weight of body in air	=	N
Weight of body in water	=	N
Weight of empty beaker	=	N
Weight of beaker plus displaced water	=	N
Apparent loss in weight of body	=	N
Weight of water displaced	=	N

If the apparent loss in weight of the body, or the upthrust on it, is found equal to the weight of the water displaced, then Archimedes' principle is verified in the case of water. Similar results are obtained if any other liquid is used.

To measure the relative density of a solid by using Archimedes' principle

On page 101 we explained the meaning of the term *relative density*, and its importance in the accurate measurement of *density*.

By definition,

$$\text{relative density of a substance} = \frac{\text{mass of any volume of the substance}}{\text{mass of an equal volume of water}}$$

or since, at any particular place, the mass of anything is proportional to its weight,

$$\text{relative density of a substance} = \frac{\text{weight of any volume of the substance}}{\text{weight of an equal volume of water}}$$

The weighings required on the right hand side of this equation are best obtained by an experiment using Archimedes' principle. We weigh a sample of the solid, first in air, and then in water. The apparent loss in weight, obtained by subtraction, is equal to the weight of a volume of water equal to that of the sample.

Thus,

$$\text{relative density of a substance} = \frac{\text{weight of a sample of the substance}}{\text{the apparent loss in weight of the sample in water}}$$

Weighings may be carried out using a spring balance calibrated in newtons. More accurately, a beam or lever balance may be used. However, these measure mass, not weight, so the mass readings must be multiplied by g, the acceleration due to gravity, at the place of the experiment (see page 68). This creates no problem since as we shall see, the value of g disappears from the final calculation, so we do not have to know its value.

Methods of weighing a body in water are shown in Fig. 12.3 or 12.4. A wooden or aluminium bridge is placed over the left-hand pan of the balance, care being taken to see that the pan does not touch it as it swings. The sample, e.g., a piece of brass, is then tied by thin thread to the lower hook of the balance stirrup so that it hangs just clear of the top of the bridge. The brass is then weighed.

A beaker containing water is then placed on the bridge so that the brass is completely immersed and does not touch the side of the beaker. Having made certain that there are no air bubbles clinging to the brass, it is weighed in water.

If m_1 and m_2 are the mass readings obtained in these two weighings, then

Weight of brass in air	$= m_1 g$
Weight of brass in water	$= m_2 g$
Apparent loss in weight of brass	$= (m_1 - m_2)g$
Relative density of brass	

$$= \frac{\text{weight in air}}{\text{apparent loss in weight in water}} = \frac{m_1 g}{(m_1 - m_2)g} = \frac{m_1}{m_1 - m_2}$$

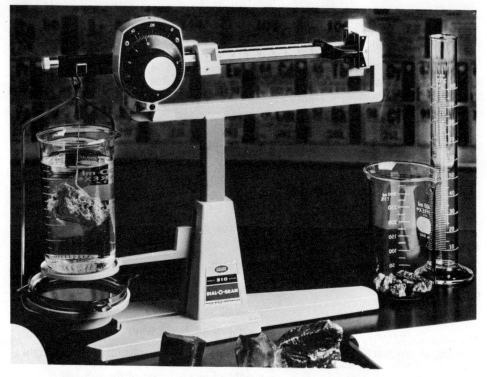

Fig. 12.4. Geological specimens being weighed in water for relative density determination

Relative density of a liquid by using Archimedes' principle

Using the same procedure as in the previous experiment, a sinker is weighed first in air, then in liquid, e.g., white spirit, and finally in water. The sinker is any convenient solid body, e.g., a piece of metal or a glass stopper.

Since the same sinker is used in both liquids, the two apparent losses in weight will be the weights of equal volumes of spirit and water respectively.

As in the previous experiment, if m_1, m_2, and m_3 are the respective mass readings obtained in these three weighings, then

Weight of sinker in air	$= m_1 g$
Weight of sinker in white spirit	$= m_2 g$
Weight of sinker in water	$= m_3 g$
Apparent loss in weight of sinker in spirit	$= (m_1 - m_2)g$
Apparent loss in weight of sinker in water	$= (m_1 - m_3)g$

Relative density of white spirit

$$= \frac{\text{weight of any given volume of spirit}}{\text{weight of an equal volume of water}}$$

$$= \frac{\text{apparent loss of weight of sinker in spirit}}{\text{apparent loss of weight of sinker in water}}$$

$$= \frac{(m_1 - m_2)g}{(m_1 - m_3)g}$$

$$= \frac{m_1 - m_2}{m_1 - m_3}$$

The modern form of hydrometer shown in Fig. 12.8 (*a*) differs scarcely if at all from those made by Robert Boyle in the seventeenth century. The lower bulb is weighted with mercury or lead shot to keep it upright, and the upper stem, graduated to read the relative density of the liquid, is made thin to give the instrument a greater sensitivity. Such hydrometers are usually made in sets of four or more, each covering a different range.

In addition, others are obtainable for special purposes. One, called a *lactometer*, has a range of 1.015–1.045 and is used for testing milk. Another, enclosed in a glass tube fitted with a rubber bulb, is used for measuring the relative density of battery acid (Fig. 12.7 (*b*)). On squeezing the bulb and then releasing it, it expands causing the air pressure inside to decrease. The greater atmospheric pressure outside pushes acid up into the glass tube, and the density can then be read on the floating hydrometer. The acid in a fully charged cell should have a relative density of 1.25–1.30. A reading of less than 1.15 indicates that recharging is necessary.

To measure the relative density of solids and liquids by a balanced lever

Relative density of a solid
A sample of the test solid is attached by a loop of thread to a metre rule pivoted near its mid-point and is balanced by the weight W, of any suitable mass, hung on the opposite side (Fig. 12.8).

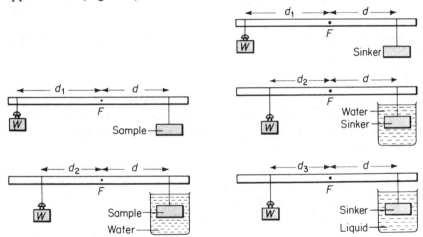

Fig. 12.8. Relative density of a solid   Fig. 12.9. Relative density of a liquid

The distance d of the sample from the pivot is kept constant and the distance of W from the pivot is noted (d_1).

The sample is then immersed in water in a beaker and the balance restored by moving W closer to the pivot (d_2).

If $\qquad\qquad W_1$ = weight of sample in air
and $\qquad\quad\; W_2$ = weight of sample in water
then, taking moments about the pivot F,

(i) $\qquad\qquad\qquad W \times d_1 = W_1 \times d$
(ii) $\qquad\qquad\qquad W \times d_2 = W_2 \times d$

therefore $\qquad\qquad W_1 = \dfrac{W \times d_1}{d}$

and $\qquad\qquad\qquad W_2 = \dfrac{W \times d_2}{d}$

Relative density of sample $= \dfrac{\text{weight in air}}{\text{loss in weight in water}}$ (see page 129)

$$= \frac{W_1}{W_1 - W_2} = \frac{\dfrac{Wd_1}{d}}{\dfrac{Wd_1}{d} - \dfrac{Wd_2}{d}} = \frac{d_1}{d_1 - d_2}$$

Several pairs of values of d_1 and d_2 may be obtained from which a mean value for the relative density may be calculated.

One advantage of this method is that no masses or weights have to be measured.

Relative density of a liquid

Two experiments are carried out exactly as described above in which the sample is replaced by any convenient sinker (e.g., a metal cylinder). These are followed by a third experiment with the sinker immersed in the liquid whose relative density is required (Fig. 12.9).

By reasoning similar to the above it may be shown that,

relative density of liquid $= \dfrac{\text{loss in weight of sinker in liquid}}{\text{loss in weight of sinker in water}}$ (see page 130)

$$= \frac{d_1 - d_3}{d_1 - d_2}$$

The steps used in arriving at this formula should be shown in full when the experiment is written up in the practical notebook.

Worked examples

1. *A ship of mass 1200 t floats in sea-water. What volume of sea-water does it displace? If the ship enters fresh water, what mass of cargo must be unloaded so that the same volume of water is displaced as before?* (*Density of fresh water* $= 1000 \text{ kg/m}^3$, *relative density of sea-water* $= 1.03$; $1 \text{ t} = 1000 \text{ kg}$.)

The ship displaces a weight of sea-water equal to its own weight and therefore a mass of sea-water equal to its own mass

$$\text{Mass of sea-water displaced} = 1200 \text{ t}$$
$$\text{Density of sea-water} = 1000 \times 1.03 \text{ kg/m}^3$$
$$\text{Volume displaced} = \frac{\text{mass}}{\text{density}} = \frac{1200 \times 1000}{1000 \times 1.03} \text{ m}^3$$
$$= 1165 \text{ m}^3 \text{ of sea-water}$$

The same volume of fresh water has a mass of $1165 \times 1000 \text{ kg} = 1165 \text{ t}$
Therefore, mass of cargo to be unloaded $\quad = 1200 - 1165 = 35 \text{ t}$

2. *What volume of brass of density* 8.5 g/cm^3 *must be attached to a piece of wood of mass 100 g and density* 0.2 g/cm^3 *so that the two together will just submerge beneath water?*

The brass and wood together will just submerge when the average density of the whole is equal to that of water (1 g/cm^3).

Let the volume of the brass $= V$ (in cm^3)

then   mass of brass $= 8.5 \, V$ (in g)

$$\text{volume of wood} = \frac{100}{0.2} = 500 \text{ cm}^3$$

and   mass of wood $= 100 \text{ g}$

For an average density of 1 g/cm³ the total mass must be numerically equal to the total volume.

Hence, $100 + 8.5\,V = 500 + V$

rearranging $7.5\,V = 400$

or $V = \dfrac{400}{7.5} = \underline{53.3 \text{ cm}^3 \text{ of brass}}$

3. *An ordinary hydrometer of mass 28 g floats with 3 cm of its stem out of water. The area of cross-section of the stem is 0.75 cm². Find the total volume of the hydrometer and the length of stem above the surface when it floats in a liquid of relative density 1.4.*

By the law of flotation, the hydrometer displaces its own weight, and hence also its own mass of any liquid.

Therefore, mass of water displaced = 28 g

and volume of water displaced = 28 cm³

Assuming top of stem to be flat and not rounded,

volume of stem above water = $0.75 \times 3 = 2.25$ cm³

therefore total volume of hydrometer = $28 + 2.25 = \underline{30.25 \text{ cm}^3}$

volume of liquid displaced = $\dfrac{\text{mass}}{\text{density}} = \dfrac{28}{1.4} = 20$ cm³

therefore volume of stem above the liquid = $30.25 - 20 = 10.25$ cm³

hence length of stem above liquid = $\dfrac{10.25}{0.75} = \underline{13.7 \text{ cm}}$

QUESTIONS: 12

(*Where required, assume* $g = 10$ m/s²)

1. Define *density* and *relative density*.

   A piece of sealing-wax weighs 0.27 N in air and 0.12 N when immersed in water. Calculate:
   (*a*) its relative density;
   (*b*) its apparent weight in a liquid of density 800 kg/m³.

2. A metal cube of side 2 cm weighs 0.56 N in air. Calculate:
   (*a*) its apparent weight when immersed in white spirit of density 0.85 g/cm³;
   (*b*) the density of the metal of which it is made.

3. A river car-ferry boat has a uniform cross-sectional area in the region of its water-line of 720 m². If sixteen cars of average mass 1100 kg are driven on board, find the extra depth to which the boat will sink in the water.

4. A rectangular metal block has a mass of 0.48 kg (480 g) and dimensions 0.05 m (5 cm) by 0.04 m (4 cm) by 0.03 m (3 cm). Calculate the density of the metal.

   The same block is now suspended from a balance so that the block is completely immersed in a liquid whose density is 1200 kg/m³ (1.2 g/cm³). What will be the reading on the balance?

5. State the *law of flotation.*

   A piece of beeswax of density 0.95 g/cm³ and mass 190 g, is anchored by a 5 cm length of cotton to a lead weight at the bottom of a vessel containing brine of density 1.05 g/cm³. If the beeswax is completely immersed, find the tension in the cotton in newtons.

6. State the principle of Archimedes and show how it may be used in determining the relative density of a liquid.

   If the relative density of ice is 0.92 and that of sea-water 1.025, what fraction of an iceberg floats above the surface? (*O.C.*)

7. The same cork is floated on mercury and water in turn. How are:
   (i) the volumes, and
   (ii) the weights of liquid displaced in each case related? (*S.*)

8. A piece of marble of weight 14 N and relative density 2.8 is supported by a light string from a spring balance and lowered into a vessel of water standing on a weighing machine. Before the stone enters the water the weighing machine reads 57.5 N. What will be the reading of both spring balance and weighing machine when the marble is completely immersed in the water?

9. A type of wood has density 0.8 times that of water. If a cube made of this wood were placed in water, what fraction of the volume would be immersed? Draw a sketch to show the effect of placing the cube in a liquid of the same density as the wood. What would you expect to happen to the position of the cube if the liquid were heated? Give a reason. (*S.*)

10. A cube made of oak and of side 15 cm floats in water with 10.5 cm of its depth below the surface and with its sides vertical. What is the density of oak?

11. A block of wood of mass 24 kg floats in water. The volume of the wood is 0.032 m³. Find:

(*a*) the volume of the block below the surface of the water;

(*b*) the density of the wood. (Density of water 1000 kg/m³.)

(*C.*)

12. A cube of wood of volume 0.2 m³ and density 600 kg/m³ is placed in a liquid of density 800 kg/m³.

(i) What fraction of the volume of the wood would be immersed in the liquid?

(ii) What force must be applied to the cube so that the top surface of the cube is on the same level as the liquid surface?

(*A.E.B.*, 1982)

13. A light spiral (helical) spring which obeys Hooke's law has an unstretched length of 220 mm. It is attached at its upper end to a fixed support and, when a piece of metal of mass 2 kg is hung from the lower end, the spring extends to a length of 274 mm.

(*a*) Find the force in newtons needed to produce an extension of 10 mm.

(*b*) When the metal is totally immersed in water, the length of the spring becomes 247 mm. What is the upthrust of the water on the metal?

(*c*) Find the mass of water displaced by the metal.

(*d*) Calculate the volume of the piece of metal.

(Take the value of g to be 10 m/s², and the density of water to be 1000 kg/m³.)   (*O.*)

13. Some molecular properties of matter

Atoms and molecules

Many of the ancient Greek philosophers believed that all substances were composed of tiny particles or atoms; but it was not until the nineteenth century that this idea developed into a useful theory for explaining some of the chemical and physical properties of matter.

In 1808, the English chemist, John Dalton, produced experimental evidence to show that chemical compounds consist of molecules which are groups of atoms of various elements united in the same simple numerical proportion.

An element is a substance which cannot be split into simpler substances, while an atom is the smallest portion of an element which can take part in a chemical change.

At the time of which we are speaking, scientists thought of atoms as being like tiny billiard-balls, but since then we have learned a great deal about the nature of the atoms themselves. The structure of atoms is discussed in chapter 45. In the present chapter we shall show how the molecular theory is used in physics to explain some of the elementary properties of gases, liquids, and solids.

Brownian movement. Kinetic theory of matter

One day in the year 1827 the botanist, Robert Brown, was using a microscope to examine some pollen particles in water when he was surprised to notice that they were in a continuous state of vigorous haphazard movement. At the time Brown could offer no explanation, but many years afterwards it came to be realized that the motion of the particles is caused by the impact of moving water molecules.

The same kind of movement can be seen in the case of smoke particles in air. Observations of this kind strongly suggest that the molecules of liquids and gases are in a state of continuous motion. This idea of molecular motion and the mathematical calculations which have been applied to it is called the *kinetic theory of matter*.

To study Brownian movement

The Brownian movement may be observed by the apparatus shown in Fig. 13.1. A small glass cell contains water with some tiny graphite particles suspended in it and these are strongly illuminated by light focused from a line filament lamp by a glass or perspex rod. This very efficient optical system was devised by Leo O'Donnell. The particles scatter the light so that when viewed through a microscope they appear as bright points moving with the same irregular motion as Brown's pollen grains.

To see the same effect in air, the glass cell is replaced by one containing a little smoke from smouldering string and covered by a thin glass slip. The smoke consists of tiny particles of combustion products and these show a similar movement.

It is important to note that the particles must be very small if their movement is to be seen. Simple kinetic theory has established the result that the mean kinetic energy

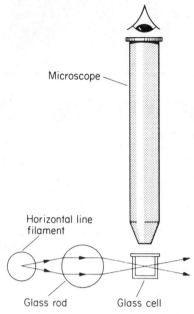

Microscope

Horizontal line
filament

Glass rod   Glass cell

Fig. 13.1. Brownian movement apparatus

is the same for all particles. Consequently, a large particle will have a low average speed which is too small to be observed.

Nature of the force between atoms and molecules

Newton's law of universal gravitation which works so well in calculations of the force between two pieces of matter whose distance apart is large compared with their size, fails to give the right answer when applied to two molecules which are very close together. *This does not mean that gravitational attraction no longer acts but that incomparably greater forces of a different kind come into action.* Now a full account of these forces belongs to very advanced physics, so we shall deal with the subject very briefly.

Later on we shall learn about electricity and see that it can produce electric and magnetic forces of attraction and repulsion. Atoms themselves contain particles of electricity in motion and so we get electric and magnetic forces between them. We shall, therefore, sum up the situation by saying that, *when atoms are very close together, the forces between them are electromagnetic in nature.*

The net result is that, when their centres are a certain distance apart, the resultant force between two atoms or molecules is zero. When closer than this they repel one another and when further apart they attract one another. Furthermore, these electromagnetic forces differ from one kind of atom to another and even between atoms of the same kind depending on whether a substance is in the liquid state or some particular kind of solid state. A good example of a substance which can exist in more than one kind of solid state is carbon.

Three states or phases of matter

Matter commonly exists in either the solid, liquid or gaseous state or phase.

In a solid substance the molecules vibrate about their zero resultant force positions, alternately attracting and repelling one another. All true solids have a crystalline structure in which the atoms are arranged in a regular pattern called the *lattice* (Fig. 13.2 to 13.4). There is, however, a borderline class of materials which appear to be solids but actually are very viscous liquids. Pitch is a good example. When struck

Fig. 13.2. This model, showing the position of the atoms in a molecular structure, was used by British biochemist Professor Dorothy Crowfoot-Hodgkin who won the Nobel prize for Chemistry in 1964. She constructed it from information obtained from the X-ray diffraction patterns formed by the crystal

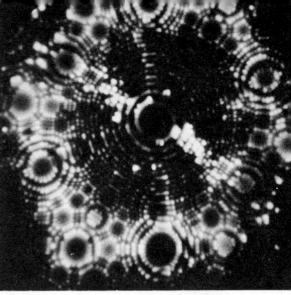

Fig. 13.3. The surface of tungsten as revealed by a field ion microscope at a magnification of about 5 million times. Each spot represents an individual atom

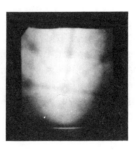

A. Polycrystalline copper

B. Tungsten (single crystal)

C. Lithium fluoride (single crystal)

D. Rock salt (single crystal)

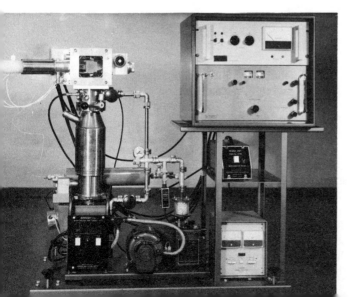

Fig. 13.4. A proton scattering microscope at Edwards High Vacuum International Research Laboratories. This instrument uses a proton beam to reveal crystal lattice structure, some examples of which are shown

with a hammer, it readily splinters, but if placed in a funnel and left for several years it slowly flows out.

In a liquid, the molecules are also vibrating to and fro alternately attracting and repelling one another with forces which can be just as strong as those in a solid (see action of a siphon, page 125). At the same time, however, the liquid molecules can move freely among one another, exchanging partners as they go. It is this freedom of movement which enables a liquid to take up the shape of any vessel in which it is placed. It is worth mentioning that experimental evidence indicates that small groups of liquid molecules can arrange themselves for very short periods of time into the same kind of regular pattern found in solids.

In a gas the molecules are much further apart than those in solids and liquids. They move at high velocities colliding with one another and with the walls of their containing vessel. Except at the moment of collision, the short-range intermolecular forces we have been describing do not come into action. Unless the gas is highly compressed, the molecules are, for the greater part of the time, so far apart that the attractive force is effectively negligible. Consequently, a gas is perfectly free to expand and completely fill the vessel containing it.

The average distance moved by a molecule between collisions is called its *mean free path*. Rudolf Clausius applied the laws of mechanics to these collisions of the molecules with the walls of the containing vessel and showed how they explained the relation between the pressure and volume of a gas.

To measure the approximate length of a molecule

A rough idea of the length of a molecule of a substance which will form a film on water may be obtained by a method originally carried out by Lord Rayleigh who was awarded the Nobel prize for physics in 1904.

From chemical studies we know that molecules of oils and fatty acids are longer than they are broad and that one end has an affinity for water. If a drop of such a substance is placed on a water surface it spreads in the form of a circular film with the molecules standing upright rather like the pile on a piece of velvet. When the spreading stops, it is reasonable to assume that the film is one molecule thick.

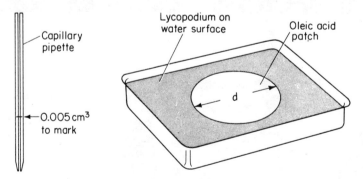

Fig. 13.5. Oil-film experiment

A suitable substance to use for this experiment is oleic acid prepared for use as a 0.5 per cent solution in methanol. 1 cm³ of oleic acid is run into a 200 cm³ measuring flask half full of methanol. More methanol is added to fill the flask to its graduation mark, ensuring thorough mixing during the process. We now have a solution which contains 0.005 cm³ of oleic acid in 1 cm³ of solution.

A large plastic dish full of water is allowed to stand until the water is at rest and is then lightly dusted with lycopodium powder from a pepper caster (Fig. 13.5). Using a capillary pipette, 0.005 cm³ of the oleic acid solution is then run into the centre of the water surface. The methanol in the solution partly evaporates and partly dissolves in the water leaving the oleic acid to spread in the form of an approximately circular patch. Several experiments are performed, using fresh water each time and a value for the mean diameter, d in cm, of the oil film calculated from measurements taken in different directions across it with a mm scale.

Calculation

Values for film diameter   cm.

$$\text{Mean diameter} = \qquad \text{cm}$$

Volume of oleic acid in $0.005 \cdot \text{cm}^3$ of solution $= 0.005^2 \text{ cm}^3$.

Assuming the film to be a flat cylinder of thickness l where l in cm is the length of a molecule, then,

$$\text{volume in cm}^3 = \pi r^2 l = \frac{\pi d^2 l}{4}$$

$$= 0.005^2 \text{ cm}^3$$

From which,

$$\text{length of molecule} = \frac{0.005^2 \times 4}{3.14 \times d^2} = \qquad \text{cm} = \qquad \text{m}$$

Only a rough value is to be expected since, in addition to other assumptions, we have ignored the space between the molecules. Experiments of this kind have been found to give values in the region of $0.000\,000\,001$ or 10^{-9} m. This is called 1 nanometre (nm).

Diffusion

The speed with which molecules move about inside a gas depends on their mass and temperature. Thus, at a given temperature the heavy molecules of carbon dioxide move more slowly than the light molecules of hydrogen. This has an important bearing on the rate at which one gas will diffuse into, or mix with another.

Thomas Graham found that, **at constant temperature gases diffuse at rates which are inversely proportional to the square roots of their densities.** This is known as **Graham's law of diffusion.**

The difference in the rates of diffusion of two gases through a porous partition may be demonstrated by the experiment shown in Fig. 13.6 (a). An unglazed earthenware pot is fitted with a rubber bung and a length of glass tubing and set up vertically with the tube dipping into a beaker of water. When the porous earthenware pot is surrounded by gas from the laboratory supply in the manner shown the gas diffuses through into the pot more rapidly than the air diffuses outwards. The pressure in the pot therefore increases, and bubbles make their appearance at the end of the glass tube.

If the inverted beaker containing the gas is taken away, the outward diffusion of gas already inside the pot is more rapid than the inward diffusion of air. Consequently, there is a reduction of pressure inside, and water is now forced up the tube (Fig. 13.6 (b)). A device based on this principle has been used for detecting the presence of the dangerous gas methane in coal-mines.

Diffusion in a liquid can be shown by placing some blue copper sulphate crystals at the bottom of a tall beaker containing water. If the beaker is placed where it will be undisturbed the crystals will dissolve and form a dense blue solution at the bottom. During succeeding weeks the blue coloration gradually extends upwards. In spite of their comparatively large mass, the copper sulphate molecules have sufficient energy to enable them to overcome the force of gravity and diffuse upwards through the water.

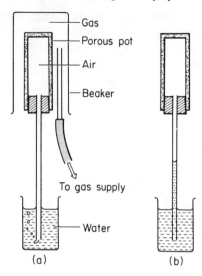

(a)   (b)

Fig. 13.6. Diffusion experiment

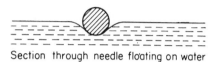

Section through needle floating on water

Fig. 13.7.

Surface tension

Most people are aware that an ordinary sewing needle can be made to float on water. The experiment works best if the needle is slightly greasy. Usually, the needle is placed on a small piece of filter paper, which is then gently placed on the water surface. Within a few seconds the paper sinks to the bottom and the needle is left floating.

Close examination reveals that the needle rests in a slight depression (Fig. 13.7). The surface of the water behaves as though covered with an elastic skin. This property of a liquid is called *surface tension*. A piece of gauze cut and bent into the form of a rectangular box will also float, but if a few drops of alcohol, soap solution or detergent are added to the water the surface tension is decreased and the gauze and needle will sink.

A tent keeps out water owing to the tension which exists in the lower surface of the rain-water in contact with the canvas. Campers soon learn that it is important not to touch the inside of the canvas, or the surface film may be broken. When this occurs, the water spreads round the strands of the material and, having nothing to support it, begins to drip through.

Further illustrations of surface tension

We have already seen that soap solution, detergents, and alcohol will lower the surface tension of water. Camphor has a similar effect. When a piece of camphor is thrown on to the surface of water it slowly dissolves and causes a reduction of surface tension in its immediate neighbourhood. The stronger pull exerted by the surrounding uncontaminated water sets up a movement of the surface and the camphor is carried along with it. After a time the concentration of the surface solution becomes uniform and all movement ceases.

The above experiment illustrates the principle of a toy duck, made of light plastic material, which swims about on the surface of water when a small piece of camphor is attached to it. A similar toy in the shape of a small metal boat has a reservoir of methylated spirit. A short wick dips in the spirit and trails over the stern. Reduction of surface tension occurs where the spirit contaminates the water and the boat moves forward.

Quantitatively, **the surface tension is defined as the tangential force in the surface acting normally per unit length across any line in the surface.** (The word "normally" here means "perpendicularly".)

The tension in the surface of a liquid is well illustrated by a soap film. A wire frame with a piece of cotton tied across it is dipped into soap or detergent solution so that a film is formed. When the film on one side of the cotton is removed by touching it with a filter-paper the tension in the film on the opposite side pulls the cotton into an arc of a circle. A similar frame has a length of cotton with a loop in it. The loop is pulled into the form of a circle when the film is removed from its centre (Fig. 13.8).

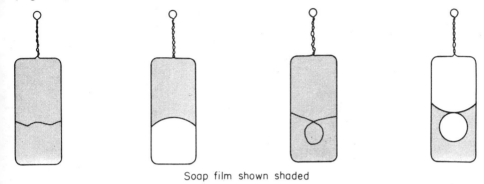

Soap film shown shaded

Fig. 13.8. Surface tension experiment

Molecular explanation of surface tension

On page 140 we explained that the molecules of a liquid are constantly oscillating to and fro alternately exerting repulsive and attractive forces on one another, yet free to move in the body of the liquid, exchanging partners as they go.

Application of the kinetic theory, together with knowledge of how the force between the molecules varies with their distance apart, leads to the conclusion that molecules which have moved into the surface of the liquid become more widely spaced than those inside it. Consequently, for most of the time they are exerting attractive forces on one another. The surface of a liquid is therefore under tension, which explains why liquids will form spherical drops, soap films spherical bubbles, and so on.

Thus, however much the surface of a liquid may expand or contract it behaves as though covered by an elastic skin under a constant tension, the value of which depends on the temperature.

Adhesion and cohesion

The force of attraction between molecules of the same substance is called *cohesion*, as distinct from *adhesion* or the force between molecules of different substances.

The adhesion of water to glass is stronger than the cohesion of water. Hence, when water is spilled on a clean glass surface it wets the glass and spreads out in a thin film. On the other hand, the cohesion of mercury is greater than its adhesion to glass. Therefore, when mercury is spilled on glass it forms small spherical droplets or larger flattened drops.

The difference between the adhesive and cohesive properties of water and mercury explains why the meniscus of water curves upwards and that of mercury curves downwards when these liquids are poured into clean glass vessels (Fig. 13.9).

Capillary attraction

If a piece of ordinary glass tubing is heated in a gas flame until it softens it may be drawn out into a tube of very fine bore. This is called a capillary tube (from the Latin *capilla* = hair).

On dipping the tube into water, alcohol, or any other liquid which wets glass it is noticed that the liquid rises in the tube to a height of several centimetres. Mercury, however, gives a capillary depression (Fig. 13.9).

In the case of liquids which wet glass the adhesion of the molecules for glass is greater than their cohesion. Consequently, the meniscus curves upwards and the liquid rises in the tube.

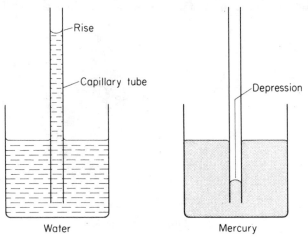

Fig. 13.9. Capillarity

Conversely, the cohesion of mercury molecules is greater than their adhesion to glass. The meniscus therefore curves downwards, and this is accompanied by a capillary depression

Blotting-paper owes its absorbent properties to the capillary rise which occurs in the narrow interstices between the soft fibres from which it is made. Oil rises up a lamp wick for the same reason.

All well-constructed houses possess a damp-course. This is a layer of slate, bitumenized lead foil or other suitable material inserted between the wall bricks just above ground, but below floor level. Being impervious to water, the damp-course prevents the capillary attraction of moisture from the ground, which would otherwise make the walls damp and unhealthy.

Osmosis

Before cooking dried fruits such as prunes and apricots it is customary to leave them soaking in cold water for some hours so that they swell up and become restored, as far as possible, to their original condition. During this time water passes through the cell walls of the fruit by a process called *osmosis*.

Osmosis can be demonstrated by the following experiment (Fig. 13.10). A narrow glass tube, more than a metre long, ends in a large thistle funnel having its mouth covered with parchment. Care must be taken to tie the parchment very securely and tightly with stout thread so that the joint is entirely free from leaks. The funnel only is then nearly filled with strong sugar solution and the whole is supported with the stem vertical and the funnel mouth downwards in a beaker of water. Very soon, the liquid level in the stem of the funnel begins to rise, showing that water is passing through the parchment into the sugar solution. The *osmotic pressure* of the solution is measured by the hydrostatic pressure, h, of the liquid column in the tube.

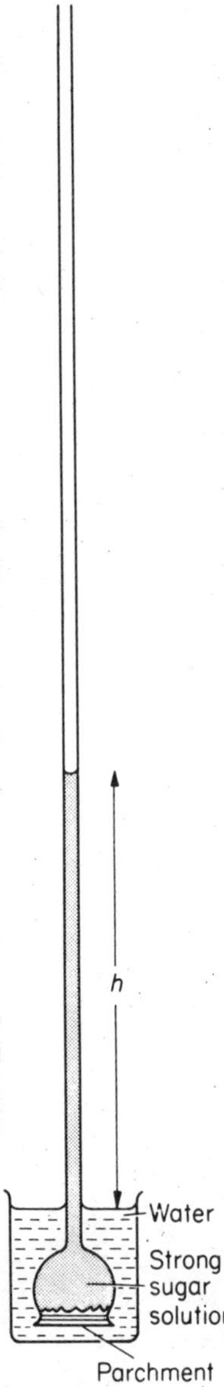

Parchment is described as a *semi-permeable membrane*, i.e., it permits water molecules to pass through its pores but obstructs the passage of the larger sugar molecules. Water molecules therefore pass through in both directions, but owing to the space occupied by the sugar molecules the concentration of water molecules is less on the sugar side of the membrane than on the water side. Consequently, the rate of passage of water molecules is greater from the water side to the sugar side.

h

Water

Strong sugar solution

Parchment

Fig. 13.10. Osmosis demonstrated

The reason why a prune swells in water now becomes apparent. The cells of the fruit contain a high concentration of sugar and are surrounded by a semi-permeable membrane.

Most animal and vegetable membranes are semi-permeable, and so osmosis plays an important part in the general functioning of living things. For example, water is absorbed through the roots of plants by this process. Osmosis also plays a part in the elimination of waste products from the blood through our kidneys. Cellophane, which is another semi-permeable material, has been successfully employed in the construction of an artificial kidney. This is connected between an artery and a vein and maintains the essential purification of the blood while the natural kidneys are out of action through illness.

Strength of materials. Elasticity

Before various materials are used in the construction of machinery and buildings, tests are carried out to ensure that they are able safely to stand up to the stresses to which they are likely to be subjected. Industrial laboratories have special apparatus for carrying out these tests.

Brittle substances such as masonry and cast iron will support large weights or forces of compression, but will easily break if stretching (tensile) forces are applied. Where tensile forces are involved an elastic material such as steel is used. *Elasticity is the ability of a substance to recover its original shape and size after distortion.*

Hooke's law

During the time London was being rebuilt after the great fire of 1666 a great deal of research went on in connection with building construction and the strength of materials.

Robert Hooke, who was Chief Surveyor to the City of London, took a leading part in this work and extended his researches to include an investigation of elasticity in general. He investigated the relationship between the deformation of a material and the force applied to it. **If the deformation of a material is proportional to the force applied, then the material is said to obey Hooke's Law.**

Hooke illustrated the law by reference to four different experiments:

(1) Loading a wire (deformation = increase in length).
(2) Loading a spiral spring (deformation = increase in length), see page 40.
(3) Loading a horizontal beam fixed at one end (deformation = depression of the free end).
(4) Tightening a watch spring (deformation = angular rotation).

Industrial applications of metallurgical studies

The behaviour of metals and alloys under stress is important in connection with the *forging* of metal components. In this process the hot or cold metal is squeezed or hammered into shape instead of being cast in a mould (Fig. 13.11 and 13.12).

Fig. 13.13 shows the crystal grain flow which takes place during a forging operation. In this condition the material becomes much tougher and more resistant to the formation of cracks which would lead to ultimate fracture. Examination of the crystal structure and behaviour of materials has led to the development of tough alloys so vital to human safety as, for example, in aircraft construction.

Fluid friction

Friction between solids was dealt with in chapter 2. We shall now say something about friction in fluids.

Bodies moving through fluids, i.e., liquids or gases, experience a retarding force

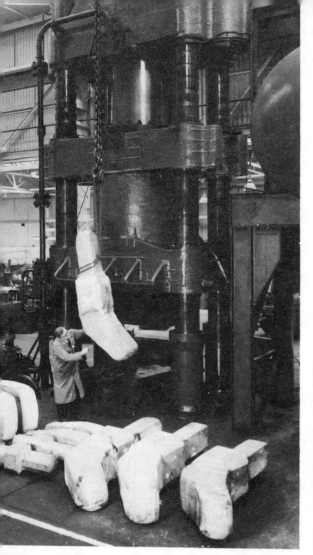

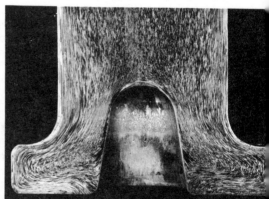

Left Fig. 13.11 40 000 000 N forging press, worked by oil pressure at the Birmingham works of Alcan Industries Ltd. The large forgings were produced in open dies, for an aircraft prototype

Bottom Fig. 13.12. Forging a propeller blade at the Birmingham works of Northern Aluminium Company

Below Fig. 13.13 Section through the hub of a forged propeller blade showing grain flow

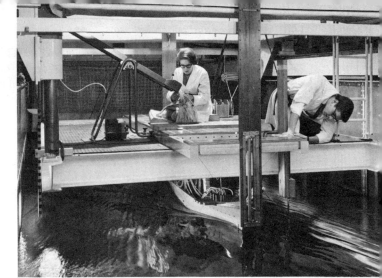

wo scientists carry out an experiment to
ure the skin friction distribution on a liner
el

Fig. 13.14. *The ship division of the National
Physical Laboratory aims to improve the hy-
drodynamic design of merchant ships and
other marine transport. This work is carried
out mainly by experiments in the ship model
laboratory at Feltham, which is one of the
largest and most modern in the world,
though analytical methods are used to an
increasing extent*

(*b*) Testing a model in the sea-keeping tank

Models (1 : 18 scale) of a 55 metre trawler
a 63 metre stern trawler in simulated storm
ditions. The models are travelling head-on
the waves at a speed equivalent to 15
h

otherwise known as *fluid friction* or *viscous drag*. This is a factor involved in the design of ships, aircraft, and other vehicles. In order to achieve a design in which the energy-wasting effects of drag are reduced to a minimum, calculations are made and small-scale models are constructed which are then tested in water tanks and wind tunnels (Fig. 13.14).

In the field of pure physics Sir George Stokes, an eminent mathematician and physicist of the nineteenth century, investigated the effects of viscous drag on small spheres falling through liquids. Unlike bodies falling in a vacuum which accelerate constantly at 10 m/s², these spheres were found to acquire a steady terminal velocity. He found that they experienced an upward retarding force, F, which depended on their radius, their velocity, and a constant called the *viscosity* of the liquid. Besides this there are two other forces acting on the sphere namely, (see Fig. 13.15)

(i) the force of gravity, W, acting downwards and

(ii) the upthrust, U, of the liquid as given by Archimedes principle (page 128) which acts upwards.

These two forces have a resultant $(W-U)$ in the downward direction. At constant temperature, $(W-U)$ is constant whilst the viscous drag, F, increases with velocity.

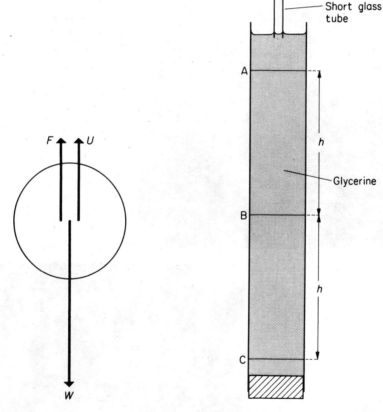

Fig. 13.15. Forces on a sphere falling through a liquid

Fig. 13.16. Measurement of terminal velocity

If, therefore, a small metal sphere is allowed to fall through the liquid it at first accelerates. As it does so its velocity increases, with consequent increase in the viscous drag. Eventually F becomes equal to $(W-U)$ and the total resultant force on the sphere is zero. In accordance with Newton's first law of motion the sphere then falls with a uniform velocity known as the *terminal velocity*.

An experiment to demonstrate terminal velocity

A tall glass jar or a tube fitted with a secure bung at its lower end about 5 cm in diameter and 50 cm in height is filled with glycerine and supported vertically (Fig. 13.16).

Small bearing balls handled with tweezers are then allowed to fall centrally down the tube by dropping them through a short piece of glass tubing which just dips into the glycerine. Balls of radius between 0.2 and 0.5 cm radius are suitable for this experiment, and they should first be wetted with glycerine contained in a small dish. This avoids the formation of small air bubbles which sometimes cling to the balls and spoil the results.

Terminal velocity will be reached after the balls have fallen some 7 or 8 cm below the level of the glycerine. A wire band A to act as a marker is therefore placed about this distance below the glycerine level on the outside of the tube. Another, C, is placed near the bottom of the tube and a third, B, midway between them, thus dividing the liquid into two sections of equal height h.

Using a stopclock, the time is noted as the balls pass each marker in turn. For a given ball it is found that the time intervals between A and B and between B and C are equal, showing that a steady terminal velocity was reached during the descent.

QUESTIONS: 13

1. What do you understand by the *Brownian movement*?

With the aid of a diagram describe an experiment to demonstrate it in either a liquid or a gas.

Give a brief explanation in terms of the simple kinetic theory.

2. What is Brownian motion? Why is it only exhibited by small particles?   (*O.C.*)

3. (*a*) Describe experiments, one in each case, to demonstrate the process of diffusion in (i) gases and (ii) liquids. For each experiment, sketch the apparatus you would use.

(*b*) For **one** of the experiments you have described, explain the diffusion process in terms of the kinetic theory of matter.
(*J.M.B.*)

4. Describe briefly two quite different experiments which indicate that liquid surfaces appear to be in a state of tension.   (*O.C.*)

5. A small steel ball which is falling in oil soon reaches its terminal velocity. What is meant by the phrase 'terminal velocity'? Explain why the ball reaches this velocity.
(*J.M.B.*)

6. Why is it that a needle may float on clean water but sinks when some detergent is added to the water?

7. Describe the differences between solids, liquids, and gases in terms of:
(i) the arrangement of the molecules throughout the bulk of the material;

(ii) the separation of the molecules, and
(iii) the motion of the molecules.

8. Describe the apparatus required and show how it would be set up in order to observe osmosis. State, with reasons, any precautions taken. Indicate the observations which would be made and the conclusions reached from them.

9. What features of the behaviour of a substance such as water lead us to believe in the existence, between molecules, of both attractive and repulsive forces? Why do we believe these forces can be very large compared with the force of gravity on a molecule?
(*O.C.*)

10. (*a*) Discuss briefly two pieces of evidence which indicate that there are attractive forces between molecules.

(*b*) Describe an experiment by means of which an estimate may be made of molecular size using a suitable liquid.   (*O.C.*)

11. A ball-bearing is released from rest from just below the surface of lubricating oil contained in a tall measuring cylinder.
(i) Describe the different stages of the motion of the ball from release until it reaches the bottom of the cylinder.

(ii) Sketch a velocity–time graph for the motion.

(iii) Describe the energy changes which are taking place during the fall.
(*J.M.B.*)

The earth's internal energy being released in the form of heat.
Two members of the Philippine Volcanology Commission inspect a fissure of
the main crater of the Taal volcano 64 km south of Manila

Internal energy and heat

14. Measurement of temperature

For several reasons, internal molecular energy which can be made available in the form of heat is the most important form of energy we have at our disposal. This has already been discussed in chapter 7. Without the heat which comes to us from the sun, life on earth would soon become extinct. At one stage or another heat plays a part in the energy transfers by which we obtain electricity, which is the most useful and convenient of all forms of energy.

Thermometers

We must be careful not to confuse the *temperature* of an object with the heat energy which can be obtained from it.

The temperature of a substance is a number which expresses its degree of hotness on some chosen scale. Thus, the temperature of a bucketful of warm water is lower than that of a hot electric lamp filament, but the water contains a much larger quantity of internal energy than the filament.

Temperature is measured by means of a *thermometer*, of which a number of different types are available. Some depend on the expansion of a liquid when heated; others on the expansion of a compound strip of two metals. Most people are familiar with the mercury or alcohol thermometers for domestic use. These usually have spherical bulbs and are mounted on plastic or metal scales. Laboratory thermometers have cylindrical bulbs for easy insertion through holes in corks and have their scales engraved directly on the stem. When the bulb is heated the liquid inside expands and the temperature is read at the level of the thread in the stem.

For some industrial purposes thermoelectric thermometers are used. These are based on the electric current which is generated when a junction of two different metals is heated (Fig. 17.16). For accurate scientific work temperatures are often measured by a platinum resistance thermometer. This consists of a fine platinum wire contained in a silica tube. Temperature changes may be calculated from the change in electrical resistance of the platinum. Details of these thermometers will be found in more advanced textbooks.

To construct a mercury thermometer

One end of a length of clean capillary tubing is heated in a bunsen flame until the glass softens and seals the end of the tube. The tube is then withdrawn from the flame and a small bulb is blown at the end. By repeating this operation the size of the bulb may be increased as required.

The size of bulb needed will depend on two things, the bore of the tube and the desired temperature range. Often, in commercial practice, bulbs of a correct size are made separately and afterwards fused on to the stem.

The next stage is the filling of the thermometer. It is placed with its open end beneath the surface of some mercury in a jar and the bulb gently heated. The air inside expands and bubbles through the mercury. On cooling, the air contracts and some mercury runs up into the bulb. The thermometer is then taken out and the bulb heated to boil the mercury. When the mercury vapour has expelled all the air the open end is quickly inverted once more in mercury. On cooling, the mercury rises and completely fills the bulb and stem. *Remember that mercury vapour is poisonous, so that all these operations must be done in a fume chamber or under an extraction hood.*

The thermometer is now taken out and heated to a temperature somewhat higher than the maximum for which it is to be used. While at this temperature the end of the stem is rotated in a small blowpipe flame, drawn out and sealed off.

The thermometer is now ready to be graduated. However, owing to the heat treatment it has received, the glass goes on contracting slowly for a considerable time. It is therefore advisable to put it aside for several months before graduation is finally carried out.

The fixed temperature points

The principle underlying the graduation of all types of thermometer is to choose two fixed and easily obtainable temperatures called the *upper and lower fixed points* and then to divide the interval between them into a number of equal parts or *degrees*.

The upper fixed point is the temperature of steam from water boiling under standard atmospheric pressure of 760 mmHg.

The temperature of the boiling water itself is not used as a fixed point for two reasons. First, local overheating may occur, accompanied by "bumping" as the water boils; secondly, any impurities which may be present will raise the boiling-point. The temperature of the steam just above the water will always be constant, and depends only on the barometric pressure at the time.

The lower fixed point is the temperature of pure melting ice.

The ice must be pure, since the presence of impurities will lower the melting-point.

Celsius temperature scale

The difference in temperature between the two fixed points is called the *fundamental interval* (Fig. 14.1). This is divided into 100 equal degrees, the ice point being called 0 °C and the steam point 100 °C. This method of subdividing the interval was suggested by a Swedish astronomer named Celsius, and is now called the *Celsius scale*. Temperatures on it are called "degrees Celsius" (°C).*

Determination of the upper fixed point

This is done by pushing the thermometer through a hole in a cork and placing it inside a double-walled copper vessel called a *hypsometer* (Fig. 14.2). Water is steadily boiled in the lower part of the hypsometer, thus keeping the bulb surrounded by pure water vapour at atmospheric pressure. For reasons already mentioned, it is important that *the bulb should not be allowed to dip into the boiling water*. The thermometer is adjusted so that the mercury thread is visible just above the top of the cork. When the thread has remained steady for some minutes its level is marked on the stem by a light scratch. The double walls reduce loss of heat and consequent cooling of the vapour surrounding the thermometer, while the manometer seen in the diagram gives warning should the pressure inside the hypsometer differ from atmospheric pressure. If the barometric pressure at the time is not equal to 760 mmHg, then the true boiling-point for the prevailing pressure must be ascertained from a table giving the variation of boiling-point with pressure. Due allowance is then made when marking the stem.

* Actually the Celsius scale as we know it today was first suggested by Linnaeus in 1745. Earlier, Celsius used a similar scale in which the ice point was marked 100° and the steam point 0°.

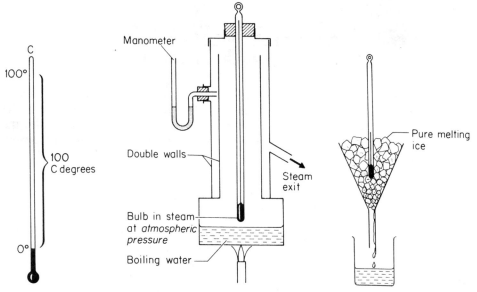

Fig. 14.1. Celsius scale   Fig. 14.2. Hypsometer (upper fixed point)   Fig. 14.3. Lower fixed point

Determination of the lower fixed point

The thermometer is placed in a glass funnel kept full of pure ice shavings, having a beaker underneath to catch the water. The mercury thread is allowed to show just above the top of the ice. When the level of the thread has remained steady for some time its position is marked as before (Fig. 14.3).

To measure temperatures with the thermometer

When a Celsius thermometer is graduated in a factory the stem between the two fixed points is divided into 100 equal parts by means of a special dividing engine.

This is generally beyond the resources of the ordinary laboratory, but the thermometer may still be used to measure an unknown temperature if the following procedure is adopted.

The distance (y) on the stem between the upper and lower fixed points is measured and compared with the length (x) of the mercury thread above the lower fixed point at the unknown temperature θ.

Then
$$\theta = \frac{x}{y} \times 100\,^{\circ}\text{C}$$

Note. The Greek letter θ (theta) is the usual symbol for temperatures in °C.

The clinical thermometer

This is a thermometer specially designed for measuring the temperature of the human body, and so it is only necessary for it to have a range of a few degrees on either side of the normal body temperature (Fig. 14.4). The thermometer is generally placed beneath the patient's tongue and left there for at least two minutes to ensure that it fully acquires the body temperature. When the thermometer is taken from the mouth the mercury thread does not contract back into the bulb, but remains in the stem. The reason for this is that the stem has a narrow constriction in its bore just

Constriction

Celsius

Fig. 14.4. Clinical thermometer

above the bulb. When the thermometer is removed from the mouth the sudden cooling and contraction of the mercury in the bulb causes the thread to break at the constriction, and so it stays in the stem at its original reading. This enables the temperature to be read at leisure. Before being used again, the mercury in the stem must be returned to the bulb by shaking.

The average body temperature of a healthy person is taken as 37 °C.

Six's maximum and minimum thermometer

This thermometer is popular among gardeners for use in greenhouses. Its purpose is to record the maximum and minimum temperatures reached since the thermometer was last read. Generally speaking, a minimum temperature occurs during the night and a maximum during the day.

Invented by James Six towards the end of the eighteenth century, the thermometer consists of a fairly large cylindrical bulb *A* which originally contained alcohol,

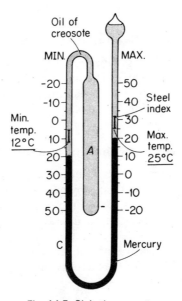

Fig. 14.5. Six's thermometer

although oil of creosote is now more generally used. This is connected by a U-shaped stem to a second bulb nearly full of the same liquid (Fig. 14.5). The bend of the U contains a thread of mercury. Two scales are provided, one against each limb of the tube so that the temperature may be read against either of the mercury levels. Resting on each of the mercury surfaces are small steel indexes provided with light springs to hold them in position in the stem. Expansion or contraction of the liquid in *A* causes a movement of the mercury thread. Consequently, one or other index is pushed forward by the mercury and left in the extreme position reached. Thus, the lower end of the index on the left indicates the minimum and that on the right the maximum temperature attained.

After readings have been taken a small magnet is used to bring the indexes back into contact with the mercury.

Choice of liquid for thermometers

Whether mercury or alcohol is used in a thermometer depends on the range over which temperature is to be measured. Mercury freezes at $-39\ °C$ and boils at $357\ °C$, while alcohol freezes at $-115\ °C$. It is therefore essential to use alcohol thermometers in places such as northern Canada and Russia, where winter temperatures of $-40\ °C$ are not uncommon. Alcohol also possesses the advantage of having an expansivity of about six times that of mercury.

Apart from these advantages, mercury is to be preferred to alcohol as a thermometric liquid for the following reasons:

(1) It does not wet glass. Alcohol tends to cling to the wall of the tube, and this leads to low readings when the thread is falling.
(2) It does not, like alcohol, vaporize and distil on to the upper part of the bore.
(3) It is opaque and easily seen, whereas alcohol has to be coloured.
(4) It is a better conductor of heat than alcohol, and therefore responds more rapidly to changes of temperature.

For extra low-temperature work (down to $-200\ °C$) pentane is used instead of alcohol. Water is unsuitable for use in thermometers, not only because it freezes at $0\ °C$ but also because of its irregular expansion (see page 166).

The thermodynamic temperature scale

One of the difficulties in the accurate measurement of temperature is that the various types of thermometer, e.g., mercury-in-glass, platinum resistance, thermoelectric, and so on give different readings when used to measure the same temperature, because the value obtained depends on the properties of the substance used in the thermometer.

It was not until the middle of the nineteenth century that the theory of temperature measurement was placed on a firm basis by Lord Kelvin, who devised an absolute scale called the thermodynamic scale, which is quite independent of the properties of any substance used in thermometers. In practice, however, the thermometer whose readings come closest to the thermodynamic scale is the constant-volume gas thermometer (Fig. 16.7), a simple form of which is shown on page 176.

Temperatures on the thermodynamic scale are not measured in degrees but in units called kelvins. They are so-named in honour of Lord Kelvin and are denoted by K *not* °K. Moreover, the thermodynamic scale has been defined in such a way as to make kelvins exactly the same size as degrees Celsius. On the thermodynamic scale the melting point of ice is 273 K, so the boiling point of water is 373 K.

A full discussion of the theory of the thermodynamic scale is outside the scope of this book, but we shall have more to say about it in chapter 16 when it will be explained why $0\ °C$ is equal to 273 K.

QUESTIONS: 14

1. List the advantages and disadvantages of mercury and alcohol as thermometric liquids.

The ice and steam points on an ungraduated thermometer are found to be 192 mm apart. What temperature is recorded in °C when the length of the mercury thread is 67.2 mm above the ice point mark?

2. A mercury thermometer with only the $0\ °C$ and $100\ °C$ markings on it was given to a student, and the student was asked to use it to estimate the temperature of a block of ice cream. Explain how the student could do this.

Draw a labelled diagram of a clinical thermometer and state **two** ways in which it differs from a normal laboratory thermometer. (A.E.B., 1982)

3. Why should a clinical thermometer not be sterilized in boiling water?

4. Describe with the aid of a clear diagram,

the construction and action of a combined maximum and minimum thermometer.

(*W.*)

5. (i) How would you determine an error, at the 100 °C mark, in a given mercury thermometer? Your answer should consist of a labelled sketch of the apparatus you would use, and brief notes on procedure. Why is it important, at the same time, to note the atmospheric pressure? If the thermometer registers 103 at 100 °C and has no zero error, what will it register at 50 °C? (*C.*)

6. Describe how you would graduate an unmarked mercury-in-glass thermometer, no other thermometer being available.

State *three* desirable physical properties of a thermometric liquid.

Mention *two* ways in which the design of a thermometer may be altered so as to increase its sensitivity. (*W.*)

15. Expansion of solids and liquids

With few exceptions, substances expand when heated, and very large forces may be set up if there is an obstruction to the free movement of the expanding or contracting bodies.

If concrete road surfaces were laid down in one continuous piece cracks would appear owing to expansion and contraction brought about by the difference between summer and winter temperatures. To avoid this, the surface is laid in small sections, each one being separated from the next by a small gap which is filled in with a compound of pitch. On a hot summer day, expansion often squeezes this material out of the joints.

In the older methods of laying railway tracks gaps have to be left between successive lengths of rail to allow for expansion. Even when such gaps have been left

Fig. 15.1. Railway lines distorted by expansion during hot weather. This condition is brought about by mechanical and thermal creep which closes the gaps left between rail sections

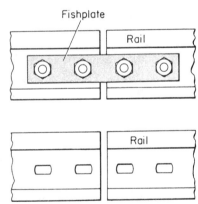

Fig. 15.2. Expansion joint

the rails may sometimes "creep" and close up the gaps. If this happens a rise in temperature may lead to buckling of the track (see Fig. 15.1). Free movement at the rail joints is allowed for by making the bolt holes slotted as shown in Fig. 15.2.

In modern practice, however, railway lines are welded together to form long, continuous lengths. With this method, it is only the last fifty to one hundred metres of any length which show expansion, usually of a few centimetres. This movement is taken up by planing the ends of the rails and overlapping them as shown in Fig. 15.3. The remainder of the rails are unable to expand and so the forces set up develop internal potential energy in the metal. To keep this internal energy to a minimum, it is best to lay the track at a time when the temperature is midway between the

Fig. 15.3. Overlapping joint to allow for expansion on continuous welded rails

Fig. 15.4. Damage to the lead on the roof of Westminster Abbey prior to its restoration in 1954. Thermal creep of this kind occurs when large sheets of lead are used on a steeply hipped roof

summer and winter averages. This technique has been made possible by the use of concrete sleepers and improved methods of fixing the rails so that the track may withstand the thermal stresses set up in it without buckling.

Allowance also has to be made for the expansion of bridges and the roofs of buildings made of steel girders. Various methods are used to overcome the difficulty, a common one being to have one end only of the structure fixed while the other rests on rollers. Free movement is thus permitted in both directions.

Over a very long period of years, expansion and contraction causes "creeping" of lead on the sloping roofs of buildings. Fig. 15.4 shows a portion of the roof of Westminster Abbey before it was repaired in 1954. When heated by the sun, the lead expands and tends to move down the roof under the force of gravity. On cooling and contracting, the force of contraction is opposed by the force of gravity on the lead and friction between it and the roof planking. This sets up a strain in the lead and gives it a very slight permanent stretch. After many years the lead stretches more and more, and eventually it forms into folds and may even break.

This trouble has been aggravated by the all too common practice in the past of using lead in very large sheets. When restorations are carried out it is now usual to replace the lead in much smaller sections.

Useful applications of thermal expansion

Although expansion can be troublesome, it often proves very useful.

The wheels of rolling-stock, particularly the driving wheels of locomotives, are fitted with steel tyres, which have to be renewed from time to time owing to wear. To ensure a tight fit the tyre is made slightly smaller in diameter than the wheel. Before being fitted the tyre is heated uniformly by special gas burners arranged in a ring. The resulting expansion enables the tyre to be slipped easily over the wheel, and on cooling it contracts and makes a tight fit.

The force of contraction when hot metal cools is also utilized in riveting together the steel plates and girders used in shipbuilding and other constructional work. The rivets are first made hot. This softens them so that they are easily burred into a head by pneumatic hammers, and their contraction on cooling serves to pull the plates tightly together.

Expansion of various substances

When rods of the same length but of different substances are heated through the same range of temperature, experiment shows that their expansions are not equal. Brass, for example, expands about one and a half times as much as steel; aluminium expands about twice as much as steel. An alloy of steel and nickel known as *invar* has an exceptionally small expansion when its temperature rises, and is used in watches and thermostats. Glass has a smaller expansion than iron. Fused silica and Pyrex glass have very low expansions.

The bimetallic strip

The difference in the expansion of brass and iron may be shown by riveting together a strip of brass and an equal strip of iron (Fig. 15.5). On heating this bimetallic strip it bends so that the brass is on the outside of the curve. The brass thus becomes longer than the iron, showing that it expands more than iron for the same temperature change.

The bimetallic strip has many useful applications, of which one of the most important is the electric *thermostat*. A thermostat is a device for maintaining a steady temperature. Fig. 15.6 shows the principle of a thermostat used for controlling

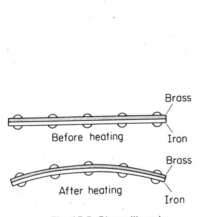

Fig. 15.5. Bimetallic strip

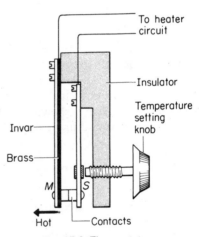

Fig. 15.6. Thermostat

the temperature of a room warmed by an electric heater. The heater circuit is completed through the two silver contacts of the thermostat, one of which is attached to a metal strip S and the other to a bimetallic strip M. If the room becomes too warm the bimetallic strip bends, separates the contacts and cuts off the current. On cooling, contact is remade and the heater switched on again. The temperature at which the contacts open is controlled by a screw which presses against an insulating sleeve on the strip S. If S is moved to the left the bimetallic strip will have to bend further before the contacts open, and so a higher temperature will be maintained in the room.

Thermostats working on the bimetal principle are also used to control the temperature of laundry irons, hot-water storage tanks, aquaria for tropical fish and for many other purposes.

Fig. 15.7 shows how a strip of bimetal is used in an automatic flashing unit suitable for direction indicator lamps on motor cars or for electric advertising signs. When the switch is closed current passes through a heating coil wound round a bimetallic strip. The heating coil is in series with the lamp, but owing to the resistance of the heating coil the current which flows in the circuit is insufficient to light the lamp. It is, however, sufficient to warm the strip, which bends upwards and closes the contacts shown. This shorts out the heating coil and applies the full voltage to the lamp through the bimetallic strip itself. The lamp therefore lights up and the strip immediately starts to cool. After a short interval the contacts open and the cycle is repeated.

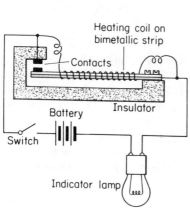

Fig. 15.7. Flasher unit

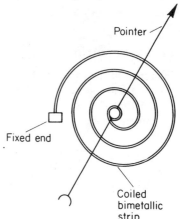

Fig. 15.8. Bimetallic thermometer

Fig. 15.8 shows the principle of the bimetallic thermometer. One end of a thin bimetallic spiral is fixed, the other end being attached to the spindle of a pointer which moves over a scale of degrees. The metals used are brass and invar, and the spiral tends to curl in a clockwise direction as the temperature rises. For simplicity, the coil in the diagram is shown as a flat spiral, but in practice it is much longer than is shown and is wound in a cylindrical form.

Use of kelvins in SI units and definitions

The thermodynamic scale which has already been mentioned on page 157 is preferred for use in SI units and definitions rather than the Celsius scale. This causes us no problem, since, as we have explained, the thermodynamic temperature unit, the kelvin (K), is exactly the same size as a temperature interval of one degree Celsius (°C).

In the ordinary course we use thermometers graduated in °C, so we record actual temperatures in °C, but express temperature differences (or intervals) in kelvins (K).

For example, in a particular experiment we may have something like this:

$$\begin{aligned}
\text{Initial temperature} &= 15\,°\text{C} \\
\text{Final temperature} &= 22\,°\text{C} \\
\text{Rise in temperature} = 22 - 15 &= 7\,\text{K}
\end{aligned}$$

To measure the linear expansivity of a metal

We have seen that various substances do not expand equally when their temperatures are raised. Their properties in this respect are expressed by the *linear expansivity*, which is defined as follows.

The linear expansivity of a substance is the fraction of its original length by which a rod of the substance expands per kelvin rise in temperature.

Various forms of apparatus are available for measuring the linear expansivity of a substance. One of these is shown in Fig. 15.9. The length of a metal rod, about 50 cm long, is carefully measured with a metre rule and then enclosed in a steam jacket, between a fixed stop *S* and a micrometer *M*. Before passing steam the rod is pushed against the stop *S* and the micrometer screw advanced until it just touches the end of the rod. The micrometer reading is recorded and also the temperature of the rod is noted from a thermometer placed inside the jacket.

The micrometer is then unscrewed several turns and a current of steam from a boiler passed through the jacket for some minutes. The micrometer is once more adjusted until it again touches the end of the rod and its new reading taken. As a precaution, the micrometer should be unscrewed again and the steam flow continued for a further few minutes. A final reading of the micrometer will make certain whether or not the rod had fully acquired the steam temperature in the first instance. Lastly, the temperature of the steam is noted. A typical set of readings for this experiment, in the case of a brass rod, is given below.

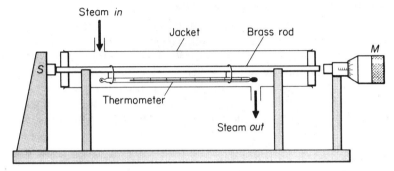

Fig. 15.9 Measurement of thermal expansion

Original length of brass rod = 50.2 cm
Initial temperature of rod = 16.6 °C
Final temperature of rod = 99.5 °C
1st micrometer reading = 4.27 mm
2nd micrometer reading = 3.48 mm

Calculation
Rise in temperature of rod = 99.5 − 16.6 = 82.9 K (see previous section)
Expansion of rod = 4.27 − 3.48 = 0.79 mm
= 0.079 cm

$$\text{Linear expansivity} = \frac{\text{expansion}}{\text{original length} \times \text{rise in temp.}}$$

$$\therefore \text{ linear expansivity of brass} = \frac{0.079}{50.2 \times 82.9}$$

$$= 0.000\,019/\text{K} = 1.9 \times 10^{-5} \text{ K}^{-1}$$

The table below gives the values of the linear expansivities for a number of different substances. The third column shows how these are written using index notation.

Linear expansivities (range 0–100 °C)

Iron	0.000 012/K	1.2 $\times 10^{-5}$ K^{-1}
Brass	0.000 019	1.9 $\times 10^{-5}$
Aluminium	0.000 026	2.6 $\times 10^{-5}$
Invar	0.000 001	0.1 $\times 10^{-5}$
Glass	0.000 008 5	0.85 $\times 10^{-5}$
Silica	0.000 000 42	0.042 $\times 10^{-5}$
Concrete	0.000 011	1.1 $\times 10^{-5}$

Thermal expansion in the kitchen

Most people know that hot water should never be poured into thick glass tumblers or dishes, since they are liable to crack. Glass is a poor conductor of heat, so that when hot liquid is poured into the glass the inside becomes hot while the outside remains cold. Expansion of the inside thus sets up a strain which cracks the glass.

Very often, cold jam-jars will crack when hot jam is poured into them. However, this may be prevented if the jars are first heated slowly in a warm oven before filling them.

Special glassware and ceramics of low expansivity are obtainable for kitchen and laboratory use. Pyrex and Pyrosil are examples. Even when very hot, dishes or beakers made of these materials may be plunged into cold water without cracking.

How to remove a tight glass stopper

Glass stoppers in bottles often become tightly fixed, and attempts to remove them by force generally result in breakage. The following treatment is invariably successful.

Two pairs of hands are required. One person holds the bottle firmly on the table, while another rapidly pulls to and fro a strip of cloth which has been wrapped once round the neck of the bottle. Work done against friction between cloth and glass becomes transferred to internal energy in the glass. This raises the temperature and causes the neck to expand sufficiently for the stopper to be withdrawn easily.

The gas thermostat

Oven temperature control in a gas-cooker is effected by utilizing the exceedingly low thermal expansion of invar, an alloy of steel with 36 per cent of nickel. From the table on page 163 it will be seen that invar expands by only one-millionth of its length per kelvin rise in temperature.

The flow of gas to the oven burners passes through a valve *A* having an invar stem. This stem is attached to the closed end of a brass tube projecting into the top of the oven (Fig. 15.10). When the burners are lit the oven begins to warm up and the

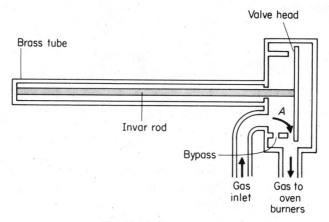

Fig. 15.10. Gas thermostat

brass tube expands. The expansion of the invar is negligible, and so it moves to the left, partially closing the valve opening and reducing the gas flow. Should the temperature of the oven fall, the brass tube contracts and the invar rod moves to the right, thus increasing the gas supply. The thermostat is provided with a rotating knob, not shown in the diagram, which varies the minimum opening of the valve and so controls the steady temperature to be maintained.

Equation for expansion

Suppose we let l_1 be the original length of a rod of material and α its linear expansivity.

For a rise in temperature of θ the expansion is $l_1\alpha\theta$

The new length, l_2 = original length + expansion

i.e.,
$$l_2 = l_1 + l_1\alpha\theta$$
or
$$l_2 = l_1(1 + \alpha\theta)$$

Expansion in area and volume

Corresponding with the equation above, we have similar equations to give the new area and volume when thermal expansion takes place.

Thus,
$$A_2 = A_1(1 + \beta\theta)$$
and
$$V_2 = V_1(1 + \gamma\theta)$$

where A_1 and A_2, and V_1 and V_2 represent the areas and volumes before and after a temperature rise of θ and β and γ the coefficients of superficial (or areal) and cubical expansion respectively.

EXPANSION OF LIQUIDS

We have already seen that liquids expand, in connection with our study of liquid-in-glass thermometers. The expansion of a liquid may be shown by means of a flask fitted with a rubber bung and a length of glass tubing (Fig. 15.11). The flask is filled with water or other liquid and the bung pushed in until the level of the liquid comes a short distance up the tube. On plunging the flask into a can of hot water it is noticed that the level of the liquid at first falls slightly and then starts to rise steadily.

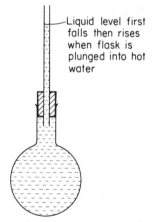

Liquid level first falls then rises when flask is plunged into hot water

Fig. 15.11. Expansion of a liquid

The initial fall in level is caused by the expansion of the glass which becomes heated and expands before the heat has had time to be conducted through the glass into the liquid.

To compare the expansions of various liquids

Different liquids have different thermal expansions. To demonstrate this, several fairly large glass bulbs with glass stems are filled to a short distance above the bulb with various liquids (Fig. 15.12). In order to make a fair comparison, the bulbs and stems must all be of the same size. The bulbs are immersed in a metal trough containing cold water and left until they have reached a steady temperature. A little extra liquid should now be added, where necessary, to make all levels the same. The bath is now heated and well stirred to ensure a uniform temperature. When the bulbs and their contents have acquired the new temperature of the bath it will be seen that the liquid levels have risen by different amounts. Thus, for a given rise in temperature, equal volumes of different liquids show different expansions in volume.

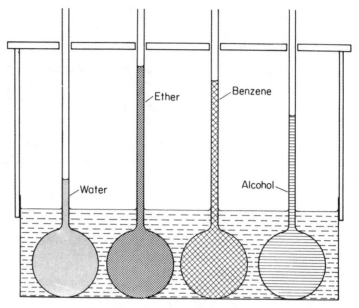

Fig. 15.12. Comparison of expansion

Real and apparent expansion of a liquid

Unlike solids, liquids have no fixed length or surface area but always take up the shape of the containing vessel. Therefore, in the case of liquids we are concerned only with volume changes when they are heated.

The real (or absolute) expansivity of a liquid is the fraction of its volume by which it expands per kelvin rise in temperature.

Any attempt at direct measurement of the expansion of a liquid is complicated by the fact that the containing vessel itself expands. However, since liquids must always be kept in some kind of vessel, it is just as useful to know the apparent expansion of a liquid, which is the difference between its real expansion and the expansion of the vessel.

The apparent expansivity of a liquid is the fraction of its volume by which the liquid appears to expand per kelvin rise in temperature when heated in an expansible vessel.

In this book we shall not be concerned with methods of measuring these expansivities.

The unusual expansion of water

Some substances do not always expand when heated. Over certain temperature ranges they contract. Water is an outstanding example.

If we start with some ice at $-10\ °C$ and supply it with heat, it expands just like any other solid until it reaches $0\ °C$. After this it begins to melt while the temperature remains constant at $0\ °C$. This melting is accompanied by a contraction in volume of about 8 per cent. Between 0 and $4\ °C$ the water contracts still further, reaching its minimum volume at about $4\ °C$. *This means that water has a maximum density at 4 °C.* Beyond 4 °C the water expands. This behaviour is described as anomalous (= irregular).

The changes in the water volume between 0 and 15 °C are shown graphically in Fig. 15.13. Unfortunately, on the scale of this graph, we cannot show the contraction in volume when ice melts, since this is nearly 700 times greater than the contraction of water between 0 and 4 °C.

Incidentally, the contraction in volume when ice melts is matched by a corresponding expansion when water freezes to form ice. This explains why pipes sometimes burst during frosty weather though the damage does not become apparent until a thaw sets in.

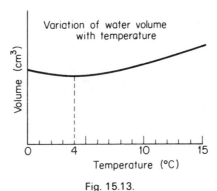

Fig. 15.13.

Frost heave

One problem brought about by the expansion of water on freezing has occurred in cold store buildings. This is *frost heave*, the name given to the damage caused to the buildings when the water in the subsoil beneath the site freezes and expands, causing upward bulging of the floor and damage to foundations and walls. This reached serious proportions about the mid-fifties as storage conditions went lower in temperature.

The trouble is now prevented by installing a heater grid of insulated stainless steel wires on the concrete site of the building during construction. Fed from the low-voltage output of a step-down mains transformer, a power of about 13.5 watts per square metre in the grid keeps the concrete a few degrees above freezing and provides for the inevitable slow continuous flow of heat upwards into the cold store. The heater grid itself is covered with a 25 mm layer of cement and sand above which is a thick layer of thermal insulation on which the cold store floor is laid.

Density changes in water

The density changes which occur when a piece of ice at − 10 °C is gradually heated up to 100°C are shown in Fig. 15.14. The maximum density region on this graph should be compared with the corresponding volume changes shown in Fig. 15.13.

Note how much greater is the density change when ice melts to form water at 0 °C compared with the density change of water between 0 °C and 4 °C.

Biological importance of the anomalous expansion of water

The peculiar expansion of water has an important bearing on the preservation of aquatic life during very cold weather (Fig. 15.15). As the temperature of a pond or lake falls, the water contracts, becomes denser and sinks. A circulation is thus set up until all the water reaches its maximum density at 4 °C. If further cooling occurs any water below 4 °C will stay at the top owing to its lighter density. In due course, ice

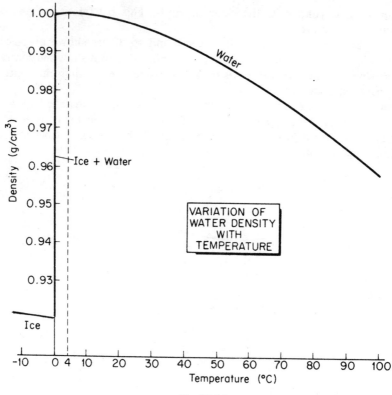

Fig. 15.14.

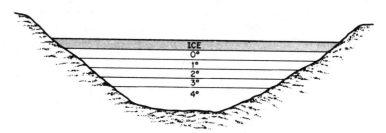

Fig. 15.15. Temperatures in an ice-covered pond

forms on the top of the water, and after this the lower layers of water at 4 °C can lose heat only by conduction (see chapter 17). Only very shallow water is thus liable to freeze solid. In deeper water there will always be water beneath the ice in which fish and other creatures can live.

QUESTIONS: 15

1. In an experiment to measure the linear expansivity of a metal, a rod of this metal 800 mm long is found to expand 1.36 mm when heated from 15 °C to 100 °C.

(a) Calculate the value of the linear expansivity.

(b) State how the expansion of the rod may be measured in such an experiment with reasonable accuracy.

(c) State how the rod may be heated up to 100 °C, and how the person conducting the experiment can ensure that it has really reached this temperature.　　　　(O.)

2. Draw a diagram of a useful device which involves the expansion of two different metals and indicate how the device works. Make clear which metal has the greater expansivity.　　　　(C.)

3. An iron tyre of diameter 50 cm at 15 °C is to be shrunk on to a wheel of diameter 50.35 cm. To what temperature must the tyre be heated so that it will slip over the wheel with a radial gap of 0.5 mm? (Linear expansivity of iron = 0.000 012/K.)   (*O.C.*)

4. Define *linear expansivity*.

Describe how you would find by experiment the linear expansivity of a metal in the form of a rod or tube. State the precautions you would take and explain your calculations.

A metal rod has a length of 100 cm at 200 °C. At what temperature will its length be 99.4 cm if the linear expansivity of the material of the rod is 0.000 02/K?

Name and describe briefly one use of the different expansions of metals.   (*A.E.B.*)

5. State whether each of the following characteristics of an aluminium washer increases, decreases, or remains unchanged when the washer is heated:
(*a*) internal diameter;
(*b*) volume;
(*c*) mass;
(*d*) density.   (*C.*)

6. The difference in length between a brass and an iron rod is 14 cm at 10 °C. What must be the length of the iron for this difference to remain at 14 cm when both rods are heated to 100 °C? (Linear expansivity of brass = 19 × 10⁻⁶/K; of iron = 12 × 10⁻⁶/K:)
    (*O.C.*)

7. The diagram, Fig. 15.16 shows an aluminium tube containing a silica cylinder. The silica cylinder acts as a thermal tap to control the flow of liquid down the tube. At 0 °C, the cylinder has a diameter of 20 mm, there is an exact fit, and no liquid flows. Calculate the size of the gap between cylinder and tube at 100 °C.

linear expansivity of aluminium
$$= 0.000\ 026/\text{K}$$
linear expansivity of silica $= 0.000\ 008/\text{K}$
    (*A.E.B.*, 1982)

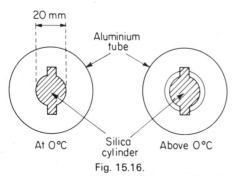

At 0 °C   Silica   Above 0 °C
            cylinder
Fig. 15.16.

8. *The linear expansivity of brass = 0.000 018 K^{-1}.** Explain the meaning of this statement.

A hole in a brass plate has diameter 20.00 mm. Through what temperature rise must the plate be heated for the diameter to become 20.09 mm?

Describe a practical device which makes use of the fact that two metals have different expansivities.   (*S.*)

9. Some thick glass vessels crack when a hot liquid is poured into them. Why do the following not crack when so treated:
(i) thin glass beakers;
(ii) thick jars made of certain kinds of glass?   (*O.C.*)

10. A copper tube has a length l_{25} at 25 °C and l_{100} at 100 °C. Use these values to write an expression for the linear expansivity of copper.   (*W.A.E.C.*)

11. A steel tape of correct length at 15.0 °C is used to measure distance on a day when the temperature is 10.0 °C. Is the result too large or too small?

If the linear expansivity of steel is 11.0 × 10⁻⁶/K, what is the error in measuring a distance of 20 m?   (*O.C.*)

12. Make a labelled diagram of an experiment to show that alcohol expands more than an equal volume of water, for the same rise in temperature.   (*S.*)

13. How does the density of water change as the temperature is lowered from, say, 15 °C to freezing-point?

Describe an experiment in support of your answer and illustrate the account by a diagram and a graph.

How would you expect the temperature of the water in a deep pond to vary with distance below the surface during a long period of hard frost?   (*O.C.*)

14. A flask filled with liquid at room temperature is sealed with a bung fitted with a length of open glass tubing, so that the level of the liquid appears a short distance up the tube. The flask is held upright and is then plunged into a beaker containing hot water. Describe and account for what is observed.

Why is mercury used in thermometers and water is never used?   (*L., part qn.*)

15. In a thermometer, the level of the mercury first <u>falls</u> and then <u>rises gradually</u> until a <u>steady level</u> is reached. Explain the three underlined observations.   (*W.A.E.C.*)

* K^{-1} is an example of index notation. It means the same as /K (see also pp. 23 and 163).

16. The gas laws

Galileo's air thermometer

The thermal expansion of air was first put to practical use in 1592 by Galileo Galilei, who used it as a crude method of indicating temperature changes. Galileo's air thermometer consisted of a glass bulb about the size of a hen's egg attached to a long, thin tube which dipped into a vessel of water.

A model of this thermometer may easily be made by fitting a glass flask with a rubber bung and a length of glass quill tubing. This is set up vertically with the end of the tube below the surface of some water in a vessel (Fig. 16.1). On placing a warm

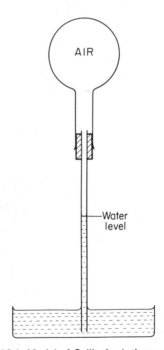

AIR

Water
level

Fig. 16.1. Model of Galileo's air thermometer

hand over the flask, the air inside becomes heated and expands. Bubbles are seen to emerge from the lower end of the tube and, when the hand is removed, the air in the flask contracts and the water rises in the tube. Subsequent changes of temperature will cause the water either to rise or fall in the tube. Galileo attached a scale to his instrument and marked it in *degrees*, but he made no attempt to standardize the scale by the use of fixed temperature points.

An air thermometer of this type has a very serious fault. Its readings are affected by changes in atmospheric pressure as well as by temperature. This defect was pointed out in 1665 by Robert Boyle, who had recently made a study of the way in which the volume of a gas alters with pressure when its temperature is kept constant.

Experiments to measure the thermal expansion of a gas are, therefore, bound to be complicated by the fact that the volume can be altered by a change in pressure as well as by a change in temperature. This difficulty does not arise in the case of solids or liquids, as these are very much less compressible.

In order to make a full study of the behaviour of a gas as regards volume, temperature and pressure, three separate experiments have to be carried out to investigate respectively:

(1) the relation between volume and pressure at constant temperature (*Boyle's law*);
(2) the relation between volume and temperature at constant pressure (*Charles's law*);
(3) the relation between pressure and temperature at constant volume (*Pressure law*).

Robert Boyle's work on air

The law relating the volume and pressure of a fixed mass of air kept at constant temperature was first investigated experimentally by Robert Boyle about the middle of the seventeenth century. In the first instance he used a glass tube in the form of a

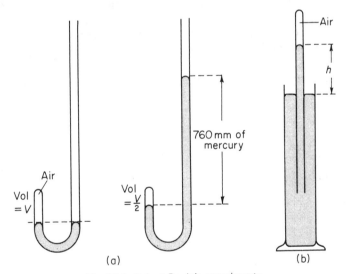

Fig. 16.2. Robert Boyle's experiments

letter J having its shorter arm closed. Mercury was added so as to trap air in the short arm and the quantity of air adjusted so that the levels were the same on both sides. More mercury was then poured in until the difference in levels was equal to the barometric height (about 760 mmHg). Having thus doubled the pressure, it was noticed that the volume of the air had been halved (Fig. 16.2 (*a*)). This kind of relationship is called *inverse proportion*.

Expressed mathematically,
$$\text{volume} \propto \frac{1}{\text{pressure}}$$

or
$$V \propto \frac{1}{p}$$

It follows that
$$V = \text{constant} \times \frac{1}{p}$$

or
$$\underline{pV = \text{constant}}$$

Altogether, Boyle took about twenty-five pairs of readings for different pressures and volumes and found that they were in very close agreement with the above equation.

The pressure was found by adding the difference in the mercury levels to the barometric height. The volume was taken as being proportional to the length of the air column in the closed tube.

The J-tube, of course, gave only readings for pressures above atmospheric. For lower pressures he used a straight tube closed at its upper end and dipping into a long jar of mercury (Fig. 16.2 (*b*)). This was raised to various heights out of the mercury and the pressure found by *subtracting* the difference in mercury levels, *h* in mm, from the barometric height.

Later experiments showed that all the so-called permanent gases behave in the same way as air, and so the result may be expressed as a universal law.

Boyle's law

The volume of a fixed mass of gas is inversely proportional to the pressure, provided the temperature remains constant.

To investigate Boyle's law for air

Fig. 16.3 illustrates a modern method of investigating Boyle's law for pressures both above and below atmospheric. Nowadays we have the advantage of flexible tubing which was not available in the seventeenth century. The apparatus consists of a burette *B* connected by a length of rubber or plastic tubing to a glass reservoir containing mercury.

It is first necessary to remove any water vapour which may be present with the air

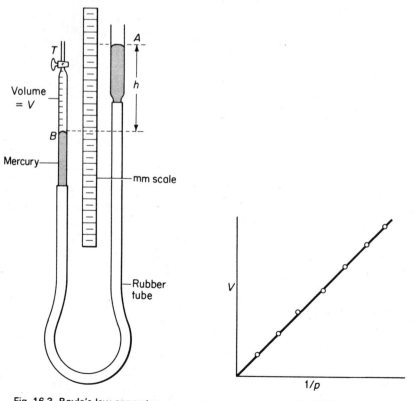

Fig. 16.3. Boyle's law apparatus

Fig. 16.4.

in the burette. The tap *T* is opened and the reservoir raised until the burette is full of mercury. A calcium chloride drying tube is attached to the tap and the reservoir lowered until the burette is about half full of dry air. The tap is then closed and the drying tube removed.

Starting with the reservoir *A* in its lowest position, the volume *V*, in cm³, of air in the burette is noted together with the readings of the mercury levels *A* and *B* on a vertical millimetre scale. The reservoir is then raised a few centimetres at a time and a series of such readings obtained.

Now, if the volume of a gas is altered, the temperature will change. It is therefore necessary to wait a short while after each adjustment of the reservoir in order to give the air time to revert to room temperature (see under *Temperature rise resulting from compression*, page 201).

The atmospheric pressure *H*, in mmHg, is read from a Fortin barometer. If the difference in the mercury levels is *h* in mm, then the absolute pressure of the enclosed air is given by $p = (H + h)$ when *A* is above *B* and $p = (H - h)$ when *A* is below *B*.

All the measurements made should be entered in a table as shown.

Barometric height = *H* =   mmHg

Volume *V* (cm³)	Level *A* (mm)	Level *B* (mm)	*h* (mm)	Pressure $p = H \pm h$ (mmHg)	$p \times V$	$\dfrac{1}{p}$

Provided the room temperature has remained constant during the experiment, we should expect the results to show that $pV = $ constant.

Graphical treatment of results

In mathematics the equation $y = mx$ gives a straight line through the origin when *y* is plotted against *x* on graph paper. In this equation *m* is a constant and a series of values of *y* are calculated by giving various values to *x*.

Now, if both sides of the equation, $pV = c$, are divided by *p* it will be converted into an equation of the form $y = mx$.

Thus,
$$V = c \times \frac{1}{p}$$

Here, *V* is the dependent variable (corresponding with *y*) and $\dfrac{1}{p}$ is the independent variable (corresponding with *x*). The constant *c* plays the same part as *m*.

Hence, if Boyle's law is true we ought to get a straight line through the origin by plotting *V* against $\dfrac{1}{p}$ (see Fig. 16.4).

Note. In more simple forms of the apparatus shown in Fig. 16.3 the graduated burette is replaced by a plain closed tube of uniform cross-section. In such cases the volume of the enclosed gas is taken as being proportional to the *length* of the air column.

Cubic expansivity of a gas at constant pressure

It may be remembered that when the linear expansivity of a solid was defined on page 162 it was not specified that the original length of the rod of material should be measured at 0 °C. Owing to the smallness of the linear expansion of most substances, it makes very little difference to the result obtained for the expansivity whether we start with the rod at 0 °C or at ordinary temperature.

In the case of a gas, however, the situation is different. The expansion of gases is very much larger than that of solids, and consequently the value of the cubic expansivity obtained will depend on the starting temperature. Hence, in order to make a satisfactory comparison between different gases, the cubic expansivity *is always calculated in terms of an original volume at 0 °C.* The cubic expansivity of a gas is thus defined as follows:

The cubic expansivity of a gas at constant pressure is the fraction of its volume at 0 °C by which the volume of a fixed mass of gas expands per kelvin rise in temperature.

To measure the cubic expansivity of air at constant pressure

A glass tube about 30 cm long and about 1.5 mm bore has one end sealed by heating in a bunsen flame. A short pellet of concentrated sulphuric acid is then introduced a little more than a third of the way down the tube by means of a quill tube drawn out into a capillary and used as a pipette as shown in Fig. 16.5 (*a*).

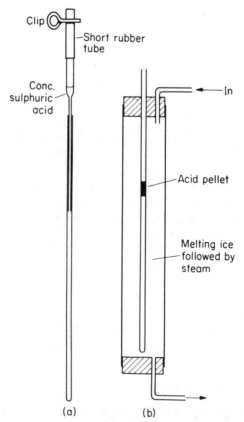

Fig. 16.5. Charles's law apparatus

Some people use a mercury pellet, but sulphuric acid has the advantage of drying the air inside the tube. The process of sealing the tube invariably introduces moisture into it, and it is essential in this experiment that the air is dry, since the presence of water vapour will give results which are abnormally high.

Care must be taken to avoid spilling concentrated sulphuric acid or getting it on one's fingers: it is advisable to enlist the help of someone experienced in handling it.

The tube is now inserted, with its open end upwards, through a hole in a bung and placed inside a glass tube with a bung at its lower end and packed with ice shavings (Fig. 16.5 (b)). When the air inside the experimental tube has fully acquired the temperature of the melting ice it is raised until the top of the pellet is just visible above the top of the bung, and the distance from the open end of the tube to the top of the pellet is measured. This distance plus the length of the pellet is afterwards subtracted from the total length of the tube from the open end to the inside of the closed end in order to find the length of the air column at 0 °C.

The ice and water are now removed from the tube and the experimental tube pushed well down. Steam is now passed in through the top and out at the bottom for several minutes. The experimental tube is again adjusted until the pellet is just visible above the top of the bung and measurements are taken as before to find the length of the air column at the steam temperature. A final check reading is taken to make certain that the air has fully reached the steam temperature, assumed to be 100 °C.

Assuming the bore of the tube to be uniform, and neglecting the small error arising from the curved inside end of the bore and acid pellet, we take the lengths l_0 and l_{100} of the air column at the ice and steam temperatures as being proportional to their volumes v_0 and v_{100}. The cubic expansivity of the air at constant pressure, γ, is given by,

$$\gamma = \frac{\text{change in volume}}{\text{volume at 0 °C} \times \text{change in temperature}}$$

$$= \frac{V_{100} - V_0}{V_0 \times 100} \text{ per K}$$

or

$$\gamma = \frac{l_{100} - l_0}{l_0 \times 100} \text{ per K}$$

The value obtained is approximately $\frac{1}{273}$ or 0.003 66/K and, moreover, practically the same result is obtained for all gases.

The original experimental work on this subject was carried out towards the end of the eighteenth century by the distinguished French scientist, Jacques Charles, whose researches were doubtless inspired by an interest in flight by hot-air balloons (Fig. 16.6). He later decided in favour of hydrogen balloons, since these had greater lifting power.

The result obtained in the experiment just described is generally known as **Charles's law**.

The volume of a fixed mass of gas at constant pressure expands by $\frac{1}{273}$ of its volume at 0 °C per kelvin rise in temperature.

Pressure expansivity of a gas at constant volume

It now remains to investigate the manner in which the pressure of a gas varies when it is heated at constant volume.

The pressure expansivity of a gas at constant volume is defined as the fraction of its pressure at 0 °C by which the pressure of a fixed mass of gas increases per kelvin rise in temperature.

To measure the pressure expansivity of air

The apparatus for this experiment was designed by the German professor of physics, Philipp Jolly, who lived during the nineteenth century. It is generally called the *simple constant volume air thermometer* (Fig. 16.7).

A glass bulb *B* joined to a capillary tube is connected to a mercury reservoir *R* by a length of rubber tubing. By means of a three-way tap air may be pumped out of the bulb and dry air afterwards admitted through a calcium chloride drying tube.

Fig. 16.6. *An old sport revived.* This modern hot-air balloon uses a propane gas burner with a heat output of 730 kilojoules per second. It has a capacity of 1840 cubic metres, and when heated to 100 °C, can lift about 500 kilograms. *Note.* The first hot-air balloon passengers were a cock, a duck and a sheep in 1783. This otherwise successful flight was marred only by the sheep which kicked the cock en route

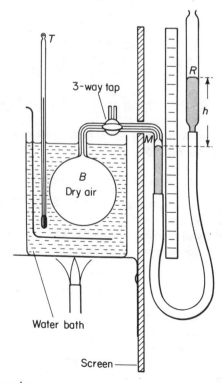

Fig. 16.7. Pressure law apparatus (simple constant volume gas thermometer)

The bulb is placed in a large beaker or can and surrounded by melting ice. The height of the reservoir is then adjusted until the mercury level reaches a fixed mark *M* on the connecting tube, and when the air in the bulb has fully acquired the temperature of the melting ice the mercury levels at *M* and *R* are read on a vertical millimetre scale.

The barometric height *H* in mmHg is read on a Fortin barometer. If the difference in the mercury levels at *M* and *R* is *h* in mm, then the pressure of the air in the bulb is given by $(H \pm h)$ according as *R* is above or below *M*.

Water is now added to the can so as to cover the bulb and the temperature of the ice-and-water mixture raised with a bunsen burner until it boils steadily. When the bulb has acquired the temperature of the boiling water the mercury level is again adjusted to the mark *M* and the new level at *R* is read off. Now the pressures in the bulb are respectively equal to the appropriate values of $(H \pm h)$ at the ice and boiling-water temperatures. If we call these p_0 and p_{100}, and assume the temperatures to be 0 °C and 100 °C respectively, then the pressure expansivity at constant volume, β, may be calculated from:

$$\beta = \frac{\text{change in pressure}}{\text{pressure at 0 °C} \times \text{temp. rise}}$$

or

$$\beta = \frac{p_{100} - p_0}{p_0 - 100} \text{ per K}$$

The result is found to approximate very closely to $\frac{1}{273}$ or 0.003 66/K. All other gases besides air give very nearly the same result.

We therefore arrive at the important conclusion that the pressure expansivity is equal to the volume expansivity.

The results obtained in the above experiment may be expressed by the **Pressure law.**

The pressure of a fixed mass of gas at constant volume increases by $\frac{1}{273}$ of its pressure at 0 °C per kelvin rise in temperature.

Note. Strictly speaking, we should have enclosed the bulb in steam and applied a boiling-point correction, instead of using boiling water (see page 154). We have, however, described the apparatus commonly found in elementary laboratories, which does not lend itself to this refinement.

The absolute or thermodynamic scale of temperature

If volume–temperature and pressure–temperature graphs for a gas are plotted on extra large sheets of graph paper and then produced or *extrapolated* backwards they cut the temperature axis at −273 °C (Fig. 16.8 and 16.9). This suggested to the early experimenters that −273 °C might be the lowest temperature attainable or the *absolute zero*. This must, however, be only an assumption, since we know that gases liquefy in most cases long before such a temperature is reached.

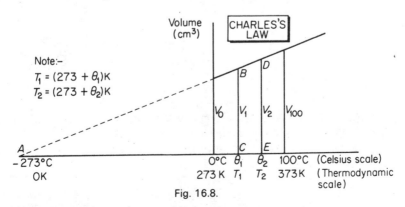

Fig. 16.8.

Thus by shifting the vertical axis 273 °C to the left and renumbering the temperature scale, −273 °C becomes zero, the ice point 273 and the steam point 373 on what is called an *absolute scale* of temperature. This comes very close to the **thermodynamic scale** of temperature proposed on theoretical grounds by Lord Kelvin about the middle of the nineteenth century. Temperatures on this scale are given in units called kelvins, instead of degrees. They are represented by T and denoted by K. Also the scale is defined in such a way that kelvins are equal to Celsius degrees in size.

It follows that temperatures on the Celsius scale can be converted to kelvins simply by adding 273. Thus 0 °C = 273 K, θ °C = (273 + θ) K and so on.

It can be shown that the thermodynamic scale can be related to the products pV for a perfect or ideal gas, i.e., one which obeys Boyle's law exactly. In the case of practical gas thermometers it is difficult to measure both p and V at the same time, so it is usual to keep V constant, in which case the temperature will be proportional to the pressure of the gas. This explains why Jolly's apparatus of the previous experiment is also called a constant-volume air thermometer.

Now air is not a perfect gas any more than any other gas, but the one which comes closest to this requirement is hydrogen. An improved form of Jolly's apparatus is called the **standard gas thermometer,** descriptions of which are to be found in more advanced textbooks. This thermometer needs laborious corrections and is far too difficult and cumbersome for ordinary day-to-day use. When very high temperatures are to be measured it is filled with nitrogen and is employed only for the purpose of obtaining accurate values for a number of other fixed points both above 373 K and below 273 K, e.g. the freezing-point of gold (1336 K) and the boiling-point of oxygen (90 K).

In practice, therefore, such instruments as the platinum resistance thermometer, thermo-couples, and so on are standardized by the fixed points so obtained and used to measure temperature over a wide range where mercury thermometers are unsuitable.

Other ways of expressing Charles's law and the Pressure law

If we assume that the two graphs in Figs. 16.8 and 16.9 have been plotted using Celsius and thermodynamic temperatures then we can use them to express Charles's

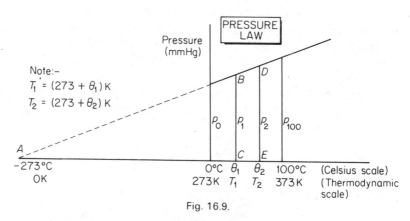

Fig. 16.9.

law and the Pressure law in a form which is usually more convenient for calculation purposes than the statements of the laws given on pages 175 and 177.

In the *volume–temperature* graph, any pair of triangles such as *ABC*, *ADE* are similar and therefore,

$$\frac{BC}{AC} = \frac{DE}{AE}$$

But the lengths of these sides represent volumes and corresponding absolute or thermodynamic temperatures to scale. Hence,

$$\frac{V_1}{T_1} = \frac{V_2}{T_2}$$

This applies to all values of volume and thermodynamic temperature and so we can say,

$$\frac{V}{T} = \text{constant}$$

Similarly, from the *pressure–temperature* graph we may show that

$$\frac{p}{T} = \text{constant}$$

These two equations enable us to state the gas laws in the following forms, which should be compared with those given previously on pages 175 and 177.

Charles's law

The volume of a fixed mass of gas at constant pressure is proportional to its thermodynamic temperature.

Pressure law

The pressure of a fixed mass of gas at constant volume is proportional to its thermodynamic temperature.

The relation between pressure, volume and thermodynamic temperature

Summarizing the gas laws dealt with in this chapter we have

$$pV = \text{constant } (Boyle's\ law)$$

$$\frac{V}{T} = \text{constant } (Charles's\ law)$$

$$\frac{p}{T} = \text{constant } (Pressure\ law)$$

These three equations can be combined into the single equation,

$$\frac{pV}{T} = \textbf{constant}$$

since,   if T is constant, then $pV = \text{constant}$

or   if p is constant, then $\dfrac{V}{T} = \text{constant}$

or   if V is constant, then $\dfrac{p}{T} = \text{constant}$

It follows that $\dfrac{pV}{T} = \text{constant}$ includes all three gas laws.

Reduction of gas volumes to standard temperature and pressure

In certain chemical experiments a volume of gas is collected at room temperature and pressure, and it is necessary to calculate the volume which the gas would occupy at 0 °C and 760 mmHg pressure. 0 °C and 760 mmHg are called standard temperature and pressure; usually abbreviated, **s.t.p.**

Worked examples

1. *125 cm³ of gas are collected at 15 °C and 755 mm of mercury pressure. Calculate the volume of the gas at s.t.p.*

This type of problem is solved by applying the relation

$$\frac{pV}{T} = \text{constant}$$

We therefore write down the initial values and the required values of the volume, pressure and temperature, remembering that temperatures must be in kelvins.

Initial values

$p_1 = 755 \text{ mmHg}$
$V_1 = 125 \text{ cm}^3$
$T_1 = (273 + 15)$
$\quad = 288 \text{ K}$

Required values

$p_2 = 760 \text{ mmHg}$
$V_2 = ?$
$T_2 = 273 \text{ K}$

Since $\dfrac{pV}{T} = \text{constant}$, it follows that

$$\frac{p_2 V_2}{T_2} = \frac{p_1 V_1}{T_1}$$

or

$$V_2 = V_1 \times \frac{p_1}{p_2} \times \frac{T_2}{T_1}$$

hence, substituting numerical values, we obtain

$$\text{volume of gas at s.t.p.} = V_2 = 125 \times \frac{755}{760} \times \frac{273}{288}$$

$$= 118 \text{ cm}^3$$

2. *An empty barometer tube, 1 m long, is lowered vertically, mouth downwards, into a tank of water. What will be the depth of the top of the tube when the water has risen 20 cm inside the tube? (Atmospheric pressure may be assumed equal to 10.4 m head of water.)*

Assuming the temperature remains constant, Boyle's law may be applied, i.e.,

$$pV = \text{constant}$$

or
$$p_2 V_2 = p_1 V_1$$

Let h = depth, in m, of water-level in tube below surface, then

$$\begin{cases} p_2 = (10.4 + h) \text{ in m of water} \\ V_2 = (0.8 \times A) \text{ in m}^3 \end{cases}$$

where A = area of cross-section of tube in m².

$$\begin{cases} p_1 = 10.4 \text{ in m of water} \\ V_1 = (1 \times A) \text{ in m}^3 \end{cases}$$

Substituting in the above equation

$$(10.4 + h)(0.8\, A) = 10.4 \times 1 \times A$$

$$10.4 + h = \frac{10.4}{0.8} = 13$$

therefore
$$h = 2.6 \text{ m.}$$
Hence top of tube is
$$2.6 - 0.8 = 1.8 \text{ m below surface}$$

3. *When tested in a cool garage at 12 °C a motor tyre is found to have a pressure of 190 kPa. Assuming the volume of the air inside remains constant, what would you*

expect the pressure to become after the tyre has been allowed to stand in the sun so that the temperature rises to 32 °C? Atmospheric pressure = 100 kPa (1 kPa = 1000 N/m²).

In this problem it must be understood that 190 kPa is the *excess pressure* above atmospheric pressure and hence the absolute pressure inside the tyre is (190 + 100) = 290 kPa.

Applying the Pressure law

$$\frac{p}{T} = \text{constant}$$

or

$$\frac{p_2}{T_2} = \frac{p_1}{T_1}$$

We have,

$$T_2 = 273 + 32 = 305 \text{ K}$$
$$p_1 = 190 + 100 = 290 \text{ kPa}$$
$$T_1 = 273 + 12 = 285 \text{ K}$$

Substituting in the equation,

$$\frac{p_2}{305} = \frac{290}{285}$$

whence

$$p_2 = \frac{290 \times 305}{285} = 310 \text{ kPa}$$

Therefore, new pressure as given by pressure gauge = 310 − 100 = 210 kPa.

Gas laws and the simple kinetic theory of matter

An introduction to the kinetic theory of matter has been given in chapter 13. On this theory a gas consists of a vast number of molecules moving with random high velocities, colliding with one another, and bouncing off the walls of the containing vessel. A force is thus set up on the walls, which is given by the rate of change of momentum as they bounce off. The pressure of the gas is the value of this force per unit area. Calculations based on the kinetic theory in conjunction with the experimental results known as Boyle's Law, Charles's Law, and the Pressure Law show that the absolute thermodynamic temperature of a perfect gas is proportional to the average kinetic energy of its molecules. These calculations are dealt with in more advanced texts, so a brief description only will be given here.

Boyle's law. At constant temperature, the average kinetic energy of the molecules is constant. If the volume of a fixed mass of gas is halved it can be shown by geometry that the number of molecular impacts per second per unit area of wall surface is doubled: in other words, the pressure is doubled. This confirms the experimental result that the pressure of a fixed mass of gas at constant temperature is inversely proportional to the volume.

Pressure law. Raising the temperature of a fixed mass of gas at constant volume increases the average kinetic energy of the molecules so that they make more frequent impacts with the walls at higher velocity. Thus the rate of change of momentum on impact is increased with consequent increase in pressure.

Charles's law. If the pressure of a fixed mass of gas is to remain constant as the temperature is raised the rate of change of momentum of the molecules on impact with the walls must remain constant. Since the velocity of the molecules will increase with temperature, the change in momentum will be greater, and to keep the rate of change the same it will be necesssary to make fewer impacts per second. Thus the volume must increase so that the molecules have to travel further between collisions with the walls. As we have seen, this condition applies if the volume varies as the thermodynamic temperature.

QUESTIONS: 16

1. Describe with a diagram and any desirable precautions how you would investigate the relationship between the pressure and volume of a given sample of air at constant temperature. Show on a sketch how you would express your results in the form of a straight-line graph.

An air bubble at the bottom of a lake 90 m deep has a volume of 1.5 cm³. What will be its volume just below the surface if the atmospheric pressure is equivalent to a height of 10 m of water? (S.)

2. State Boyle's law and describe an experimental method of verifying it.

A uniform, narrow-bored tube, closed at one end, contains some dry air which is sealed by a thread of mercury 15.0 cm long. When the tube is held vertically, with the closed end at the bottom, the air column is 20.0 cm long, but when it is held horizontally, the air column is 24.0 cm long. Calculate the atmospheric pressure. (C.)

3. State Boyle's law and explain how you would attempt to verify it. Sketch your apparatus.

A narrow uniform glass tube contains air enclosed by a thread of mercury 15 cm long. When the tube is vertical, with the open end uppermost, the column is 30 cm long. When the tube is inverted the length of the column becomes 45 cm.

Explain why the length of the column changes and calculate the pressure of the atmosphere. (O.C.)

4. State *Boyle's law* and apply it in order to solve the following problem.

A U-tube of uniform cross-section has limbs of unequal length and is held vertically with its longer limb open to the atmosphere so that air may be trapped in the shorter limb, which is closed, by adding mercury. The length of the air column in the closed limb is 16 cm when the *difference* in the vertical heights of the meniscus levels in the two limbs is 20 cm. When the difference is increased to 77 cm, by adding more mercury, the air column is only 10 cm long. Use these data to calculate the height of the mercury barometer, which may be considered to have remained constant during the experiment.

Explain why the assumption of Boyle's law may not be strictly valid in this case. (L.)

5. Describe an experiment to examine the relation between the volume and the temperature of a constant mass of dry air, at constant pressure. Draw a labelled diagram of the apparatus used in such an experiment

and describe clearly how the experiment is performed. In a similar experiment the following readings were obtained:

Volume in m³	7.0×10^{-6}	7.6×10^{-6}	8.2×10^{-6}	8.6×10^{-6}	8.8×10^{-6}
in cm³	7.0	7.6	8.2	8.6	8.8
Temperature in °C	15.0	40.0	65.0	80.0	90.0

Plot a graph of volume on the y-axis and temperature on the x-axis. Use the graph or information from the graph to determine values for:

(i) the volume of the air at 0 °C;

(ii) the gradient. Explain what information this gives about the way the gas expands. In each case show clearly how the values were obtained. (J.M.B.)

6. The mercury in a barometer of a cross-sectional area 1 cm² stands at 75 cm, and the space above it is 9 cm in length. What volume of air, measured at atmospheric pressure, would have to be admitted into the space to cause the column of mercury to drop to 59 cm? (W.)

7. State Charles's law for the expansion of a gas heated at constant pressure. Describe how you would test the truth of the law experimentally, and show how it leads to the idea of an absolute zero of temperature.

The density of hydrogen at 0 °C and standard pressure is 0.009 g/litre. Find the volume occupied by 1 g of hydrogen at 200 °C and standard pressure. (O.)

8. The results shown in the table below were obtained in an experiment to verify Boyle's law.

Pressure (kN/m²)	400	320	160	80
Volume (mm³)	2.0	2.5	5.0	10.0
$\frac{1}{\text{volume}}$ (mm⁻³)	0.5			

(a) Copy the table into your answer book and complete it.

(b) Plot a graph of pressure on the y-axis against $\frac{1}{\text{volume}}$ on the x-axis.

(c) State the relationship which this graph shows between pressure and volume.

(d) **From your graph** calculate the volume when the pressure was 240 kN/m².

(e) State which **two** physical properties of the gas were kept constant.

9. 1000 cm³ of air at 10 °C is heated to 80 °C. What will be the new volume if the pressure remains atmospheric? (J.M.B.)

10. Before starting a long journey a motorist checked her tyre pressures and found them to be 3×10^5 Pa. At the end of the journey, the

pressures were found to be 3.3×10^5 Pa. The temperature of the tyres and contained air at the start of the journey was 17 °C. Assuming the volume of the tyres remains constant, determine the temperature of the air in the tyres at the end of the journey.

(*A.E.B.*, 1982)

11. Describe fully the apparatus to be used, the precautions which should be taken, and the observations to be made to establish the relationship at constant pressure between the volume of a fixed mass of gas and its temperature over the range approximately 0 to 100 °C.

Sketch the graph you would expect to obtain and indicate the expected relationship.

A thick-walled steel cylinder used for storing compressed air is fitted with a safety valve which lifts at a pressure of 1.0×10^6 N/m^2. It contains air at 17 °C and 0.8×10^6 N/m^2. At what temperature will the valve lift? (*L.*)

12. To what temperature must 2 litres of air at 17 °C be heated at constant pressure in order to increase its volume to 3 litres?

(*J.M.B.*)

13. State the law which indicates how the pressure of a fixed mass of air varies with its temperature as recorded on a mercury thermometer, when the volume remains constant.

Describe an experiment to verify the law.

80 cm^3 of hydrogen are collected at 15 °C and 75 cm of mercury pressure. What is its volume at s.t.p.? (*L.*)

14. Draw and label a constant-volume air thermometer. State two advantages of such a thermometer as compared with a mercury thermometer.

A capillary tube, sealed at both ends, contains a small thread of mercury which divides the enclosed air into two parts, the ratio of the lengths being 3 to 1, the whole tube is initially at 0 °C. If the tube is heated to 273 °C, how will the air inside be affected?

(*S.*)

15. What is meant by "the absolute zero of temperature"? What is this temperature measured on the Celsius scale?

A car tyre was tested before being driven on a motorway. The reading on a tyre-pressure gauge which reads the pressure above atmospheric pressure was 120 000 N/m^2 and the temperature was 7 °C. At the end of the journey the temperature of the tyre was found to be 35 °C. What reading would you expect on the same tyre-pressure gauge if the volume of air in the tyre

remained constant and atmospheric pressure throughout the journey was 100 000 N/m^2?

(*J.M.B.*)

16). (*a*) A fixed mass of gas is enclosed in a vessel. Explain in terms of molecular theory:

(i) how the pressure measured at the wall of the vessel is produced by the gas molecules;

(ii) how the pressure would be affected if the temperature increased, keeping the volume constant;

(iii) how the pressure would be affected if the volume of the vessel were reduced, keeping the temperature constant.

(*b*) Also explain how you would expect the pressure to change if a quarter of the mass of gas were allowed to escape, the volume and temperature of the gas remaining in the vessel being unchanged. (*O.*)

17. According to the kinetic theory, gases exert pressure by repeated perfectly elastic collisions between gas molecules and the walls of the container. Imagine such a gas container to be a cube, each edge of which has a length l, containing N gas molecules each of mass m. Consider **one** molecule which is moving towards one wall of the cube in a direction perpendicular to the wall with a velocity c.

(*a*) (i) What is the momentum of the molecule just before it collides with the wall?

(ii) What is the momentum of the molecule just after it has collided with the wall?

(iii) What is the change in the momentum of the molecule for each collision with that wall?

(*b*) (i) Still assuming that the gas molecule is moving perpendicular to the wall, how much later will the **same** molecule collide with the **same** wall, assuming that no time is lost in collisions on the way?

(ii) How many times will this molecule hit the same wall in each second?

(*c*) (i) If, on average, one-third of all the gas molecules move in this way, state the change in momentum occurring at this wall in each second.

(ii) State the law which relates force to change of momentum. Hence deduce the force exerted on this same wall by the molecules.

(iii) Hence deduce the pressure exerted by the gas in terms of N, m, c and the volume of gas V.

(*d*) If the temperature of the gas is measured by mc^2 and is kept constant, what does the equation derived in (*c*) predict about the result of multiplying pressure and volume for a fixed mass of gas maintained at constant temperature?

Explain your answer. (*J.M.B.*)

17. Transmission of heat

If a steel rod is held by one end and the opposite end placed in a bunsen flame it is noticed, very soon, that the rod becomes warm to the fingers. Heat travels through the metal by a process called *conduction*.

This process is complex. It differs between metals and non-metals, and only a brief explanation can be attempted here.

When a metal is heated the free electrons (see page 367) which it contains begin to move faster, i.e., their kinetic energy increases. These electrons then drift towards the cooler parts of the metal, where their energy is transferred by collision to the metal molecules there. At the same time there is a drift of slower-moving (cooler) electrons in the reverse direction.

To a much less extent, heat energy is transmitted through a metal by vibrations of the atoms themselves which pass on energy from one to the other in the form of waves (see chapter 26). These waves are of very high frequency and are transmitted in tiny energy packets called "phonons".

In non-metals which have no free electrons heat energy is conducted entirely by phonons.

Good and bad conductors of heat

Most metals are good conductors of heat; silver and copper are exceptionally good. On the other hand, substances such as cork, wood, cotton and wool are bad conductors. Both good and bad conductors have their uses. The "bit" of a soldering iron is made of copper, so that when its tip is cooled through contact with the work, heat is rapidly conducted from the body of the bit to restore the temperature of the tip and maintain it above the melting-point of solder.

Bad conductors have a very wide application. Beginning with our own personal comfort, we prevent loss of heat from ourselves by a covering of poorly conducting material. Textiles are bad conductors of heat, since they are full of tiny pockets of air enclosed by the fibres of the material. Air, in common with all gases, is a very bad conductor of heat. It is usual to say that wool is warmer than cotton. Technically, of course, we imply that it has a lower thermal conductivity than cotton. Double glazing in windows reduces heat loss from a building owing to the poorly conducting layer of air between the two sheets of glass.

A stone floor feels cold to the bare feet, but a carpet on the same floor feels warm. This difference arises from the fact that stone is a better conductor of heat than a carpet.

To begin with both the stone floor and the carpet are at the same temperature. This may be verified by placing a thermometer in contact with each in turn. Since the feet are warmer than either, heat tends to flow from the feet. Stone, being the better conductor, conveys heat away from the feet more rapidly than the carpet. Consequently, the feet feel cold on the stone but warm on the carpet.

Precisely the same effect is experienced when handling a garden fork in winter. The iron part of the fork feels cold, but the wooden handle warm.

Fig. 17.1. *Keeping the heat in*. A layer of thermally insulating material (fibreglass) reduces loss of heat from the oven by conduction. As an extra precaution, the lagging is encased in bright aluminium foil which helps to prevent further heat loss by radiation

Fig. 17.2. *Keeping the heat out*. Refrigeration hold of a cargo vessel during construction. Note the use of thick layers of mineral fibre to prevent the entry of heat by conduction

Lagging

Loss of heat by conduction through the walls of an oven is reduced by constructing it with double walls. The space between is packed with slag wool or glass fibre. These substances are not only very poor conductors but also have the merit of being non-flammable. Material of low thermal conductivity used for the purpose of preventing heat loss is called *lagging*. Extensive use has been made in the past of asbestos mixed with plaster for the lagging of steam boilers, hot-water storage tanks, and pipes. Where such installations are dealt with today, precautions are taken to avoid health hazards from asbestos dust, and increasing use is being made of other materials. Similarly, cold-water pipes are lagged with strips of felt or sacking to prevent freezing during very cold weather. (See also Fig. 17.1 and 17.2.)

Ignition point of a gas

A flammable gas will burn if its temperature reaches a value known as the "ignition point". The effect of a good conductor in the neighbourhood of a flame can be shown by placing a wire gauze about 5 cm above a bunsen-burner. If the gas is turned on and lighted underneath the gauze it is found that the flame does not pass through the gauze. The wires of the gauze conduct the heat of the flame away so rapidly that the hot gases passing through the gauze are cooled below the ignition temperature (Fig. 17.3).

The gas is now turned out and after the gauze has cooled, the gas is again turned on and lit above the gauze. This time the flame continues to burn above the gauze. As in the previous case, the wires conduct heat rapidly away, with the result that the

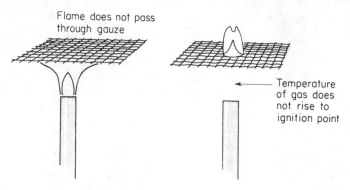

Flame does not pass
through gauze

Temperature
of gas does
not rise to
ignition point

Fig. 17.3. Gauze experiment

temperature of the gas in contact with the underneath surface of the gauze is not
raised to its ignition point. The flame will pass through the gauze only if it should
become red hot. As we shall now show, this experiment illustrates the principle of
the Davy safety lamp.

The Davy safety lamp

The enormously increased output of coal for industrial purposes towards the end of
the eighteenth century brought with it a corresponding increase in the number of
fatal mine accidents. A flammable gas called methane is often found in coal-mines.

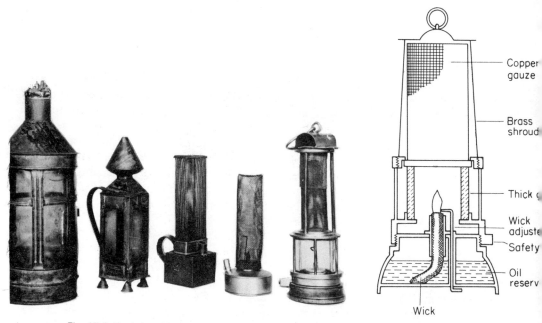

Fig. 17.4. Evolution of the Davy lamp

Copper
gauze

Brass
shroud

Thick

Wick
adjuste

Safety

Oil
reserv

Wick

Fig. 17.5. Davy safety lamp

This, when mixed with the air of the mine, exploded when it came into contact with the
naked flames of the candles which, at that time, were used for illumination.

In 1813 a society was formed to study methods for preventing these explosions,
and Sir Humphry Davy was approached for advice. Davy investigated the problem

and eventually found a remedy in the *safety lamp* Fig. 17.4. In its original form this consisted of a simple oil burner completely surrounded by a cylinder of wire gauze. The gauze, however, threw undesirable shadows, and later a thick cylindrical glass window was added, still keeping the gauze above, but encased in a brass shroud to protect it from damage (Fig. 17.5).

Should the atmosphere surrounding the lamp contain methane, its presence will be indicated by the flame becoming surrounded by a bluish haze. This is caused by the methane burning when it comes into contact with the flame. The flame cannot extend beyond the gauze and cause an explosion, since the wires of the gauze rapidly conduct the heat away. The temperature of the gauze, therefore, never rises to the ignition point of the gas–air mixture in the mine.

Although Davy lamps have long been replaced by electric lamps they will always be remembered as an important application of science in the interests of human safety.

Comparison of thermal conductivities

Rods of different materials but having the same length and diameter are passed through corks inserted in holes in the side of a metal trough (Fig. 17.6). The rods are first dipped into molten paraffin wax and withdrawn to allow a coating of wax to

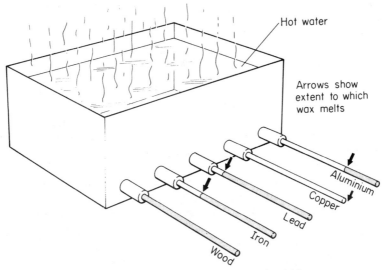

Fig. 17.6. Comparison of thermal conductivities

solidify on them. Boiling water is then poured into the trough so that the ends of the rods are all heated to the same temperature. After some minutes have elapsed it is noticed that wax has melted to different distances along the rods, indicating differences in their thermal conductivities.

Comparison of wood and copper

The difference between the conductivity of a poor conductor such as wood and a good conductor such as copper can be shown by turning down one end of a wooden rod in the lathe so that it will just fit into a copper tube of the same diameter. The two materials together will then form a single rod of uniform diameter. A piece of paper is stuck round the joint so that it covers the wood and copper equally (Fig. 17.7).

On passing the rod several times through a bunsen flame, the paper chars where it covers the wood, but remains unharmed where it covers the copper. Copper conducts the heat from the paper so rapidly that it remains comparatively cool. On the other hand, the wood conducts the heat very slowly from the paper, with the result that its temperature quickly rises to the point at which it begins to burn.

Conduction of heat through liquids

All ordinary liquids, with the exception of mercury and other metals in the liquid state, are poor conductors. Nevertheless, heat can be transmitted very quickly through liquids by a different process called *convection*. This will be described later.

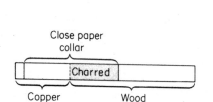

Fig. 17.7. Relative conductivity of copper and wood

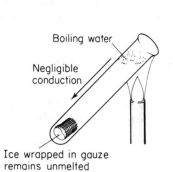

Fig. 17.8. Showing that water is a bad conductor of heat

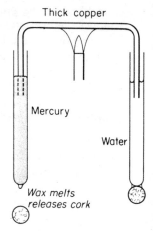

Fig. 17.9. Mercury conducts heat better than water

To prevent convection, and to confine the process of heat transmission to conduction only, it is necessary to heat a liquid at the top. Thus we may show water to be an extremely bad conductor of heat by wrapping a piece of ice in gauze to make it sink and placing it at the bottom of a test-tube nearly full of water. By holding the top of the tube in a bunsen flame, the water at the top may be boiled vigorously while the ice at the bottom remains unmelted (Fig. 17.8).

Mercury may be shown to be a better conductor than water by taking two test-tubes containing mercury and water respectively and attaching a cork to the bottom of each with melted wax (Fig. 17.9). A piece of thick copper wire bent twice at right angles is then placed with a leg in each of the two liquids. On heating the centre of the wire with a bunsen flame, heat is conducted through the metal equally into the water and mercury. In a very short time the wax on the mercury-filled tube melts and the cork falls off. Very prolonged heating is necessary before the same occurs with the water-filled tube.

It may be noted that *gases* are far worse conductors of heat than liquids.

CONVECTION

Convection in liquids

When a vessel containing a liquid is heated at the bottom a current of hot liquid moves upwards and its place is taken by a cold current moving downwards. Unlike conduction, where heat is passed on from one section of the substance to another as described on page 184, the heat is here actually carried from one place to another in the liquid by the movement of the liquid itself. This phenomenon is called *convection*. The same process occurs when a gas is heated.

Convection currents in water may be shown by filling a large spherical flask with water and dropping a single large crystal of potassium permanganate to the bottom of it through a length of glass tubing. A finger is placed over the end of the tube, which is then removed, together with the coloured water it contains. This method of introducing the crystal ensures getting it in the centre and also prevents it from colouring the water before it is required. On heating the bottom of the flask with a very small gas flame, as shown in Fig. 17.10, an upward current of coloured water will ascend from the place where the heat is applied. This coloured stream reaches the top and spreads out. After a short time it circulates down the sides of the flask, showing that a convection current has been set up.

Pattern of windmill

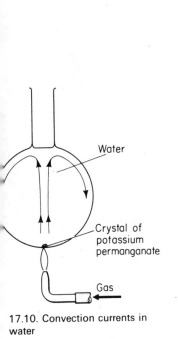

17.10. Convection currents in water

Fig. 17.11. Convection windmill

Hot upward air current from lamp

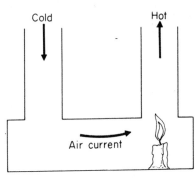

Fig. 17.12. Ventilation by convection

Explanation of convection currents

When a portion of liquid near the bottom of a vessel is heated it expands. Since its mass remains unaltered, it becomes less dense, and therefore rises. Thus a warm convection current moves upwards; for the same reason a cork rises in water or a hydrogen-filled balloon rises in air. In effect, convection is an application of Archimedes' principle. See chapter 12.

If, on the other hand, some liquid in a vessel is heated at the top, the liquid there expands and remains floating on the denser liquid beneath. No convection current is set up, and the only way in which heat can travel downwards under these conditions is by conduction.

Convection in air

The air convection current rising from an electric lamp may be shown with the aid of a small windmill. A suitable windmill may be cut with scissors from thin card or aluminium foil to the pattern of Fig. 17.11. The vanes are slightly bent and the mill, pivoted on a piece of bent wire, is held over the top of an electric lamp. When the lamp is switched on the windmill rotates in the upward hot air current. A device similar to this is often used to produce a flickering effect in domestic electric heaters of a type which are disguised to resemble glowing coal fires.

During the eighteenth century coal-mines were ventilated by sinking two shafts to the workings, known as the upcast and downcast shafts respectively. A fire was lit at the bottom of the upcast shaft, which caused the air in it to become heated and rise. Fresh air entered the downcast shaft and passed through the passages of the mine workings before it, in turn, became heated and passed out through the upcast shaft.

In this way a constant flow of fresh air was maintained through the mine. Fig. 17.12 shows a laboratory model to illustrate this method of ventilation. It consists of two wide glass tubes projecting from the top of a rectangular wooden box with a removable glass front. A short piece of candle is lit at the base of one of the tubes. When a piece of smouldering brown paper is held over the top of the other tube, the direction of the convection currents will be rendered visible by the passage of smoke through the box.

The domestic hot water supply system

The domestic hot water supply system consists of a *boiler*, a *hot water storage tank* and a *cold supply tank* interconnected by pipes arranged as shown in Fig. 17.13. When the system is working a convection current of hot water from the boiler rises up the flow pipe *A* while cold water descends to the boiler through the return pipe *B*, where

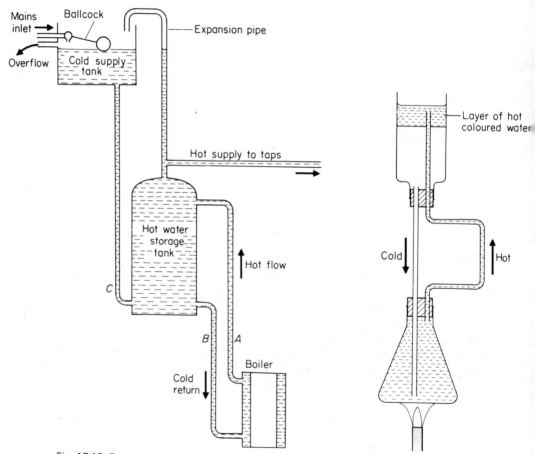

Fig. 17.13. Domestic hot water supply system

Fig. 17.14. Circulation by convectic

it becomes heated in turn. In this way a circulation is set up, with the result that the hot water storage tank gradually becomes filled with hot water from the top downwards. It is important to notice that the flow pipe *A* leaves the boiler at the top and enters the top of the hot tank, while the return pipe *B* connects the bottom of the hot tank to the bottom of the boiler.

Hot water for use in kitchen and bathroom is taken from a pipe leading from the top of the hot tank. When hot water is run off an equal volume of water from the cold supply tank enters the hot storage tank at the bottom through the pipe *C*. The whole system is thus kept constantly full of water and no air can enter. The water-level in the cold tank is maintained by a supply from the mains which enters through a ball-cock.

An expansion pipe rises from the top of the hot tank and is bent twice at right angles so that its end is over the cold tank. This is a safety precaution; if the fire is allowed to burn so fiercely that the water boils, steam and hot water are discharged harmlessly into the cold tank and no damage results. The expansion pipe also permits the escape of dissolved air which comes out of the water when it is heated, as otherwise this might cause troublesome air-locks in the pipes.

Fig. 17.14 shows a glass model which demonstrates the convection currents described above. To begin with, the water in the lower flask is coloured with blue ink, while the tubes and upper vessel are full of colourless water. When the flask is steadily heated with a bunsen-burner a hot convection current of coloured water rises up the bent tube. After a short time a visible layer of coloured hot water collects at the top of the upper vessel. Eventually this layer increases in depth until the upper vessel is entirely filled with coloured hot water. Finally, coloured warm water will be seen descending the straight tube.

Land and sea breezes

At places on the coast in summer-time it is noticeable that a breeze generally blows from the sea during the day, while at night the direction of the wind is reversed. These breezes are local convection currents.

During the day the land is heated by the sun to a higher temperature than the sea. There are two reasons for this. First, water has a higher specific heat capacity than earth; secondly, the surface of the sea is in constant motion, leading to mixing of the

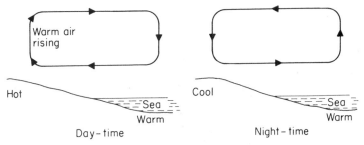

Fig. 17.15. Land and sea breezes

warm surface water with the cooler layers below. Air over the land is therefore heated, expands and rises while cooler air blows in from the sea to take its place. The circulation is completed by a wind in the upper atmosphere blowing in the opposite direction (Fig. 17.15).

At night the land is no longer heated by the sun and cools very rapidly. On the other hand, the sea shows practically no change in temperature, since it has been heated to a greater depth than the land, and consequently acts as a larger reservoir of heat. By comparison the sea is now warmer than the land, so that the air convection current is reversed.

RADIATION

Both conduction and convection are ways of conveying heat from one place to another which require the presence of a material substance, either solid, liquid, or gas.

There is a third process of heat transmission which does not require a material medium. This is called *radiation*, and is the means by which energy travels from the sun across the empty space beyond the earth's atmosphere.

This radiant energy consists of invisible *electromagnetic* waves which are able to pass through a vacuum. The waves are partly reflected and partly absorbed by objects on which they fall. The part which is absorbed becomes transferred to internal energy. Radiation which has passed through a vacuum can be easily felt by holding the hand near to a vacuum-filled electric lamp when the current is switched on. A more general discussion on the subject of electromagnetic waves will be found at the end of chapter 26.

The detection of radiation. The thermopile

Radiant energy from a hot body may be detected by converting it into electric energy. A simple experiment serves to show how this is done. A copper and an iron wire are twisted together to form a junction, while the free ends of the wires are connected to the terminals of a sensitive galvanometer (Fig. 17.16). On warming the

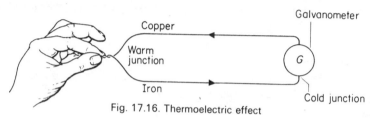

Fig. 17.16. Thermoelectric effect

junction an electric current is produced in the circuit and the galvanometer gives a deflection. This is called the *thermoelectric* effect.

Bismuth and antimony are two metals which show the thermoelectric effect in a marked degree, and they are used for the detection of radiation in an instrument

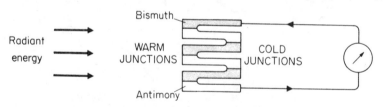

Fig. 17.17. Thermopile construction

called a *thermopile*. In order to magnify the effect, as many as 64 pairs of antimony and bismuth bars are joined in series, to give 64 junctions on which the radiation is allowed to fall. The bars are placed side by side, insulated from one another by paper, and their ends are soldered together. The whole is mounted in plaster of paris set in a short brass cylinder, and provided with two terminals connected to the free ends of the two end bars (Fig. 17.17).

To compare the radiation from different surfaces

The rate at which a body radiates energy depends on its temperature and the nature and area of its surface. It is found that, for a given temperature, a body radiates most energy when its surface is dull black and least when its surface is highly polished.

A comparison of the radiating powers of different surfaces was first described by Sir John Leslie of Edinburgh early in the nineteenth century. Leslie used a hollow copper cube, each side of which had a different surface. One may be highly polished metal, another coated with lamp black by holding it in the flame of a candle, while the remaining two surfaces may be painted in a light and dark colour respectively (Fig. 17.18).

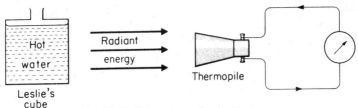

Fig. 17.18. Comparison of radiating powers

The cube is filled with hot water and a thermopile placed at the same distance from each face in turn. In each case the steady deflection obtained on the galvanometer is recorded. The results show that the dull black surface produces the largest, and the polished metal the smallest deflection. Of the painted surfaces, the darker one is usually better, but this is not always the case. The texture of the surface appears to be a more important factor than its colour.

Absorption of radiation by a surface

As stated earlier in this chapter, radiation falling on a surface is partly absorbed and partly reflected. The absorbing powers of a dull black and a polished surface may be compared by using two sheets of tinplate, one polished and the other painted dull black. On the reverse side of each plate, a cork is fixed by means of a little melted paraffin wax. The plates are then set up vertically, a short distance apart, with a bunsen burner midway between (Fig. 17.19). When the burner is lit, both surfaces receive equal quantities of radiation. In a very short time the wax on the dull black plate melts and the cork slides off. The polished plate, however, remains cool and the wax unmelted.

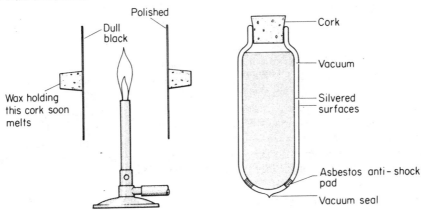

Fig. 17.19. Comparison of absorbing powers

Fig. 17.20. Vacuum flask

This experiment shows that the dull black surface is a much better absorber of radiation than the polished surface. The polished surface is therefore a good *reflector* of radiation.

The experiment should be repeated with other types of surface whose radiating powers have been previously compared by the Leslie cube experiment. In every case it is found that the better radiator is also the better absorber of radiation.

Practical applications of radiation

The investigations on radiation and absorption described above have a number of useful applications. Buildings which are whitewashed or painted in light colours keep cooler in summer, since the light surfaces reflect radiation from the sun. Many factory roofs are now aluminium-painted. The bright surface reduces the heat lost in winter, and keeps the interior cool in summer. We ourselves choose light-coloured clothing in summer for the same reason, and in very hot countries white clothing is generally the rule.

Brightly polished objects retain their internal energy for a long period. This is one reason why a silver teapot is to be preferred to others. See also Fig. 17.1.

The so-called "radiators" of a hot water central heating system do, in fact, emit most of their heat by convection. Nevertheless, in order to increase the proportion of energy radiated they are sometimes painted a dark colour.

The vacuum flask

The vacuum flask is also commonly known as a Thermos flask, which is the trade-name used by a large manufacturing firm. Originally, it was devised by Sir James Dewar for the purpose of storing liquefied gases. Liquid oxygen, for example, boils at the very low temperature of -183 °C (90 K), so that if it is placed in an ordinary flask it rapidly boils away. It is necessary therefore to keep it in a vessel through which heat cannot pass.

The vacuum flask consists of a double-walled glass vessel having a vacuum between the walls. Both walls are silvered on the vacuum side (Fig. 17.20). No heat can enter or leave the inner flask by conduction or convection across the vacuum. A certain amount of heat can be gained by the flask through radiation, but this is reduced to a minimum owing to the silvering. In addition, there will be a little heat transmitted by conduction through the thin glass walls at the neck, and through the poorly conducting cork. The sum total of this heat transfer is very small, so that a cold liquid inside remains cold for a very long period.

The vacuum flask is equally suitable for keeping liquids hot.

The greenhouse

Anyone who walks into a greenhouse, even on a day when the sunlight is rather dull, realizes how efficiently it acts as a radiation trap.

Very hot bodies such as the sun emit most of their radiation in the form of visible light and short wavelength *infrared* rays which easily pass through glass without being absorbed. (The term *wavelength* is explained on page 287). These rays are absorbed by the earth and objects inside the greenhouse which, in turn, raise the temperature of the air by conduction and convection. The warm air, being enclosed, cannot escape. The warm objects inside also radiate energy, but, owing to their comparatively low temperature the infrared rays they emit are of *long* wavelength and cannot penetrate the glass (see pages 282, 305, 308).

The global greenhouse effect

The earth's atmosphere behaves like a greenhouse. The sun's radiation easily passes through it and is absorbed by the earth's surfaces. As it warms up, the earth re-radiates energy of longer wavelengths. This becomes absorbed by water vapour, carbon dioxide and other gases in the atmosphere. The warm atmosphere, in turn, radiates back to the earth, so making it warmer than it would be otherwise.

However, over the past century the amount of carbon dioxide has increased, mostly from burning coal. Indeed, the concentration of carbon dioxide is expected to be doubled by the end of the next century. This will increase the greenhouse effect and consequently the average temperature of the earth should increase.

In all probability this will lead to shrinking of the world's ice caps with a consequent rise in mean sea level. In addition, there will almost certainly be changes in the world's wind and weather pattern. Finally, it may help to delay or ward off the possibility of another ice age.

This is necessarily a very brief account. Interested readers are referred to several books on the subject, details of which may be obtained from a good library.

QUESTIONS: 17

1. Describe briefly an experiment to illustrate each of the following:

(a) water is a bad conductor of heat,

(b) copper is a better conductor of heat than iron,

(c) convection currents in gases,

(d) a rough surface is a better emitter of radiation than a polished surface. (A.E.B.)

2. A stone floor feels very cold to bare feet in the winter, but a carpet in the same room feels comfortably warm. Why is this? (L.)

3. Draw a labelled diagram of a domestic hot water system and explain its action.

Distinguish the processes by which a room becomes warmed by a hot-water radiator. Which of these processes is usually most effective, and how does this affect the design of the radiator? (L.)

4. Explain the transfer of thermal energy by conduction and convection. Describe:

(i) a laboratory experiment which illustrates convection;

(ii) a large-scale example of convection in nature.

A beaker of hot water is placed on a bench. Describe all the ways in which it loses heat and suggest a simple way minimizing each. (C.)

5. What is meant by conduction of heat? Sketch the apparatus you would use to show that water is a poor conductor of heat.

Why are woollen materials bad conductors of heat? (L.)

6. Distinguish between conduction and radiation of heat. Describe an experiment to show that a dull black surface is a better absorber of heat than a polished one. Give two ways by which the heat lost by a body can be assisted and show how this is effected in ONE practical example. (S.)

7. Give an account of the transfer of heat by conduction. How would you show that copper is better than iron as a conductor of heat?

Describe the Davy safety lamp, and explain what happens when it is surrounded by a flammable mixture of gas and air. (O.)

8. Describe briefly how the heat insulation of the windows of a house can be improved. Give one other form of thermal insulation available for houses. (A.E.B.)

9. Two similar kettles containing equal masses of boiling water are placed on a bench; the surface of one is highly polished and the surface of the other is covered with soot. Compare their rates of cooling, giving reasons for your answer.

Give a labelled diagram of a Dewar (or Thermos) flask, and explain the principles on which its action depends. (W.)

10. An electric filament lamp with a clear glass bulb is switched on and gives a bright white light. The bulb contains a small quantity of argon (an inert gas). Give an account of the parts played by conduction, convection, and radiation in the loss of heat from the lamp filament. (C.)

11. Describe a simple experiment to show how the heat radiated from a hot object depends on the nature of the surface. Indicate the result you would expect and state two practical applications of this effect. (S.)

18. Development of the concept of energy

Heat and internal energy

In chapter 7 we talked about energy in general and noticed that heat is involved as a step in practically all the various energy transfers which lead to the production of that most useful of all forms of energy, namely electricity.

Our present concept of energy and its measurement arose out of a search for an understanding of what it was that made things hot. As this quest took more than two and a half centuries, it is well that we should spare a little time in this chapter saying something about the scientists who were responsible for it.

On page 78 we explained that, in physics, the word *heat* is the name given to energy in the process of transfer from one body to another as the result of a temperature difference between them. *Once heat has been transferred to a body or substance, it ceases to be heat and instead becomes internal molecular energy.*

In the past, and this must be remembered when reading older textbooks, the word heat has been used to mean the same thing as internal energy. Indeed, the early scientists who themselves were responsible for the development of the idea of energy used the word heat in the same way.

Early ideas on the nature of internal energy

It had long been known that nails became hot when hammered into a piece of wood. In the seventeenth century Robert Boyle explained this by saying that the hammer blows set the particles of metal into violent vibration and so concluded that the rise in temperature was simply caused by *motion*. But there the matter rested, and by the beginning of the eighteenth century a new theory became popular. Heat, as it was called, came to be considered as a weightless fluid known as *caloric*. This idea appealed to men such as the Scottish chemist, Joseph Black, who found it easier to talk about caloric rather than a vague quantity of motion. And so the caloric theory gained ground and was destined to occupy a prominent place in the study of the subject for the next 150 years.

The observations of Count Rumford

In spite of the popularity of the caloric theory during the eighteenth century, there were some who did not find it altogether satisfactory. Its main critic was an American named Benjamin Thompson, who for his services to the Elector of Bavaria was later given the title of Count Rumford.

Towards the end of the century, when he was in charge of the arsenal at Munich, Rumford noticed that a big rise in temperature was produced during the boring of brass cannon. In order to study this effect he immersed one of the cannon in a wooden box filled with water and then subjected it to the action of a blunt borer worked by horses. As the operation proceeded the temperature of the water gradually rose until after about 2½ hours it actually boiled. From this, together with observations on a number of other experiments, Rumford became convinced that such a thing as caloric did not exist. As he saw the situation, *motion* only had been supplied to the brass cannon and borer. He therefore affirmed his belief in the older idea that the rise in temperature was caused by motion.

Rumford's experiments aroused considerable interest at the time and many people went to see this strange phenomenon of water boiling without fire. But although he referred to the "heat" as being "generated by friction" and "communicated by motion", *he measured neither the friction nor the motion and failed to grasp the connection between the two, namely, the performance of work which was transferred to internal energy in the brass and water.*

Progress during the nineteenth century

During the years which followed Rumford's experiments the concepts of work and energy began to emerge in scientific thought. Moreover, the rapid development of steam power, which was then taking place, played its part in stimulating a general interest in the transfer of internal energy from steam into mechanical work.

At first the discussion was confined to the theoretical level. In England Thomas Young drew attention to the confusion which existed between force as such and the work which is done when a force is exerted through a distance. He was also the first to use the word *energy* to refer to the capacity of a moving body for doing work. This was in 1807. The German physicist, Robert Mayer, calculated the work done when a gas is compressed and assumed that the whole of this work became transferred to internal energy which increased the temperature of the gas.

Generally speaking, the main body of scientific opinion was hostile to the new ideas. Eventually, however, during the period 1840–50 a series of experiments was carried out by James Joule which established that, when the temperature of a body was raised by doing work on it, there was an exact equivalence between the work done and the internal energy acquired.

Joule's experiments

James Joule was the son of a Salford brewer and became interested in the subject of energy at a very early age. He spent many years making careful experiments to show that mechanical and electrical energy could be transferred to internal energy in water which produced a rise in temperature.

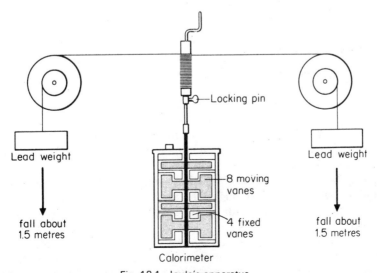

Fig. 18.1. Joule's apparatus

His best-known apparatus is illustrated in Fig. 18.1. Two heavy lead weights were attached to string wound round the spindle of an eight-vaned paddle wheel which rotated inside a copper vessel containing water. Inside the vessel were four fixed vanes which prevented the water from being carried round bodily. The work done by the paddle against the resistance offered by the water was transferred to internal energy in the water and copper. By means of the handle the weights could be wound up and allowed to fall several times. The internal energy acquired by the water and copper was calculated from their masses and temperature rise and Joule was able to show that this was exactly equivalent to the transfer of potential energy from the falling weights.

Potential energy transferred to internal energy in a waterfall

From the results of his experiments, Joule came to the conclusion that the water at the bottom of a waterfall ought to be slightly warmer than that at the top.

At the top of the fall the water possesses potential energy, which becomes transferred to kinetic energy as it descends. Part of this kinetic energy becomes transferred to internal molecular energy when the motion of the water is arrested at the bottom. Joule decided to carry out a test on a particular waterfall in Switzerland. Unfortunately, even with a very sensitive thermometer he failed to detect the small temperature rise expected, since the water was too broken by spray.

Measurement of heat and internal energy

We shall now deal with some of the calculations involved in heat and internal energy measurements. Since heat is a form of energy transfer, it is measured in joules the same as any other kind of energy. Now, the easiest way to produce heat, and at the same time be able to measure it accurately in joules, is by an electrical method, but the explanation of this must wait until after we have studied electricity and its units (chapters 41 and 42).

In the meantime we shall explain some of the terms used in our calculations.

Heat capacity

The heat capacity of a body of any kind is defined as the heat required to raise its temperature by 1 K.

The SI unit of heat capacity is therefore the joule per kelvin (J/K).

For the meaning and use of the kelvin in SI units and definitions see page 162.

Specific heat capacity

If we take equal masses of water and oil and warm them in separate beakers by the same gas burner it is found that the oil temperature may rise by 10 K in 3 min, but the water may rise by only 5 K. Since the rate of supply of heat is the same in both cases, it is clear that oil has a smaller heat capacity than an equal mass of water. When comparing the heat capacities of various substances we talk of their *specific heat capacities*. The word *specific* is used in physics when we refer to unit quantity of a physical property.

The specific heat capacity of a substance is defined as the heat required to raise the temperature of unit mass of it through 1 K (Symbol used $= c$.)

It follows that the unit of specific heat capacity in the SI system is the joule per kilogram kelvin (J/kg K).

In the following table it will be seen that water has the unusually high specific heat capacity of 4200 J/kg K. Very few substances have a higher value than this, the most notable being hydrogen at constant volume, and mixtures of certain alcohols with water. Incidentally, the high specific heat capacity of hydrogen, coupled with its high thermal conductivity renders it a very efficient cooling gas for enclosed electric generators.

Table of specific heat capacities in J/kg K

Aluminium	900	Lead	130
Brass	380	Mercury	140
Copper	400	Methylated spirit	2400
Glass (ordinary)	670	Sea-water	3900
Ice	2100	Water	4200
Iron	460	Zinc	380

Methods for measuring specific heat capacities are fully dealt with in chapter 42.

Heat calculations

1. *How many joules of heat are given out when a piece of iron of mass 50 g and specific heat capacity 460 J/kg K, cools from 80 C to 20 C?*

To say that the specific heat capacity of iron is 460 J/kg K means that 1 kg of iron gives out or takes in, as the case may be, 460 J when its temperature changes by 1 K.

It follows that 50 g (= 0.05 kg) of iron in cooling through 1 K gives out,

$$0.05 \times 460 \text{ J}$$

Hence 50 g of iron in cooling from 80 °C to 20 °C i.e., through (80 − 20) K gives out

$$0.05 \times 460 \times (80 - 20) = \underline{1380 \text{ J}}$$

If we write the above expression in words we get a useful formula: heat energy given out (or received) .

$$= \text{mass} \times \text{specific heat capacity} \times \text{temperature change}$$

Putting this equation into symbols we have,

$$\text{heat energy in joules} = mc(\theta_2 - \theta_1)$$

where m = mass in kg
c = specific heat capacity in J/kg K
θ_2 = higher temperature in °C
θ_1 = lower temperature in °C

so that $(\theta_2 - \theta_1)$ = change in temperature in kelvins.

Note. The Greek letter θ (theta) is the accepted symbol for temperature in °C. If t is used instead of θ it may lead to confusion in equations where t is used to represent time in seconds.

2. *What is the final temperature of the mixture if 100 g of water at 70 C is added to 200 g of cold water at 10 °C and well stirred? (Neglect heat absorbed by the container.)*

From the table above we note that the specific heat capacity of water is 4200 J/kg K.

Heat given out by hot water = heat received by cold water.

Let the final temperature of the mixture = θ in °C
Then, change in temperature of hot water = $(70 - \theta)$ in kelvins
and, change in temperature of cold water = $(\theta - 10)$ in kelvins

Thus, using the formula $mc(\theta_2 - \theta_1)$ explained in the previous example, we substitute values in the equation.

Heat given out by hot water = heat received by cold water which gives (remembering to convert g to kg),

$$0.1 \times 4200 \times (70 - \theta) = 0.2 \times 4200 \times (\theta - 10)$$

dividing both sides, by 4200

$$7 - 0.1\,\theta = 0.2\,\theta - 2$$

rearranging

$$9 = 0.3\,\theta$$

whence

$$\theta = \underline{30\,°C}$$

$$= \text{final temperature of mixture}$$

3. *The temperature of a piece of copper of mass 250 g is raised to 100 °C and it is then transferred to a well-lagged aluminium can of mass 10.0 g containing 120 g of methylated spirit at 10.0 °C. Calculate the final steady temperature after the spirit has been well stirred. Neglect the heat capacity of the stirrer and any losses from evaporation and use the table of specific heat capacities on page 199 for any data required.*

Let the final steady temperature $= \theta$

The copper cools from 100 °C to θ
The aluminium and spirit both warm up from 10.0 °C to θ
Proceeding as in the previous examples, remembering to work in kg,

Heat in J given out by copper $= 0.25 \times\ \ 400 \times (100 - \theta)$
Heat in J received by aluminium $= 0.01 \times\ \ 900 \times (\theta - 10)$
Heat in J received by spirit $= 0.12 \times 2400 \times (\theta - 10)$

$$\text{Heat given out} = \text{heat received}$$

therefore

$$100 \times (100 - \theta) = 9 \times (\theta - 10) + 288 \times (\theta - 10)$$

or,

$$10\,000 - 100\,\theta = 297\,\theta - 2970$$

Rearranging,

$$12\,970 = 397\,\theta$$

whence

$$\theta = \frac{12\,970}{397} = \underline{32.7\,°C}$$

$$= \text{final steady temperature.}$$

4. *The temperature of 500 g of a certain metal is raised to 100 °C and it is then placed in 200 g of water at 15 °C. If the final steady temperature rises to 21 °C, calculate the specific heat capacity of the metal.*

Let the specific heat capacity of the metal $= c$ in J/kg K
Heat in J given out by metal $= 0.5 \times c \times (100 - 21)$
Heat in J received by water $= 0.2 \times 4200 \times (21 - 15)$
Equating, $0.5 \times c \times 79 = 0.2 \times 4200 \times 6$

whence

$$c = \frac{5040}{39.5} = \underline{128\ \text{J/kg K}}$$

From the table of specific heat capacities we infer that the metal is probably lead.

To measure the specific heat capacity of lead by the shot-tube method

Whilst, as its name implies, the term specific heat capacity is used in internal energy calculations where the energy transfer is by heat, it is also used when the energy transfer is by mechanical means, i.e., by doing work on a body.

The experiment we shall now describe illustrates the transfer of potential energy to internal molecular energy, and may be used to find a rough value for the specific heat capacity of lead.

A cardboard tube about a metre long contains a quantity, m in kg, of lead shot and is fitted with corks at both ends. One of the corks has a small hole plugged with a wooden peg. This allows for the insertion of a thermometer to take the temperature of the shot (Fig. 18.2).

If the tube is inverted, the shot falls through a distance h in m. In so doing its potential energy, *mgh* in J, becomes transferred to kinetic energy which, in turn, becomes internal molecular energy when the shot is brought to rest.

The temperature, θ_1, of the shot is taken immediately before the experiment starts and the tube is then inverted 100 times in order to obtain a measurable temperature rise. The final temperature, θ_2, of the shot is noted.

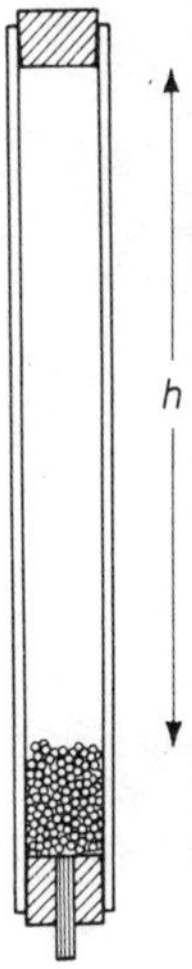

Let the specific heat capacity of lead = c
$$\text{and } g = 10 \text{ m/s}^2$$

then the specific heat capacity may be calculated as follows:

Force in newtons on lead in falling	$= mg$
Total distance fallen in metres	$= 100 \times h$
Work done in joules	$= 100\,mgh$
Rise in temperature of lead	$= (\theta_2 - \theta_1) = \theta$
Equivalent heat input	$= mc\theta$
Hence,	$mc\theta = 100\,mgh$
therefore	$c = \dfrac{100\,gh}{\theta} \text{ (in J/kg K)}$

Note that the mass of the lead disappears from the final calculation, and therefore *the mass of the shot does not have to be known*. This experiment is worth trying as a matter of interest, but the reader is warned not to expect a very accurate result.

There are two main sources of error. First, the whole of the shot does not fall through *h*, as some of it inevitably begins to slide before the tube reaches a vertical position. Secondly, the shot cools in falling through the air. The air thus gains some of the internal energy produced, and this, in turn, is lost to the cardboard.

Plug

Fig. 18.2. Shot tube apparatus

Temperature rise resulting from compression

Anyone who has pumped up a bicycle tyre knows that the lower part of the pump barrel may become quite warm. Erroneously, this is often attributed to work done against friction. On reflection, however, one must come to the conclusion that the friction of an oiled plunger against the smooth barrel wall is far too small to do any appreciable amount of work. The increase in internal energy which raises the temperature comes, of course, from the work done in compressing the air.

Conversely, if compressed air or any other gas is allowed to expand it performs external work, and the energy required comes from the internal energy of the gas itself. Consequently, the gas cools. See also page 181.

Importance of Joule's work

When an account of Joule's work became known in the middle of the nineteenth century it aroused but little interest, as the concept of work and energy was new to science. At the time it was not generally realized that Joule's experiments provided the first reliable experimental evidence for the truth of the principle of the conservation of energy.

This principle was put forward by the German physicist, Hermann von Helmholtz, in a book published in 1847, but it had earlier been accepted by other far-seeing scientists, particularly Sadi Carnot, Robert Mayer, and Sir William Grove.

The conservation of energy has already been discussed in chapter 7. Joule's experiments had shown that internal molecular energy could be put into a substance by mechanical work and that there was *an exact equivalence between these two forms of energy*. Later it was demonstrated that the same exact relationship existed between other forms of energy, for example, electric energy, chemical energy, and heat. One can readily appreciate why Joule's memory has been honoured by giving his name to the SI unit of energy.

Once the principle of conservation of energy had thus been established, the way became open for great advances in science. It formed the basis of a new branch of the study of heat and energy known as *thermodynamics*. Calculations could now be made regarding certain problems in physics with a certainty that the answer would be correct. In the field of applied physics the same can be said with regard to calculations on the design of steam turbines, internal combustion engines, rockets and jet engines, electric motors, generators, and power installations.

The first law of thermodynamics

The work of Joule and others may be summed up in a statement known as the first law of thermodynamics:

Change in internal energy of any system = heat inflow + work done on the system
or, in symbols $U = Q + W$ (see page 78)

This is simply a way of stating the law of conservation of energy as applied to heat and mechanical work changes. The equation must, of course, be treated algebraically, i.e., heat *outflow* and work done *by* the system would be written with minus signs.

The word *system* in the above equation refers to any body or device which is involved with heat, work, and energy changes. The steam turbine discussed on page 79 can be taken as a practical example of such.

Before concluding this chapter it will be useful to add a final comment. Work done on a body may or may not change its internal energy. For example, a frictional force acting through a distance on the body will do work which is transferred to internal molecular energy and this will have the same effect as heat transfer, i.e., it will produce a rise in temperature.

By contrast, the work done by a force in lifting a body above the earth's surface will be transferred to gravitational potential energy of the body as a whole without affecting the internal energy of its molecules.

If, however, the body is now allowed to fall, its potential energy will be transferred to internal energy in the body itself; to the ground at the place of impact; to the air through which it fell, and lastly to a little sound energy which is mechanical molecular energy transmitted by longitudinal waves (page 314). Finally, it must be noted that while the bulk of the energy will be transferred to internal energy in the body, the sum total of this together with the other parts mentioned will be conserved and equal to the original potential energy.

QUESTIONS: 18

1. Where necessary, use the values of specific heat capacity given on page 199 to calculate the following.

(a) the heat given out when 50 g of iron cools from 45 °C to 15 °C;

(b) the specific heat capacity of gold if 108 J of heat raise the temperature of 9 g of the metal from 0 °C to 100 °C;

(c) the heat required to raise the temperature of 1000 kg of sea-water through 40 K.

2. A piece of lead of mass 500 g and at air temperature falls from a height of 25 m. What is:

(a) its initial potential energy;

(b) its kinetic energy on reaching the ground? Assume $g = 10$ m/s².

Assuming that all the energy becomes transferred to internal energy in the lead when it strikes the ground, calculate the rise in temperature of the lead if its specific heat capacity is 130 J/kg K.

State the energy changes which occur from the moment the lead strikes the ground until it has cooled to air temperature again.

3. A waterfall is 100 m high and the difference in temperature between the water at the top

Development of the concept of energy 203

and that at the bottom is 0.24 K. Obtain a value for the specific heat capacity of water in J/kg K explaining the steps in your calculations. Mention any assumptions you make. ($g = 10$ m/s².)

4. Explain the following:
    (i) when the brakes of a moving car are applied for an appreciable time, they get hot;
    (ii) when the tyre of a car is pumped up, the pump gets warm. (C.)

5. A car of mass 1000 kg travelling at 72 km/h is brought to rest by applying the brakes. Assuming that the kinetic energy of the car becomes transferred to internal energy in four steel brake drums of equal mass, find the rise in temperature of the drums if their total mass is 20 kg, the specific heat capacity of steel is 450 J/kg K, and the work done is equal on all four drums.

6. Some hot water was added to three times its mass of water at 10 °C and the resulting temperature was 20 °C. What was the temperature of the hot water? (S.)

7. A bath contains 100 kg of water at 60 °C. Hot and cold taps are then turned on to deliver 20 kg per minute each at temperatures of 70 °C and 10 °C respectively. How long will it be before the temperature in the bath has dropped to 45 °C? Assume complete mixing of the water and ignore heat losses.
(O.C.)

8. The temperature of a brass cylinder of mass 100 g was raised to 100 °C and transferred to a thin aluminium can of negligible heat capacity containing 150 g of paraffin at 11 °C. If the final steady temperature after stirring was 20 °C, calculate the specific heat capacity of paraffin. (Neglect heat losses, and assume specific heat capacity of brass = 380 J/g K.)

9. A piece of copper of mass 40 g at 200 °C is placed in a copper calorimeter of mass 60 g containing 50 g of water at 10 °C. Ignoring heat losses, what will be the final steady temperature after stirring? (Specific heat capacity of copper = 400 J/kg K.)

10. Explain why the bit of an electric drill becomes hot during use. (S.)

11. In an experiment, a cardboard tube closed at both ends and containing a quantity of lead shot was inverted a number of times. The following readings were taken:

Length of tube	= 1.10 m
Mass of lead shot	= 50 g
Initial temperature of lead shot	= 15 °C
Final temperature of lead shot	= 22 °C
Room temperature	= 16 °C
Number of inversions of tube	= 120

Calculate the specific heat capacity of the lead shot. State, giving a reason for each answer:
    (i) which of the experimental readings taken did not affect the experiment;
    (ii) whether you would expect the experimental result to be too high or too low. ($g = 10$ m/s².) (A.E.B.)

12. The temperature of a gas contained in a cylinder increases when it is compressed. Use the kinetic theory to explain this observation. (W.A.E.C.)

19. Latent heat

Latent heat of vaporization

When a kettle is put on to boil the temperature of the water steadily rises until it reaches 100 °C. At this temperature it starts to boil, that is to say bubbles of vapour form at the bottom and rise to the surface, where they burst and escape as steam.

Once the water has begun to boil, the temperature remains constant at 100 °C. But at the same time, heat is being steadily absorbed by the water from the gas flame or heating element. This heat, which is going into the water but not increasing its temperature, is the energy needed to convert the water from the liquid state to the vapour state.

Experiment shows that 2 260 000 J are required to convert 1 kg of water at its boiling-point to steam at the same temperature. This is known as the *specific latent heat of steam*. "Latent" means hidden or concealed. This extra heat goes into the vapour but does not indicate its presence by producing a rise in temperature.

When the steam condenses to form water the latent heat is given out. This is one reason why a scald from steam does more harm than one from boiling water. Other liquids besides water absorb latent heat when they turn into vapour. For example, 860 000 J are required to convert 1 kg of alcohol at its boiling-point to vapour at the same temperature. These quantities of heat are called the *specific latent heats of vaporization*.

The specific latent heat of vaporization of a substance is the quantity of heat required to change unit mass of the substance from the liquid to the vapour state without change of temperature. (Symbol = l.)

The SI unit of specific latent heat of vaporization is the **joule per kilogram (J/kg).** However, in order to avoid having to write very large numbers the alternative units kJ/kg or MJ/kg may be used instead.

$$1 \text{ kJ} = 1\,000 \text{ J}$$
$$1 \text{ MJ} = 1\,000\,000 \text{ J}$$

so we may express the specific latent heat of vaporization of water as 2260 kJ/kg or 2.26 MJ/kg.

Latent heat of fusion

Just as latent heat is taken in when water changes to vapour at the same temperature, so the same thing occurs when ice melts to form water. But in this case the latent heat is not so great. It requires only 336 000 J to convert 1 kg of ice at 0 °C to water at the same temperature. Likewise, when water at 0 °C freezes into ice, the same quantity of heat is given out for every 1 kg of ice formed. This is called the *specific latent heat of ice*.

As already mentioned, the phenomenon of latent heat is not confined to water alone. Other substances also absorb latent heat when they melt; conversely, they give out latent heat on solidifying. This heat is called *latent heat of fusion*.

The specific latent heat of fusion of a substance is the quantity of heat required to

convert unit mass of the substance from the solid to the liquid state without change of temperature. (Symbol $= l$.)

The same units, J/kg, or alternatively kJ/kg or MJ/kg, are used for fusion as for vaporization.

The measurement of specific latent heat is best done by electrical methods which are fully dealt with in chapter 42.

Latent heat calculations

Fig. 19.1 is a self-explanatory illustration of the heat required at various stages when 2 g of ice at $-6\,°C$ are completely converted into steam at $100\,°C$.

The following data are used:

Specific heat capacity of ice	$=$	$2\ 100$ J/kg K $(= 2.1$ J/g K$)$
Specific heat capacity of water	$=$	$4\ 200$ J/kg K $(= 4.2$ J/g K$)$
Specific latent heat of ice	$=$	$336\ 000$ J/kg $\quad(= 336$ J/g$)$
Specific latent heat of steam	$= 2\ 260\ 000$ J/kg	$\quad(= 2\ 260$ J/g$)$

To measure melting point from a cooling curve

The latent heat given out when a molten substance freezes to the solid state may be shown by the following experiment with naphthalene. Naphthalene is a white crystalline solid obtained from coal-tar. It has a pungent smell and is often used by gardeners as a soil fumigant.

A test-tube containing naphthalene is held vertically by a clamp and stand (Fig. 19.2). The naphthalene is heated gently by a very small bunsen flame until it just melts. A thermometer is inserted in the naphthalene and the heating continued until the temperature of the melted naphthalene is about $100\,°C$. The bunsen flame is then removed, and readings of the thermometer are taken at minute intervals as the tube and its contents are cooling. It is noticed that when the freezing point, or what is the same thing, the melting point, of the naphthalene is reached the temperature remains constant at $80\,°C$ until all the naphthalene has solidified. After this the temperature begins to fall again. The temperature changes are illustrated most strikingly by plotting a graph of temperature against time. The flat portion of the graph represents the time during which the naphthalene is solidifying. At this stage its temperature remains constant at $80\,°C$ although heat is steadily being lost by convection and radiation all the time. The heat lost is exactly compensated by the latent heat of fusion of the naphthalene, which is being given out during the change from the liquid to the solid state.

Other substances, for example paradichlor benzene, can be used in this experiment and their melting points found from the flat portion of the cooling curve (see also Fig. 19.3).

Cooling produced by evaporation

Some liquids have a low boiling-point, and thus change from liquid to vapour quite easily at ordinary temperatures. These are called *volatile liquids*. Methylated spirit and ether are examples.

If a little methylated spirit or eau-de-Cologne is spilt on the hand it evaporates rapidly and the hand feels very cold. To change from liquid to vapour, the spirit requires latent heat. This it obtains from the hand, which thus loses heat and cools. Water also causes the hand to become cold, but not so noticeably as methylated spirit. The spirit has a lower boiling-point than water, and so it evaporates more quickly at the temperature of the hand.

Campers are well aware that milk can be cooled more efficiently by wrapping the

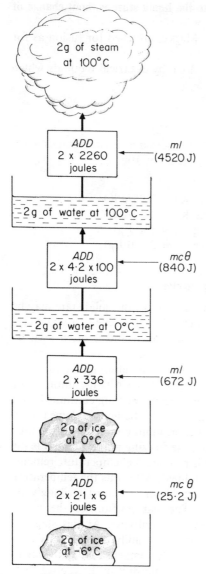

2g of steam
at 100°C

ADD
2 x 2260
joules

ml
(4520 J)

2g of water at 100°C

ADD
2 x 4·2 x 100
joules

$mc\theta$
(840 J)

2g of water at 0°C

ADD
2 x 336
joules

ml
(672 J)

2g of ice
at 0°C

ADD
2 x 2·1 x 6
joules

$mc\theta$
(25·2 J)

2g of ice
at −6°C

Fig. 19.1. Heat calculations

Fig. 19.3. *Cooling curve technique in the ironfounding industry.* As part of routine quality control procedure, molten cast-iron samples are poured into a crucible and allowed to cool. Temperature/time data are recorded automatically. The thermometer used here is a platinum/platinum-rhodium thermocouple (see page 153). The shape of the cooling curve provides information on the composition and quality of the metal

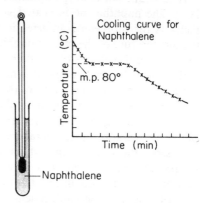

Temperature (°C)

Cooling curve for Naphthalene

m.p. 80°

Time (min)

Naphthalene

Fig. 19.2. Melting point determination

bottle in a wet cloth than by standing it in a bucket of cold water. If the rate of evaporation can be speeded up by placing the wet bottle in a draught, so much the better.

Perspiration is the body's method of maintaining a constant temperature. When perspiring heavily after exercise it is unwise to stand about in a draught, or overcooling may result from evaporation. The resulting chill may lower the resistance of the body to infection.

Dogs, who do not perspire from the skin, hang out their tongues during hot weather in order to achieve a cooling effect.

To make ice by the evaporation of ether

A beaker about one-third full of ether is stood in a small pool of water on a flat piece of wood (Fig. 19.4). A current of air is then bubbled through the ether by means of a rubber tube attached to bellows. The ether evaporates into the bubbles, and the vapour is carried quickly away as the bubbles rise to the surface and burst, thus increasing the rate of evaporation. The rapid change from the liquid to the vapour state requires latent heat. This comes from the internal energy of the liquid ether

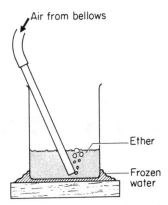

Fig. 19.4. Cooling by evaporation

itself, with the result that it soon cools well below 0 °C. At the same time heat becomes conducted through the walls of the beaker from the pool of water below it, and eventually the water cools to 0 °C. After this it begins to lose latent heat, and freezes.

Cooling by evaporation explained by the kinetic theory

The molecules of a liquid have an average kinetic energy which increases with temperature.

Molecules near the surface which happen to be moving faster than average can escape from the attraction of their neighbours and escape out of the liquid. Some of these may collide with other molecules above the liquid and so bounce back into it. But many others may escape altogether and their escape will be assisted if a current of air is passed over or through the liquid. Bubbling air through a liquid also increases the rate of evaporation by increasing the surface area from which molecules may escape.

In this way the liquid loses its most energetic molecules while the less energetic ones are left behind. The average kinetic energy of the remaining molecules is therefore reduced and this results in a fall in temperature.

Fusion and vaporization in relation to the kinetic theory of matter

Fusion
We explained on page 138 that the molecules of solids vibrate to and fro alternately attracting and repelling one another. Their total energy can be looked on as consisting of two parts: kinetic energy which depends on the temperature; and potential energy which depends on the force between the molecules and their distance apart.

When a substance changes from the solid to the liquid state the molecules have a wider range of movement than in the solid, thus going into extra close and extra distant positions. Their potential energy is therefore increased and the additional energy required is the latent heat of fusion.

Vaporization when heat is supplied
When vaporization occurs the vapour occupies a much larger volume than the liquid so that energy is required to separate the molecules against their mutual attractions

(internal work). In addition extra energy is required to enable the vapour to expand against the atmospheric pressure (external work). The heat required to provide the sum of internal and external work without change of temperature is the latent heat of vaporization.

The refrigerator

Cooling in a domestic refrigerator takes place when a volatile liquid, Freon,* evaporates inside a copper coil surrounding the freezing box (Fig. 19.5). As fast as the vapour is formed it is removed by an electric pump, and so under the reduced pressure the liquid evaporates rapidly and may even boil. The necessary latent heat

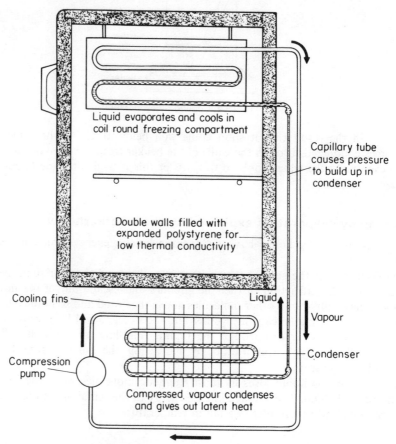

Fig. 19.5. Domestic electric refrigerator

of vaporization is provided at the expense of the liquid's own internal energy. Consequently the liquid cools. An experiment to demonstrate the cooling which occurs when a liquid boils under reduced pressure is described on page 215.

The vapour which has been pumped off passes into a second coil (the condenser) outside the cabinet where it is compressed by the pump and condenses back to liquid. Here latent heat is given out, and to enable this heat to be dissipated quickly the condensing coil may be fitted with copper fins. Heat is removed by conduction into the fins and thence by convection and radiation to the surroundings.

From the condensing coil the liquid is passed back into the evaporator coil round the freezing box. In this manner a continuous circulation of liquid and vapour is set up. The rate of vaporization and the consequent degree of cooling is controlled by a

* Freon is a collective term for suitable refrigerants, an example being dichlorodifluoromethane (boiling-point about −30 °C or 243 K).

thermostat switch (not shown in the diagram), which switches the pump motor on and off at intervals. The thermostat is adjustable and is provided with a dial which may be set to give the desired low temperature inside the cabinet.

A different type of refrigerator is also in common use which, instead of a pump, employs a gas flame or electric heater to provide the energy necessary to maintain the circulation of liquid and vapour. The method of producing the circulation is somewhat more complex than that described above. Nevertheless the basic principle, namely, cooling by vaporization under reduced pressure, remains the same.

Change of volume on solidification

When water freezes to form ice expansion occurs and the ice takes up a bigger volume than the water. For this reason, water pipes are liable to burst during very cold weather, although the leaks do not occur until a thaw sets in.

Sometimes the expansion of a substance on solidification serves a useful purpose. Molten type-metal, for example, expands very slightly when it solidifies, and so takes up a sharp impression of the mould.

Most substances, however, contract in volume when they solidify. Paraffin wax is a typical example. When some molten paraffin wax is allowed to solidify in a test-tube the shrinkage in volume is shown by a deep cleft in the surface (Fig. 19.6).

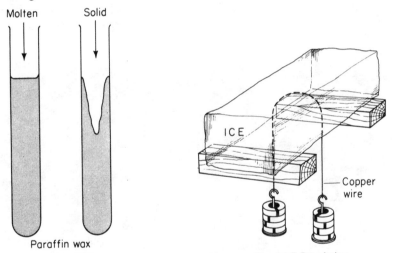

Fig. 19.6. Change in volume on solidification

Fig. 19.7 Regelation

Effect of pressure on melting point. Regelation

If a substance expands on solidifying, then the application of pressure lowers the melting point. Conversely, substances which contract in volume on solidifying have their melting points raised by pressure. Thus, the freezing point of water is lowered by just over 0.007 K per atmosphere increase in pressure, while the freezing point of paraffin increases by about 0.04 K per atmosphere.

The experiment illustrated in Fig. 19.7 shows the effect of pressure on the melting point of ice in a rather striking manner. A block of ice rests on two supports, and a thin copper wire with heavy weights at each end is hung over it. After an hour, more or less, depending on the size of the block, the wire cuts right through it and falls to the floor, leaving the ice still in a solid block. This phenomenon is called *regelation* (= refreezing).

Several factors are involved here. The pressure of the wire lowers the melting point of the ice in contact with it, and so the ice melts and flows above the wire. The latent heat required for the melting comes, in the first instance, from the copper wire. As soon as the water passes above the wire it is no longer under pressure and therefore refreezes. In so doing it gives out latent heat, and *this heat is conducted down through the wire to provide heat for further melting of the ice beneath.*

It must be realized that rapid conduction of heat down through the copper is an important factor in the process. An iron wire of smaller thermal conductivity cuts through much more slowly. A thin string of very low conductivity will not pass through at all.

Regelation is a factor in the making of snowballs. Compression of the snow by hand causes slight melting of the ice crystals, and when the pressure is removed refreezing occurs and binds the snow together. In very cold weather the pressure exerted is insufficient to melt the snow, and so it fails to bind (see also Fig. 19.8).

Why ice is slippery

The ease with which a skater glides over the ice depends on the formation of a thin film of water between the blade of the skate and the ice. At one time this was generally believed to be caused by melting under pressure. However, a simple calculation shows that the pressure exerted by a skater is about 1000 kPa and this would lower the melting-point by less than 0.1 K. Yet skating is possible even when the temperature is several degrees below zero.

It was first pointed out by F. P. Bowden, and later by J. Fremlin that the water film in this case is more likely to be brought about by the work done against friction. This work becomes transferred to internal energy and the ice melts in consequence.

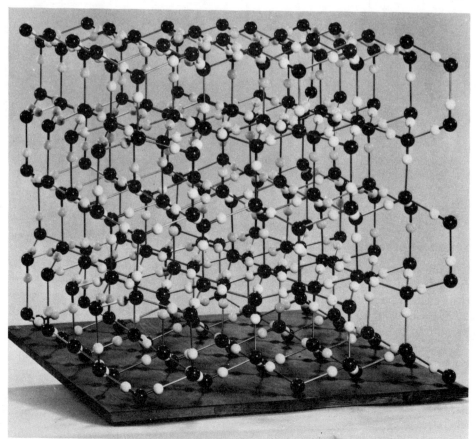

Fig. 19.8. *Model showing arrangement of atoms in a hexagonal ice crystal.* This model was constructed at the National Physical Laboratory from theoretical and experimental studies designed to give information on the forces between atoms in ice and so lead to a better understanding of the behaviour and properties of ice and liquid water

Worked example

Dry steam is passed into a well-lagged copper can of mass 250 g containing 400 g of water and 50 g of ice at 0 °C.

The mixture is well stirred and the steam supply cut off when the temperature of the can and its contents reaches 20 °C. Neglecting heat losses, find the mass of steam condensed.

(Specific heat capacities: water, 4.2 J/g K; copper, 0.4 J/g K. Specific latent heats: steam, 2260 J/g; ice, 336 J/g.)

Using the principle of conservation of energy, we may say,

heat given out by steam = heat received by ice, water and can.

Let mass of steam condensed = m (in grams)

Heat in joules given out by:

Steam condensing to water at 100 °C = $m \times 2260$

Condensed steam cooling from 100 °C to 20 °C
$$= m \times 4.2 \times 80$$
$$\text{Total} = m \times 2596$$

Heat in joules received by:

Ice melting to water at 0 °C	= 50×336
Melted ice warming from 0 °C to 20 °C	= $50 \times 4.2 \times 20$
Water warming from 0 °C to 20 °C	= $400 \times 4.2 \times 20$
Calorimeter warming from 0 °C to 20 °C	= $250 \times 0.4 \times 20$

$$\text{Total} = 56\,600 \text{ J}$$

Hence $m \times 2596 = 56\,600$

or $m = 21.8$ g

Answer: mass condensed = 21.8 g

QUESTIONS: 19

Note. When required in any question, values for the specific heat capacity and specific latent heats of water given on page 205 should be used.

1. Calculate the quantity of heat required to melt 4 kg of ice and to raise the temperature of the water formed to 50 °C. Take the specific latent heat of ice to be 3.4×10^5 J/kg and the specific heat capacity of water to be 4.2×10^3 J/kg K. *(O.C.)*

2. Why are pieces of ice at 0 °C, added to a drink at room temperature, more effective in cooling the drink than an equal mass of water at 0 °C? *(C.)*

3. Define *specific heat capacity* and *specific latent heat of vaporization.*

Calculate the heat required to convert 2 kg of ice at -12 °C to steam at 100 °C.

4. Define *specific latent heat of ice.*

A refrigerator can convert 400 g of water at 20 °C to ice at -10 °C in 3 hours. Find the average rate of heat extraction from the water in joules per second.

5. What do you understand by *specific latent heat of fusion?* Find the quantity of heat required to melt completely 200 g of lead initially at 27 °C given that, for lead: melting-point = 327 °C; mean specific heat capacity = 0.14 J/g K; specific latent heat of fusion = 270 J/g.

If the heat is supplied to the lead at the rate of 30 joules per second, find:

(a) the time taken to bring the lead to its melting-point;

(b) the additional time required to melt it.

6. 160 g of molten silver at its melting-point, 960 °C, is allowed to solidify at the same temperature and gives out 16 800 J of heat. What is the specific latent heat of silver?

If the mean specific heat capacity is 230 J/kg K how much additional heat does it give out in cooling to -40 °C?

7. (a) The results in the table were obtained when a hot liquid in a test tube was allowed to cool in a laboratory.

Temperature (°C)	85	61	56	56	56	50	40
Time (minutes)	0	1	2	3	4	5	6

(i) Draw a graph of temperature on the y-axis against time on the x-axis. On the graph show clearly the freezing point of the liquid.

(ii) Describe what could be seen happening inside the test tube during the cooling process.

(iii) Explain the shape of the graph, by

stating what is happening in the test tube at each stage of the cooling process.

(b) Calculate the total quantity of heat required to change 0.01 kg of ice at $-10\,°C$ completely into steam at $100\,°C$.

(Specific heat capacity of ice = 2100 J/kg K. Specific heat capacity of water = 4200 J/kg K. Specific latent heat of fusion of ice = 336 000 J/kg. Specific latent heat of vaporization of water = 2 260 000 J/kg.)   (*J.M.B.*)

8. Define *specific latent heat of steam*.

Use the kinetic theory of matter to explain how a liquid absorbs energy when it boils at constant temperature and turns into vapour.

An electric kettle contains 1.5 litres of water at $20\,°C$. Find:

(a) the heat required to bring it to boiling-point;

(b) the additional heat necessary to boil half the water away, assuming that all the heat from the element goes into the water. If the element is rated at 1000 W, how long does it take the water to come to the boil?

9. A copper can together with a stirrer of total heat capacity 60 J/K contains 200 g of water at $10\,°C$. Dry steam at $100\,°C$ is passed in while the water is stirred and until the whole reaches a temperature of $30\,°C$. Calculate the mass of steam condensed.

10. After a period of hard frost with temperatures several degrees below zero, snow may fall and it is noticed that the air temperature often rises to $0\,°C$. Suggest an explanation in terms of the latent heat of ice.

11. Explain why vegetables stored in a cellar can be protected from damage by frost by having tubs of water placed in the cellar.   (*O.C.*)

12. What is meant by *specific latent heat of fusion*? Give a simple explanation, by reference to internal molecular energy, of the changes which occur in a substance when it is supplied with heat at constant temperature and goes from the solid to the liquid state.

13. 0.5 kg of naphthalene contained in an aluminium can of mass 0.4 kg is melted in a water bath and raised to a temperature of $100\,°C$. Calculate the total heat given out when the can and its contents are allowed to cool to room temperature, $20\,°C$. Neglect losses by evaporation during the heating process and give your answer to the nearest kilojoule. (For naphthalene: melting-point = $80\,°C$; specific heat capacity for both liquid and solid = 2100 J/kg K; specific latent heat of fusion = 170 000 J/kg. For aluminium: specific heat capacity = 900 J/kg K.)

14. Describe briefly with two reasons, two ways (other than direct heating) by which a quantity of liquid may be made to evaporate more quickly.   (*S.*)

15. With the aid of a labelled diagram, describe a simple laboratory experiment in which evaporation is shown to produce cooling.

Explain this effect in terms of the kinetic energy of the molecules.   (*A.E.B.*)

16. A metal can which is efficiently lagged with expanded polystyrene contains a mixture of water and crushed ice at $0\,°C$. It is weighed, and immediately afterwards the mixture is well stirred while dry steam at $100\,°C$ is passed in. When all the ice has just melted, the steam supply is stopped and a second weighing reveals a gain in mass of 15 g. Calculate the mass of ice originally present.

17. A copper cylinder of mass 90 g, supported by a length of thread is lowered into a vacuum-walled vessel containing liquid nitrogen at its boiling-point, $-196\,°C$. The nitrogen boils vigorously for a short time. Explain this.

When the boiling has ceased, the copper is transferred to a large vessel containing water at $0\,°C$ and it is observed that a sheath of ice forms round the copper. Explain why this occurs and calculate the mass of ice formed. (Mean specific heat capacity of copper = 300 J/kg K.)

18. State what is meant by the *kinetic theory of matter*, and employ the theory to explain cooling by evaporation.   (*C.*)

19. (a) With the aid of a clearly labelled diagram, describe the operation of the cooling system of a domestic refrigerator.

(b) Refrigerators can also provide background heating for a flat. How is it possible for one piece of equipment to perform the apparently contradictory functions of heating and cooling?   (*A.E.B.*)

20. A beaker contains 200 g of water at $15\,°C$. 25 g of ice at $0\,°C$ is added to the water which is stirred until the ice is completely melted.

(a) How much heat is needed to melt all the ice?

(b) What is the mass of water produced by melting all the ice?

(c) Calculate the lowest temperature of the mixture, assuming that all the heat to melt the ice is taken from the water and that no heat enters or leaves the system.

(Assume: Specific heat capacity of water = 4200 J/kg K. Specific latent heat of fusion of ice = 336 000 J/kg.)   (*J.M.B.*)

20. Vapours

It is a matter of common knowledge that some water in an open vessel eventually dries up through evaporation. Liquids vary in the rate at which they evaporate at ordinary temperatures. Methane and ether, for example, disappear rapidly. They are said to be *volatile*. On the other hand, lubricating oil and mercury never *seem* to evaporate, however long they are allowed to stand.

How liquids evaporate

When discussing the process of evaporation (page 207), we used the kinetic theory to explain how molecules escape from a liquid.

If a liquid is heated, the energy which goes into it becomes mechanical energy in the molecules. More and more of the molecules gain enough kinetic energy to enable them to escape from the attraction of their neighbours and eventually escape from the liquid. A rise in temperature is, therefore, accompanied by an increase in the rate of evaporation.

Vapour pressure

Let us suppose that some liquid is poured into a bottle which is then corked up. Owing to evaporation, the space above the liquid begins to fill with vapour. The vapour molecules move about in all directions and exert pressure when they bounce off the walls of the bottle. They also strike the surface of the liquid and many re-enter it. Eventually *a state of dynamic equilibrium is reached in which the rate at which molecules leave the liquid is equal to the rate at which others return to it*. The use of the word *dynamic* to describe the equilibrium stresses the fact that the molecules are in continuous motion with equal two-way traffic at the surface of the liquid.

Under these conditions the space above the liquid is said to be saturated with vapour, and the pressure exerted is called the *saturated vapour pressure* (s.v.p.).

For a given temperature the saturated vapour pressure of a liquid is always the same whether there is air in the space above it or not.

Before equilibrium has been reached in the manner described the vapour is said to be *unsaturated*.

A saturated vapour is one which is in a state of dynamic equilibrium with its own liquid or solid.

To measure the saturated vapour pressure of a liquid

The vapour pressure of a liquid may be studied and measured by the aid of an ordinary simple barometer as shown in Fig. 20.1 (*a*). Having set up the barometer as described on page 112, a very small quantity of the chosen liquid is introduced into the lower end of the tube by means of a special bent pipette. In this way a small drop of liquid is allowed to rise up the mercury column. On reaching the top it evaporates and the pressure exerted by the vapour depresses the column.

If sufficient liquid is added so that a small quantity remains on top of the mercury the space above will become saturated and the saturated vapour pressure may be measured by the total depression, *h*, of the column. In order that this may be done accurately it is advisable to have a second barometer tube set up in case the barometric pressure should alter during the course of the experiment. Also, the liquid used in the pipette should be freshly boiled so as to remove any dissolved air which would spoil the results.

Saturated vapour pressure does not depend on volume

It is important to note that saturated vapour pressure at any given temperature is independent of the volume of the vapour so long as there is some free liquid present to ensure saturated conditions.

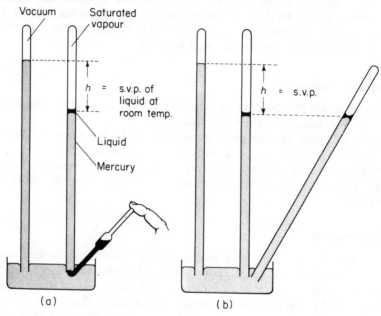

Fig. 20.1. S.v.p. of a liquid at room temperature

This may be demonstrated by tilting the barometer tube containing the vapour (Fig. 20.1 (*b*)). The mercury rises up the tube but keeps at the same level. As the volume decreases, the excess vapour condenses back to liquid and the vapour pressure remains constant.

Variation of saturated vapour pressure with temperature

The effect of temperature on the saturated vapour pressure of a liquid may be studied by means of two barometer tubes placed side by side in a water bath (Fig. 20.2).

If water is the liquid to be investigated a little freshly boiled water is introduced into one of the tubes in the manner already described. The temperature of the water bath is then noted and the saturated vapour pressure found by measuring the difference, *h* in mm, in the mercury levels.

The water bath is stirred continuously and its temperature raised by passing in steam from a boiler. At 10 K intervals readings of temperature and saturated vapour pressure are taken.

From the results, a graph may be plotted, similar to that shown, although with this apparatus it is inadvisable to attempt to reach a temperature as high as 100 °C.

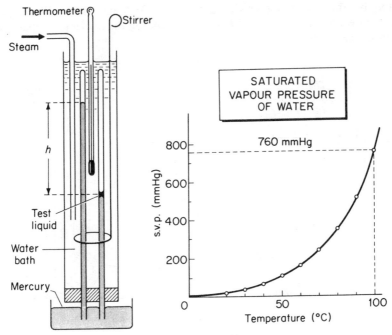

Fig. 20.2. Variation of s.v.p. with temperature

What happens when a liquid boils

If a liquid is heated its temperature begins to rise, and therefore the saturated vapour pressure will increase. Ultimately, the saturated vapour pressure becomes equal to the external atmospheric pressure. At this stage the further addition of heat will cause bubbles of vapour to form inside the body of the liquid and rise to the surface. This process is called boiling or *ebullition*.

The boiling point (b.p.) of a substance is defined as the temperature at which its saturated vapour pressure becomes equal to the external atmospheric pressure.

Variation of the boiling point of water with pressure

Since the atmospheric pressure does not remain constant, the boiling point of water is liable to vary from day to day. Water boils at exactly 100 °C only when the barometric pressure is at the standard value of 760 mmHg. Tables are available which give the boiling point at other pressures. These tables show that the boiling point of water changes by approximately 0.037 K per mmHg change of pressure in the region of 760 mmHg.

Boiling under reduced pressure

Water can be made to boil without heating it simply by reducing the atmospheric pressure above it to a value less than the saturated vapour pressure. This may be done with the aid of a filter pump.

A stout, round-bottomed flask containing warm water is fitted with a two-holed rubber bung through which passes a thermometer and a short glass tube (Fig. 20.3). When the flask is connected to a good filter pump the water begins to boil as soon as the pressure becomes less than the saturated vapour corresponding to the temperature of the water at the time.

Since no heat is being supplied from outside, the necessary latent heat of vaporization has to come from the water itself. It therefore cools and the temperature indicated by the thermometer drops rapidly. To some extent this experiment illustrates how cooling occurs in the coils surrounding the freezing compartment of a refrigerator (see page 208).

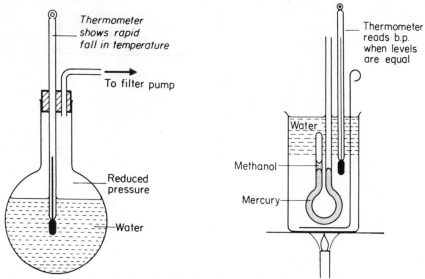

Fig. 20.3. Water boiling under reduced pressure

Fig. 20.4. To measure b.p., using a small quantity of liquid

Fresh water supply on ships. Desalination

Rapid distillation of sea-water under reduced pressure can be used to provide fresh water on steamships. The reduction in pressure in the condenser is brought about by a vacuum pump very similar in principle to a laboratory filter pump, except that it works by a steam jet instead of a water jet. The water is heated by waste steam from the turbines.

Desalination (salt-removing) plants working on this principle can produce over 600 t of distilled water per day which is used not only to feed the boilers but also for all domestic uses on board. For the latter purpose a small quantity of a mixed salts solution is added to make it like naturally occurring water.

Determination of boiling point

We may use the fact that the saturated vapour pressure of a liquid at its boiling point is equal to atmospheric pressure as a means of finding the boiling point.

A small glass U-tube, closed at one end and bent as shown in Fig. 20.4, is filled with clean mercury, with the exception of a small space at the top. The space is then filled with a little freshly boiled liquid, e.g., methanol.

By suitably tilting the tube while a finger is held over the open end, the methanol may be caused to pass into the closed arm and remain on top of the mercury when the tube is vertical. The excess mercury in the open limb may then be removed by holding the tube over a beaker and slowly inserting a glass rod.

Having thus prepared the tube, it is placed in a large beaker of water together with a thermometer. The water is slowly heated and stirred continuously. As the temperature rises the liquid begins to vaporize, and when the mercury levels are the same on both sides the vapour pressure is atmospheric. The thermometer now indicates the boiling point of the liquid. A mean value of the boiling point is obtained from two readings, one taken when the temperature is rising and the other when it is falling through the boiling point.

The pressure cooker

The time required for cooking vegetables and other food can be greatly reduced if the boiling point of the water is raised. This can be done by the use of a pressure cooker.

The pressure cooker, or "digester" as it used to be called, dates from the seventeenth century and was invented by a Frenchman named Denys Papin. In an account written in 1681, Papin describes how he used the digester to extract gelatin from beef bones. He goes on to say that he afterwards added lemon juice and sugar to the gelatin and found the resulting lemon jelly much to his liking.

The modern counterpart of the digester commonly takes the form of a stout aluminium container fitted with a lid having a rubber sealing ring (Fig. 20.5). Steam is allowed to escape through a loaded pin valve which can be set to blow at excess pressures varying from a third to one atmosphere. By this means it is possible to cook at temperatures up to 120 °C or more, with a corresponding saving of time and fuel.

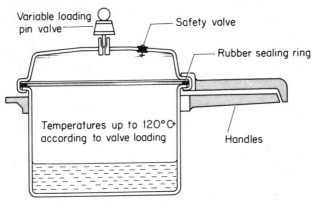

Fig. 20.5. Pressure cooker

Explorers in mountainous regions find pressure cookers indispensable, and they are useful in other places where the atmospheric pressure is low. For example, Quito, the capital of Ecuador, is over 2700 m above sea-level, and here water normally boils at about 90 °C.

ATMOSPHERIC MOISTURE

Owing to evaporation which goes on continuously from the sea and other water surfaces, the atmosphere always contains water vapour. The quantity of vapour which air can hold depends on the temperature. If warm air containing water vapour is cooled it can hold less vapour, so that below a certain temperature the excess vapour condenses out.

Dew point

Those who wear spectacles know that when they enter a warm room from the cold outside air a film of moisture is often deposited on the spectacle lenses. Similarly, a cold mirror may become dulled when brought into a warm atmosphere.

In such cases the cold glass surface cools the air in its vicinity to a temperature below that for which the water vapour present is sufficient to saturate the air. Excess vapour then condenses out, and the highest temperature at which this can occur is called the *dew point*.

The dew point is defined as the temperature at which the water vapour present in the air is just sufficient to saturate it.

Relative humidity

On page 206 mention was made of the part played by perspiration in controlling bodily temperature. We shall now look into this a little more fully.

The rate at which moisture evaporates from the skin depends on three factors, namely, the temperature, the amount of water vapour present in the atmosphere and the rate of movement of air over the skin.

On occasions when the air is still and its vapour content is near to saturation the rate of evaporation from the body slows up. Under these conditions even moderate exercise induces profuse perspiration, and the skin remains moist or clammy. The atmosphere is described as being close or humid. Matters can be improved if the rate of evaporation is increased by setting the air in motion. Where there is no natural breeze an artificial one can be created by fans. This is often done in rooms where humid conditions frequently occur.

At other times, when the air is comparatively dry, evaporation from the skin takes place more easily, and faster cooling of the body ensues. Thus, in hot continental climates, where the atmosphere tends to be dry, high temperatures can be tolerated more comfortably than in tropical regions, where heavy rainfall leads to near-saturation conditions.

From the point of view of personal comfort and for certain industrial purposes to be mentioned later, the actual quantity of water vapour present in the air is not so important as its nearness to saturation or *relative humidity*.

$$\text{Relative humidity} = \frac{\text{mass of water vapour in a given volume of air}}{\substack{\text{mass of water vapour required to saturate the same volume} \\ \text{of air at the air temperature}}}$$

We need not here go into detail but, as a matter of interest, this ratio may also be expressed in the more convenient form.

$$\text{Relative humidity} = \frac{\text{s.v.p. of water at the dew point}}{\text{s.v.p. of water at the original air temperature}}$$

There are a number of different methods available for measuring relative humidity by instruments called *hygrometers*. We shall describe one of the simpler and more common types in use.

Wet- and dry-bulb hygrometer

For meteorological purposes the relative humidity may be measured quickly with a reasonable degree of accuracy by means of the wet- and dry-bulb hygrometer (Fig. 20.6). This instrument has two thermometers mounted side by side, one of which has its bulb covered with a muslin wick dipping into a small vessel of water. Owing to evaporation of water from the muslin, the wet bulb is cooled to an extent which depends on the dryness of the surrounding air. Tables have been drawn up relating the relative humidity to the dry-bulb temperature and the wet-bulb depression.

Humidity control

In certain manufacturing processes the relative humidity is an important factor. Cotton fibres, for example, must not become too dry, or they become brittle and difficulties arise through electrification by friction. For this reason the British cotton spinning industry came to be established in the damp climate of Lancashire on the west side of the Pennines. In this part of the country moisture-laden winds from the Atlantic are forced upwards over the high ground where the atmospheric pressure is less. In the resulting expansion the air does work, and the energy for this is provided

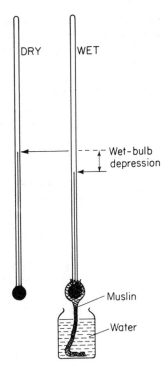

DRY WET

Wet-bulb depression

Muslin

Water

Fig. 20.6. Wet- and dry-bulb hygrometer

from the internal energy of the air (see page 201). Consequently, the wind cools and excess moisture is precipitated as rain.

In contrast, a dry atmosphere is needed in factories where the assembly of certain electrical components is carried out. This requirement also applies to warehouses for the storage of food and for the seasoning of wood. In such cases special equipment is installed to dry the air and recirculate it through the building.

Air conditioning

At the present time increasing use is being made of air-conditioning plant in ships, theatres, concert halls and other buildings. Large fans draw air from outside and pass it through water sprays to wash out dust and other pollution. Afterwards it is heated and humidified to the correct value before being conveyed through large pipes or ducts to the various rooms.

QUESTIONS: 20

1. What do you understand by the terms *saturated vapour, unsaturated vapour,* and *boiling point?*

Describe a method of measuring the saturated vapour pressure of alcohol between room temperature and a few degrees below the boiling point.

Sketch roughly the type of curve you would expect to obtain. (Saturated vapour pressure at 15 °C = 3 cm. Boiling point of alcohol = 78 °C.)   (*O.C.*)

2. Describe briefly, with reasons, two ways (other than by direct heating) by which a quantity of liquid may be made to evaporate more quickly.   (*S.*)

3. Describe an experiment to show that the boiling point of water is decreased by a reduction in pressure.

Describe one practical application of an increase in the boiling point of water under increased pressure.   (*C.*)

4. A simple barometer has a thin layer of water on top of the mercury column. When the temperature is 20 °C the height of the mercury column is 74.1 cm. What is the

atmospheric pressure? Assuming that this remains constant, what will the barometer read if the temperature falls to 10 °C? Ignore the change in density of the mercury. (Saturated vapour pressures of water at 10 °C and 20 °C are 9 mm and 17 mm of mercury respectively.) (*A.E.B.*)

5. What do you understand by a *saturated* vapour?

Describe how you would measure the saturated vapour pressure of water at room temperature.

Explain as fully as you can what happens when a liquid boils. Why would you expect the boiling point of a liquid to be lowered when the pressure above the free surface is vapour? (*O.*)

6. The following table gives the pressure of saturated water vapour at various temperatures:

Temperature (°C)	40	50	60	70
S.v.p. (mm of mercury)	55.3	92.5	149.4	234.0

Find at what temperature water in a flask would boil if the pressure above the water surface were reduced to 100 mm of mercury. (*O.C.*)

7. Describe an experiment to show that a liquid boils when its saturated vapour pressure is equal to the pressure on its surface.

Define relative humidity, and explain the action of a wet- and dry-bulb hygrometer. (*S.*)

8. Explain the distinction between *saturated* and *unsaturated* vapour, and between *evaporation* and *boiling*.

Heat is supplied at a rate of 500 W to a pressure cooker containing water and fitted with a safety valve. Steam escapes at such a rate that the loss of water is 10.4 g/min. If heat is supplied at the rate of 700 W 15.6 g of water is lost per minute.

Suggest an explanation of these figures and deduce:

(*a*) the latent heat of steam in joules per gram at the temperature of the cooker, and

(*b*) the rate of loss of heat from the cooker at this temperature by other processes than evaporation. (*O.C.*)

9. Explain the statement: "The saturation vapour pressure of water at 60 °C is 149.4 mm of mercury."

Describe an experiment to determine the saturation vapour pressure of water in the temperature range 25–70 °C. Sketch the curve you would expect. (*W.A.E.C.*)

The Isaac Newton reflecting telescope, formerly at the Royal Greenwich Observatory when it was at Herstmonceaux in Sussex, England.
This telescope, fitted with a new mirror, and computer-controlled by satellite link from England, has now been installed in the new observatory at La Palma in the Canary Islands

Optics

21. Light rays and reflection of light

The sharp edges of shadows make us realize that light travels in straight lines. We can also demonstrate this fact by a simple experiment with three cardboard screens having small holes in their centres. These are set up so that the holes are in a straight line by threading string through the holes and pulling it taut (Fig. 21.1). Light from a candle or lamp placed at A can then be received by an eye at B. If, however, one of

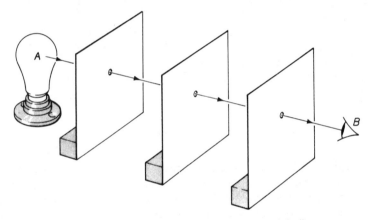

Fig. 21.1. Showing that light travels in a straight line

the screens is moved so that the holes are no longer in a straight line the light is cut off.

Later on, however, it will be seen that light is a form of wave motion and shows the same kind of behaviour as water waves. We shall show that light energy *can* spread out through a narrow opening to produce *diffraction* and *interference* effects (chapter 26).

Rays and beams of light

Ordinarily, the term *ray* refers to a narrow stream of light energy, e.g., that coming through a small hole in a screen, but scientifically it has a more precise meaning.

A ray is the direction of the path taken by light. In diagrams rays are represented by lines with arrows on them.

A beam is a stream of light energy, and may be represented by a number of rays which may be either diverging, converging or parallel. Examples of each of these will be met with in due course.

The pinhole camera

This device is believed to be an Arabic invention of the eleventh century and was used for viewing eclipses of the sun without danger to the eyesight. It was described by Leonardo da Vinci in the sixteenth century, when it was known as the *camera obscura* (= dark room). It took the form of a small darkened room into which light was admitted through a single small hole in one wall. The result was that an inverted image of the scene outside was formed on the whitened wall opposite. Artists of the time used it as an aid to achieve correct perspective in their drawings and paintings.

The present-day successor of the camera obscura is called a *pinhole camera* and consists of a box with a small hole in a metal plate at one end and a screen of tracing-paper or frosted glass at the other (Fig. 21.2). An image is formed on the screen which will be seen more clearly if external light is excluded by covering head

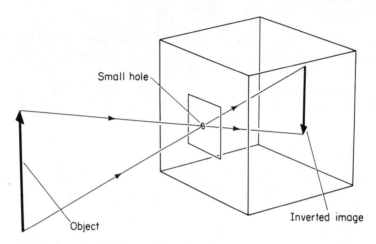

Fig. 21.2. Image formation by a pinhole camera

and camera with a dark cloth. Since light travels in straight lines, it follows that a given point on the screen will be illuminated solely by light coming in a straight line through the pinhole from a certain point on an object outside. Rays of light from the various parts of an object outside will thus travel in straight lines through the pinhole, and form a multitude of tiny patches on the screen. These tiny patches combine to form an inverted image of the object.

A blurred image results if the hole is made much larger than a pinhole. This may be explained if we think of the larger hole as being equivalent to a group of small holes close together, each of which produces its own image on the screen. These images overlap, and the resultant effect is an image which is brighter but very blurred.

If a line is drawn through the pinhole perpendicular to both object and image it may be proved by similar triangles that

$$\frac{\text{height of image}}{\text{height of object}} = \frac{\text{distance of image from pinhole}}{\text{distance of object from pinhole}}$$

This fraction is called the *magnification* of the camera. For large magnification it is obvious that the distance of the object from the pinhole must be small compared with the distance between pinhole and screen.

If the screen is replaced by a photographic plate or film very satisfactory pictures of still subjects may be taken with this camera using time exposures of suitable length (Fig. 21.3). Snapshots can be taken only with a lens camera, since the larger aperture of the lens admits more light energy per second than a pinhole.

On the other hand, a pinhole camera requires no focusing as does a lens camera (page 269). Although, according to the size of the pinhole, the image may be slightly

Fig. 21.3. Photograph taken with a pinhole camera

blurred, all parts of the image are in equally sharp focus whatever their distances from the camera. In the language of photographers, the pinhole camera has an *infinite depth of focus*.

Shadows

We have already mentioned that the sharp edge of a shadow indicates that light travels in straight lines.

When an obstacle is placed in the path of light coming from a *point source* the shadow formed on a screen is uniformly dark (Fig. 21.4). The point source of light used here has been made by putting an electric lamp inside a tin with a small hole in it.

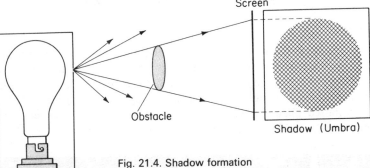

Fig. 21.4. Shadow formation

If an *extended source* is used the shadow is seen to be edged with a border of partial shadow called *penumbra* to distinguish it from total shadow or *umbra* (Fig. 21.5). The extended source shown in this diagram has been made by putting a pearl electric lamp inside a tin with a large hole in it. Points inside the umbra receive no light at all from the source. The penumbra receives a certain amount of light from the source, but not as much as it would receive if the obstacle were removed.

Eclipses

An eclipse of the sun by the moon occurs when the moon passes between the sun and the earth and all three are in a straight line. Fig. 21.6 (*a*) shows how the umbra and penumbra are produced, and the appearance of the sun as seen from various positions on the earth's surface (see also Fig. 21.7). Of course, owing to the vast distance between sun and earth our diagram cannot be drawn to scale.

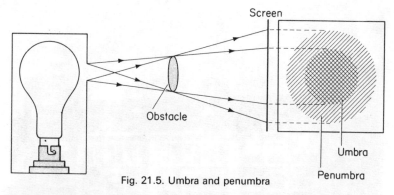

Fig. 21.5. Umbra and penumbra

On occasions when the distance of the moon from the earth is such that the tip of the umbra fails to reach the earth's surface an *annular eclipse* occurs, and from one place on the earth the sun presents the appearance of a ring of light (Fig. 21.6 (*b*)).

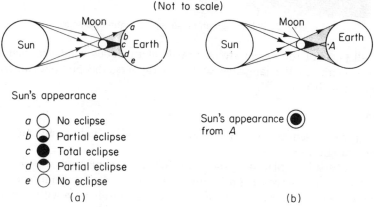

Fig. 21.6. Eclipse of the sun by the moon

The nature of light

Scientists have always been puzzled by the nature of light. In the seventeenth century there were two schools of thought concerning it. Sir Isaac Newton regarded light as a stream of corpuscles or tiny particles travelling in straight lines. The Dutch physicist, Huygens, held that light consisted of waves in a substance called the *ether*, which he supposed filled the whole of space, including that between the atoms of matter, and which could not be removed even from a vacuum.

Fig. 21.7. *Stages towards a total eclipse of the sun as seen from Virginia on 7th March 1970.* The photographs show the moon moving across the sun and finally reaching total eclipse

As time went on and more became known about the behaviour of light, Huygens's wave theory came to be accepted as the better one. At the present day, however, we have reason to believe that light consists of streams of tiny wave-like packets of energy called "photons", which travel at a speed of 3×10^8 m/s or 3×10^5 km/s.

Atoms emit light at the high temperatures produced by chemical reaction in a flame, by the heating of thin tungsten wire in the ordinary electric lamp or by the bombardment of gas molecules by electrons in a discharge lamp tube.

The sun and sources as described above are said to be *self-luminous*, since they emit light of their own accord. The common objects around us are not self-luminous, but we are able to see them because they reflect light from the sun or other sources in all directions. Mirrors and highly polished surfaces reflect light strongly, and we shall now deal with the laws governing the reflection.

Reflection of light

Fig. 21.8 illustrates the terms we use in the study of reflected light. MM' represents the surface of a plane mirror. AO, called the *incident ray*, is the direction in which light falls on to the reflecting surface. O is the *point of incidence* and OB the *reflected ray*. The angles i and r which the incident and reflected rays make with ON, the *normal* or perpendicular to the reflecting surface at the point of incidence, are called the *angles of incidence* and *reflection* respectively.

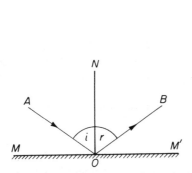

Fig. 21.8. Reflection from a plane mirror

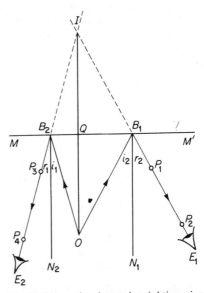

Fig. 21.9. Locating image by sighting pins

Laws of reflection

(1) **The incident ray, the reflected ray and the normal at the point of incidence all lie in the same plane.**

(2) **The angle of incidence is equal to the angle of reflection.**

To verify the laws of reflection of light

A strip of plane mirror is set up vertically, with its silvered surface on a line MM' drawn on a sheet of white paper on a drawing-board (Fig. 21.9). A pin O, to serve as an object, is stuck into the paper about 7 or 8 cm from MM'.

With the eye in some convenient position E_1, two pins P_1 and P_2 are stuck into the paper so as to be in a straight line with the image I of the pin O seen in the mirror.

For accuracy, these sighting pins must be placed as far apart as the paper will allow. The two sighting pins are next removed, and their positions marked by small pencil crosses and lettered P_1 and P_2. The same procedure is carried out with the eye in several other positions, such as E_2, at least two positions on either side of the object O being taken. When this has been done the mirror is removed.

The points P_1 and P_2, etc., are then joined by pencil lines to cut MM' at B_1, etc., and these lines are produced backwards behind the mirror. Here they ought all to intersect at I.

The lines OB_1, OB_2, etc., represent incident rays and B_1P_2, B_2P_4, etc., are the corresponding reflected rays. Normals B_1N_1, B_2N_2 and so on are constructed and the angles of incidence and reflection for each pair of rays noted in a table. In each case it is found that $i = r$, thus verifying the second law.

Angle of incidence (i)	Angle of reflection (r)

This experiment can be performed on a flat drawing-board only if the mirror is set up at right angles to the board. This verifies the first law of reflection.

Note. If IO is joined, cutting MM' at Q, it is found that $IQ = OQ$ and IO is perpendicular to MM'. We therefore infer that the line joining object and image is at right angles to the mirror and the image is as far behind the mirror as the object is in front.

We shall learn more about the image from a later experiment.

Parallax

When we look out of the window of a moving train, trees, chimneys, towers and other objects in the landscape appear to be moving relatively to each other. Thus, at one moment a tree may appear to be to the right of a church spire and a few seconds later to the left of it. Their actual positions in space have, of course, remained fixed. This apparent relative movement of two objects owing to a movement on the part of the observer is called *parallax*.

No parallax is observed between the cross surmounting a spire and the spire itself, since these two objects coincide in position.

To locate images by no-parallax

The method of no-parallax is frequently used in light experiments to locate the positions of the images of pins (Fig. 21.10). The eye is moved from side to side while

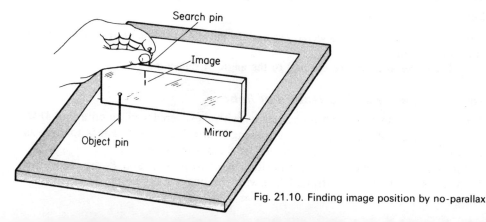

Fig. 21.10. Finding image position by no-parallax

viewing a *search* pin placed in the neighbourhood of the image. A position is found for which both pin and image appear to coincide in the same straight line. When this condition of no-parallax holds the search pin gives the position of the image.

To study the image formed in a plane mirror

A straight line MM' is drawn across the centre of a sheet of drawing paper to represent a reflecting surface and a large letter E to serve as an object (Fig. 21.11). A strip of plane mirror is then stood vertically with its silvered surface over MM' and the image of the object letter E seen in the mirror is found in the following manner. An object pin is stuck into the paper at the various points O_1, O_2, etc., in turn on the

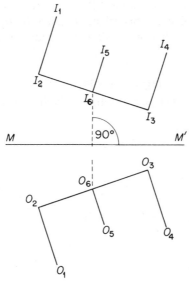

Fig. 21.11. Image in a plane mirror

object letter, and each time the images, I_1, I_2, etc., are located by the method of no-parallax, using a search pin as described in the previous section. When adjusting the search pin to coincidence with the image it is useful to keep in mind that, of the two, *the further one moves with the eye.*

Suitable measurements should now be taken on the diagram and observations recorded in the notebook to verify the following facts.

The image in a plane mirror is:
(1) *The same size as the object.*
(2) *The same distance behind the mirror as the object is in front.*
(3) *Laterally inverted* (see next page).
(4) *Virtual* (*it cannot be formed on a screen*).

Finally we see from this experiment that the line joining any point on the object to its corresponding point on the image cuts the mirror at right angles. It is important to remember this, as we use it when making graphical constructions of images in plane mirrors.

Parallax in pointer instruments

Electrical and other instruments which have a pointer moving above a scale are liable to parallax errors if the eye is not vertically above the pointer when taking a reading.

Good-class instruments have a plane mirror on the scale. Parallax error is avoided if the eye is positioned so that, when taking a reading, the pointer exactly covers its own image in the mirror.

Looking into a plane mirror

We were, of course, already familiar with some of the above facts from our everyday experience with mirrors. Looking into a mirror, we see an image of the face situated apparently behind the mirror. If we now move backwards the image will recede so that it is always the same distance behind the mirror as the object is in front. Unlike the images formed on a screen by a projector, which are said to be *real* in the sense that they are formed by the actual intersection of real rays, the image we see in the mirror cannot be formed on a screen, It is said to be *virtual* and is produced at the place *where the reflected rays appear to intersect* when their directions are produced backwards behind the mirror. This is further explained in the next paragraph. It is also to be noticed that the left ear of the image is formed from our own right ear as object. This effect is called *lateral inversion*: it is even more strikingly demonstrated when we look at the image of a printed page in a mirror (Fig. 21.12).

How the eye sees an image in a plane mirror

Let us consider how the eye sees the tip of the image of a candle flame in a plane mirror (Fig. 21.13).

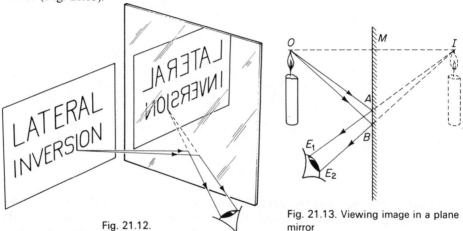

Fig. 21.12.

Fig. 21.13. Viewing image in a plane mirror

We know now that the position of the image I can be constructed by drawing a line through O perpendicular to the mirror at M and making $OM = IM$. On looking into the mirror the eye sees I apparently by the cone of rays IE_1E_2, which enter the eye pupil. The portion IAB of this cone obviously does not exist: there is no light behind the mirror. The real portion ABE_1E_2 results from light from O travelling down the cone OAB and reflected from the mirror at AB in accordance with the laws of reflection.

The complete real cone of rays $OABE_1E_2$ is called the "pencil" of light by which the eye sees the image I.

Images formed in two mirrors inclined at 90°

When two mirrors are inclined at right angles we have not only the images I_1 and I_2 formed by a single reflection but in addition two extra images produced by *two reflections*. The pencil of light by which the eye sees one of these, $I_{1.2}$, is shown in Fig. 21.14. The subscript $_{1.2}$ in the symbol $I_{1.2}$ signifies the order in which the reflections take place from the mirrors 1 and 2.

The other image $I_{2.1}$ may be seen by looking into mirror 1. Actually the images $I_{1.2}$ and $I_{2.1}$ are superimposed on one another.

The images I_1 and I_2 themselves act as objects for the formation of images $I_{1.2}$ and $I_{2.1}$, and the positions of these images are found in the usual way, i.e., a perpendicular to the mirror is drawn through the object and the object and image distances from the mirror are made equal.

Geometrically, the object and all the images lie on a circle whose centre is at the intersections of the mirrors. It is useful to remember this when drawing ray diagrams.

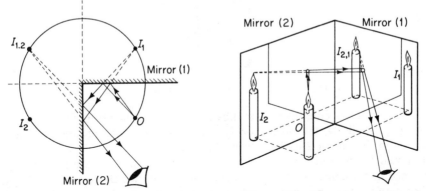

Fig. 21.14. Reflection from two mirrors at 90°   Fig. 21.15. Images formed by two mirrors at 90°

Fig. 21.15 shows, in perspective, the pencil of rays by which a point on the image $I_{2.1}$ is seen.

The kaleidoscope

The kaleidoscope consists of two strips of plane mirror M_1 and M_2 about 15 cm long, placed at an angle of 60° inside a tube (Fig. 21.16). At the bottom of the tube is a ground-glass plate to admit light, on which is scattered small pieces of brightly

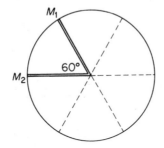

Fig. 21.16. Mirrors in a kaleidoscope

coloured glass. These pieces of coloured glass act as objects, and on looking down the tube five images are seen, which together with the object form a symmetrical pattern in six sectors.

This may be compared with Fig. 21.14, where the mirrors are at 90° and there are only three image spaces.

The number of different patterns obtained is unlimited, as a fresh one is produced every time the tube is shaken to rearrange the pieces.

Parallel mirrors

An infinite number of images are formed of an object placed between two parallel mirrors. These all lie on a straight line through the object perpendicular to the mirrors (Fig. 21.17).

The positions of the images may be found by the usual construction, remembering that each image seen in one mirror will act as a virtual object and produce an image

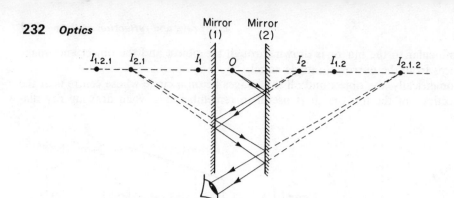

Fig. 21.17. Images formed by two parallel mirrors

in the other mirror. Thus, using our subscript notation, the object for image $I_{2.1.2}$ is image $I_{2.1}$. Similarly, the object for $I_{2.1}$ is I_2, while I_2 was produced by a single reflection in the mirror 2 by light from the object O.

It should be noticed that $I_{1.2}$, which is the image of I_1 in the mirror 2, is not concerned in the construction of the pencil of light by which the eye sees $I_{2.1.2}$.

Two parallel plane mirrors should be set up to examine the images of a candle or other object placed between them. The more remote the images, the fainter they become, since some of the light energy is absorbed by the mirrors at each successive reflection.

The periscope

The simple periscope consists of two plane mirrors, fixed facing one another at an angle of 45° to the line joining them. The user is enabled to see over the heads of a crowd or over the top of any obstacle. The upper mirror M_1 produces an image I_1, which may then be regarded as an object for the lower mirror M_2 (Fig. 21.18). The

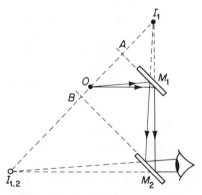

Fig. 21.18. Mirrors in a periscope

diagram shows the pencil of light by which the eye sees a point on the final image $I_{1.2}$, which has been formed by reflection of light from each mirror in turn.

When constructing this diagram it should be noticed that $OA = I_1A$, and $I_1B = I_{1.2}B$, and, furthermore, that the line $I_1I_{1.2}$ is perpendicular to the mirrors.

Periscopes used in submarines are more elaborate than the simple type described here. Prisms are used instead of mirrors, and the tube supporting them incorporates a telescope to extend the range of vision (see page 257).

Pepper's ghost

This method of producing the illusion of a ghost on the theatrical stage was invented by John Pepper, director of the Royal Polytechnic Institution, London, in the mid-nineteenth century. A large sheet of polished plate glass placed diagonally across the stage acts as a mirror, but at the same time permits objects on the stage to be seen through it. An actor attired to represent the ghost is concealed in the wings and is strongly illuminated. All else surrounding him is painted or draped dull black so that an image of the actor only is formed by the plate glass. Nowadays, perspex or a stretched sheet of clear polythene is used instead of glass. The illusion is rendered complete by the fact that light from objects on the stage behind the image causes the ghost to appear transparent. Headless ghosts result simply from enclosing the actor's head in dull black cloth.

A practically identical arrangement causes a candle to appear to be burning inside a bottle full of water. How this may be done is shown in Fig. 21.19.

Diffuse reflection

Mirrors and sheets of glass are highly polished surfaces. The surfaces of most objects, however, are found to have tiny irregularities when they are examined under a microscope. Paper is an example. Thus when a parallel beam of light falls on such

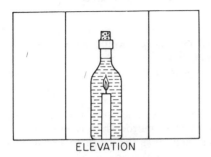

ELEVATION

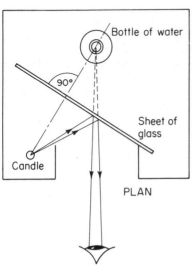

PLAN

Fig. 21.19 Model to illustrate Pepper's ghost

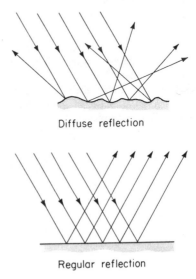

Diffuse reflection

Regular reflection

Fig. 21.20.

a surface the individual rays strike it at different angles of incidence. The rays are therefore reflected in different directions from the surface (Fig. 21.20). We say that the light is *diffusely reflected* from the surface.

Rotation of reflected ray

If a fixed ray is incident on a plane mirror at an angle i the angle of reflection is also equal to i, and so the angle between the two rays will be $2i$.

Suppose now that the mirror is rotated though an angle θ. The angle of incidence will increase by θ, and so also will the angle of reflection. The total angle between the two rays is now $2i + 2\theta$. Consequently, the angle of reflection will rotate through an angle of 2θ. Thus, *the reflected ray rotates through twice the angle through which the mirror rotates.*

QUESTIONS: 21

1. Draw a diagram of the apparatus you would use to show by a simple experiment that light travels in straight lines. (*S.*)

2. Describe the construction of a *pinhole camera*. Upon what fact does it rely for its action?

Discuss the effect on the size, sharpness, and brightness of the image of:

(*a*) doubling the object distance;

(*b*) doubling the diameter of the pinhole. Illustrate your answer by sketches. (*O.C.*)

3. Explain, with appropriate diagrams in each case, Two of the following:

(*a*) how an eclipse of the sun occurs, distinguishing clearly between total and partial eclipses;

(*b*) why three images are formed when an upright pin is placed between the reflecting surfaces of two vertical plane mirrors meeting at right angles;

(*c*) the use of plane mirrors to make a simple periscope – why are lenses usually fitted to periscopes? (*W.*)

4. Describe a simple experiment to locate the image of an object as seen in a plane mirror. State the result expected.

Describe and explain the use of a plane mirror under the scale of some accurate pointer-reading instruments. (*S.*)

5. State the laws of reflection of light.

Show that the image of an object observed in a plane mirror is as far behind the mirror as the object is in front.

Show that when a ray of light is reflected from a plane vertical mirror and the mirror is turned round a vertical axis through an angle θ, the reflected ray is turned through 2θ.

(*O.C.*)

6. State the laws of reflection of light, and describe how you would verify them experimentally.

Explain, with the help of a ray diagram, how the eye sees the image of a bright point formed by a plane mirror.

A man sits in an optician's chair, looking into a plane mirror which is 2 m away from him, and views the image of a chart which faces the mirror and is 50 cm behind his head. How far away from his eyes does the chart appear to be? (*O.*)

7. Draw a ray diagram to show that a vertical mirror need not be 180 cm long in order that a man 180 cm tall may see a full-length image of himself in it.

If the man's eyes are 12 cm below the top of his head find the shortest length of mirror necessary and the height of its base above floor level.

22. Spherical mirrors

In this chapter we shall study the way in which images are formed by curved mirrors. The kind we use are generally made by silvering a piece of glass which would form part of the shell of a hollow sphere. Silvering the glass on the outside gives a concave or *converging* mirror, while silvering on the inside gives a convex or *diverging* mirror.

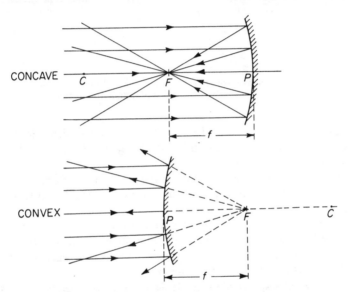

Fig. 22.1. Reflection of parallel beams from curved mirrors

The *principal axis* of a spherical mirror is the line joining the *pole P* or centre of the mirror to the *centre of curvature C* which is the centre of the sphere of which the mirror forms a part (Fig. 22.1).

The *radius of curvature r* is the distance *CP*. In the case of a concave mirror the centre of curvature is in front of the mirror; in a convex mirror it is behind.

Principal focus

When a parallel beam of light falls on a plane mirror it is reflected as a parallel beam; but in the case of a concave mirror the rays in a parallel beam are all reflected so as to converge to a point called a *focus*.

If the incident rays are parallel to the principal axis the point through which all the reflected rays pass is on the principal axis just midway between the pole and the centre of curvature and is called the *principal focus, F* (Fig. 22.1).

The same figure also shows what happens when a beam of light parallel to the principal axis falls on a convex mirror. In this case the rays are reflected so that they all appear to be coming from a principal focus midway between the pole and centre of curvature *behind* the mirror.

A concave mirror, therefore, has a real principal focus, while a convex mirror has a virtual one.

Mirrors of large aperture

The remarks in the previous paragraph apply only to mirrors which are small in size or aperture compared with their radii of curvature. This is not the case with a hemispherical mirror, for instead of producing a point focus from a parallel beam, the reflected rays intersect to form a surface called a *caustic* (Fig. 22.2).

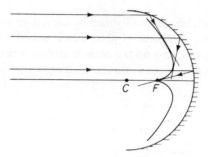

Fig. 22.2. Reflection from a hemispherical mirror

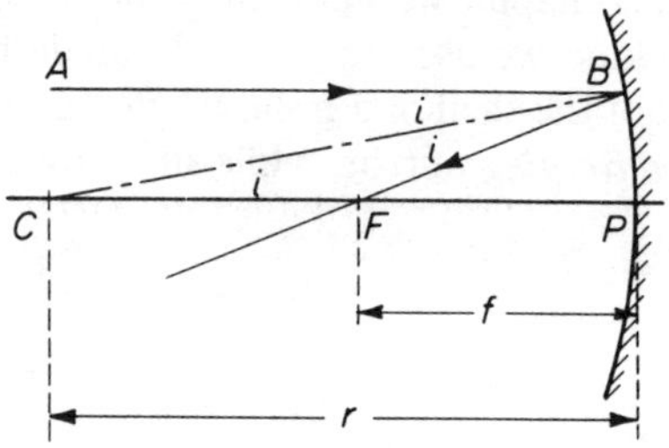

Fig. 22.3. Relation between focal length and radius of curvature

A bright caustic curve can often be seen on the surface of tea in a cup. This is formed when light from a distant lamp is reflected from the inside of the cup, which acts as a curved mirror of large aperture.

If the mirror is small or, what amounts to the same thing, if only rays close to the principal axis are used, only the apex of the caustic is formed, and this becomes the principal focus.

With these facts in mind we therefore have the following definitions.

The principal focus of a spherical mirror is that point on the principal axis to which all rays originally parallel and close to the principal axis converge, or from which they appear to diverge, after reflection from the mirror.

The focal length *f* of a spherical mirror is the distance between the pole of the mirror and principal focus.

The focal length of a spherical mirror is half its radius of curvature

In Fig. 22.3 *AB* is an incident ray parallel and close to the principal axis *PC* of a concave mirror.

By definition, the reflected ray passes through the principal focus *F*.

Since by geometry, the radius *CB* is normal to the mirror surface at *B*, it follows from the laws of reflection that,

$$\angle ABC = \angle CBF$$

But
$$\angle ABC = \angle BCF \text{ (alternate angles)}$$
$$\angle CBF = \angle BCF$$
$$\therefore \quad BF = FC \text{ (sides of isosceles } \Delta)$$

Now *B* is very close to *P*

$$\therefore \quad BF = PF \text{ very nearly}$$
$$\therefore \quad PF = FC$$

or
$$f = \frac{r}{2}$$

Construction of ray diagrams

Since a point on an image can be located by the point of intersection of two reflected rays, we have to consider which are the most convenient rays to use for this purpose.

Remembering that, by geometry, the normal to a curved surface at any point is the radius of curvature at that point, one very useful ray to draw will be one which is incident along a radius of curvature. Since this is incident normally on the mirror, it will be reflected back along its own path.

Another useful ray is one which falls on the mirror parallel to the principal axis. By definition, this will be reflected through the principal focus. Conversely, any incident ray passing through the principal focus will be reflected back parallel to the principal axis. The same observations also apply to convex mirrors, so we may briefly sum them up into a set of rules for constructing images formed by spherical mirrors.

(1) Rays passing through the centre of curvature are reflected back along their own paths.
(2) Rays parallel to the principal axis are reflected through the principal focus.
(3) Rays through the principal focus are reflected parallel to the principal axis.
(4) (Useful when using squared paper.) Rays incident at the pole are reflected, making the same angle with the principal axis.

Images formed by a concave mirror

Fig. 22.4–22.9 are ray diagrams showing the image formed by a concave mirror for different positions of an object placed on the principal axis. In all of these diagrams the object is represented by a vertical arrow OA and the image by IB.

Starting with Fig. 22.4 in which the object is placed between F and P, it will be seen that rules 1 and 2 have been used to locate the image of the point A on the object. On drawing the two reflected rays they diverge from a point B behind the mirror. B is therefore a virtual image of A. If the same construction is carried out for a series of other points along the object OA a corresponding set of virtual image points will be formed along a vertical line IB. The image formed in this case is thus larger than the object, virtual, erect and behind the mirror. This type of image is formed by a shaving or make-up mirror and also by the small concave mirror used by dentists for examining teeth.

In Fig. 22.5 the object has been moved to the principal focus. All rays given from any point on the object are reflected parallel to one another. It is usual, in such cases, to regard the rays as intersecting to form an image at infinity.

Fig. 22.6–22.9 are of particular interest, since in these cases *real images* are formed. If a white screen is placed at the image position the image will be formed on it and thus rendered visible from all directions. It is important to be able to distinguish clearly between real and virtual images. **A real image is formed by the actual intersection of rays, whereas a virtual image is one formed by the apparent intersection of rays when their directions have been produced backwards.** The practical distinction is that a real image can be formed on a screen while a virtual image cannot.

It must be understood that the use of full and dotted lines in ray diagrams is not a matter of personal preference but is an accepted convention. Full lines are used to represent real rays, objects and real images, while dotted lines are used for virtual rays and images. Also, *all rays should be arrowed* to show the direction in which the light is travelling. If these rules are followed the diagrams will be made more informative and there will be less likelihood of confusion.

In Fig. 22.6–22.8 the object and image positions are examples of *conjugate foci*. **Conjugate foci are any pair of points such that an object placed at one of them gives rise to a real image at the other.**

The fact that an object and its real image can be interchanged follows from the *principle of reversibility of light.* See also page 252.

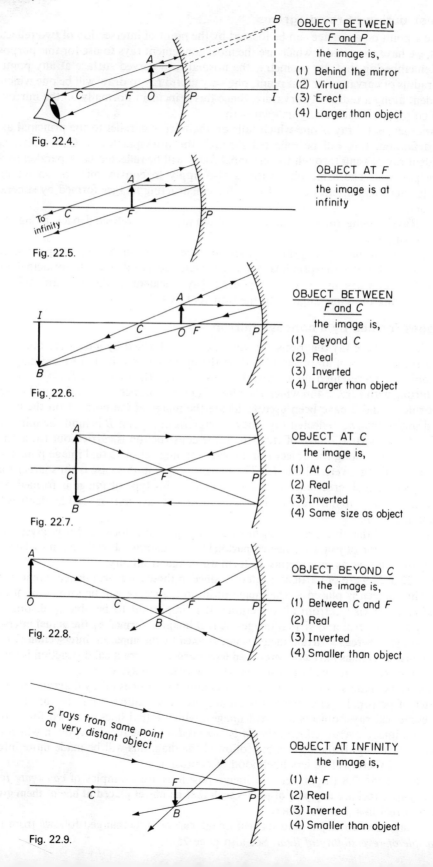

OBJECT BETWEEN
F and P

the image is,

(1) Behind the mirror
(2) Virtual
(3) Erect
(4) Larger than object

Fig. 22.4.

OBJECT AT *F*

the image is at infinity

Fig. 22.5.

OBJECT BETWEEN
F and C

the image is,

(1) Beyond *C*
(2) Real
(3) Inverted
(4) Larger than object

Fig. 22.6.

OBJECT AT *C*

the image is,

(1) At *C*
(2) Real
(3) Inverted
(4) Same size as object

Fig. 22.7.

OBJECT BEYOND *C*

the image is,

(1) Between *C* and *F*
(2) Real
(3) Inverted
(4) Smaller than object

Fig. 22.8.

2 rays from same point on very distant object

OBJECT AT INFINITY

the image is,

(1) At *F*
(2) Real
(3) Inverted
(4) Smaller than object

Fig. 22.9.

Parallel beam from curved mirrors

In connection with the reversibility of light it is worth noting that a *narrow* parallel beam of light may be obtained from a point source of light placed at the principal focus of a concave mirror of *small* aperture. We can regard the image in this case as being at infinity. If a *wide* parallel beam is required as from a car headlamp then the section of the mirror must be in the form of a *parabola* (Fig. 22.10).

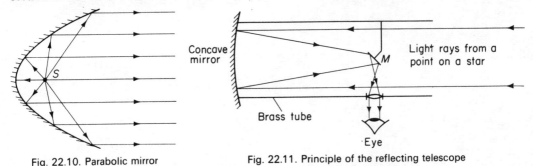

Fig. 22.10. Parabolic mirror

Fig. 22.11. Principle of the reflecting telescope

Reflecting telescopes

The case shown in Fig. 22.9 illustrates the principle of the reflecting telescope. When the object is a very long distance away from the mirror the rays from any particular point on it are practically parallel when they reach the mirror. Consequently, an image is formed at the principal focus.

The first telescope of this type was made by Sir Isaac Newton in the seventeenth century (Fig. 22.11). In order to see the image conveniently a small plane mirror, *M* is placed at 45° to the axis of the concave mirror and just in front of the principal focus. This reflects the rays to one side, and the image may now be viewed through a lens.

Newton's first telescope had a mirror of diameter about 25 mm. One of the largest reflecting telescopes in use today, called the *Hale Telescope*, is at Mount Palomar Observatory in California. It has a mirror of diameter of just over 5 m which is made of special glass coated with aluminium. The mirror took several years to make and was installed at the top of Mount Palomar, where the cloud- and dust-free atmosphere gives excellent visibility. Such a large mirror collects enough light energy to make it possible to see or photograph very distant stars and nebulae.

The world's largest reflecting telescope, with a mirror 6 metres in diameter, is in the Soviet Union. At present (1983), plans have been prepared at the Royal Greenwich Observatory for a telescope with six mirrors which together have a light-collecting capacity equivalent to a mirror 18 metres in diameter. See also Fig. 22.13.

Images formed by a convex mirror

Unlike the concave mirror, which can produce either real or virtual images according to the position of the object, the convex mirror gives virtual images only. These are always erect and smaller than the object and are formed between *P* and *F* (see Fig. 22.12).

Fig. 22.12. Image formed by diverging mirror

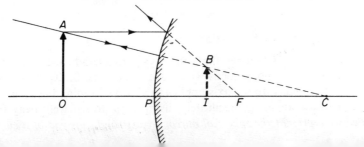

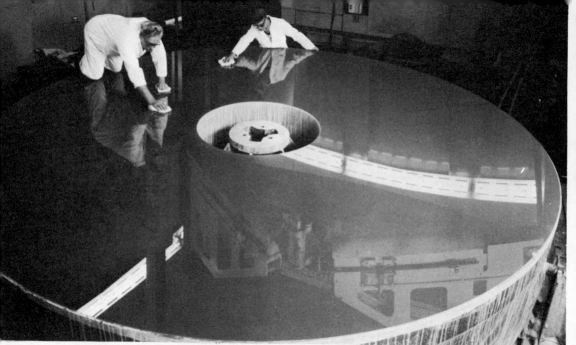

Fig. 22.13. *Polishing the 17 tonne primary mirror of the British William Herschel 4.2 metre telescope.*
The telescope is to be sited at the new International Observatory on La Palma in the Canary Islands.
Using satellite communication links, the telescope will be controlled from a computer console at the
Royal Greenwich Observatory at Cambridge in England.

La Palma was chosen as the site of the new observatory for its exceptionally fine weather record. The
whole project is a co-operative venture between the governments of Britain, Spain, Sweden, Denmark
and Holland

Convex mirrors are very convenient for use as car driving mirrors, since they
always give an erect image and a wide field of view. Fig. 22.14 shows why a convex
mirror has a wider field of view than a plane mirror of the same size.

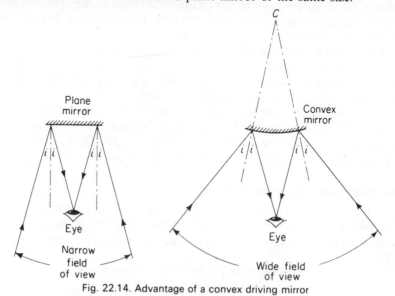

Fig. 22.14. Advantage of a convex driving mirror

Accurate construction of ray diagrams

Earlier in this chapter we mentioned that only rays parallel and *close* to the principal
axis are brought to a true point focus. If we are to have an undistorted image the
same condition of closeness to the axis applies to all rays forming the image. Nevertheless, we can locate an image accurately, using rays drawn well away from the
axis, *provided we represent the spherical mirror by a straight instead of a curved line.*

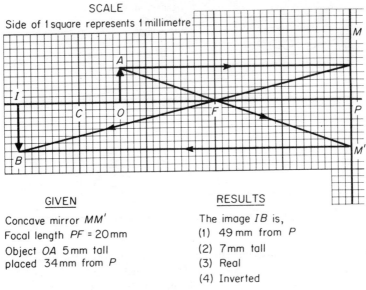

SCALE
Side of 1 square represents 1 millimetre

GIVEN

Concave mirror *MM'*
Focal length *PF* = 20mm
Object *OA* 5mm tall
placed 34mm from *P*

RESULTS

The image *IB* is,
(1) 49mm from *P*
(2) 7mm tall
(3) Real
(4) Inverted

Fig. 22.15.

Worked example

By means of an accurate graphical construction, determine the position, size and nature of the image of an object 5 mm tall, standing on the principal axis of a concave mirror of focal length 20 mm and 34 mm from the mirror.

The construction is best done on squared paper, using the side of one square as a unit to represent 1 mm to scale (Fig. 22.15).

A straight line *CP* is drawn to represent the principal axis with a line *MPM'* at right angles to it to represent the concave mirror. The principal focus *F* is marked 20 units from *P*.

The object is represented by a line *OA*, 5 units long, perpendicular to the axis *PC* and 34 units from *P*.

Two incident rays are now drawn from *A*, one parallel to the axis and the other passing through *F*. The corresponding reflected rays are drawn through *F* and parallel to the axis respectively, and these will be seen to intersect to give a real image at *B*. The image is completed by drawing *IB* perpendicular to the axis *PC*.

It will be observed that *IB* is 7 units tall and is 49 units from *P*. According to the scale used, it therefore follows that the image is 7 mm tall and is situated 49 mm from the mirror.

To measure the focal length of a concave mirror

We have already seen (Fig. 22.7) that when an object is placed at the centre of curvature of a concave mirror a real image is formed at the same place as the object. This fact is used in the following two methods for finding the focal length of a concave mirror.

Method (1). *Using an illuminated object at C*. The object used in this experiment consists of a hole cut in a white screen made of sheet metal and illuminated from behind by a pearl electric lamp. Sharpness of focusing will be greatly assisted if a thin cross-wire is placed across the hole (Fig. 22.16).

A concave mirror, mounted in a holder, is moved to and fro in front of the screen until a sharp image of the object is formed on the screen adjacent to the object. When this has been done both object and image are at the same distance from the

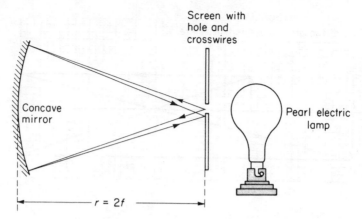

Fig. 22.16. Finding radius of curvature of a mirror

mirror, and hence both must be situated in a plane passing through the centre of curvature and at right angles to the axis. The distance between mirror and screen is measured. Half this distance is the focal length of the mirror.

Method (2). *Using a pin at C.* The concave mirror is supported vertically in a suitable holder, and a pin, stuck in a cork held in a clamp and stand, is adjusted so that the tip of the pin is at the same level as the centre of the mirror. The pin is then moved along the bench in front of the mirror until a real inverted image of the pin is seen somewhere in front of the mirror. If necessary, the tilt of the mirror should be adjusted until the tip of the image is at the same level as the tip of the pin. The pin is then moved to and fro until there is no parallax between the pin and its image. When this adjustment has been properly made the pin and its image will remain in the same straight line when the eye is moved from side to side (see Fig. 22.17).

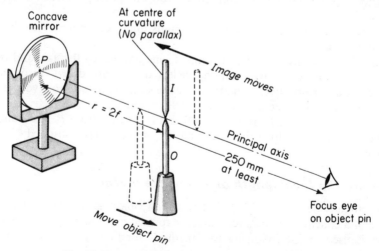

Fig. 22.17. Location of image by no-parallax method

Sometimes the image is a little difficult to find, but this can usually be overcome by making quite sure that the principal axis of the mirror passes through the tip of the object pin.

In experiments of this type one must resist a natural inclination to look *into* the mirror. *The eye should be fixed on the pin*, and the image will be seen to move backwards and forwards as the pin is moved to and fro. The pin is halted just when the image is exactly above it.

Method (3) *u and v method.* For spherical mirrors it can be proved that

$$\frac{1}{v} + \frac{1}{u} = \frac{1}{f}$$

where u = object distance from mirror
       v = image distance from mirror
       f = focal length

Hence, if u and v are measured we can calculate f.

Two pins are required, one to act as an object and the other as a search pin. The object pin is placed in front of the mirror between F and C so that a magnified real image is formed beyond C. The search pin is then placed so that there is no parallax between it and the real image (Fig. 22.18).

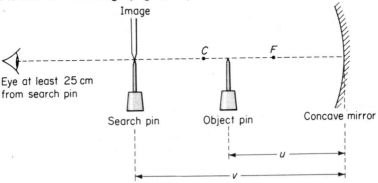

Fig. 22.18. The u and v method

The distance of the object pin from the mirror will then give the object distance u, and that of the search pin the image distance v.

Several pairs of object and image distances are obtained in this way and the results entered in a table as shown. A mean value for the focal length f may then be calculated. Care should be taken to apply correct arithmetical signs according to the sign convention used (see end of chapter).

u (cm)	v (cm)	$\dfrac{1}{u}$	$\dfrac{1}{v}$	$\dfrac{1}{v} + \dfrac{1}{u}$
			Mean value of $\dfrac{1}{f}$	
			Mean value of f	

The above method of calculation requires a table of reciprocals, but if this is not available it will be found easier to calculate the focal length from

$$f = \frac{uv}{u + v}$$

Magnification formula

Since the image formed by a spherical mirror varies in size, we refer to the linear or transverse magnification, m, which is defined as,

$$m = \frac{\text{height of image}}{\text{height of object}}$$

and in all cases it can be shown that

$$m = \frac{\text{image distance from mirror}}{\text{object distance from mirror}}$$

As a typical case, consider Fig. 22.19, in which the image has been constructed using rules 1 and 4 (page 237).

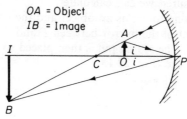

OA = Object
IB = Image

Fig. 22.19.

The right-angled triangles IBP and OAP are similar, and hence

$$\frac{IB}{OA} = \frac{IP}{OP}$$

or $\dfrac{\text{height of image}}{\text{height of object}} = \dfrac{\text{image distance}}{\text{object distance}} = \text{magnification} = m$

Mirror formula. Sign convention

It is shown in more advanced textbooks that, for both concave and convex mirrors,

$$\frac{1}{v} + \frac{1}{u} = \frac{1}{f}$$

where u = object distance from mirror
v = image distance from mirror
f = focal length

We have already seen that an image is formed sometimes in front of a curved mirror and sometimes behind it. This makes it necessary to have a sign convention so that we may distinguish between the two cases and obtain the correct answer when substituting in the formula.

Examples showing the use of the two sign conventions in common use are given below.

NEW CARTESIAN CONVENTION

(1) All distances are measured from the mirror as origin.
(2) Distances measured against the incident light are negative.
(3) Distances measured in same direction as incident light are positive.

One advantage of this system is that, if the object is placed to the left of the mirror the ordinary graphical (Cartesian) convention of signs comes into operation, i.e.,

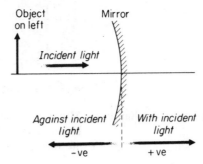

Fig. 22.20. New Cartesian convention applied to mirrors

distances measured to the left are negative; those to the right, positive (Fig. 22.20).

On the new Cartesian convention the focal length of a concave mirror is negative; that of a convex mirror, positive.

REAL-IS-POSITIVE CONVENTION

(1) All distances are measured from the mirror as origin.
(2) Distances of real objects and images are positive.
(3) Distances of virtual objects and images are negative.

On the real-is-positive convention a concave mirror has a real principal focus and hence a positive focal length. A convex mirror has a virtual principal focus and a negative focal length.

Worked examples

1. *An object is placed (a) 20 cm, (b) 4 cm, in front of a concave mirror of focal length 12 cm. Find the nature and position of the image formed in each case.*

New Cartesian

(a) $u = -20$ cm (object on left)
  $f = -12$ cm

Substituting in the formula,

$$\frac{1}{v} + \frac{1}{-20} = \frac{1}{-12}$$

$$\frac{1}{v} = \frac{1}{-12} + \frac{1}{20} = \frac{-2}{60}$$

whence $v = -30$ cm.

Minus sign means image on same side as object.

Hence a real image is formed 30 cm from mirror on same side as object.

(b) $u = -4$ cm

Substituting in the formula,

Real-is-positive

$u = +20$ cm (real object)
$f = +12$ cm

Substituting in the formula,

$$\frac{1}{v} + \frac{1}{20} = \frac{1}{12}$$

$$\frac{1}{v} = \frac{1}{12} - \frac{1}{20} = \frac{2}{60}$$

whence $v = 30$ cm.

Plus sign means a real image.

Hence a real image is formed 30 cm from mirror on same side as object.

$u = 4$ cm

Substituting in the formula,

$$\frac{1}{v} + \frac{1}{-4} = \frac{1}{-12}$$

$$\frac{1}{v} = \frac{1}{-12} + \frac{1}{4} = \frac{2}{12}$$

whence $v = 6$ cm.

Plus sign means image on right hand side, i.e., behind mirror.

Hence a virtual image is formed 6 cm from mirror.

$$\frac{1}{v} + \frac{1}{4} = \frac{1}{12}$$

$$\frac{1}{v} = \frac{1}{12} - \frac{1}{4} = \frac{-2}{12}$$

whence $v = -6$ cm.

Minus sign means virtual image.

Hence a virtual image is formed 6 cm from mirror.

2. *A concave mirror produces a real image* 1 cm *tall of an object* 2.5 mm *tall placed* 5 cm *from the mirror. Find the position of the image and the focal length of the mirror.*

New Cartesian

Using the magnification formula,

$$m = \frac{\text{image distance}}{\text{object distance}}$$

we have, $\dfrac{1}{0.25} = \dfrac{\text{image distance}}{5}$

whence image distance = 20 cm on same side as object.

thus $\left.\begin{array}{l} v = -20 \text{ cm} \\ \text{and } u = -\ 5 \text{ cm} \end{array}\right\}$ both on left

Substituting in the formula,

$$\frac{1}{-20} + \frac{1}{-5} = \frac{1}{f}$$

$$\frac{-1-4}{20} = \frac{1}{f}$$

whence $\underline{f = -4 \text{ cm}}$

Real-is-positive

Using the magnification formula,

$$m = \frac{\text{image distance}}{\text{object distance}}$$

we have, $\dfrac{1}{0.25} = \dfrac{\text{image distance}}{5}$

whence image distance = 20 cm on same side as object.

thus $\left.\begin{array}{l} v = +20 \text{ cm} \\ \text{and } u = +\ 5 \text{ cm} \end{array}\right\}$ both real

Substituting in the formula,

$$\frac{1}{20} + \frac{1}{5} = \frac{1}{f}$$

$$\frac{1+4}{20} = \frac{1}{f}$$

whence $\underline{f = 4 \text{ cm}}$

3. *A convex mirror of focal length* 18 cm *produces an image on its axis,* 6 cm *away from the mirror. Calculate the position of the object.*

New Cartesian

$v = +\ 6$ cm (image on right)
$f = +18$ cm (convex mirror)

Substituting in the formula,

$$\frac{1}{6} + \frac{1}{u} = \frac{1}{18}$$

$$\frac{1}{u} = \frac{1}{18} - \frac{1}{6} = \frac{-2}{18}$$

whence $u = -9$ cm.

Minus sign means object is to left of mirror, i.e., 9 cm in front of it.

Real-is-positive

$v = -\ 6$ cm (virtual image)
$f = -18$ cm (virtual focus)

Substituting in the formula,

$$\frac{1}{-6} + \frac{1}{u} = \frac{1}{-18}$$

$$\frac{1}{u} = \frac{1}{-18} + \frac{1}{6} = \frac{2}{18}$$

whence $u = +9$ cm.

Plus sign means real object and, therefore, 9 cm in front of mirror.

QUESTIONS: 22

1. An object is placed on the axis of a converging (concave) mirror of focal length 200 mm. The image produced is inverted and has a magnification of 1.5. By calculation or by scale drawing on graph paper determine the position of the object. (*A.E.B.*, 1982)

2. Distinguish between a *real* image and a *virtual* image. Explain how a concave mirror can give either kind of image. Illustrate each case by a ray diagram.

A concave mirror has a radius of curvature of 20 cm. Find the position, magnification, and nature of the image of a small pin placed on the axis and at right angles to it, and 15 cm from the pole. (*O.*)

3. Show on a ray diagram the *centre of curvature, axis, pole,* and *principal focus* for a concave spherical mirror.

Describe how you would find by an optical method the radius of curvature of a concave mirror.

Draw a ray diagram to show how a concave mirror can produce a virtual image.

An object 2 cm long is placed 40 cm in front of a concave mirror of focal length 15 cm so that it is perpendicular to, and has one end resting on, the axis of the mirror. Find by means of a ray diagram, drawn to suitable scales, the size and position of the image. (*A.E.B.*)

4. Describe and explain how you would measure the radius of curvature of a concave spherical mirror by experiment.

Such a mirror has a radius of curvature 20 cm. Using a scale diagram, determine full details of the image when an object 1.5 cm high is placed at a distance of 4 cm from the mirror.

By means of an appropriate diagram, show how a suitable curved mirror may be used to produce an accurately parallel beam of light from a small source. (*S.*)

5. A concave mirror is used to form an image of an object pin. Where must the object be placed to obtain: (*a*) an upright, enlarged image; (*b*) an image the same size as the object? (*J.M.B.*)

6. State the laws of reflection of light.

Draw a semicircle, radius 10 cm, to represent the cross-section of a spherical concave mirror. Two rays of light, parallel to the principal axis, are incident on the mirror, and each is 6 cm distant from the axis. By applying the laws of reflection, find the point where these rays cross.

Repeat this process on the same diagram, for two other rays each 2 cm from the principal axis. What deduction can you make from your result? (*S.*)

7. Explain with the aid of diagrams, the formation of
(i) virtual, and
(ii) real images. Give one example of each.

Show on a ray diagram the *centre of curvature, axis, pole,* and *focus* of a *convex mirror.*

Give two reasons why convex mirrors are frequently used as driving mirrors.

An object 1 cm tall is placed 30 cm in front of a convex mirror of focal length 20 cm so that it is perpendicular to, and has one end resting on, the axis of the mirror. Find the size and position of the image formed by the mirror. (*A.E.B.*)

8. A small object is placed on the principal axis of a convex mirror of focal length 10 cm. Determine the position of the image when the object is 15 cm from the mirror. (*J.M.B.*)

9. Draw a ray diagram to show how a spherical mirror can produce a real diminished image of an object. (*J.M.B.*)

10. Explain what is meant by the terms *centre of curvature, principal focus,* and *focal length of a spherical mirror.* Illustrate your answers with labelled diagrams of:
(*a*) a concave mirror;
(*b*) a convex mirror.

Describe an experiment to determine the radius of curvature of a concave mirror.

A concave spherical mirror of radius of curvature 20 cm forms an erect image 30 cm from the mirror and 5 cm high. Find the position and size of the object and show with a scale diagram how the image is formed. (*A.E.B.*)

23. Refraction at plane surfaces

A pond or a swimming bath both appear much shallower than they actually are; a straight stick appears bent when partly immersed in water; and the landscape "shimmers" on a hot summer's day. These, and many similar effects are caused by *refraction*, or the change in direction of light when it passes from one medium to another.

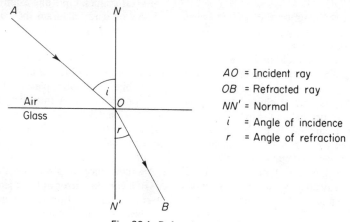

AO = Incident ray
OB = Refracted ray
NN' = Normal
i = Angle of incidence
r = Angle of refraction

Fig. 23.1. Refraction of light

The terms used in connection with refraction are illustrated in Fig. 23.1, which represents the passage of a ray of light from air to glass. The *angle of incidence, i*, is the angle between the incident ray and the normal at the point of incidence. The *angle of refraction, r*, is the angle between the refracted ray and the normal.

It is important to remember that. when a ray passes from one medium to a more optically dense medium, the ray bends towards the normal. Conversely, a ray passing from glass or water into air is bent away from the normal.

Before going on to the next section it should be mentioned that refraction can be explained in terms of a *change in the speed of light* when it passes from one medium to another. This important aspect of refraction is dealt with in chapter 26 as part of our study of wave motion.

The laws of refraction

Although many scientists worked on the problem, the laws governing the refraction of light when it passes from one substance to another resisted discovery for centuries. It was not until 1621 that Willebrord Snell, Professor of Mathematics at Leyden University, discovered the exact relationship between the angles of incidence and refraction.

The laws of refraction are now stated as follows:

Law (1) **The incident and refracted rays are on opposite sides of the normal at the point of incidence and all three are in the same plane.**

Law (2) (Also known as Snell's law.) The ratio of the sine of the angle of incidence to the sine of the angle of refraction is a constant for a given pair of media.

To investigate Snell's law of refraction

A straight line *SS'*, to represent the surface of separation between air and glass, is drawn on a sheet of drawing paper on a drawing-board, together with a normal *ON* and several lines at various angles to *ON* to represent incident rays (Fig. 23.2).

A ruler is placed along *SS'* and a rectangular glass block carefully placed in contact with it in the position shown. The ruler is now transferred to the lower edge of the block and the block is then removed.

A line *TF'*, to represent the lower edge of the block, is now drawn. Without moving the ruler, the block is now placed carefully in contact with the ruler. The two lines *SS'* and *TT'* should now coincide exactly with the upper and lower vertical faces of the block.

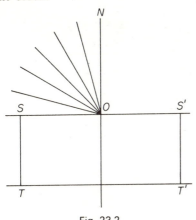

Fig. 23.2.

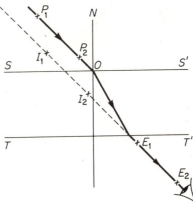

Fig. 23.3. Refraction through a glass block

This is a better method than simply drawing round the block with a pencil. The block is thick and its edges are usually bevelled, which renders it difficult to draw lines to coincide exactly with the block edges.

Method. Two pins P_1 and P_2 are stuck in the paper as far apart as possible along one of the lines drawn to represent an incident ray (Fig. 23.3). Then, looking through the block, the emergent ray is located by sticking two sighting pins E_1 and E_2 in the paper exactly in line with the images I_1 and I_2 of the pins P_1 and P_2. This procedure is carried out for all the incident rays, each time marking the positions of the pins with fine pencil crosses.

When this has been done the block is removed and the points E_1 and E_2 joined. Finally, the refracted rays are drawn in by joining *O* to the points where the emergent rays leave the block.

The ratio sin *i*/sin *r* may now be found for each pair of rays by looking up the sines in tables (at the end of the book). The ratio should be found to be practically constant.

i	*r*	sin *i*	sin *r*	$\dfrac{\sin i}{\sin r}$

Also, a graph of sin i against sin r should be plotted and a mean value for $\dfrac{\sin i}{\sin r}$ found from its gradient. See also the section entitled, Refractive Index.

Alternative treatment of results. Instead of measuring the angles i and r and finding the sines from tables, the ratio $\dfrac{\sin i}{\sin r}$ may be found by the following method which Snell himself used.

The largest convenient circle centre O is drawn and the incident and refracted rays (not the emergent rays) are produced, if necessary, to cut the circle at A and C. Perpendiculars AB and CD are drawn to the normal ON (Fig. 23.4).

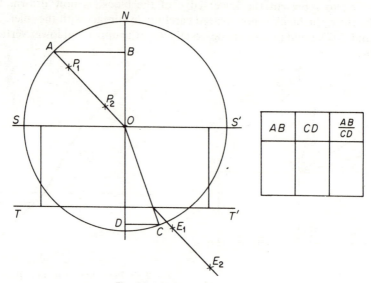

AB	CD	$\dfrac{AB}{CD}$

Fig. 23.4. Verifying Snell's law

Referring to Fig. 23.4 in which, for simplicity, only one pair of rays has been shown, we see that

$$\sin i = \frac{AB}{AO}, \text{ and } \sin r = \frac{CD}{CO}$$

therefore

$$\frac{\sin i}{\sin r} = \frac{AB}{AO} \times \frac{CO}{CD}$$

but

$$AO = CO \text{ (radii of circle)}$$

therefore

$$\frac{\sin i}{\sin r} = \frac{AB}{CD}$$

For each pair of rays, AB and CD are measured and recorded in a table. If the value of $\dfrac{AB}{CD}$ is constant, then Snell's law is true in this case.

Refractive index

The value of the constant $\dfrac{\sin i}{\sin r}$ for a ray passing from one medium to another is called the refractive index of the second medium with respect to the first; and is denoted by the symbol n.

If, however, the first medium is air it is usual to speak of n simply as the refractive index of the second medium but, strictly speaking, the *absolute* refractive index of a medium is the value of n when the first medium is a vacuum.

Thus, if a ray passes from air into water, then

$$\text{refractive index of water} = n = \frac{\sin i}{\sin r}$$

It follows that the mean value of the ratio $\dfrac{AB}{CD}$ obtained in the experiment to verify Snell's law will be equal to the refractive index of glass.

Manufacturers have to know the refractive index of the glass used for making lenses of telescopes, microscopes, cameras and other optical instruments. For crown glass, n is about 1.52; for flint glass about 1.65. The refractive index of water is 1.33.

Geometrical construction for refracted ray

The alternative method of treating the results of the last experiment suggests a way of finding the path of a ray through a glass block or some other medium, when i and n are given.

Example. *Trace the path of a ray through a glass block of refractive index 1.52, for light entering at an angle of incidence of 45°. Measure the angle of refraction.*

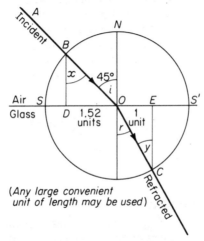

Fig. 23.5. Finding path of refracted ray

Draw SS' to represent the surface of separation between glass and air and AO an incident ray making an angle of 45° with the normal NO (Fig. 23.5). Then, using the largest possible scale for best accuracy, mark off OD equal to 1.52 units of length.

At D erect a perpendicular to cut AO at B. With centre O and radius OB, draw a circle. Along OS', mark off OE equal to 1 unit of length. Through E draw a perpendicular to SS' to cut the circle at C. Then OC is the required refracted ray and the angle of refraction r is found to be 28°.

Proof.
$$OB = OC \text{ (radii)}$$
$$i = x \quad \text{and} \quad r = y \text{ (alternate angles)}$$
$$\frac{\sin i}{\sin r} = \frac{\sin x}{\sin y} = \frac{1.52}{OB} \,\bigg/\, \frac{1}{OC} = 1.52 = n \text{ (given)}$$

Some effects of refraction

The apparent upward bending of a stick when placed in water is shown in Fig. 23.6. Rays of light from the end B of the stick pass from water to air, and are bent away from the normal, since they are passing to a less optically dense medium. Entering the eye, the rays appear to be coming from a point C above B. C is thus the image of

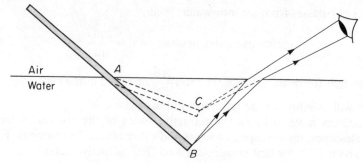

Fig. 23.6. A stick appears bent in water

B as a result of refraction. The same reasoning applies to any point on the immersed portion of the stick *AB*, so that the observer sees an image apparently in the position *AC*.

To avoid confusion in this and similar ray diagrams one should always be careful to use the accepted convention of drawing *real rays, real images and objects in full lines, and virtual rays and images in dotted lines.* In addition, an arrow should be placed on a ray to show the direction in which the light travels.

Fig. 23.7 illustrates the appearance of print viewed through the top of a piece of thick glass placed over it. Since the rays are refracted away from the normal when they pass from glass to air, the print and the bottom of the glass appear raised. The apparent raising of the print thus occurs for the same reason that a stick appears to be bent upwards when placed in water, as explained above.

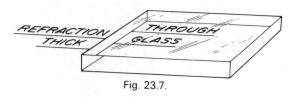

Fig. 23.7.

The principle of reversibility of light

The principle of reversibility of light, which states that the paths of light rays are reversible, has already been mentioned in connection with reflection on page 237. The same holds true in the case of refraction and, indeed, for rays passing through any optical system.

The refractive indices for a ray passing fron air to glass and from glass to air may be distinguished by using the symbols $_an_g$ and $_gn_a$ respectively.

Thus, using Fig. 23.1,

$$_an_g = \frac{\sin i}{\sin r}$$

and by the principle of reversibility of light,

$$_gn_a = \frac{\sin r}{\sin i}$$

hence

$$_gn_a = \frac{1}{_an_g}$$

The equation will be used later on in the chapter in connection with critical internal reflection. If $_an_g = 1.5$ it follows that $_gn_a = 1/1.5 = 0.67$.

Real and apparent depth

A thick slab of glass appears to be about two-thirds of its real thickness when viewed from vertically above. Similarly, water in a pond appears to be only three-quarters of its true depth.

Fig. 23.8 shows how this comes about. Rays from a point *O* at the bottom of the slab are refracted away from the normal where they leave the glass and then enter the pupil of the eye as though coming from a virtual image *I* above *O*. For clarity, the horizontal scale of this diagram has been exaggerated. The diameter of the eye pupil is, of course, so small that only rays very close to the normal will enter it.

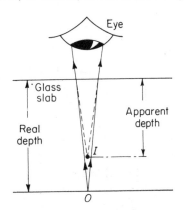

Fig. 23.8. Real and apparent depth

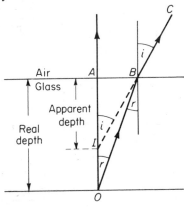

Fig. 23.9.

Refractive index related to real and apparent depth

The way refractive index is related to real and apparent depth is illustrated in Fig. 23.9. Once again the horizontal scale is exaggerated, but it is to be understood that *OBC* is a ray very close to the normal which enters the eye from a point *O* at the bottom of the slab. The emergent ray *BC* appears to be coming from a virtual image *I*, so that *AI* is the apparent depth of the slab.

By using the *principle of reversibility of light*, the refractive index of glass is given by

$$n = \frac{\sin i}{\sin r}$$

but $\qquad \angle AIB = i \text{ (corresponding angles)}$

and $\qquad \angle AOB = r \text{ (alternate angles)}$

therefore $\qquad n = \dfrac{\sin \angle AIB}{\sin \angle AOB}$

$$= \frac{AB}{BI} \bigg/ \frac{AB}{BO} = \frac{BO}{BI} = \frac{AO}{AI} \text{ when } B \text{ is very close to } A$$

or $\qquad n = \dfrac{\text{real depth}}{\text{apparent depth}}$

To measure refractive index by the real and apparent depth method

(*a*) *Glass.* A glass block is placed vertically over a straight line ruled on a sheet of paper (Fig. 23.10 (*a*)). A pin on a sliding cork adjacent to the block is then moved up or down until there is no parallax between it and the image of the line seen through the block. Measurements of the real and apparent depth are then taken as indicated in the diagram and the refractive index of glass calculated.

(*b*) *Water or other liquid.* The same method as that described above may be used except that the glass block is replaced by a tall beaker containing the liquid and having a pin lying on the bottom.

However, it will be found easier to adjust for no parallax if we use the fact that the image in a plane mirror is as far behind the mirror as the object is in front (Fig. 23.10 (*b*)). A slip of mirror rests across the top of the beaker, and a sliding pin above it is

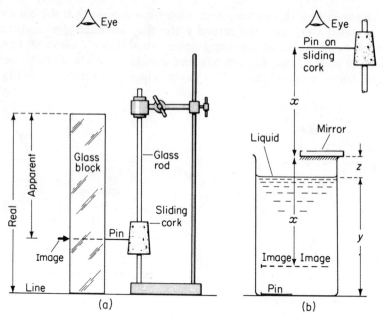

Fig. 23.10. Refractive index by apparent depth method

adjusted until there is no parallax between the image of this pin in the mirror and the image of the pin lying on the bottom of the beaker.

Fig. 23.10 (b) makes it clear that $n = \dfrac{\text{real depth}}{\text{apparent depth}} = \dfrac{y}{x-z}$

Several readings may be obtained, each time varying the depth of the liquid. A graph of real depth against apparent depth is plotted and a mean value of n obtained from its gradient.

Total internal reflection. Critical angle

When light passes from one medium to a more optically dense medium there will always be both reflection and refraction for all angles of incidence (Fig. 23.11). But this is not always so when light passes from one medium to a less optically dense medium.

Suppose we consider a ray passing from glass to air. Starting with a small angle of incidence (Fig. 23.12 (a)), we get a weak internally reflected ray and a strong refracted ray.

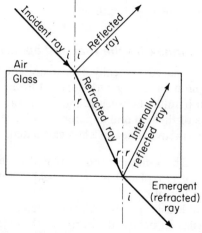

Fig. 23.11. Refraction and reflection in a glass block

As the angle of incidence increases the angle of refraction also increases, and at the same time the intensity of the reflected ray gets stronger and that of the refracted ray weaker.

Finally, at a certain critical angle of incidence *c*, the angle of refraction becomes 90° (Fig. 23.12 (*b*)).

Since it is impossible to have an angle of refraction greater than 90°, it follows that for all angles of incidence greater than the critical angle *c* the incident light undergoes what we describe as *total internal reflection* (Fig. 23.12 (*c*)).

See also optical fibres, page 272.

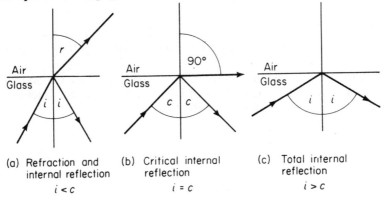

(a) Refraction and internal reflection
$i < c$

(b) Critical internal reflection
$i = c$

(c) Total internal reflection
$i > c$

Fig. 23.12. What happens when light passes from glass to air

Relation between critical angle and refractive index

We have already seen that $_gn_a = \dfrac{1}{_an_g}$, and hence, from Fig. 23.12 (*b*).

$$\frac{\sin c}{\sin 90°} = \frac{1}{_an_g}$$

But

$$\sin 90° = 1$$

therefore

$$\sin c = \frac{1}{_an_g}$$

or

$$_an_g = \frac{1}{\sin c}$$

For crown glass of refractive index $n = 1.5$, the critical angle *c* is about 42°. In the case of water ($n = 1.33$) it is about 49°. These values are obtained from the above equation, using a table of sines.

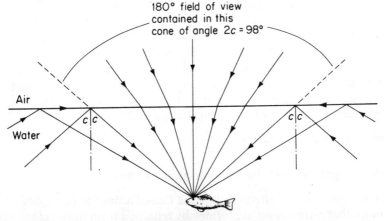

180° field of view contained in this cone of angle $2c = 98°$

Fig. 23.13. The fish's-eye view

The fish's-eye view

At whatever depth a fish happens to be it has a full view of everything above the water, provided, of course, that the water surface is unruffled. Fig. 23.13 explains this and shows how the fish enjoys a 180° field of view apparently all squeezed into a cone of angle about 98° (i.e., twice the critical angle for water). Outside this range the fish will see objects in the water and on the bottom which are mirrored in the surface of the water by total internal reflection.

To measure the critical angle and refractive index of the material of a prism

An equilateral glass prism *ABC* is placed on a sheet of paper and its outline drawn on the paper with the aid of a ruler in a manner similar to that described for the glass block on page 249.

An object pin *O* is then stuck in the paper touching the side *CA* and its image is viewed through the side *BC* (Fig. 23.14). Light enters the eye after internal reflection from *AB*, followed by refraction through *BC*.

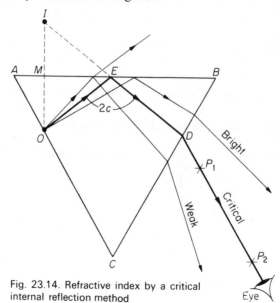

Fig. 23.14. Refractive index by a critical internal reflection method

If the eye is well to the left the image seen is weak owing to the fact that the light has undergone only partial internal reflection. If, however, the eye is well to the right the image is bright, since total internal reflection has occurred. Therefore, as the eye moves from left to right through the critical position the image suddenly becomes brighter. The emergent ray for this position is located by pins P_1 and P_2. The prism is then removed and $P_1 P_2$ joined to cut *BC* in *D*.

Because *AB* acts as a reflecting surface the ray *ED* appears to be coming from *I*, which is the image of *O* in *AB*. Rays *OE* and *ED* can therefore be located as follows. Draw *OMI* perpendicular to *AB*, making *IM = OM*. Join *ID* to cut *AB* in *E*. Join *OE*. Then angle *OED* = twice the critical angle *c*. Hence *c* can be found and the refractive index calculated from $n = \dfrac{1}{\sin c}$.

Multiple images formed by a thick glass mirror

Ordinary mirrors, made by silvering the back face of a sheet of glass, suffer from the disadvantage that extra images are formed by reflection from the front surface of the glass and also by multiple internal reflection inside the glass. Fig. 23.15 shows how

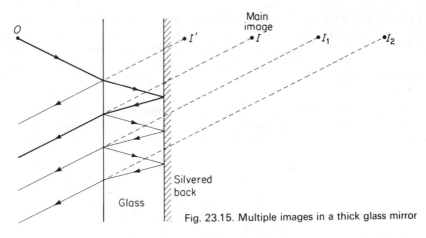

Fig. 23.15. Multiple images in a thick glass mirror

the images are formed. Ordinarily these images are very much weaker than the image formed by the silvered surface and go unnoticed. But if silvered glass mirrors are used in certain optical instruments, the extra images are a nuisance. For this reason plane mirrors in reflex cameras, and the curved mirrors of telescopes are aluminized on the front surface. Care should be taken when handling these, as the metallized surface is easily scratched.

Total internal reflection in prisms

The problem of providing an untarnishable mirror which gives one image only has been solved in the submarine periscope by using 45° right-angled glass prisms, as shown in Fig. 23.16 (*a*). Actually a submarine periscope is a combined periscope and telescope, but for simplicity in the diagram the lenses have been omitted. Light enters the faces of the prisms normally and falls on the hypotenuse face internally at an angle of incidence of 45°. Total reflection occurs here, since the critical angle for ordinary glass is about 42° (page 255).

For reasons explained in chapter 24, it is necessary to place slides or films upside down in a projector in order to obtain a picture the right way up on the screen. It is not always possible to do this, since it is occasionally necessary to project an image of a thin glass cell containing a liquid. For example, polarization on the copper plate

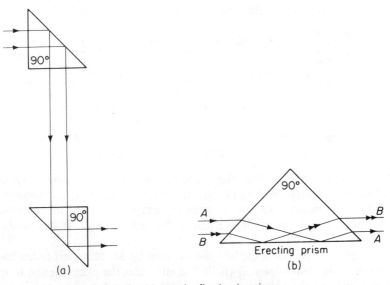

Fig. 23.16. Total internal reflection in prisms

of a simple cell (see page 397) may be shown on a screen by this method. The image is obtained the correct way up by placing an *erecting prism* in front of the projection lens. Fig. 23.16 (*b*) shows how a right-angled prism is used for this purpose. Light enters the face of the prism approximately parallel to the base or hypotenuse face. Total reflection occurs at the base, with the result that the rays passing through are inverted.

A third example of employing total internal reflection in right-angled prisms is found in prism binoculars. A prism binocular is simply a pair of telescopes conveniently shortened in length by causing the light to traverse the tube three times instead of once.

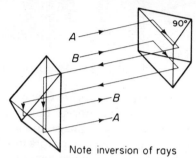

Note inversion of rays

Fig. 23.17. Binocular prisms

If light is incident perpendicular to the hypotenuse face of the prism it will undergo two internal reflections, finally emerging parallel to its original path but travelling in the opposite direction. Fig. 23.17 indicates how two prisms are used in this manner in the binocular. Note that the rays A and B are inverted, giving the added advantage that the final image is the right way up (see also page 272).

The mirage

Mirages are usually associated with hot deserts. The traveller in a desert often sees what appears to be a sheet of water a short distance ahead of him. This he is never able to reach, since it is an optical illusion.

It is not necessary to make a journey to the Sahara in order to see this phenomenon. Mirages are quite common elsewhere. On a hot summer's day the surface of a hot roadway in the near distance may appear wet and shiny as if after rain. Since we are accustomed to seeing the sky mirrored in the surface of still water, the natural inference made is that the road surface is wet (Fig. 23.18).

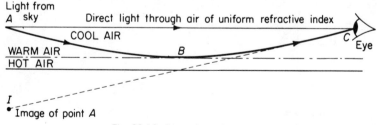

Fig. 23.18. Formation of a mirage

Two explanations have been suggested. Traditionally, it has been explained by the progressive refractive bending of light from the sky as it passes through successive layers of less optically dense air of increasing temperature near the hot ground until it finally meets a layer at an angle equal to the critical angle. It then undergoes total reflection.

A rather more probable explanation is given by M. Minnaert in his book entitled *Light and colour in the open air* (Bell). He attributes the phenomenon to the progressive bending of light as it passes through the warmer layers of decreasing refractive index until it becomes parallel to the ground. After this it proceeds to bend upwards. The process is thus regarded as one of continuous and progressive refraction rather than total reflection.

QUESTIONS: 23

1. Explain the statement "the refractive index of glass is 1.5". *(J.M.B.)*

2. In an experiment carried out with a rectangular block of perspex, the following readings were taken:

Angle of incidence (i)	22°	28°	33°	38°	41°
sin i	0.374 6		0.544 6	0.615 7	0.656 1
sin r	0.258 8		0.358 4	0.454 0	0.438 4
Angle of refraction (r)	15°	18°	21°	27°	26°

Complete the table and then plot a graph of sin i against sin r.

Use your graph to determine:
(a) which reading was measured or recorded incorrectly;
(b) the refractive index of perspex (air to perspex);
(c) the angle of refraction (to the nearest degree) when the angle of incidence is 45°.
(A.E.B.)

3. By a graphical construction, and without using a table of sines, find the angle of refraction for a ray of light incident at an angle of 55° on the plane surface of a slab of perspex of refractive index 1.49.

4. State the *laws of refraction* and define *refractive index*.
Describe in detail an experiment to find the refractive index of glass.

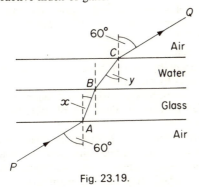

Fig. 23.19.

Fig. 23.19 shows a slab of glass of uniform thickness, lying horizontally. Above it is a layer of water. A ray of light PQ is incident upwards on the lower surface of the glass and is refracted successively at A, B, and C, the points where it crosses the interfaces. Calculate
(i) angle x,
(ii) angle y, and
(iii) the refractive index for light passing from the water to glass. (Refractive index of glass $= \frac{3}{2}$; refractive index of water $= \frac{4}{3}$.)
(A.E.B.)

5. A microscope is focused on a mark on a table. When the mark is covered by a plate of glass 2 cm thick the microscope has to be raised 0.67 cm for the mark to be once more in focus. Calculate the refractive index of the glass. *(O.C.)*

6. Answer any TWO of the following, illustrating by means of diagrams:
(a) Explain why a stick, partly immersed in water and placed obliquely to the surface, appears bent at the surface.
(b) When looking obliquely into a mirror at the image of a candle several images are seen. Explain this. Which of the images is the brightest? Why?
(c) What is meant by: (i) *total internal reflection*; (ii) *critical angle*?
Explain, with the aid of a diagram, how a suitable glass prism may be used to turn a ray of light through 180°. *(L.)*

7. (i) Draw a ray diagram to show why a pool of water appears to be only three-quarters of its real depth when viewed vertically from above.
(ii) Find the critical angle of a medium of refractive index 1.65. *(J.M.B.)*

8. A ray of light is incident on the plane surface of a transparent material at such an angle that the reflected and refracted rays are at right angles to each other. Draw a diagram to illustrate this and calculate the refractive index if the angle of refraction = 30°. *(S.)*

9. Define *refractive index* and explain briefly a method of measuring its value for water.
A small flat object, e.g., a penny, lies at the bottom of a tank containing water to a depth of 16 cm. Find by drawing or by trigonometry the apparent position of the object when viewed from directly above.
Does the object appear to be the same size as it does when the tank is empty and it is viewed from the same point? If not, does it seem larger or smaller? Give a reason for your answer. (Refractive index of water $= \frac{4}{3}$.)
(O.C.)

10. A small object is placed at the bottom of a beaker containing a liquid of refractive index $\frac{4}{3}$ to a depth of 10 cm. A pin placed

horizontally above the beaker is adjusted until there is no parallax between the image of the pin formed by reflection at the liquid surface and the image of the object. What is the height of the pin above the liquid surface? (*L.*)

11. What is meant by the critical angle of a medium? How would you find the refractive index and critical angle of a given specimen of glass?

Show by a ray diagram how a right-angled glass prism may be used:
(i) to turn a ray through 90°;
(ii) to turn a ray through 180°;
(iii) to invert a beam of light. (*J.M.B.*)

12. What is meant by *total internal reflection* of light? State the conditions under which this occurs. Define critical angle and explain how the critical angle for a material in air depends on its refractive index.

Name one practical application of total internal reflection.

Find the angle of incidence of a ray of light on one face of a 60° prism if the ray is just totally internally reflected on meeting the next face. (Take the refractive index of glass to be 1.5.) (*O.*)

13. State the laws of refraction of light, defining the angles of incidence and refraction and showing them on a diagram.

A slab of glass 1 cm thick is silvered underneath and rests on a table. A speck of dust is on the top face. Find, by a drawing to scale or by calculation, the position of the image of the speck by reflection at the silvered surface and refraction at the top. (Refractive index of glass = 1.5.) (*O.C.*)

14. Define *refractive index* and describe ONE method by which its value for glass may be determined.

ABC is a triangular prism, made of glass of refractive index 1.5, in which the angle *A* is 30° and the angle *C* is 60°. A ray of light is incident normally on the face *AB*. By graphical construction, or otherwise, determine the path of the ray and show it clearly on a diagram. (*L.*)

15. Give the meaning of the terms *refractive index, critical angle,* and *total internal reflection,* illustrating your answers by diagrams.

Describe briefly how you can determine the refractive index of water. If a pinhole camera pointing vertically upwards is filled with a liquid and taken to an open space the image of the sky is found to be a circular disc of definite radius when the receiving screen is large enough. Explain this and deduce the refractive index of the liquid given that the perpendicular distance of the pinhole from the image is 10 cm and the radius of the image is 8 cm. (*O.C.*)

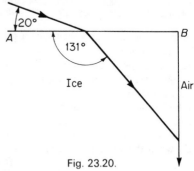

Fig. 23.20.

16. Fig. 23.20 shows the path of a ray of light through one corner of a cube of ice. Find:
(*a*) the angle of incidence on the face *AB*;
(*b*) the angle of refraction at this face.

Mark, on the diagram, the critical angle for ice, and mark also the direction of any one additional ray which can occur owing to partial reflection. (*C.*)

17. An insect hovers in a fixed position above the still water of a pond. Draw a diagram to show approximately where it appears to be to a fish vertically below it. (*O.C.*)

18. Define *refractive index* and explain what is meant by *critical angle,* illustrating your answers by diagrams.

Describe how you would find the refractive index of glass in the form of a triangular prism.

On warm sunny days tarmac roads often appear to be covered with pools of water some distance ahead, which disappear when approached. How do you explain this?

(*S.*)

24. Lenses and optical instruments

Magnifying glasses or lenses have been in use for centuries and were well known to the Greeks and medieval Arabs. Lenses of many different types play an important part in our own everyday life. Apart from the benefit of spectacles which enable millions of people to read in comfort, our lives would be vastly changed if we had no cameras, projectors, microscopes or telescopes, all of which function by means of lenses.

Not all lenses can be used as magnifying glasses. There are some, used in opera glasses and in spectacles for short-sighted persons, which always give a diminished, erect virtual image. These are referred to as concave or *diverging* lenses, while mag-

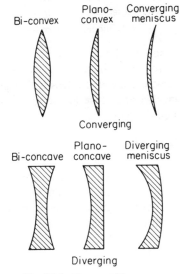

Fig. 24.1. Shapes of lenses

nifying glasses are called convex or *converging* lenses. The two types can be readily distinguished from one another; converging lenses are thickest in the middle while diverging lenses are thinnest in the middle. Fig. 24.1 illustrates some of the more common types of lenses.

Technical terms

A simple lens is usually a piece of glass bounded by spherical surfaces.

The *principal axis* of a lens is the line joining the centres of curvature of its surfaces.

If a parallel beam of light, parallel to the principal axis, is incident on a converging lens, the rays, after passing through the lens, all converge to a point on the axis which is called the *principal focus, F.* In the case of a diverging lens the rays will spread out after passing through the lens, as if diverging from a focus behind the lens. The principal focus is thus real for a converging lens and virtual for a diverging lens (Fig. 24.2).

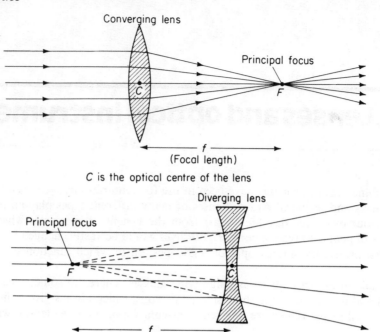

C is the optical centre of the lens

Fig. 24.2. Principal foci of lenses

When we were discussing spherical mirrors on page 236 it was explained that a true point focus is obtained only for rays which are very close to the axis. The same holds true for lenses. Thus, **the principal focus of a lens is that point on the principal axis to which all rays originally parallel and close to the axis converge, or from which they diverge, after passing through the lens.**

Lenses compared with prisms

In chapter 25 we shall see how a triangular glass prism causes light to be *deviated* on passing through it. Now a lens may be regarded as being made up of a very large number of portions of triangular prisms, the angles of which decrease from the edges of the lens to its centre. As a result, light is deviated more at the edges of a lens than at the centre. This will explain how a beam of light parallel to the principal axis of a

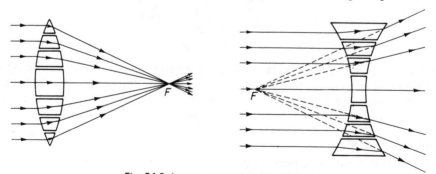

Fig. 24.3. Lenses compared with prisms

lens is either brought to a real focus in the case of a converging lens or diverges from a virtual focus in the case of a diverging lens (see Fig. 24.3).

Optical centre of a lens. Focal length

The central portion of a lens may be regarded as a small part of a parallel-sided slab, so that rays passing through it are not deviated but only slightly displaced parallel to their original direction. When the lens is thin this displacement is sufficiently small to be ignored, so that in all our diagrams rays going through the centre, C, of the lens are drawn straight. The centre of the lens is thus called the *optical centre*.

The focal length of a lens is the distance between the optical centre and the principal focus (see Fig. 24.2).

A lens has two principal foci

Since light may pass through a lens in either direction, there will be two principal foci equidistant from the optical centre, one on either side of the lens. These are denoted by the symbols F and F'.

Apart from their use in locating the principal axis of a lens, the centres of curvature of the faces of a lens are not of particular importance in elementary work. We shall see later that two points on the axis which are important are the points $2F$ and $2F'$. As their names indicate, these are situated respectively at a distance of twice the focal length from the optical centre on either side of the lens.

Construction of ray diagrams

Three particular classes of rays are used in geometrical constructions to locate the image formed by a converging lens:

(1) Rays parallel to the principal axis which pass through the principal focus after refraction through the lens.
(2) Rays through the principal focus which emerge parallel to the principal axis after refraction through the lens.
(3) Rays through the optical centre which are undeviated.

Two of these rays only are sufficient to locate an image, and which particular pair is chosen is merely a matter of convenience.

Fig. 24.4–24.9 are a series of diagrams to show the type of image formed as the object is moved progressively along the principal axis, starting at a point between the lens and the principal focus. As is usual in optical diagrams, the object is represented by a vertical arrow OA standing on the principal axis, and IB represents the image. Arrows are placed on the rays to show the direction of the light.

Formation of images by a converging lens

Fig. 24.4 illustrates the use of a lens as a magnifying glass. It will be noticed that the image formed is erect, virtual and magnified and on the same side of the lens as the object. Used in this way, a magnifying glass is sometimes called a *simple microscope* to distinguish it from the compound microscope, which is a more powerful instrument consisting of two or more lenses mounted in a brass tube (page 271).

Fig. 24.6 shows how a lens is used as a projection lens for the purpose of throwing a magnified real image of a slide or film on a screen. In this case the image is inverted, and so the slide must be put in the projector upside down.

Fig. 24.8 shows the action of a simple camera lens in producing a small real inverted image on a sensitive plate or film.

Image formed by a diverging lens

Fig. 24.10 illustrates the formation of an image by a diverging lens. For all positions of the object, the image is virtual, erect and smaller than the object, and is situated between the object and the lens.

Solution of problems by graphical construction

In ray diagrams a lens is represented by a straight line so that the actual paths of the rays through the lens itself are ignored. This is justified, since the lenses used are thin.

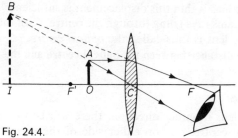

Fig. 24.4.

OBJECT BETWEEN
LENS and F'
the image is,
(1) Behind the object
(2) Virtual
(3) Erect
(4) Larger than object

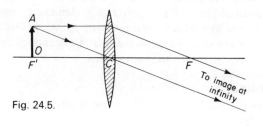

Fig. 24.5.

OBJECT AT F'
the image is
at infinity

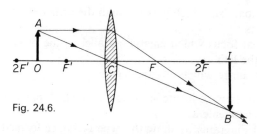

Fig. 24.6.

OBJECT BETWEEN
F' and $2F'$
the image is,
(1) Beyond $2F$
(2) Real
(3) Inverted
(4) Larger than object

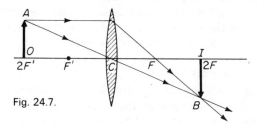

Fig. 24.7.

OBJECT AT $2F'$
the image is,
(1) At $2F$
(2) Real
(3) Inverted
(4) Same size as object

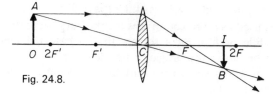

Fig. 24.8.

OBJECT BEYOND
$2F'$
the image is,
(1) Between F and $2F$
(2) Real
(3) Inverted
(4) Smaller than object

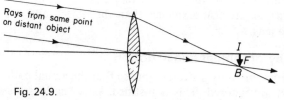

Fig. 24.9.

OBJECT AT INFINITY
the image is,
(1) At F
(2) Real
(3) Inverted
(4) Smaller than object

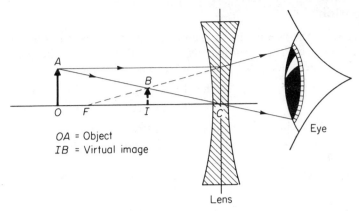

OA = Object
IB = Virtual image

Fig. 24.10. Formation of an image by a diverging lens

Example. *An object* 10 mm *tall stands vertically on the principal axis of a converging lens of focal length of* 10 mm, *and at a distance of* 17 mm *from the lens. By graphical construction find the position, size and nature of the image.*

Using squared paper, the ray diagram is drawn to scale, taking the side of one square as a unit to represent 1 mm (Fig. 24.11). A horizontal straight line *OCI*

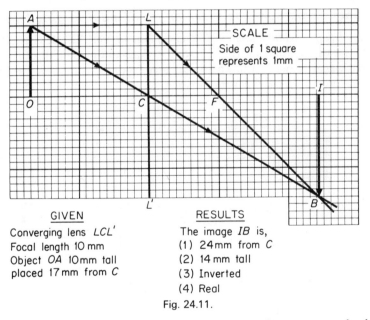

SCALE
Side of 1 square
represents 1mm

GIVEN

Converging lens *LCL'*
Focal length 10 mm
Object *OA* 10mm tall
placed 17mm from *C*

RESULTS

The image *IB* is,
(1) 24mm from *C*
(2) 14 mm tall
(3) Inverted
(4) Real

Fig. 24.11.

represents the principal axis. *LCL'* at right angles to it represents the lens. The object is represented by *OA*, 10 units long, drawn perpendicular to *OC* and 17 units from *C*. The two rays used to locate the image are:

(1) *ACB* through the optical centre (undeviated), and
(2) *AL* parallel to the principal axis which emerges through the focus *F*.

The intersection *B* of the two emergent rays is the image of *A*.
IB, perpendicular to the axis, will therefore represent the image of *OA*.
Giving the result to two significant figures, we see from the construction that the image is a real one, 24 mm (24 units) from the lens on the opposite side to the object and is 14 mm (14 units) tall.
All data given and measurements made should be recorded on the squared paper as shown.

Magnification

As we have seen, the size of the image produced by a lens varies according to the position of the object. The *linear magnification* is the ratio of the height of the image to the height of the object, and is usually denoted by the letter *m*. Thus

$$m = \frac{\text{height of image}}{\text{height of object}}$$

It is useful to know how *m* depends on the distances of the image and object from the lens. Referring to Fig. 24.6 and 24.10, it will be found easy to prove that the angles of triangle *CIB* are respectively equal to the angles of triangle *COA*.

Hence the triangles *CIB* and *COA* are similar, and it follows that

$$\frac{IB}{OA} = \frac{IC}{OC}$$

or

$$m = \frac{\text{distance of image from lens}}{\text{distance of object from lens}}$$

It is customary to denote the image and object distances from the lens by the letters *v* and *u* respectively, so that

$$m = \frac{v}{u} \text{ numerically}$$

Experiments to measure the focal length of a converging lens

Method (1). Using an illuminated object and plane mirror. The lens is set up in a suitable holder with a plane mirror behind it so that light passing through the lens is reflected back again (Fig. 24.12). The object used is a hole and cross-wire in a white screen illuminated by a pearl electric lamp.

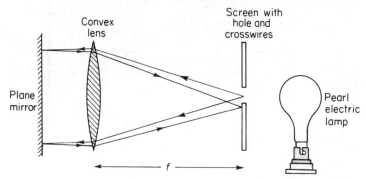

Fig. 24.12. Finding focal length of a converging lens

The position of the lens holder is adjusted until a sharp image of the object is formed on the screen alongside the object itself. The object will now be situated in the focal plane of the lens, i.e., a plane through the principal focus at right angles to the principal axis. Under these conditions, rays from any point on the object will emerge from the lens as a parallel beam. They are therefore reflected back through the lens and brought to a focus in the same plane as the object. The distance between lens and screen now gives the focal length of the lens.

Method (2). Using a pin and plane mirror. In this method the object consists of a pin set vertically in a special carrier or simply stuck in a cork held by a clamp and stand. Having adjusted the pin so that its tip is at the same horizontal level as the centre of the lens, a position is found for which there is no parallax between it and the real image formed (Fig. 24.13). For best results, attention should be given to the tilt of the plane mirror so that the tip of the image and the tip of the object pin appear to touch

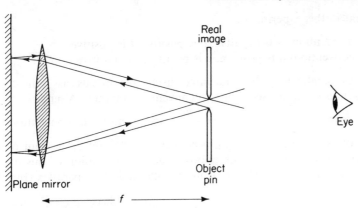

Fig. 24.13. Focal length of a converging lens by pin method

at the same level as the centre of the lens. The distance between pin and lens will then be equal to the focal length of the lens.

Method (3). *Measurement of object and image distances.* The lens is set up in front of an illuminated object so that a real image is formed on a white screen placed on the opposite side (Fig. 24.14). Having adjusted the lens so that the image is sharply in focus, the object distance u and the image distance v from the lens are measured.

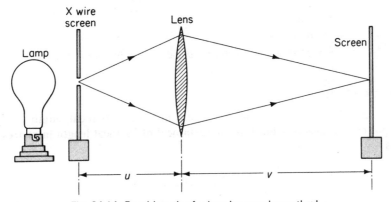

Fig. 24.14. Focal length of a lens by u and v method

It can be proved that u, v and the focal length f are related by the formula

$$\frac{1}{v} - \frac{1}{u} = \frac{1}{f} \quad \text{(New Cartesian sign convention)}$$

or

$$\frac{1}{v} + \frac{1}{u} = \frac{1}{f} \quad \text{(Real-is-positive sign convention)}$$

For details of these conventions see pages 273–74.

Several pairs of values of u and v are found and the results entered in a table as shown. A mean value for the focal length f may then be calculated.

u (cm)	v (cm)	$\dfrac{1}{u}$	$\dfrac{1}{v}$	$\dfrac{1}{v} \pm \dfrac{1}{u}$
			Mean value of $\dfrac{1}{f}$	
			Mean value of f	

Note that in this experiment:

(N.C. convention) u is negative, v is positive, f is positive.
(R.P. convention) u is positive, v is positive, f is positive.

The focal length may be calculated using reciprocals as indicated, with − or + sign in last column according to convention adopted. Alternatively, it may be calculated from $f = \dfrac{uv}{u - v}$ or $f = \dfrac{uv}{u + v}$ according to the convention used.

Instead of using an illuminated object, a pin may be set up in front of the lens so that it forms a real image on the opposite side. The position of this image is then located by the aid of a search pin, using the method of no-parallax (Fig. 24.15).

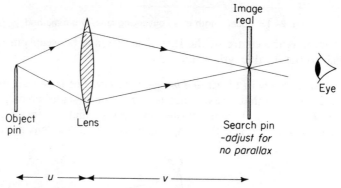

Fig. 24.15. Focal length from u and v by no-parallax method

Power of a lens

Opticians usually refer to the *power* of a lens instead of its focal length.
The power of a lens is defined as the reciprocal of its focal length in metres. Thus,

$$\text{Power} = \frac{1}{\text{focal length in metres}}$$

The unit of power is the **dioptre (D)** which is the power of a lens of focal length one metre.

The shorter the focal length, the greater the power. For example, in the case of a lens of focal length 20 cm (0.2 m)

$$\text{Power} = \frac{1}{0.2} = 5\,\text{D}$$

The projector

Fig. 24.16 shows the arrangement of lenses in a cinema or slide projector. (See also Fig. 24.6.)

The illuminant is either a carbon electric arc or a *quartz iodine* lamp to give a small but very high intensity source of light. This is situated at the centre of curvature of a small concave mirror which reflects back light otherwise wasted at the back of the projector housing. In order to obtain a brilliant picture on the screen, a *condenser* consisting of two plano-convex lenses collects light which would otherwise spread out and be wasted, and causes it to converge through the slide on to the projection lens.

The projection lens is mounted in a sliding tube, so that it may be moved to and fro to focus a sharp image on the screen.

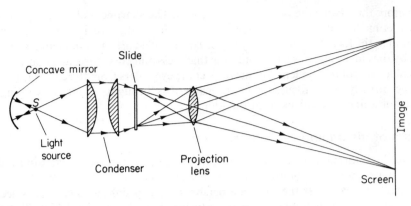

Fig. 24.16. Optical projection system

The camera

In its simplest form, a lens camera consists of a lens and a sensitive film mounted in a light-tight box, with provision for adjusting the distance between lens and film. A shutter of variable speed, and a diaphragm of variable aperture, regulate the amount of light energy admitted through the lens.

These features of the camera are shown in Fig. 24.17. A sharp image of the scene being photographed is focused on the film by turning the screw mount of the lens. This varies the distance between lens and film. The correct setting of the lens for an object at any given distance from the camera is obtained from a scale engraved on the lens mount.

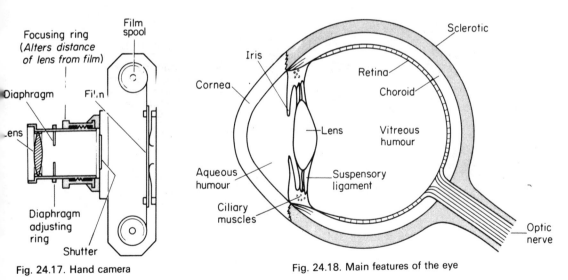

Fig. 24.17. Hand camera

Fig. 24.18. Main features of the eye

The eye

Fig. 24.18 is a simplified diagram of the human eye. In many respects it is similar to the camera. It has a tough, white wall called the *sclerotic* of which the front portion, the *cornea*, is transparent. Situated in the *aqueous humour* in front of the eye *lens* is the *iris* or coloured part of the eye which automatically adjusts the size of the *pupil* or circular opening in the centre according to the intensity of the light falling on it. Light passing through the eye lens crosses the *vitreous humour* to form an image on the *retina*. From here, electrical signals are transmitted to the brain via the optic nerve.

The main refraction of the light entering the eye occurs at the cornea which has rather more than twice the power of the eye lens. The sharpness of the image formed on the retina for objects at different distances from the eye is controlled by an alteration in the focal length of the eye lens. This is called *accomodation* and is brought about by the *ciliary muscles* which vary the thickness and curvature, and consequently the focal length, of the eye lens. In this respect the eye differs from the camera, since, as we have already seen, focussing in the camera is brought about by varying the distance of a fixed focus lens from the film.

Defects of vision and their correction

The so-called normal eye can accommodate for clear vision of objects from infinity (the far point) down to about 25 cm (the near point).

We shall now consider the case of a person with **long sight** who can see objects at infinity but whose near point is somewhat further than 25 cm (Fig. 24.19). True long

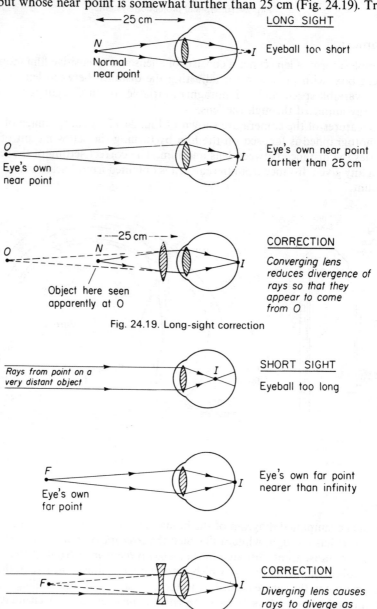

Fig. 24.19. Long-sight correction

Fig. 24.20. Short-sight correction

sight, as distinct from loss of accommodation due to advancing age, is caused by the eye-ball being too short.

Thus if a printed page is held at the normal 25 cm from such an eye it will appear blurred since, for this distance, the eye lens can form a sharp image only at a point behind the retina.

Correction is effected by a converging spectacle lens which reduces the divergence of the rays entering the eye just sufficiently to make them appear as though coming from the eye's own more distant near point.

In a case of simple **short sight** the eye-ball is too long so that the effectively parallel rays from a point for such an eye on a very distant object are focused in front of the retina. The actual far point for such an eye may be only a metre or less away (Fig. 24.20).

Correction is obtained by a diverging spectacle lens which diverges the rays entering the eye so that they appear to be coming from the eye's own far point.

The compound microscope

This instrument employs two lenses of short focal length arranged as shown in Fig. 24.21. The first of these, called the objective, produces an enlarged, real, inverted image I_1 of a small object O.

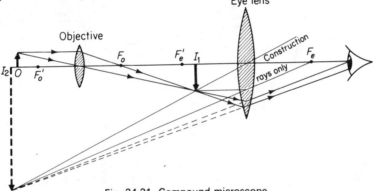

Fig. 24.21. Compound microscope

The image I_1 then acts as an object for the *eye lens*, which gives a virtual, image I_2 which is still further enlarged. This is the image which is seen by the eye.

It should be noticed that the thin rays drawn without arrows in the diagram are only the necessary construction rays to locate the position of the image I_2. The actual rays from the object by which the eye sees the tip of the image I_2 are the thicker arrowed rays.

Angular magnitude and apparent size

It is well known that street lamp-posts appear to be shorter the further they are away, although, in fact, all are of the same height.

The angular magnitude of an object is the angle it subtends at the eye. This angle

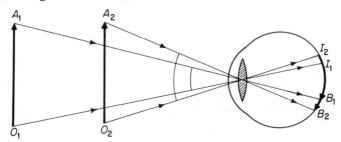

Fig. 24.22. Apparent size depends on angle subtended at the eye

determines the size of the image formed on the retina and hence governs the *apparent size* of the object (Fig. 24.22).

The astronomical telescope

The function of a telescope is to increase the angle which a distant object appears to subtend at the eye, and therefore produces the same effect as if the object were either larger or else closer to the eye.

Fig. 24.23 shows how this is done by an *astronomical telescope*. For a very distant object, e.g., a star which is effectively at infinity, rays coming from any point on it are sensibly parallel on reaching the telescope. A real image *I* is therefore formed in the focal plane of the objective. The focal plane of the objective coincides with the focal plane of the eye lens. Consequently, the image *I* acts as an object for the eye lens

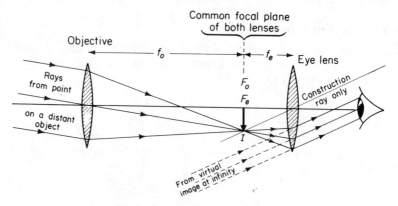

Fig. 24.23. Astronomical telescope

and, in normal adjustment, a final virtual, greatly magnified image is formed at infinity. Clearly the telescope owes its magnifying power to the fact that the angle subtended at the eye by the final image at infinity is very much greater than that subtended by the distant object.

For astronomical purposes it does not matter that the final image is inverted.

For high magnifying power the objective should have a long focal length f_o, and the eye lens a short one f_e. In this respect, a telescope differs from a compound microscope.

Optical fibres

Total internal reflection (page 255) can be used to convey light through a thin glass fibre (Fig. 24.24), with very little loss of energy, provided the angle of incidence is greater than the critical angle. A bundle of such fibres forms a flexible *light guide*.

The glass cladding of slightly lower refractive index than that of the core prevents surface damage to the core fibre which otherwise would allow light to escape.

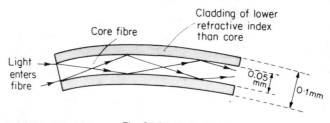

Fig. 24.24. Optical fibre

Light guides can be used to illuminate an otherwise inaccessible space since the light is kept inside the guide even if the guide is slightly curved. Conversely, if an objective is placed at one end of a light guide and an eyepiece at the other, the arrangement forms an *endoscope*. This is a collective name for a group of such instruments which are used in medicine and internal surgery to examine various body cavities. Endoscopes also have industrial applications.

Optical fibres in telecommunications

One application of optical fibres is their increasing importance in telecommunications.

Research is already far advanced and production of the necessary equipment has begun. It is expected that, by the end of the century, if not before, optical fibres will have largely replaced copper cables, not only for telephone conversations but also for transmitting pictures, television programmes, computer information and so on.

Briefly, in the case of a telephone conversation, the energy of sound waves entering the microphone is transferred to electric wave energy. This is transferred to electric pulses and the energy of these controls light pulses which travel through the optical fibre. At the receiving end, the light energy is transferred back to sound energy.

One big advantage of an optical fibre is its small size and weight compared with a copper cable of equivalent signal carrying capacity. This is an important factor in a spacecraft or an aircraft. See also Fig. 24.25 (*a*).

Sign conventions

When using the lens formula mentioned on page 267 correct signs have to be applied to object and image distances and focal lengths, according to the convention chosen.

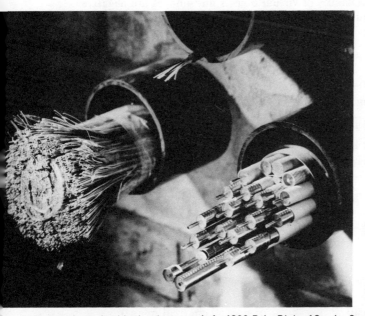

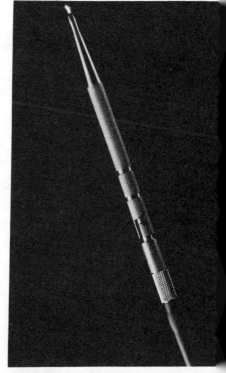

Fig. 24.25 (*a*). *Optical cable development. Left:* 4800 Pair. *Right:* 18-tube 9 mm coaxial cable. *Centre:* 8 core optical cable which is capable of a voice circuit capacity equal to that of the coaxial cable

Fig. 24.25 (*b*). A precision diamond scalpel for use in eye surgery. The illumination is obtained by light passing through a fibre optic light guide

NEW CARTESIAN CONVENTION

The lens equation is

$$\frac{1}{v} - \frac{1}{u} = \frac{1}{f}$$

All distances are measured from the optical centre, being positive if in the same direction as the incident light and negative if against it.

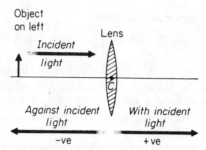

Fig. 24.26. New Cartesian convention applied to lenses

This convention is more easily applied if the object is always placed to the left of the lens, in which case all distances to the left of the lens are negative and all those to the right are positive. *This automatically brings all signs into agreement with the ordinary Cartesian graphical convention used in mathematics; hence its name* (Fig. 24.26). The focal length of a converging lens is +ve; a diverging lens −ve.

REAL-IS-POSITIVE CONVENTION

The lens equation is

$$\frac{1}{v} + \frac{1}{u} = \frac{1}{f}$$

All distances are measured from the optical centre. Distances of real objects and real images are positive. Distances of virtual objects and images are negative.

The focal length of a converging lens is +ve (real focus).
The focal length of a diverging lens is −ve (virtual focus).

Worked examples

1. *An object is placed* (a) 20 cm, (b) 5 cm *from a converging lens of focal length* 15 cm. *Find the nature, position and magnification of the image in each case.*

New Cartesian	*Real-is-positive*
(a) $u = -20$ cm (object left of lens)	$u = +20$ cm (real object)
$f = +15$ cm (converging lens)	$f = +15$ cm (converging lens)

Substitute in $\dfrac{1}{v} - \dfrac{1}{u} = \dfrac{1}{f}$ Substitute in $\dfrac{1}{v} + \dfrac{1}{u} = \dfrac{1}{f}$

$$\frac{1}{v} - \frac{1}{-20} = \frac{1}{15} \qquad\qquad \frac{1}{v} + \frac{1}{20} = \frac{1}{15}$$

$$\therefore \frac{1}{v} = \frac{1}{15} - \frac{1}{20} = \frac{4-3}{60} = \frac{1}{60} \qquad \therefore \frac{1}{v} = \frac{1}{15} - \frac{1}{20} = \frac{4-3}{60} = \frac{1}{60}$$

or $v = 60$ cm (on right, real)   or $v = 60$ cm (real image)

$$m = \frac{\text{image distance}}{\text{object distance}} = \frac{60}{20} = 3 \qquad m = \frac{\text{image distance}}{\text{object distance}} = \frac{60}{20} = 3$$

Answer. (*a*) A real image is formed 60 cm from lens on side opposite to object, of magnification 3.

(*b*) $u = -5$ cm (object left of lens)

$$\frac{1}{v} - \frac{1}{-5} = \frac{1}{15}$$

$$\therefore \frac{1}{v} = \frac{1}{15} - \frac{1}{5} = \frac{1-3}{15} = \frac{-2}{15}$$

or $v = -7.5$ cm (on left, virtual)

$$m = \frac{\text{image distance}}{\text{object distance}} = \frac{7.5}{5} = 1.5$$

$u = 5$ cm (real object)

$$\frac{1}{v} + \frac{1}{5} = \frac{1}{15}$$

$$\therefore \frac{1}{v} = \frac{1}{15} - \frac{1}{5} = \frac{1-3}{15} = \frac{-2}{15}$$

or $v = -7.5$ cm (virtual image)

$$m = \frac{\text{image distance}}{\text{object distance}} = \frac{7.5}{5} = 1.5$$

Answer. (*b*) A virtual image is formed 7.5 cm from lens on same side as object, of magnification 1.5.

2. *Find the nature and position of the image of an object placed* 10 cm *from a diverging lens of focal length* 15 cm.

$u = -10$ cm (left of lens)
$f = -15$ cm (diverging)

Substitute in $\dfrac{1}{v} - \dfrac{1}{u} = \dfrac{1}{f}$

$$\frac{1}{v} - \frac{1}{-10} = \frac{1}{-15}$$

$$\frac{1}{v} = \frac{1}{-15} - \frac{1}{10} = \frac{-2-3}{30}$$

whence $v = -6$ cm.

$u = +10$ cm (real object)
$f = -15$ cm (diverging)

Substitute in $\dfrac{1}{v} + \dfrac{1}{u} = \dfrac{1}{f}$

$$\frac{1}{v} + \frac{1}{10} = \frac{1}{-15}$$

$$\frac{1}{v} = \frac{1}{-15} - \frac{1}{10} = \frac{-2-3}{30}$$

whence $v = -6$ cm.

Answer. A virtual image is formed 6 cm from lens on same side as object.

QUESTIONS: 24

1. Define *principal focus*, *focal length* of a converging lens, and explain the meaning of *real image* and *virtual image*.

A small object is placed 6 cm away from a converging lens of focal length 10 cm. By means of a carefully drawn scale diagram find the nature, position, and magnification of the image. (*O.C.*)

2. Describe the various types of spherical converging lenses.

Draw ray diagrams to show how
(*a*) a real diminished image, and
(*b*) a virtual magnified image, can be formed by a converging lens. Mention examples of applications of the converging lens where these two types of image are produced.

A simple magnifying glass produces an enlarged erect image when an object is situated 10 cm from the lens. If the length of

the image is twice that of the object, calculate:
(*a*) the distance of the image from the lens;
(*b*) the focal length of the lens. (*O.C.*)

3. Draw and name the THREE types of converging lenses.

Describe an experiment to find accurately the focal length of a converging lens.

Where must an object be placed so that the image formed by a converging lens will be
(i) at infinity,
(ii) the same size as the object,
(iii) erect,
(iv) inverted and enlarged, and
(v) as near the object as possible?

An object 1 mm tall is placed 2.9 cm from a convex lens of focal length 3 cm so that it is perpendicular to, and has one end resting on, the axis of the lens. What is the size and position of the image? (*A.E.B.*)

4. Explain the terms real image and virtual image.

Draw a ray diagram to scale showing how a thin converging lens of focal length 10 cm forms a real image twice as large as the object, and write down the distance of object and image from the lens. Explain the action of a projection lantern.   (*O.*)

5. What is meant by a *real* image? Explain, with the help of a ray diagram, how a convex lens forms a real image of a small object close to its axis.

Describe the construction of a photographic camera. If the focal length of the camera lens is 20 cm, how far away from the film must the lens be set in order to photograph an object 100 cm from the lens?   (*O.*)

6. Describe how you would determine the focal length of a convex lens, using a plane mirror and one pin.

A convex lens has a focal length of 10 cm. Draw ray diagrams to show how it may be used to obtain:

(*a*) a magnified real image;

(*b*) a magnified virtual image. In each case find, by calculation or construction, where the object must be placed to obtain a magnification of 3.   (*C.*)

7. What do you understand by the term "parallax"?

Describe an accurate method of determining the focal length of a convex lens.

How will the focal length of a plano-convex lens be affected by:

(i) increased refractive index of the glass;

(ii) increased radius of curvature of the surface? Consider only rays which are normal to the plane surface and give, with diagrams, your reasons in general terms.   (*S.*)

8. Draw a diagram of a slide projector and use it to explain the function of the condensing lens and of the concave mirror placed behind the light source.

What advantage has a matt white surface for the screen over a glossy surface?

A slide projector using a slide 5 cm × 5 cm produces a picture 3 m × 3 m on a screen placed at a distance of 24 m from the projection lens. How far from the lens must the slide be? Make an approximate estimate of the focal length of the projection lens. If this lens gets broken and the only substitute available is one of about half its focal length, what would you do to arrange that the picture is still the same size?   (*O.C.*)

9. Explain the main features of the human eye, illustrating your answer by a simple labelled diagram.

Determine graphically the position, size, and nature of the image of an object 2 cm high placed 40 cm away from a diverging lens of focal length 20 cm.   (*W.*)

10. When is a person said to be suffering from long sight? Draw a diagram of the eye to show how this defect may be corrected by the use of a suitable type of lens.

Mention **two** ways in which a photographic camera is similar to the human eye and **one** way in which it is different.   (*W.A.E.C.*)

11. A converging (convex) lens has a focal length of 5 cm.

(i) What is the power of the lens?

(ii) If this lens were used in an astronomical refracting telescope, for which part of the telescope would it be most suitable?

Describe the other lens which would be used in the telescope, stating a suitable value for its focal length.

(iii) What would be the distance between the two lenses if the telescope were in normal adjustment (i.e. with the final image at infinity)?   (*J.M.B.*)

12. (i) Given two converging lenses of focal lengths 50 cm and 5 cm, indicate (on a ray diagram) how the lenses would be arranged, with other accessories, to obtain a magnified image of a distant object which is not on the axis of the lens system.

(ii) Describe how the final image would be formed.

(iii) What alteration in the arrangement would you suggest for a greater magnification?

(iv) If the final image is to be formed at infinity, what would be the distance between the lenses of the system?   (*W.A.E.C.*)

13. (*a*) (i) Draw a ray diagram showing how a lens may be used as a magnifying glass.

(ii) Name the type of lens used.

(iii) Describe the image formed by the lens used in this way.

(iv) Suggest a suitable value for the focal length of the lens you might usefully use in this way.

(v) Calculate the power of the lens referred to in part (iv).

(*b*) A lens used as in (*a*) is often used, together with another lens, to form an astronomical telescope.

(i) Draw a ray diagram of such a telescope, name the types of lens used and show the positions of the principal foci of the lenses and the positions of the images formed.

(ii) Describe the images formed by the lenses used in this way.

(iii) Suggest a suitable value for the focal length of the objective lens you would use.

(*J.M.B.*)

25. Dispersion and colour

It had been known for centuries that small fragments of colourless glass and precious stones glittered in bright colours when white light passed through them, but it was not until the middle of the seventeenth century that Sir Isaac Newton investigated the problem systematically.

Newton's work on this subject arose out of the need for finding a way of removing coloration from the images seen through a telescope. At that time ·this instrument had recently been invented by a Dutch spectacle-maker named Lippershey.

Newton's experiment with a prism

Newton began his experiments by making a small circular hole in one of the window shutters of his room at Cambridge. Light from the sun streamed through this hole and made a circular white patch on the opposite wall. On placing a triangular glass prism before the hole, an elongated coloured patch of light was formed on the wall. Newton called this a *spectrum*, and noted that the colours formed were in the order Red, Orange, Yellow, Green, Blue, Indigo, and Violet (Fig. 25.1). Most people, however, are unable to distinguish the indigo from the rest of the blue.

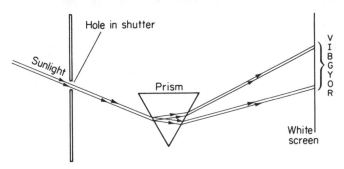

Fig. 25.1. Newton's experiment with a prism

The theory which he put forward to explain the spectrum was that white light consists of a mixture of seven different colours. The refractive index of glass is different for each colour, so that when white light falls on the prism each colour in it is refracted at a different angle, with the result that the colours are spread out to form a spectrum. It should be noted that when the light is incident on the prism as shown in Fig. 25.1, *it is refracted towards the base of the prism*, the violet being deviated most and the red least. The separation of white light into its component colours by a prism is called *dispersion.* Strictly speaking, there are many shades of each colour in the spectrum, each shade merging gradually into the next.

Improvement on Newton's original experiment

The spectrum formed in Newton's first experiment was impure. This resulted from the fact that it consisted of a series of circular coloured images all overlapping.

Later Newton devised the arrangement shown in Fig. 25.2, which produced a fairly pure spectrum. A converging lens L is placed so as to form an image I of a narrow slit S brightly illuminated by white light. A prism is then placed in the path of the light from the lens. Dispersion of the light on passing through the prism leads to the

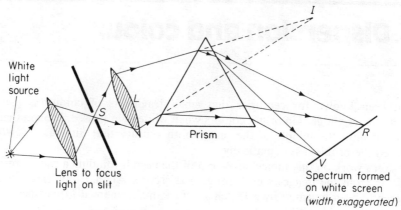

Fig. 25.2. Production of a fairly pure spectrum

formation of a spectrum VR on a white screen. This consists of a series of coloured images of the slit all touching one another. If the slit is made narrow, overlapping is reduced to a minimum and the resulting spectrum is fairly pure.

Production of a pure spectrum. Spectrometer

A better method of obtaining a pure spectrum than the one just described is to use a parallel beam of light as in an instrument called a *spectrometer*. This is used for examining the spectra of hot gases and other sources of light (Fig. 25.3).

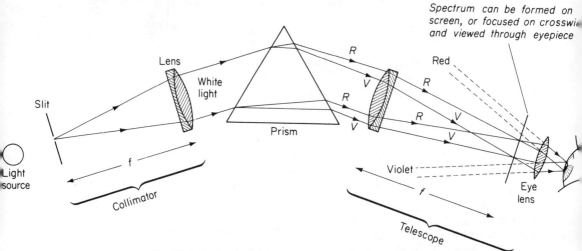

Fig. 25.3. Production of a pure spectrum using two lenses and a prism

An illuminated slit is placed at the principal focus of a converging lens so that a parallel beam of light emerges from it and falls on a prism. Refraction through the prism splits up the light into separate parallel beams of different colours, each of which is brought to its own focus in the focal plane of a second lens. Here the spectrum may be formed on a white screen or, alternatively, viewed through a magnifying eyepiece.

The combination of slit and first lens is called the *collimator* (to collimate means to make parallel). The second lens with cross wires in its focal plane, together with the eyepiece form a *telescope*. (See Fig. 25.4.)

Fig. 25.4. In this spectrometer the collimator with its adjustable slit is seen on the left; and the telescope with its eye-piece on the right. The rotating table in the centre supports the prism. Further details, including adjustments will be found in more advanced texts.

Recombination of the colours of the spectrum

The colours of the spectrum may be recombined to form white light by allowing the spectrum to be formed on a row of small rectangular plane mirrors (Fig. 25.5). On adjusting the angle which the mirrors make with the incident light so that they all reflect the light to the same place on the screen a white patch of light is formed.

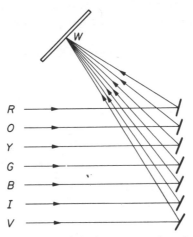

Fig. 25.5. Recombination of spectrum colours by mirrors

Newton showed the recombination of the colours to form white light in another way. He painted the colours of the spectrum in sectors on a disc. When rotated at high speed the disc appears white. It is only fair to say that the whiteness thus obtained is slightly greyish, owing to the difficulty of obtaining pigments which are pure colours. The experiment works by reason of the *persistence of vision*. The impression of an image on the retina of the eye is retained for a small fraction of a second after the image has disappeared. Consequently, the brain sums up and blends together the rapidly changing coloured images of the disc, and thus produces the sensation of a stationary white image.

It may be mentioned in passing that persistence of vision is responsible for the absence of flicker in the picture formed by a ciné projector. In this case a constant succession of images of the film is thrown on the screen at the rate of 24 per second. The eye retains the sensation of each image until it receives the next, so producing an impression of continuity.

Colour of objects in white light

When white light falls on any particular body, then either all the colours in the white light may be reflected from the body, when it appears white, or only some of them may be reflected while the others are absorbed. In the latter case the body appears coloured. The energy of the light absorbed is generally converted into internal energy so that the body becomes slightly warmer. The colour which the body presents to the eye is the colour of the light which it reflects. Thus the leaves of plants appear green, since they reflect green light and absorb the other colours.

White paper reflects all the colours of the spectrum while black paper absorbs all of them. Blackness is thus due to the absence of light of any colour.

Light filters

Interesting results are obtained when light filters in the form of sheets of gelatine coloured with various dyes are placed, in turn, in front of the slit in Fig. 25.2 or 25.3. By this means, the light transmitted by the filter can be analysed into its component colours. It is observed that certain colours, depending on the colour of the filter, are now absent from the spectrum. The missing colours are those of light which has been absorbed by the filter while the remaining colours have been transmitted.

Now one would expect red gelatine to transmit only red light, green gelatine only green light and so on. Indeed, this generally proves to be so when tested by experiment. But an unusual result is obtained with yellow gelatine. *The spectrum of the light passing through most types of yellow gelatine is found to consist of red and green as well as yellow.* What is even more striking is that this particular yellow light looks just the same to the eye as that which comes from a filter passing only pure yellow. To distinguish between the two, the former kind of yellow is called *compound yellow* light.

Experiment shows that the yellow petals of flowers and most yellow paints are examples of compound yellow.

Appearance of coloured objects in coloured light

A convenient source of light of various colours may be obtained by placing an ordinary electric lamp in a box with an opening which may be covered with gelatine sheets of different colours. By means of such a lamp the appearance of different coloured objects in different colours of light may be examined in a dark room.

Then it is found that red bodies look red in red light while green and blue ones look black, since they have absorbed the red light. In like manner a red poppy appears black in green light. On the other hand, the compound yellow petals of a daffodil appear black only in blue or violet light. The daffodil appears yellow only in yellow or white light. In red light it looks red and in green light, green.

Primary and secondary colours

Although yellow may be produced by mixing red and green lights, it is not found possible to produce either red, green, or blue by mixing two other colours. For this reason, red, green, and blue are called *primary* colours.

Yellow is called a *secondary* colour. The other two secondary colours are *cyan* made by mixing green light and blue light, and *magenta*, by mixing red light and blue light.

Mixture of coloured lights

Before proceeding further it must be pointed out that mixing coloured paints is an entirely different thing from mixing coloured lights. Paints will be dealt with later.

The effect of mixing coloured lights may be investigated by using three projectors fitted with slides of various coloured gelatine sheets, and arranged so as to produce overlapping images on a white screen.

In this way it may be shown that a mixture of the three primary colours, red, green, and blue, gives a white patch on the screen. However, a successful result is obtained only by using the right kind of red, green, and blue gelatine for producing the colour and by having each light of the correct intensity. This can only be done by experimenting with different types of gelatine and by having lamps of the appropriate brightness.

By using two projectors only the following facts may also be verified:

$$Red + Green = Yellow$$
$$Red + Blue = Magenta$$
$$Blue + Green = Cyan$$

These results, together with the knowledge that a mixture of the three primaries, red, green, and blue, gives white, lead us to expect that:

$$Red + Cyan = White$$
$$Green + Magenta = White$$
$$Blue + Yellow = White$$

A further experiment with two projectors using appropriate gelatines shows that these inferences are correct.

Two colours such as those described above which give white light when added together are called *complementary* colours.

To sum up, Fig. 25.6 shows the result obtained when three circular overlapping patches of red, green and blue light are formed on a screen by three projectors respectively.

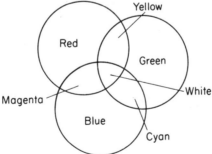

Fig. 25.6. Mixture of the three primary coloured lights

Mixing coloured pigments

One of the first things that a student of painting learns is that green paint can be made by mixing yellow and blue paint.

This would not be possible if the paints available were *pure* yellow and *pure* blue. The success of this method of making green paint depends on the fact that the pigments in common use are impure colours.

Yellow paint is a compound yellow so that, when illuminated by white light, it reflects red, yellow, and green light and absorbs the blue. Similarly, blue paint is not a pure colour: it reflects blue and green and absorbs red and yellow.

When the two paints are mixed, then between them they absorb red, yellow, and blue. The only colour they both reflect is green. Consequently, the mixture looks green. This process is called *colour mixing by subtraction* to distinguish it from the effect of mixing coloured lights by reflection from a white surface which is called *colour mixing by addition*.

Infrared and ultraviolet radiation

If a spectrum from an electric arc lamp or from the sun is produced on a screen by one of the methods described in this chapter it may be shown that invisible radiant energy is incident on the screen just beyond each of the extreme ends of the visible spectrum (Fig. 25.7).

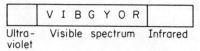

	V I B G Y O R	
Ultra- violet	Visible spectrum	Infrared

Fig. 25.7. Borders of the visible spectrum

Just beyond the red end of the spectrum is a region occupied by *infrared* radiation. This is invisible, but its presence may be demonstrated by placing a thermopile where the rays can fall on it. A galvanometer connected to the thermopile will give a deflection. Details of the thermopile are given on page 192.

Ultraviolet radiation can be detected in several ways. If a piece of photographic paper is placed on the screen it will become darkened to a variable extent where the spectrum falls on it, but maximum darkening is caused by ultraviolet radiation just beyond the violet end. Ultraviolet radiation also causes certain substances to *fluoresce*, i.e., glow with visible light. Quinine sulphate is an example. A colourless solution of this salt fluoresces with a blue light under the action of ultraviolet radiation. Paper lightly smeared with vaseline behaves similarly.

Laundry washing powders usually contain small quantities of a substance which fluoresces with a bluish white light under the action of the ultraviolet radiation in sunlight. This enhances the whiteness of linen and helps to combat the natural darkening of the material with age. More is said about the radiant energy spectrum on page 305.

QUESTIONS: 25

1. (*a*) What is a pure spectrum?

(*b*) In the formation of the spectrum of white light by a prism:
   (i) which colour is deviated least;
   (ii) which colour is deviated most?
   (*J.M.B.*)

2. Explain with the aid of a clear diagram how you would arrange apparatus to project a pure spectrum of white light on a white screen. Show on your diagram the paths of TWO rays of red light and TWO rays of blue light.

Describe what would be the appearance of the screen if:

(*a*) a pure blue filter was placed in the path of the beam;

(*b*) the filter was removed and a pure red screen substituted for the white screen;

(*c*) with the red screen remaining, the blue filter was replaced in the beam. Give brief explanations for your answers.   (*W.*)

3. A narrow beam of white light is incident upon a triangular glass prism. Draw a clear diagram to illustrate what is meant by:
   (*a*) deviation;
   (*b*) dispersion.

With the help of a well labelled diagram, describe how a pure spectrum may be obtained from white light using such a prism and any other necessary apparatus.

State and explain what form of spectrum would be obtained if:

(*c*) a red filter;

(*d*) a green filter were placed in the path of the beam.   (*L.*)

4. Describe, with the help of a diagram, how you would use a prism and two convex lenses to obtain a pure spectrum from white light. Explain how the colours of the spectrum can be recombined.

A mark is made in red ink on a strip of

white paper. How would you expect the appearance of the mark to change as the paper is moved along the spectrum? (*O.*)

5. Describe the optical arrangement for producing a pure spectrum of the light from an electric filament lamp on a white screen. Give a diagram showing the path of a beam of light through the arrangement.

How is the appearance of the spectrum altered if:

(*a*) a piece of red glass is placed in front of the lamp;

(*b*) this is removed and the white screen is replaced by a green one;

(*c*) the red glass and green screen are used together? Give reasons for your answers.

(*L.*)

6. Explain, with ray diagrams, how a glass prism:

(*a*) deviates, and

(*b*) disperses a ray of white light incident upon it.

Draw a labelled ray diagram to show how a pure spectrum of white light may be produced.

Describe and explain the appearance of the spectrum if:

(i) a sheet of blue glass is inserted in the path of the light;

(ii) sheets of red and yellow glass are inserted together in the path of the light, the blue having been removed. What is seen when a pure spectrum of white light is formed on a red screen in a dark room? (*J.M.B.*)

7. Describe and explain the appearance of a red tie with blue spots when observed in:

(*a*) red light;

(*b*) green light (*L.*)

8. Describe how a spectrum is formed on a screen, using a glass prism and a white light source.

How would you show the presence of infrared and ultraviolet radiation? (*C.*)

9. Explain two ways of reconstituting the colours of the spectrum into white.

State and account for the different effects produced when:

(i) yellow and blue pigments;

(ii) yellow and blue lights are mixed. (*S.*)

10. A plant with green leaves and red flowers is placed in:

(*a*) green;

(*b*) red;

(*c*) blue light. What colour will the leaves and flowers appear in each case? Assume that all the colours are pure. (*C.*)

Wave motion

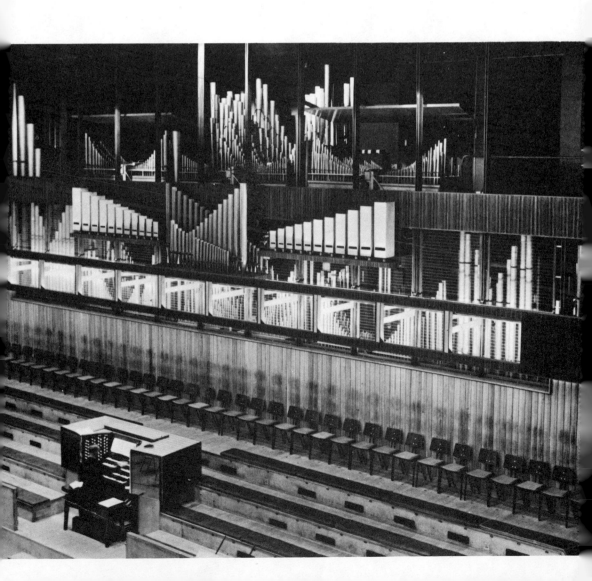

26. Transverse waves and light

In this chapter we shall deal with the experimental study of wave motion with special reference to the properties of water waves and light waves.

Transverse waves

Probably everyone has, at some time or another, thrown a stone into a pond or other smooth sheet of water and noticed the circular ripples which spread out from the spot where the stone entered the water. These ripples are an example of a wave motion travelling over a circular wavefront.

A somewhat simpler type of *transverse wave*, is seen when one end of a piece of rope or string is moved up and down in a direction perpendicular to its length. The particles of the rope near the end exert a drag on their neighbours so that these begin to oscillate as well. This process continues throughout the rope, until finally any particular particle is oscillating up and down slightly later than the one immediately before it. The net result is that the rope presents the appearance of a series of equidistant crests and troughs which travel forward with a certain velocity, called the *wave velocity* (Fig. 26.1).

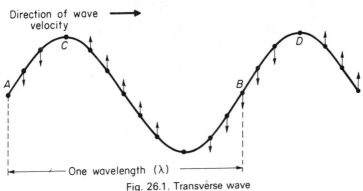

Fig. 26.1. Transverse wave

It is important to realize that it is only the shape or form of the wave which moves forward. The individual particles of the rope merely oscillate up and down with a motion similar to that of a pendulum bob. The actual motion of the particles has been indicated by small arrows in the diagram.

The maximum displacement of a particle from its rest position is called the amplitude of the wave.

The wavelength (λ) is defined as the distance between two successive particles which are at exactly the same point in their paths and are moving in the same direction. Such pairs of particles are said to be in the same *phase*. Examples are A and B or C and D (Fig. 26.1).

Any line or section taken through an advancing wave in which all the particles are in

the same phase is called the wavefront. We used this term earlier with reference to circular water ripples.

The time taken for a wave particle to make one complete oscillation is called the *periodic time.*

The number of complete oscillations made in 1 second is called the frequency (f).

The SI unit of frequency is called the hertz (Hz) and is defined as 1 cycle (or oscillation) per second.

In the time it takes the particles to make one complete oscillation the whole wave moves forward one wavelength.

Hence in 1 second the wave moves forward a distance $f\lambda$.

But the distance moved per second is the velocity, v.

Hence

$$v = f\lambda$$

To construct a transverse wave model

Much useful information about wave motion may be gained by constructing the wave model illustrated in Fig. 26.2.

A series of strips, 2.5 mm wide are shaded alternately in pencil or colour on a piece of drawing paper or thin card. A wave curve is then cut out of a longer strip,

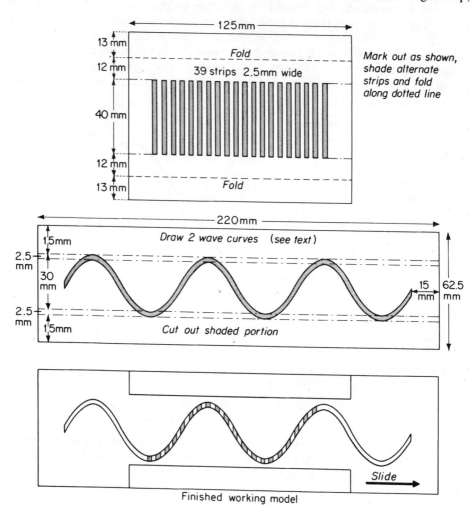

Fig. 26.2. Making a transverse wave model

for which purpose the curve in Fig. 26.1 has been made the correct size to suit the dimensions suggested, and this may be traced. Two curves should be drawn, 2.5 mm apart, and the space between cut out with a sharp penknife or single-edged razor-blade.

When the wave strip is inserted in the guide and slid along, it shows clearly how the forward motion of a wave is associated with transverse oscillation of the wave particles.

Water waves and light waves

On page 226 we mentioned the opposing views of Sir Isaac Newton and the Dutch scientist Christian Huygens regarding the nature of light. We shall now describe some experiments with water waves which show that, in some ways, their behaviour bears a striking resemblance to that of light. Afterwards we shall deal with some experiments which provide strong evidence in support of Huygens's wave theory of light.

To study water waves with a ripple tank

A ripple tank for showing the properties of water waves comprises a shallow trans-parent tray of water with a point light source above it and a white screen on the floor below (Fig. 26.3). Before adding the water the tray is levelled with a spirit-level to ensure a uniform water depth of rather less than 1 cm.

Straight parallel waves may be produced by a horizontal metal strip, or circular waves by a vertical ball-ended rod. When either of these is dipped into the water a

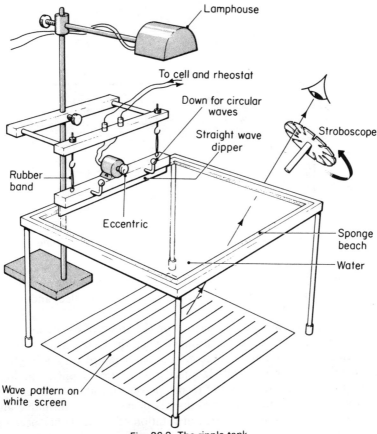

Fig. 26.3. The ripple tank

pulse of ripples is sent across the surface. Alternatively, continuous ripples may be obtained by fixing the dipper to a horizontal bar suspended by rubber bands. The bar is moved up and down by the vibrations of a small electric motor having an eccentric (=off-centre) metal disc on its rotating spindle. A rheostat in the motor circuit controls the speed and hence the frequency of the waves sent out. Owing to the lens effect of the wave crests and troughs, the light source produces a bright and dark wave pattern on the white screen below.

In all the experiments which follow it is best to try the effect of single pulses first and afterwards to study the pattern formed by continuous waves.

Use of a stroboscope. The progress of single wave pulses can be followed by the eye: so can that of continuous waves, provided that the wavelength is not too short.

If desired, however, the pattern formed by continuous waves can be made to appear stationary by the use of a stroboscope (shown in Fig. 26.3). One of the simplest forms of stroboscope is a disc about 25 cm in diameter with a number of equidistant radial slits cut in it. Pivoted on a handle, it is rotated by placing a finger in a hole near the centre.

If the wave pattern is viewed through the slits as shown in the ripple tank diagram it is found that, for a certain speed of rotation, the waves appear to be at rest. This occurs if the time taken for successive slits to cross the line of sight is exactly equal to the time taken by a wave crest to move into the position occupied by the one in front. If the speed of rotation of the stroboscope is varied, the waves will appear to move either forwards or backwards. Precisely the same effect is in operation when the wheel spokes of a vehicle in the cinema or television screen first come to rest and then move backwards as the wheel slows up.

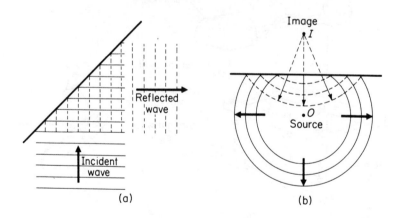

Fig. 26.4. Reflection of straight and circular waves from a plane surface

While on the subject of stroboscopes, it is worth mentioning a different type consisting of a lamp which flashes on and off at a controlled and measurable rate.

This type is useful for examining and measuring the speed of oscillating or rotating machine components, which appear stationary when the lamp flashes at the correct frequency. Incidentally, neon and fluorescent discharge lamps vary in brightness 100 times per second when run off a 50 Hz alternating current supply. The light from them may be used to check the speed of record player turntables. A disc on which is drawn the appropriate number of radial lines is placed on the turntable and these lines appear to be at rest when the speed of rotation of the table has been correctly adjusted.

Reflection of straight and circular waves

Using the straight dipper, parallel waves are set up and reflected from a plane metal strip stood upright in the water to act as a reflecting surface. Various angles of incidence should be tried. Fig. 26.4 (*a*) shows the wave pattern obtained for an angle of incidence of 45°. Note that the direction of propagation of the wave energy is at right angles to the waves, and that the angle of incidence is equal to the angle of reflection.

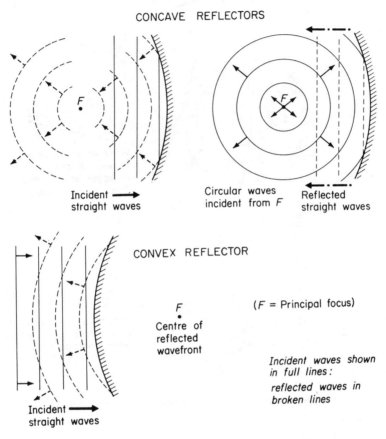

Fig. 26.5. Reflection of straight waves from a curved reflector

Fig. 26.4 (*b*) shows the pattern obtained with circular waves. The reflected waves are a set of circles whose common centre is a point as far behind the reflector as the source is in front. We are reminded of the corresponding case in optics where an object in front of a plane mirror gives rise to a virtual image situated at the same distance behind the mirror.

The plane reflector should now be replaced by concave and convex reflectors in turn and copies made of the wave patterns obtained with both straight and circular waves (Fig. 26.5 and 26.6).

Owing to interference between incident and reflected waves, to be discussed later, it will be found that a much clearer picture of what is happening may often be gathered from single wave pulses rather than from continuous ones. Circular pulses may be generated either by the finger or dipper, or better still, by drops of water from a bulb pipette.

Incidentally, the sponge beach of the ripple tank absorbs the energy of the waves and so prevents unwanted reflections. Some tanks achieve the same result by having gently shelving sides.

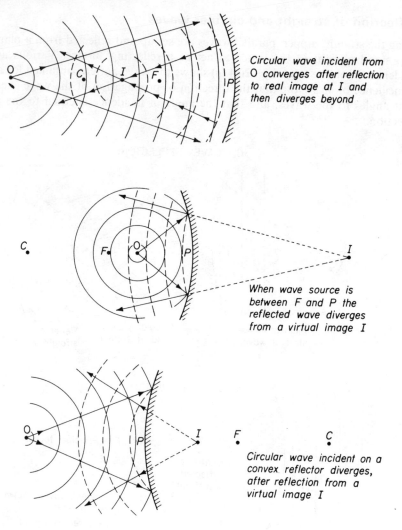

Circular wave incident from O converges after reflection to real image at I and then diverges beyond

When wave source is between F and P the reflected wave diverges from a virtual image I

Circular wave incident on a convex reflector diverges, after reflection from a virtual image I

Fig. 26.6. Reflection of circular waves from a curved reflector

Refraction of waves at plane boundaries

When straight waves pass from deep to shallow water, their wavelength becomes shorter. Both the long and the short waves, however, appear at rest when viewed simultaneously through the stroboscope. This shows that, although the wavelength λ has altered, the frequency, f, has remained the same. Now, since the velocity, $v = f\lambda$, it means that the waves travel more slowly in shallow water than in deep. This can be illustrated by placing a rectangular piece of perspex of suitable thickness in the tank to reduce the local water depth.

Furthermore, when the angle of incidence is anything other than zero (i.e., perpendicular incidence), the change in wavelength and speed automatically brings about a change in the direction of travel of the waves when they cross the boundary (see Fig. 26.7). In other words, *refraction* occurs. Now the direction in which the waves are travelling is at right angles to the wavefront, so in accordance with the usual convention we have drawn a normal and marked the angle of incidence, i, and the angle of refraction, r. This may be compared with the refraction of light when it passes from one medium to another.

It is clear from Fig. 26.7 that the wavelength has changed from λ_1 to λ_2. From the geometry of the diagram we note that there are two right-angled triangles with angles i and r, and sides λ_1 and λ_2 respectively, together with a common hypotenuse AB.

Also, we have already seen on page 250 that the refractive index, n, is defined by,

$$n = \frac{\sin i}{\sin r}$$

Therefore, the refractive index for water waves passing from deep to shallow water is,

$$n = \frac{\lambda_1/AB}{\lambda_2/AB} = \frac{\lambda_1}{\lambda_2}$$

Now the stroboscope tells us that the frequency, f, of the waves remains unaltered: hence, using the wave equation $v = f\lambda$,

velocity in deep water $= v_1 = f\lambda_1$
velocity in shallow water $= v_2 = f\lambda_2$

Therefore, $\dfrac{v_1}{v_2} = \dfrac{\lambda_1}{\lambda_2}$ so we can also say,

$$\text{refractive index} = n = \frac{v_1}{v_2} = \frac{\text{velocity in deep water}}{\text{velocity in shallow water}}$$

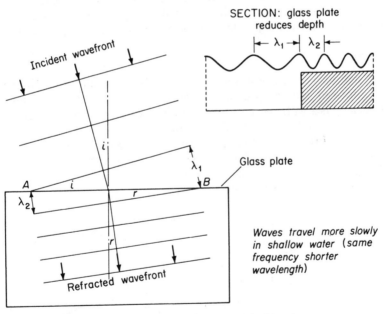

Fig. 26.7. Refraction of waves at a plane boundary

Comparison with light waves. Using a diagram similar to Fig. 26.7, Huygens considered the case of plane light waves being refracted from one medium to a more optically dense medium and obtained the same result, namely,

$$\text{refractive index} = \frac{\text{velocity of light in first medium}}{\text{velocity of light in second medium}}$$

Now, the mean refractive index of water (as far as light is concerned) is 1.33. Therefore, if Huygens was right it meant that light ought to travel 1.33 times faster in air than in water. In contrast, Newton's corpuscular theory gave a theoretical result which was the exact opposite to this.

When Huygens and Newton put forward their two different theories in the sixteenth century it was not possible to test them experimentally, since no method was available for measuring the velocity of light in anything other than free space. Nearly 150 years later, it was a triumph for the wave theory when Jean Foucault in France devised a method for measuring the velocity of light in both air and water, and found that it actually was less in water.

Also the expression $\dfrac{velocity\ in\ air}{velocity\ in\ water}$ gave the correct value, 1.33, for the refractive index of water.

Refraction of waves at curved boundaries

When carrying out experiments on refraction, using very shallow water, it will be found that a very slight trace of detergent in the water will improve the results by reducing surface tension effects.

The wave-focusing action of a shallow lens-shaped patch of water may be studied by placing a bi-convex piece of perspex in the tank. About 2 mm depth of water above the perspex coupled with a full water depth of about 8 mm gives satisfactory results.

Fig. 26.8 illustrates the focusing effect for straight waves. Having tried straight waves, the water dropper should be used to start circular pulses at various distances along the axis of the water lens and diagrams drawn in the practical notebook. The positions of the images formed by the refracted pulses in these *wave diagrams* should be compared with the corresponding *ray diagrams* for light on page 264.

Diffraction

Interesting and unexpected results are obtained when straight waves are incident on an opening formed between two vertical metal barriers placed in the ripple tank.

If the opening is a wide one compared with the wavelength of the waves, they will pass through in parallel straight lines, though we cannot fail to notice a slight bending round at the edges. Matters are entirely different when the opening is a narrow one about the same order of width as the wavelength. The wavefront now emerges with a pronounced circular shape and the waves spread out in all directions from the opening. This effect is called *diffraction* (see Fig. 26.9).

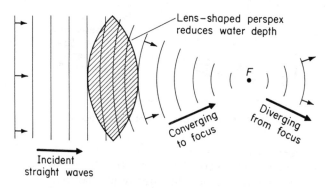

Fig. 26.8. Refraction of waves at curved boundary

Light behaves in a similar manner. If a parallel beam of light falls on a screen after passing through a wide slit the diffraction is negligible; very much less, in fact, than in the case of water waves. We get a sharp-edged patch of light on the screen so that it appears that the light is travelling in straight lines. Indeed, we have already mentioned the straight line propagation of light in chapter 21, and we saw later that it worked very well in explaining the formation of images by mirrors and lenses.

On the other hand, if light passes through a very narrow slit it can be shown to spread out in a manner similar to the water waves in Fig. 26.9. The reason why we never notice the diffraction of light in everyday life is that the wavelength of light is

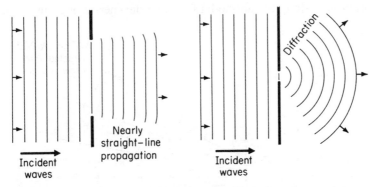

Fig. 26.9. Diffraction of waves through wide and narrow openings

exceedingly small and so the effect is unobservable except when the light passes through very narrow openings. Even in the laboratory we have to go to some trouble to demonstrate it. However, in due course we shall show how diffraction can be used to measure the wavelength of light.

Interference

When two ball-ended dippers are attached to the vibrator of the ripple tank, two sets of circular ripples are sent out which pass through one another as shown in Fig. 26.10.

Where the two waves are superposed in the same phase, e.g., crest on crest, we get lines of increased disturbance or *constructive interference*. These are called the *antinodal* lines. In between these are the *nodal* lines along which the waves are exactly out of phase, e.g., the crests of one are superposed on the troughs of the other. Here, provided the amplitudes of the two waves are the same, we now get zero resultant disturbance of the water surface, or *destructive interference*.

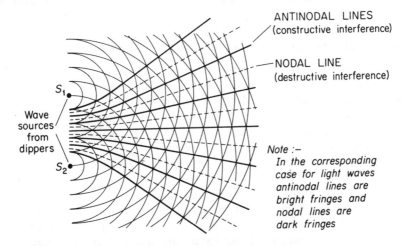

Fig. 26.10. Interference of circular waves

A similar interference pattern is obtained if either a straight or a circular wave is incident on a vertical barrier having two small apertures. In this case, interference takes place between the emerging diffracted waves (see Fig. 26.11).

How the wave nature of light was first discovered

If we imagine that the two water wave sources of Fig. 26.11 to be replaced by two point light sources then, if light is a form of wave motion, we should expect similar constructive and destructive interference to occur. In other words we ought to get increased brightness along the antinodal lines and darkness along the nodal lines.

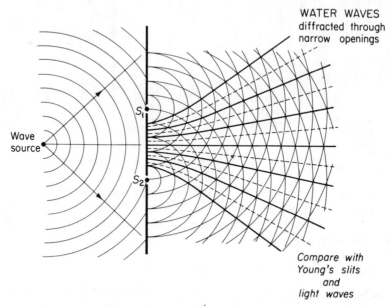

Fig. 26.11. Interference of diffracted waves

At the beginning of the nineteenth century, Thomas Young did, in fact, perform such an experiment using the light diffracted from two pinholes and obtained a series of light and dark bands or *interference fringes* on a screen placed in the path of the light. This was very strong evidence for the wave theory of light. Later he repeated the experiment using the light from two narrow parallel slits, with similar results.

Production of interference fringes using Young's slits

Fig. 26.12 shows the general scheme of a modern version of Young's experiment. Two very *narrow, close, and parallel* slits S_1 and S_2 are illuminated by the light from a single slit S parallel to them, and placed in front of a strong monochromatic (= one colour or wavelength) light source. A sodium discharge lamp giving orange light is suitable, or alternatively, a white source may be used together with a colour filter which transmits only a limited range of wavelengths.

The interference fringes can be seen by setting up a translucent screen and viewing from the side opposite to the slits. Tracing-paper makes a suitable screen. Otherwise they can be examined through a magnifying eyepiece.

Note that the fringes are formed in space. They are said to be *non-localized*. Hence, light and dark bands will be formed on a screen placed anywhere in the fringe region and the spacing of the bands will increase as the screen is moved further from the slits. The same applies when using a magnifying eyepiece which shows a section across the fringes in its image plane.

The number of fringes seen depends on the width of the slits. The narrower the slits the greater will be the number of fringes, owing to the increased angular diffraction. They will, however, be much fainter since less light energy gets through. The average wavelength of light is about 0.000 5 mm. For convenience, this is usually written as 5×10^{-4} mm or 5×10^{-7} m, the use of a negative index signifying *division* by the particular power of 10.

In order to pass sufficient light energy to give easily visible fringes, the slits have to be a good many wavelengths wide. The angular diffraction of the light passing through them is, therefore, quite small. We may think of it as being somewhere between that of the two water cases shown in Fig. 26.9. Consequently, the fringes are

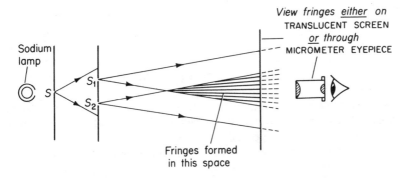

Fig. 26.12. Young's slits experiment (transverse scale exaggerated)

confined to a much smaller space than our water wave experiments might lead us to suppose. It must be borne in mind, therefore, that all our diagrams relating to Young's experiment have, for clearness, been drawn with exaggerated scales.

The slit sources must be coherent. The reader may have wondered why it is necessary to have a single slit in front of the light source. The reason is that lamps do not emit light waves in the same regular manner as the dipper sends out water waves in a ripple tank.

The atoms in a light source give out millions of wave packets all with different phases and in different directions. If the single slit were not there the two slits would receive light from different parts of the source and so the light emerging from them would be in different and constantly varying phase. The function of the single slit is to cause the light to spread out over a cylindrical diffracted wavefront on which the phase remains constant. Thus, if the two slits are equidistant from the single slit they simultaneously receive light in the same phase. If any sudden phase changes do occur in the light from the single slit, then the two slits will both be affected equally. In these circumstances the light from the two slits is said to be *coherent*.

Thomas Young was well aware of this condition. One cannot obtain a constant interference pattern from two independent slit sources.

Practical details. Good results will be obtained only if proper care is given to the preparation of the slits. One of the commonest methods is to use a pin to rule two parallel lines about a third of a millimetre apart on a piece of thin glass which has been coated with Aquadag (colloidal graphite) and allowed to dry. The graphite is removed by the pin point, thus leaving two transparent slits. This is not as easy as it sounds: several trials may be necessary in order to obtain good slits.

Good results are obtained by the following method. A hole about 1 cm in diameter is made in a thin sheet of metal and, diametrically across it a short length of copper wire about 0.4 mm diameter is fixed with adhesive. To ensure straightness the wire should be cut from a length which has been stretched slightly by clamping one end in a vice and pulling the other end with pliers. The slits are formed by sticking two pieces of razor blade on either side of the wire. This is best done under a low-power microscope when it will be found comparatively easy to push the two pieces of blade into position to form two narrow, equal, and parallel gaps before the adhesive sets. A pair of slits made in this way, about 0.06 mm wide and about 0.4 mm apart will, if used with a strong light source, give up to 18 fringes. Wider slits give fewer but brighter fringes.

The single slit can be made by the same method, omitting the wire. It is, however, a definite advantage to use a variable slit if one is available, since its width can then be adjusted to give maximum brightness combined with good definition of the fringes.

Needless to say, it is best to work in a dark room or at any rate in a dimly lit laboratory. When setting up the apparatus it is essential to see that the light source and single slit both lie on the perpendicular bisector of the line joining the two slits S_1 and S_2. *If the fringes are poor, it will probably be due to lack of parallelism between the slits.* A slight rotation of the slit S one way or the other should bring about the desired results.

The distances between the components is not critical. The slit screens may be some 15 cm apart. The eyepiece should be placed where its field of view comfortably contains all the observable fringes: 20 to 30 cm from the slits may suit a micrometer eyepiece. If, however, a translucent screen is used it should be placed 50 to 100 cm from the slits if rough measurements of the fringe spacing are to be made with a half-millimetre scale.

More accurate results will naturally be obtained from micrometer measurements. The micrometer eyepiece mentioned in this description has a vertical crosswire on a horizontal slide which is moved by a micrometer screw (not shown in Fig. 26.12). Readings are taken when the crosswire is centred over the extreme fringes visible and from these the mean distance between adjacent fringes is calculated.

We shall now discuss the ray geometry of Young's experiment and show how the wavelength of the light may be measured.

Ray geometry of Young's experiment

In Fig. 26.13 all light waves leave S_1 and S_2 in the same phase, and the rays give the directions of the wave paths.

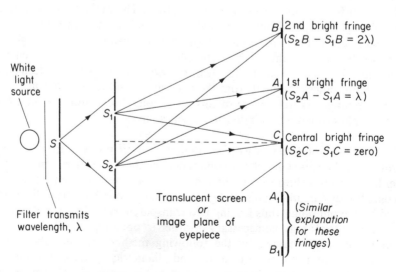

Fig. 26.13. Ray diagram for Young's experiment

If we consider a point C on the perpendicular bisector of S_1S_2, the waves travelling along the rays S_1C and S_2C have travelled equal distances. Hence they will arrive in phase and interfere constructively to make C the centre of a bright fringe.

The next bright fringe is at A where the wave path S_2A is one wavelength longer than S_1A. Once more, the waves are in phase, making A the centre of a bright fringe.

Similarly, B is the centre of a bright fringe where the path difference between S_2B and S_1B is two wavelengths. Subsequent bright fringes will be formed where the path differences are three, four, etc. wavelengths. The same explanation applies to the bright fringes formed on the side opposite to C, namely, at A_1, B_1, and so on.

In between the bright fringes we get the dark ones (not shown on the diagram). The centres of these will be situated where the wave paths differ in length by 0.5 λ, 1.5 λ, 2.5 λ, and so on, i.e., where the path difference is an odd number of half-wavelengths.

If, for convenience, we talk in terms of half-wavelengths, we may sum up the whole situation by saying:

For bright fringes: wave path difference = zero or an even number of half-wavelengths.

For dark fringes: wave path difference = an odd number of half-wavelengths.

Measurement of wavelength from Young's experiment

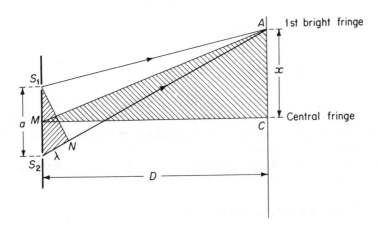

Fig. 26.14. Wavelength calculation from Young's experiment

Fig. 26.14 shows the ray geometry for the first bright fringe next to the central one. For clarity the vertical scale of this diagram has, like the others, been greatly exaggerated: actually the fringe spacing x is only about one six-hundredth of the distance D.

We saw in the previous section that the distance S_2A is one wavelength longer than S_1A. Thus, if we drop a perpendicular S_1N on to the line S_2A it will cut off a length $S_2N = \lambda$. Bearing in mind the smallness of the distance a between slits and the fringe spacing x we may, to a very close approximation, regard the two right-angled triangles S_2NS_1 and ACM as equiangular and therefore similar.

Hence
$$\frac{\lambda}{a} = \frac{x}{D}$$

or
$$\lambda = \frac{ax}{D}$$

The fringes are all effectively equidistant and so we take the fringe spacing x as equal to the average spacing of as many fringes as can be seen and measured. It has already been explained how this may be done either with a micrometer eyepiece or, more roughly with a half-millimetre scale used in conjunction with a magnifying glass.

The distance *a* between the slits may also be measured with the half-millimetre scale, though more accurate results are obtained with a travelling microscope. This microscope is fitted with a crosswire on which the slits are focused in turn. The consequent movement *a* of the microscope carriage is measured by a vernier or micrometer screw.

Owing to the smaller percentage error involved, it is sufficiently accurate to measure the distance *D* with an ordinary millimetre scale.

Wavelength and colour

The colour of light is related to the frequency and therefore to the wavelength of the light waves if they are passing through the same medium. The wavelength increases as we go from violet towards the red end of the spectrum. The wavelength of red light is approximately twice that of violet.

Returning to Young's experiment, it is clear from the equation

$$\lambda = \frac{ax}{D}$$

that if *a* and *D* are constant, then

$$\lambda \propto x$$

or, in words, *the fringe spacing is proportional to the wavelength.*

A very good way of demonstrating this is to make a triple light filter from three horizontal strips of red, green, and blue gelatine. A white light source is used in this experiment and the light filter is placed just in front of the fringe area on the screen or just in front of the eyepiece.

Red has the longest, and blue the shortest wavelength of the three colours and the resulting appearance of the fringes is as in Fig. 26.15.

Fringes formed by white light

Broadly speaking, white light consists of seven different colours, as we saw in chapter 25. So if a white light source is used in Young's experiment we get seven sets of differently spaced coloured fringes all superimposed on one another.

The previous experiment shows that the central fringes for each colour are in the same straight line. The one place where all colours constructively interfere together will therefore be at the central fringe. Consequently this will be white and the

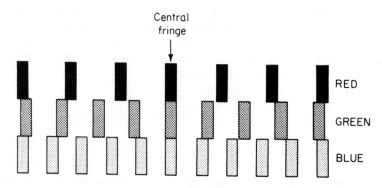

Fig. 26.15. Fringe spacing depends on wavelength

resultant appearance will be a central white band, bordered symmetrically on either side by bands of composite colours due to superposition of the other differently spaced colour fringes.

The diffraction grating

Young's demonstration of the wave nature of light was the prelude to further experimental and theoretical work on the subject which continued during the nineteenth century.

Within a few years, the German physicist, Joseph von Fraunhofer, invented a more satisfactory way of measuring the wavelength of light. Instead of two slits, he used as many close parallel slits as it was then possible to obtain. Such a device is called a *diffraction grating*. Some he made by ruling as many lines as he could on a piece of smoked glass: others were constructed from parallel wires, kept equidistant by locating them in between the threads of two fine screws.

Better gratings were made by later workers who used accurate temperature controlled dividing engines to rule parallel lines on sheets of glass. Replicas of such gratings can be made by coating them with a layer of collodion (a solution of cellulose nitrate) which, when dry, is carefully peeled off and attached to a piece of glass. Modern gratings may have 500 or more lines per millimetre. It is now possible to obtain large gratings on cellulose acetate sheet. For elementary work this may be cut up into pieces about 50 mm square and sandwiched between thin glass plates for protection and to keep it flat.

We shall first explain the action of a diffraction grating and then go on to show how wavelengths of light may be measured with it. Note that all our diagrams have been simplified by showing only a few grating elements but it must be realized that, in practice, there are some hundreds per millimetre.

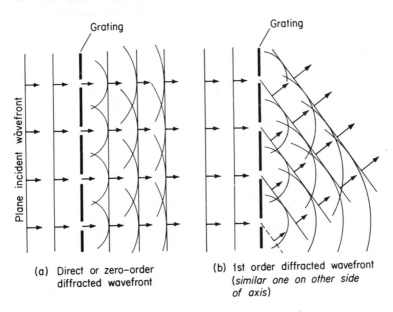

(a) Direct or zero-order diffracted wavefront

(b) 1st order diffracted wavefront (*similar one on other side of axis*)

Fig. 26.16. Light waves diffracted through a grating

Fig. 26.16 shows the cylindrical diffracted wavelets emerging from the slits of a diffraction grating when a beam of parallel light is incident upon it. The whole effect is similar to that obtained with water waves in a ripple tank when straight waves are incident on a barrier with a number of equidistant narrow openings.

Now the slits of a diffraction grating are only a few wavelengths wide: very much finer than those used in Young's experiment. This enables the light to diffract over an angle of practically 180° but the reduced light energy passing through in any particular direction is, of course, compensated by the sum-total effect of the large number of slits.

Let us suppose that the arcs of the circles drawn represent the wavefronts of the diffracted wavelets and that these are spaced one wavelength apart.

The tangent planes drawn to touch these cylindrical wavefronts suggest that there are at least three main directions along which all the wavelets combine in step with one another, and experiment shows that this is so (see Fig. 26.16). Incidentally, this way of regarding a plane wavefront as being composed of a vast number of spherical or cylindrical wavelet fronts was first used by Huygens in his original wave theory of light.

The resultant wavefront emerging parallel to the grating is the *zero order* wavefront, so-called because all the wavelets have travelled the same distance (i.e., zero path difference between them).

The two plane wavefronts emerging at an angle on either side are called the *first order* diffracted wavefronts, a name derived from the fact that each wavelet has to travel exactly one wavelength further than its next-door neighbour in order to be in step along the combined wavefront.

Relation between angle of diffraction and wavelength

Fig. 26.17 shows the usual arrangement for a diffraction grating experiment. A parallel beam of light is incident on the grating from an illuminated slit S placed at the

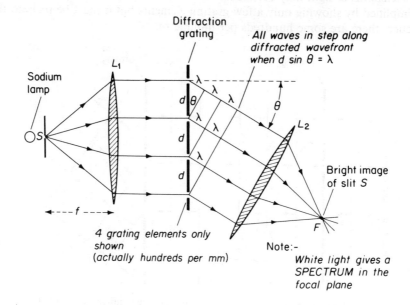

Fig. 26.17. Principle of the diffraction grating

principal focus of a converging lens L_1. The first order diffracted beam from the grating enters a second lens L_2 which brings it to a focus in the principal focal plane. Here we get an image of the slit S which may be formed either on a white screen or else viewed through an eye lens.

If we are using monochromatic (= one colour) light, then for first order diffraction, the distances between successive wavefronts leaving the grating in the same phase will be equal to the wavelength, λ.

Now, the *sine of an angle* $= \dfrac{perpendicular}{hypotenuse}$

Hence, from the ray geometry of the diagram, we see that

if θ = angle of diffraction of the beam

and d = grating space (or distance between adjacent slit centres)

then
$$\frac{\lambda}{d} = \sin \theta$$

or
$$\lambda = d \sin \theta$$

Thus, if we know d, and can measure θ, we can calculate λ.

Formation of spectrum by diffraction grating

If the monochromatic light source used previously is replaced by white light, then each wavelength in the white light will be diffracted at its own particular angle θ which satisfies the equation

$$\lambda = d \sin \theta$$

The net result is the formation of a continuous line of coloured images of the slit in the focal plane of the lens L_2. In other words, a white light spectrum is formed there. The next experiment will serve to illustrate this.

To measure the wavelength of light by a simple diffraction grating experiment

Fig. 26.18 shows how the ordinary type of ray-box used in elementary laboratories may be used in conjunction with a diffraction grating and a second cylindrical lens to form a spectrum of white light and to measure the wavelength of any particular colour in it. The cellulose acetate grating mentioned earlier is very suitable for the purpose and gives good results.

The ray-box itself consists of a vertical line filament electric lamp which can be adjusted in the usual way to give a parallel beam of light from the cylindrical lens L_1. The diffraction grating is then set up in front of the lens with a second lens L_2 immediately in front of it.

The whole arrangement is placed on a sheet of paper as shown. If a small white screen is placed on the axis of the lenses and at a distance from L_2 equal to its own focal length, f, a sharp white image of the line filament will be formed on it. This arises from the fact that the direct or zero order diffracted wavefronts emerging from the grating are all parallel to one another for all wavelengths, and consequently all colours come to a focus in the same direction.

On either side of the axis two positions can be found for which a sharp spectrum is focused on the screen, formed in the manner described in the previous section.

Suppose we wish to measure the wavelength of yellow light. The position of this is marked in pencil on the paper for both spectra in turn. The position, C, of the centre of the lens L_2 is also marked.

Having removed the ray-box, the pencil marks are then joined to form an angle $Y_1 C Y_2$. Clearly this angle $= 2\theta$ where θ is the required angle for substitution in the equation

$$\lambda = d \sin \theta$$

The number of lines on the grating, N per millimetre, is supplied by the makers, so that the value of d is equal to $\dfrac{1}{N}$.

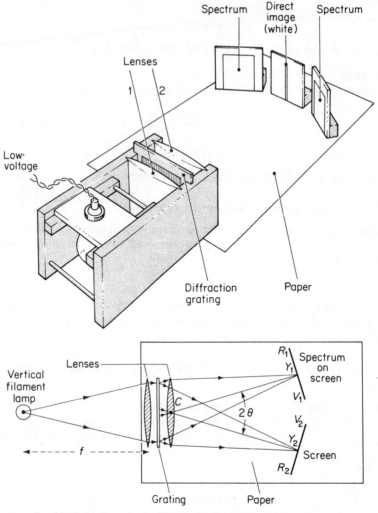

Fig. 26.18. Measurement of light wavelength by a diffraction grating

When doing this experiment it is important to observe that, owing to its longer wavelength, the red of the spectrum is diffracted through a larger angle than the blue. *This is the exact opposite of what we see with a spectrum formed by a glass prism* where the blue light is refracted through a larger angle than the red.

Higher order spectra from a diffraction grating

We have seen that reinforcement occurs for light diffracted from adjacent slits in a diffraction grating at an angle θ given by,

$$\lambda = d \sin \theta$$

Reinforcement will also occur when the angle of diffraction is such that the path difference for light from adjacent slits is 2λ, 3λ, and so on. We therefore may obtain diffracted images for any particular wavelength at larger angles given by,

$$2\lambda = d \sin \theta$$
$$3\lambda = d \sin \theta$$

These are called the *second and third order* images respectively. The actual number of orders possible for any particular colour will depend on the width of the grating space. Sin θ cannot be greater than 1 (90° diffraction), so that the maximum number of orders possible for any given wavelength cannot be greater than the whole-number value of *n* given by,

$$n\lambda = d \times 1$$

Two orders of spectra can be seen using the plastic replica described earlier in this chapter. Coarse gratings with a wide spacing give several orders as may readily be seen if a diffraction experiment is performed using a piece of fine wire gauze.

Electromagnetic waves

When the wave theory of light first gained general acceptance it was considered that light waves were conveyed through a transparent elastic medium which filled the whole of space, even a vacuum. This substance was called the *ether*.

A further step forward was made in 1845, when Michael Faraday showed that, under certain conditions, light waves passing through a material medium were affected by a magnetic field. Now by that time, it was known that there was an inseparable connection between magnetism and electricity. Faraday's experiment gave a strong hint that light might well have electrical properties.

Some years later, the eminent mathematician and physicist, James Clerk Maxwell, became very interested in Faraday's work on electricity and eventually put forward a mathematical theory suggesting that an oscillating electric current should be capable of radiating energy in the form of *electromagnetic waves* (e.m. waves). An electromagnetic wave can be visualized as an oscillating electric force travelling through space accompanied by a similar oscillating magnetic force in a plane at right angles to it. More importantly, Maxwell's equations led to the conclusion that such waves, if they existed, would travel with the same velocity as light.

Some twenty years after the publication of Maxwell's theory, the German scientist, Heinrich Hertz, showed that electromagnetic waves could indeed be produced by means of an oscillating electric spark. Moreover, he performed numerous experiments to demonstrate that the newly discovered waves underwent reflection, refraction, diffraction, and interference: in short, they behaved exactly like light waves but with a much greater wavelength.

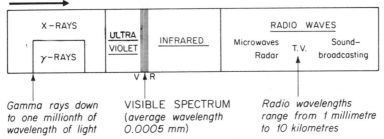

Fig. 26.19. The electromagnetic spectrum

The inference was that light waves themselves were also electromagnetic and further experimental and theoretical studies have since confirmed this belief. More will be said on the subject after we have studied electricity.

The work of Hertz was developed by Marconi and others who laid the foundations of our present-day use of electromagnetic waves in radio communication.

Fig. 26.19 shows the whole range of electromagnetic waves in order of increasing wavelengths. Any particular range of wavelengths is referred to as a *band*. It will be noticed that the visible wavelengths occupy a very small band in the complete electromagnetic spectrum.

The SI unit of frequency

In recognition of the importance of Hertz's researches into electromagnetic waves, his name has been given to the unit of frequency.

The SI unit of frequency is called the hertz (Hz) and is equal to a frequency of 1 cycle per second.

The term hertz is not restricted to wave frequencies only but is used for any regularly occurring event, e.g., the frequency of a pendulum, an alternating current, a musical note, and so on.

Larger frequency units in common use are

$$
\begin{aligned}
\text{the kilohertz} \quad (\text{kHz}) &= 1000 \text{ Hz} \\
\text{the megahertz} \,(\text{MHz}) &= 1\,000\,000 \text{ Hz} \\
\text{the gigahertz} \quad (\text{GHz}) &= 1\,000\,000\,000 \text{ Hz}
\end{aligned}
$$

Example. Calculate the frequency of a radio wave of wavelength 150 m. The velocity of all e.m. waves in free space is 3×10^8 m/s.

For a wave,

$$v = f\lambda \text{ (page 296)}$$

hence,

$$f = \frac{v}{\lambda} = \frac{3 \times 10^8}{150} = 2 \times 10^6 \text{ Hz}$$

$$= 2 \times 10^3 \text{ kHz}$$
$$= \underline{2 \text{ MHz}}$$

The inverse square law for electromagnetic waves

Let us suppose that an area of 1 m² which forms part of the surface of a sphere of radius 1 m receives wave energy from a point source placed at the centre *S* (Fig. 26.20).

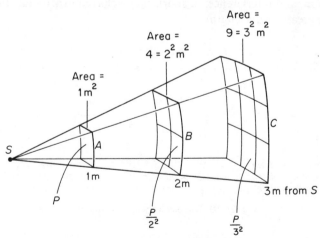

Fig. 26.20. Inverse square law for electromagnetic waves

If *P* is the energy passing through this unit area in joules per second, i.e., if *P* is the power in watts, then we say that *P* is the *wave intensity* in W/m².

If the distance is increased to 2 m it is clear from the geometry of the figure that the same power is now spread over an area of $2^2 = 4$ m². Consequently the wave intensity is reduced to $\dfrac{P}{2^2}$. Similarly, when the distance is increased to 3 m the wave intensity in W/m² is now only $\dfrac{P}{3^2}$

Wave intensity is defined as the power transmitted per unit area of the wavefront.
In general, the wave intensity at any distance x from a point source is given by

$$\frac{P}{x^2}$$

This relationship between wave intensity and distance is expressed by the inverse square law.

The electromagnetic wave intensity from a point source in free space is inversely proportional to the square of the distance from the source.

Light as a special case. The measurement of light wave energy (photometry) is complicated by the fact that light sources contain non-visible radiation as well, e.g., ultraviolet and infrared. Special units called lux (not W/m^2) therefore have to be used, but all the same, the inverse square law still holds. In this book we shall not be concerned with photometry.

Attenuation

The inverse square law for radiation as presented above holds strictly for waves through free space from a point source. If the radiation passes through a material medium of some kind then the law is modified by the fact that some of the wave energy is progressively absorbed. The loss of power from this cause is described as *attenuation*.

Origin and sources of electromagnetic waves

The whole range of electromagnetic radiation pours on to the earth from the sun and other heavenly bodies in outer space. Those frequencies which are stopped by the earth's atmosphere have been detected by instruments in man-made satellites.

Otherwise some of the main sources on earth are given in the table on page 308.

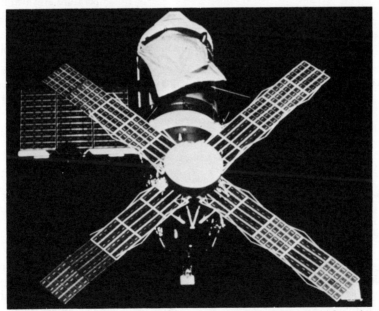

Fig. 26.21. *Skylab II*. A picture taken against a black sky background from the service module, just after the satellite was launched in May 1973. This 77 t spacecraft sent valuable information to earth over a period of six years before its final disintegration in July 1979. Its power requirements came from the solar panels which transferred energy in the sun's radiation to energy in an electric current.

Geostationary satellites which orbit the earth at a height of about 35 000 km and maintain the same position above its surface are used for communication purposes, mainly telephone and television links. Weather satellites transmit cloud cover pictures to earth for use in weather forecasting

Radar

Radar equipment used in ships, aircraft, and airfield control towers employs pulses of *very short radio waves*, for the purpose of finding the distances from the transmitter of various objects, e.g., coastline and building outlines together with other ships or aircraft.

A rotating (scanning) aerial sends out a series of high-frequency radio pulses which are received back as echoes from the objects on which they fall. The time interval between transmission and reception of the pulse depends on the distance of the reflecting object. These signals are used to produce a picture of the area scanned on the screen of a special cathode ray tube (Fig. 26.22). One form of cathode ray tube is described· in chapter 44.

WAVE-BAND	ORIGIN	SOURCES
GAMMA RADIATION	Energy changes in nuclei of atoms	Radioactive substances (see chapter 45)
X-RADIATION	(1) High energy changes in electron structure of atoms (2) Decelerated electrons	X-ray tubes (see chapters 44 and 45)
ULTRAVIOLET RADIATION	Fairly high energy changes in electron structure of atoms	(1) Very hot bodies, e.g., the electric arc (chapter 41) (2) Electric discharge through gases, particularly mercury vapour in quartz envelopes (chapter 44)
VISIBLE RADIATION	Energy changes in electron structure of atoms	Various lamps, flames and anything at or above the temperature at which it begins to emit red light
INFRARED RADIATION	Low energy changes in electron structure of atoms	*All* matter over a wide range of temperature from absolute zero upwards
RADIO WAVES	(1) High-frequency oscillatory electric currents (2) Very low energy changes in electron structure of atoms	Radio transmitting circuits and associated aerial equipment

QUESTIONS: 26

(*Note: questions dealing with sound waves may be deferred until after reading the chapters on Sound.*)

1. Give experimental details of how you would demonstrate, for water ripples,

(*a*) the reflection of a circular ripple from a plane reflector,

(*b*) the reflection of a plane (straight) ripple from a concave reflector.

Draw diagrams, to illustrate the effect of each of these reflections, showing

(i) the incident ripple and the reflected ripple just after reflection has begun,

(ii) some later positions of the reflected ripple.

When two vibrators, a short distance apart and vibrating together, touch the surface of water in a ripple tank, a stationary pattern occurs on the water surface. Draw a diagram to illustrate the pattern, and give an explanation of how it occurs. (*C.*)

2. Draw diagrams of waves as seen on the surface of the water in a ripple tank which illustrate the following cases of refraction and reflection of light:

(*a*) a double convex lens producing a real

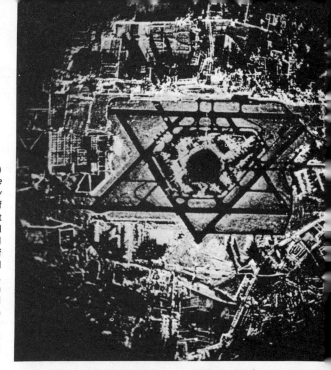

Fig. 26.22. *The Plan Position Indicator (PPI) display of the Decca Mark 3 Airfield Surface Movement Indication radar at Heathrow Airport.* To ensure safe landing and take-off operations, Air Traffic Control staff must first ensure that runways are clear of aircraft and vehicles. In conditions of low visibility, ASMI radar alone can fulfil this task. Aircraft taking off and landing at high speed are easily recognized as continuously moving targets on the display. The equipment provides coverage in the area from 200 to 4 000 metres from the radar, and can detect small vehicles at a range of up to 2 500 metres.
Note. Radar waves are electromagnetic waves (see chapter 26) and must not be confused with sound waves

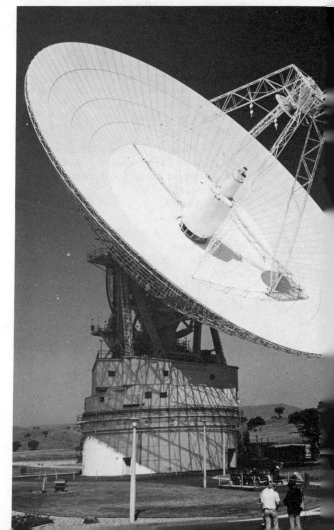

Fig. 26.23. *Parabolic reflecting radio telescope at the Australian space tracking station at Tidbinbilla near Canberra.* Radio waves being similar to light waves may be reflected in the same manner. Thus a weak radio signal is intensified when received by an antenna situated at the principal focus of a parabolic 'dish'. Such radio telescopes are in operation in many places over the world. They are used for detecting radiation from outer space as well as for sending and receiving signals from spacecraft

image of a point source on the axis;

(b) a concave mirror focusing a parallel beam.

Indicate the direction of the wave travel by arrows and label the images *I*. Show two waves that have travelled beyond *I*. What effect does the lens or mirror have on the speed of the waves striking them?   (*O.C.*)

3. Describe one type of stroboscope, explain its principle by describing an example of its use.

A disc ruled with radial lines is used to check the turn-table speed of a transcription unit for playing records at $33\frac{1}{3}$ rev/min. If the disc appears stationary when viewed in the light from a neon discharge lamp run from a 240 V 50 Hz a.c. supply what should be:

(a) the angle between the lines;

(b) the total number of lines?

4. A cart has wheels of 3 m circumference. Each wheel has 8 spokes. In a cine-film of it taken at 16 pictures per second, the wheels seem not to rotate although the cart is moving with uniform speed. What can you deduce about the speed of the cart?   (*O.C.*)

5. Radio and light waves travel at a velocity of 3.0×10^8 m/s in air. Calculate

(a) the wavelength in air of radio waves when transmitted at a frequency of 150 MHz, and

(b) the velocity of light in glass of refractive index 1.5.

6. (a) Sketch a displacement–distance graph for a transverse wave, showing two complete cycles. Mark on your graph distances to show what is meant by

(i) wavelength,

(ii) amplitude.

Label a point A anywhere on your graph and then label

(iii) a point B which is vibrating in-phase with A,

(iv) a point C which is vibrating in anti-phase (180° out of phase) with A.

(b) If a wave has a velocity of 330 m/s and a wavelength of 0.5 m, calculate the frequency of the vibrator producing the wave.

(c) Name **one** example of a transverse wave and **one** example of a longitudinal wave. (Chapter 27 may help you.)   (*J.M.B.*)

7. Water ripples are caused to travel across the surface of a shallow tank by means of a suitable straight vibrator. The distance between successive crests is 3.0 cm and the waves travel 25.2 cm in 1.2 s. Calculate the *wavelength* and velocity of the waves, and the *frequency* of the vibrator. Explain the two terms in italics.

As the ripples cross the tank they meet a shallower section, the edge of which is straight and at 45° to the initial direction of the ripples. If the wave velocity in this section is only two-thirds of the initial value, describe as fully as possible what happens to the ripples Your answer should be illustrated by suitable calculations and a clear diagram.   (*S.*)

8. A straight vibrator causes water ripples to travel across the surface of a shallow tank. The waves travel a distance of 33 cm in 1.5 s, and the distance between successive wave crests is 4.0 cm. Calculate the frequency of the vibrator.

Explain how you would use the arrangement to show

(a) refraction, and

(b) diffraction of water waves. Sketch the wave patterns you would expect to observe in each case.   (*W.*)

9. Describe with the aid of diagrams what happens when plane waves are incident on the gap between two obstacles as in a ripple tank if:

(i) the gap is wide;

(ii) the gap is narrow.

How would you expect the appearance of the resulting pattern to change if, when using the narrow gap, the wavelength of the incident plane wave became shorter?   (*L.*)

10. (a) Two small loudspeakers, a metre or so apart, are emitting pure notes of the same frequency and intensity (loudness). When a listener moves in a line parallel to the line joining the loudspeakers the intensity of the sound which is heard fluctuates. Explain, with the aid of a diagram, what is heard and name the phenomenon. Suggest one path along which he could move so that the intensity of the sound would not fluctuate.

(b) Two men stand a distance apart beside a long metal fence on a still day. One man places his ear against the fence while the other gives the fence a sharp knock with a hammer. Two sounds, separated by a time interval of 0.5 s, are heard by the first man. If the velocity of sound in air is 330 m/s and in the metal 5280 m/s, how far apart are the men?   (*W.*)

11. The velocity of light in water is 2.2×10^8 m/s and the velocity of light in glass is 2.0×10^8 m/s. Calculate

(i) the refractive index for light passing from water to glass,

(ii) the angle of incidence in the water which would produce an angle of refraction of 30° in the glass.   (*J.M.B.*)

12. In Fig. 26.24 two small vibrating sources, A and B, emit waves of the same frequency. Along the line X_1X_2, all points on which are equidistant from A and B, the intensity diminishes steadily, but along the line Y_1Y_2 there is found to be a repeated rise and fall in intensity. Account for each of these phenomena and indicate how a position of maximum or minimum intensity along Y_1Y_2 is related to its distances from A and B and the wavelength of the waves.   (*L.*)

13. State the conditions which must be fulfilled in order that two wave trains shall continuously interfere to produce no effect at a certain point.

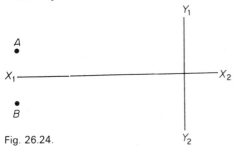

Fig. 26.24.

In an experiment using Young's slits, interference bands x (in cm) apart are produced on a screen D (in cm) from the slits by light of wavelength λ (in cm) coming from two slits d (in cm) apart. The formula for λ in terms of x, d, and D is $\lambda = \dfrac{xd}{D}$. This formula holds also for sound waves and can be applied to the following problem.

Two posts are fixed 4.0 m apart on a line which is parallel to and 100 m away from a straight road. On each post is fixed a loudspeaker and these are connected in parallel to an a.c. supply of frequency 1650 Hz (cycles/s) so as to emit pure tones.

An observer walking along the road notices that the sound he hears alternates from loud to quiet, the quiet positions being 5.0 m apart. With the aid of a diagram, explain this effect and calculate the wavelength of the sounds emitted and the velocity of sound in the air.

Suggest a path along which the observer could walk so as not to hear alternate loud and quiet sounds.   (*O.C.*)

14. In a *Young's slits* experiment, using sodium light, seven fringe spaces were found to occupy 2.8 mm when viewed through a micrometer eyepiece whose image plane was 24 cm from the slits. Calculate the separation of the slits, assuming the wavelength of sodium light is 6×10^{-4} mm.

When the sodium lamp was replaced by a white light source with a green gelatine filter, and the eyepiece moved 12 cm further from the slits, five green fringe spaces were found to occupy 2.3 mm. Find, in metres, the mean wavelength of the light passed by the filter.

Describe what you would expect to see through the eyepiece if the green filter were removed.

15. What is a diffraction grating? Explain with a simple diagram how it is used to produce a spectrum of the light from a small bright source. Mark the positions in which the red and blue colours are seen and establish a formula by which the wavelength of a coloured beam may be determined from its spectrum position.   (*O.C.*)

16. Why are the elements of a diffraction grating regularly spaced? How may a grating be used to study the spectrum from an electric lamp with a straight filament?

What is meant by different *orders* of spectrum?

A narrow detector of infrared radiation, insensitive to visible light is moved through the spectrum. It is found that a strong response occurs in the second order spectrum for green light of wavelength 4.9×10^{-5} cm yet there is no response in the first order spectrum for this wavelength. Suggest an explanation for this and deduce the wavelength of the radiation responsible for the strong response of the detector. How could you check your explanation if you were given a piece of material which absorbed green light but was transparent to infrared?   (*O.C.*)

17. (*a*) Of the following types of electromagnetic radiation: ultraviolet, gamma rays, radio waves:
    (i) which has the longest wavelength?
    (ii) which has the highest frequency?
    (*b*) What is the speed of electromagnetic radiation in free space?   (*J.M.B.*)

18. Name four types of electromagnetic radiation and briefly describe one process by which each may be produced.

19. Write down in order of increasing wavelength the principal regions of the electromagnetic spectrum. Indicate the approximate wavelengths of the radiations.
                    (*A.E.B.*)

20. Define *wave intensity* and state the *inverse square law* for electromagnetic radiation in free space.

What do you understand by *attenuation* of a wave?

21. Explain the purpose of the reflector in a radiotelescope.   (*J.M.B.*)

27. Longitudinal waves and sound

In this chapter we shall study the way in which sounds are produced and investigate some of their properties and uses.

Sound produced by vibration

The majority of musical instruments produce notes either by the vibration of stretched strings or by the vibration of air in pipes. These vibrations cause the air in the neighbourhood to vibrate also, and this disturbance of the air travels out in the form of a *longitudinal* wave.

In this chapter we shall show how sound waves differ from the *transverse* waves we talked about in the previous chapter. When a sound wave enters the ear part of its energy is converted into minute electric currents which travel along nerves to the brain, and so we are made aware of the sounds received.

For experimental purposes, musical sounds may be made by tuning forks. A tuning fork has two hard steel prongs. When struck on a piece of hard rubber the prongs vibrate and a note of definite pitch is given out.

The pitch of a note is denoted by a letter or symbol which refers to its position in the musical scale (see Fig. 28.2). Later on we shall see that pitch depends on frequency or the number of vibrations made per second.

On close examination the prongs of a sounding tuning fork are seen to present a hazy appearance owing to their state of vibration. The vibration can also be demonstrated in other ways. For example, a small pith ball suspended by thread is kicked to one side if touched with one of the prongs (Fig. 27.1). Again, the water in a beaker is violently agitated when the vibrating prongs are dipped into it.

Sound waves require a material medium

Unlike light and other electromagnetic waves, sound waves cannot travel through a vacuum. This was discovered in 1654 by Otto von Guericke, shortly after he had invented the air pump. He gradually pumped the air out of a glass bottle containing a clockwork bell and noticed that the sound steadily got weaker until it almost disappeared. Nowadays we generally use a small electric bell hung from rubber bands inside a bell-jar (Fig. 27.2). When all the air has been removed from the jar the ringing can no longer be heard, although the hammer can still be seen striking the gong. The faint burring noise which is sometimes audible comes from vibration which has travelled through the rubber bands supporting the bell.

Von Guericke found that fish were attracted by the sound of a bell rung under water, and therefore concluded that sound could travel through water as well as air. He did not, however, carry out a control experiment to see if the fish would come to the bell without the added inducement of tasty bait, so his conclusion was not truly scientific. Nevertheless, von Guericke would have been intrigued to know that, 300 years later, fishermen would be using transistor "beepers" which, when lowered into water, are claimed to attract fish over a radius of about $1\frac{1}{2}$ kilometres.

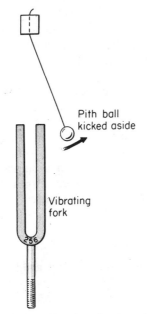

Fig. 27.1. Showing vibration of a tuning fork

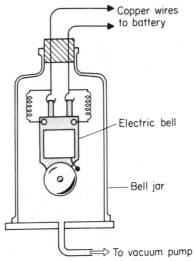

Fig. 27.2. Sound will not travel through a vacuum

Sound can travel through solids

At one time doctors used stethoscopes consisting of thin wooden rods with broadened ends. By placing one end to his ear and applying the other end to the patient's chest the physician could hear the sound of heart-beats transmitted through the wood. Motor mechanics sometimes use wooden rods as stethoscopes to assist in tracing the source of knocking noises in engines.

If a long iron fence is available the passage of sound through iron can be demonstrated. One person places his ear against the fence while another gives the fence a single tap with a stone some considerable distance away. Two sounds will be heard, one coming through the iron, followed by another through the air. Actually, sound travels about fifteen times faster through iron than through air.

Velocity of sound in water

In 1826, J. Colladon and J. Sturm measured the velocity of sound in water and found it to be about four times that in air. The experiment was carried out on Lake Geneva in Switzerland using two boats which were situated some 13.5 km apart. One boat supported a bell beneath the water surface, while the other had a trumpet with an elastic membrane stretched across it.

The hammer of the bell was connected to a lever holding a lighted match. At the same instant that the hammer struck the bell the match dipped into a small bowl containing gunpowder. The observer in the other boat listened for the sound through a tube connected to the trumpet and timed the interval between seeing the gunpowder flash and hearing the sound. The experiment was carried out at night, so that the flash was more easily visible. The result they obtained was 1435 m/s at 8.1 °C.

Velocity of sound in air

During the seventeenth century several scientists made attempts to measure the velocity of sound in air. The usual method was for two experimenters to station themselves a measured distance of several kilometres apart. One of them fired a cannon, while the other timed the interval between seeing the flash and hearing the report. It was assumed that the time taken for light to travel between the two stations was negligible. This assumption was justified, since light travels at 300 000 000 or 3×10^8 m/s, compared with sound, which has a velocity of only about 330 m/s.

In the early experiments simple pendulums or water clocks were used for timing, as pendulum clocks did not come into use until the end of the century. Later on, better results were obtained by having a cannon at each end and timing the sound in both directions. This is called *reciprocal* firing. By taking an average velocity, errors due to wind were, to some extent, eliminated.

Further improvements came in the nineteenth century, when accurate chronometers became available and electrical methods of timing were introduced. The various factors which affect the velocity of sound in air will be discussed in the next section. Its value for dry air at 0 °C is approximately 331 m/s.

Factors which affect the velocity of sound in air

Pitch and loudness. It is a matter of common experience that **the pitch and loudness of sounds have no effect on their velocity**. For instance, if we walk away from a band playing in the open air or from music coming from a loud-speaker the various notes are still heard in correct time whatever the distance from the source.

The only exception to this rule occurs in the case of the shock wave from a big explosion, when it is found that the noise travels with a higher velocity in the immediate neighbourhood of the source.

Pressure. The effect of pressure changes on the velocity of sound in air and other gases was investigated theoretically by Sir Isaac Newton, who showed that the velocity of sound is proportional to $\sqrt{\dfrac{pressure}{density}}$.

Now, in accordance with Boyle's law, if the pressure of a fixed mass of air is doubled the volume will be halved. Hence the density will be doubled. Thus, at constant temperature the ratio $\dfrac{pressure}{density}$ will always remain constant, however much the pressure may vary.

Consequently, **changes of pressure have no effect on the velocity of sound in air.**

Temperature. Anything which changes the air density without altering the pressure will cause a change in the velocity of sound. Change in temperature can bring this about. If the air temperature increases at constant pressure the air will expand according to Charles's law, and therefore become less dense.

The ratio $\dfrac{pressure}{density}$ will therefore increase, and hence **the velocity of sound increases with temperature.**

This result was first tested experimentally by an Italian scientist named Bianconi about the middle of the eighteenth century. He measured the velocity of sound by the cannon-firing method and obtained higher values in summer than in winter.

Sound waves

Sound waves in air differ from the transverse waves described in the previous chapter in that the wave particles oscillate in the same direction as the wave instead of at right angles to it. Waves of this type are called *longitudinal waves*. Owing to the longitudinal motion of the wave particles, sound waves consist of a series of *compressions* followed by *rarefactions*.

Fig. 27.3 shows how a vibrating tuning-fork sends out a sound wave. When the prong moves to the right it compresses the air particles together. This disturbance is then transmitted from particle to particle through the air, with the result that a pulse of compression moves outwards. Similarly, a reverse movement of the prong gives rise to a pulse of rarefaction of the air.

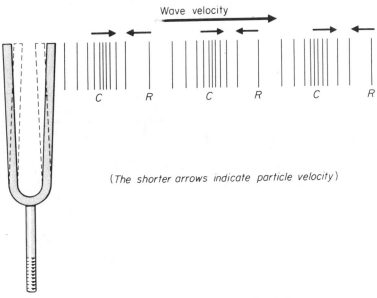

(The shorter arrows indicate particle velocity)

Fig. 27.3. Sound waves from a tuning fork

It is important to note that the particle at the centre of a compression is moving through its rest position in the *same* direction as the wave, while the particle at the centre of a rarefaction is moving through its rest position in the *opposite* direction to the wave.

As in the case of a transverse wave, the distance between two successive particles in the same phase is called the wavelength, and the same wave equation applies, namely, $v = f\lambda$.

To construct a longitudinal wave model (Crova's disc)

The manner in which oscillatory motion of the air layers gives rise to the compressions and rarefactions of a sound wave may be illustrated by a simple device known as Crova's disc.

A circle of radius about 7.5 mm is drawn on a piece of stout paper or thin card and eight equidistant points are marked round its circumference. These are numbered 1 to 8. Using these points successively as centres, a set of circles of progressively increasing radii are drawn as shown in Fig. 27.4.

When completed, the disc is cut out with scissors and pinned through its centre to a strip of card with a slit cut in it.

The disc is then rotated and viewed through the slit. The portions of the circles seen through the slit represent adjacent layers of air in a sound wave and, as the disc rotates, compressions followed by rarefactions will be seen to travel across. Each individual layer, however, simply oscillates about a mean position.

An experiment on the reflection of sound

Two metal or cardboard tubes are set up, inclined to one another in a horizontal plane, and pointing towards a vertical flat surface of any hard material, e.g., a drawing-board (Fig. 27.5).

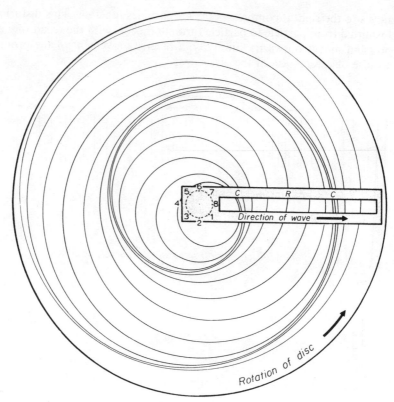

Fig. 27.4. Crova's disc illustrates longitudinal wave motion

A ticking watch is placed near the end of one tube and the ear is placed at the end of the other tube. Sound waves pass down the first tube and are reflected from the board. The loudness of the sound heard through the second tube is found to be a maximum when the board is adjusted so that the normal to it lies in the plane of the tubes and makes equal angles with their axes.

Under these conditions the incident and reflected waves and the normal are in the same plane and the angle of incidence is equal to the angle of reflection. Sound thus obeys the same laws of reflection as light.

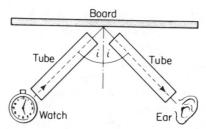

Fig. 27.5. Reflection of sound

Echoes

Echoes are produced by the reflection of sound from a hard surface such as a wall or cliff.

Let us suppose that a person claps his hands when standing some distance from a high wall and listens for the echo. The time which elapses before the echo arrives will depend on the distance away of the wall. Now, in order that the echo may be heard separately from the original clap it must arrive at least 0.1 s later. Since sound travels at about 330 m/s, the reflected wave must have travelled a total distance of at least 33 m, and consequently the minimum distance of the wall must be about 17 m.

When the reflecting surface is less than this distance the echo follows so closely upon the direct sound wave that they cannot be distinguished as separate sounds. One merely receives the impression that the original sound has been prolonged. This effect is called *reverberation.*

The echelon echo

Another effect explained by the reflection of sound is the high-pitched ring of short duration heard when one stamps on the pavement near to a fence made of equally spaced iron bars or wooden palings, or indeed, near any series of regularly spaced reflecting surfaces such as a flight of steps.

Thus, in the case of an iron fence whose bars are 20 cm apart the reflected sound pulses reach the ear at intervals equal to the time taken by the sound to travel *twice* the distance between the bars. Putting this in another way, the wavelength of the note heard will be twice the distance between the bars or 40 cm, giving a note of frequency

$$f = \frac{v}{\lambda} = \frac{330}{0.4}$$

$$= 825 \text{ Hz}$$

The description *echelon* comes from the French word for *step.*

The acoustics of buildings

Reverberation is particularly noticeable in cathedrals and other large buildings where multiple sound reflections can occur from walls, roof and floor. For example, in St. Paul's Cathedral, London, it takes about 6 seconds for the notes of the organ to die away after the organist has stopped playing.

Excessive reverberation in a concert hall is undesirable, as it causes music and speech to sound confused and indistinct. On the other hand, it is not a good thing to have no reverberation at all. Speakers and singers who have practised in an empty hall are often disconcerted by the seeming weakness of their voices when they perform in the same hall full of people. The soft clothing of the audience absorbs the sound instead of reflecting it, and consequently the music and speech appear to be weaker. Some degree of reverberation is therefore useful, as it prevents a hall from being acoustically "dead" and improves the hearing obtained in all parts of the building.

The characteristics of a building in relation to sound are called its *acoustics*, and the pioneer work in this field was carried out in the early twentieth century by Professor W. C. Sabine of Harvard University. The most important property of a concert hall is its *reverberation time*, which is defined as the time taken for sound of a specified standard intensity to die away until it just becomes inaudible. Sabine began his investigations by measuring the reverberation time of a number of empty halls and churches. For this purpose he used a small organ pipe which gave a note of the required intensity.

Further experiments were then carried out with cushions placed on the seats and with various types of wall and floor coverings in order to ascertain which reverberation time gave the best results. For the majority of halls this was found to lie between 1 and 2 seconds.

From his results Sabine obtained a formula relating the reverberation time to the volume of a hall, the surface area of its walls, ceiling and so on, and also the sound-reflecting properties of these surfaces. Later on, this formula was found very useful in planning new concert halls.

Research has been continued on the foundations laid by Sabine, and nowadays no

Fig. 27.6. *Balcony of the Royal Festival Hall, London,* showing the use of padded leather panels backed by sound-absorbent material to reduce the reverberation period and so control the acoustic properties of the hall

Fig. 27.7. *Sound power measurement of a large centrifugal fan at the N.P.L.* The sound is measured at the 20 corners of the surrounding metal framework. The ceiling, walls and area beneath the floor grating are covered with mineral-fibre wedges which absorb incident sound waves and prevent unwanted echoes

architect would draw up plans for a new hall without paying attention to its acoustic design. Fig. 27.6 shows part of a wall of the Royal Festival Hall, London, during the course of construction. Here, reverberation is controlled by covering the walls with leather panels put on over a layer of sound-absorbent glass fibre.

Anechoic and soundproof rooms

For certain special purposes, e.g., investigation of the properties of loud-speakers and other sound equipment, it is necessary to have rooms whose walls absorb completely all sound energy falling upon them. This is achieved by lining the walls, ceiling, and floor with anechoic (=without echo) wedges composed of glass fibre, or plastic foam, encased in muslin (Fig. 27.7).

Efficient sound insulation is necessary for the comfort of those who dwell in blocks of flats. In modern building practice the spaces between floors and ceilings are usually filled with some inelastic material which absorbs the sound instead of transmitting it.

To measure the velocity of sound by an echo method

In places where a clear space extends for about 100 m from a high wall the velocity of sound may be found by measuring the time taken for a sound to travel to the wall and back again. The experiment is best carried out by two persons working together, one to make the sound and the other to carry out the timing.

One observer claps together two small wooden blocks held one in each hand and listens for the echo from the wall. Having obtained some idea of the time interval, he continues to clap the two blocks together and adjusts the rate of striking until each clap is made simultaneously with the arrival of the echo from the previous clap. The time interval between successive claps is then equal to the time taken for sound to travel twice the distance between observer and wall.

When the correct rate of striking has been achieved the second observer uses a stop-clock or watch to measure the time occupied by thirty or more clap intervals. It is, of course, important to begin the timing by *counting from nought*.

The distance from the wall is measured by a tape or other means.

Let: distance from wall $= x$ in metres; no. of clap intervals $= n$; time $= t$ in seconds

therefore   sound travels a distance $2x$ in time $\dfrac{t}{n}$

hence   velocity of sound in m/s $= \dfrac{distance}{time} = \dfrac{2x}{\dfrac{t}{n}} = \dfrac{2nx}{t}$

Several determinations should be made and a mean value for the velocity of sound calculated.

Echo sounding

An echo sounder or *fathometer* is a device used on a ship for the purpose of measuring the depth of the sea. A detailed explanation of the apparatus is beyond the scope of this book, but the basic principle involved is comparatively simple.

An electrical device fitted to the bottom of the ship sends out regular sound impulses which are reflected back from the sea-bed and received by a *hydrophone*, which is a microphone designed to work in water (see page 501). Obviously the time interval between the emission of a signal and its arrival back to the ship depends directly on the depth of the sea. A special electric circuit utilizes the time intervals to control the movement of a stylus over a moving strip of graph paper marked with a depth scale. The stylus thus plots a curve which is a scale contour of the sea-bed, giving a continuous indication of depth as the ship proceeds on its way.

This information is of great value to the pilot, as it permits the safe navigation of the vessel through uncharted waters, particularly where shallows or sandbanks are likely to be encountered.

It is important to note that the sound impulses sent out by a fathometer have a frequency of the order of 50 000 Hz. High-frequency waves of this type are said to be *ultrasonic*, as they cannot be heard by the human ear. In this connection they have a twofold advantage. First, they cannot be confused with engine noises or other sounds made by the ship. Secondly, very high-frequency waves can be confined to very narrow beams, and hence can penetrate sea-water to large distances without undue loss of energy by diffraction.

Clinical and other applications of ultrasound

Ultrasonic waves are being increasingly used in medical and surgical diagnosis by a technique based on the fact that different types of tissue, e.g., bone, fat, and muscle have different reflective properties for very high-frequency waves.

When a portion of the body is scanned by an ultrasonic beam, the varying echoes are recorded electronically and used to form a picture of what lies beneath on a television-type screen. The method serves a purpose similar to that of X-radiography (page 514), but is preferable in circumstances where the use of X-rays is inadvisable for health reasons. See Fig. 27.8.

Another application of ultrasound is for testing the quality of the linings and pads used in vehicle and aircraft brakes, see Fig. 27.9. Surgical instruments, jewellery, and similar articles may be cleaned thoroughly if they are immersed in a tank of cleaning solution through which ultrasonic waves are passed. Dirt is readily removed from crevices by the high frequency vibrations which are set up.

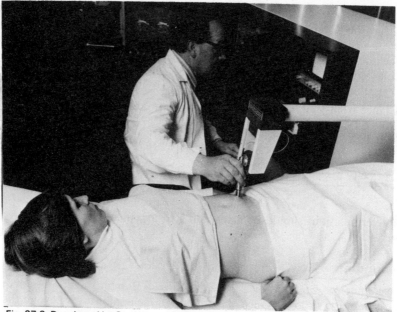

Fig. 27.8. Developed by Sonicaid, a British company, this ultrasonic probe produces a three-dimensional picture of various internal organs such as the pancreas, liver, and kidneys. It is of particular importance in obtaining information throughout the progress of pregnancy

Fig. 27.9. *Ultrasonic equipment at the Ferodo Research Laboratories.* This technician is making a control check of brake linings using ultrasonic apparatus. Any variation in quality or internal structure is detected by a change in the echo pattern displayed on a fluorescent screen. Tests of this kind are especially useful for the inspection of friction pads for vehicle and aircraft disc brakes

QUESTIONS: 27

1. What is the nature of a sound wave? How is it propagated?

Describe experiments, one in each case, to show:

(*a*) that the source of a sound is a vibrating body;

(*b*) that a material medium is necessary to transmit sound. (*J.M.B.*)

2. State two essential differences between the modes of propagation of sound and light waves. (*J.M.B.*)

3. A person standing 99 m from the foot of a tall cliff claps his hands and hears an echo 0.6 s later. Calculate the velocity of sound in air.

4. A student, standing between two vertical

cliffs and 480 m from the nearest cliff, shouted. She heard the first echo after 3 s, and the second echo 2 s later. Use this information to calculate

(i) the velocity of sound in air,
(ii) the distance between the cliffs.

(A.E.B., 1982)

5. Explain how sound waves travel through a gas and describe carefully an experiment which shows that a material medium is necessary for the transmission of sound.

A and B are two observers 1 km apart. There is a steady wind blowing. When a gun is fired at A the time interval between the flash and report observed at B is 3.04 s. When a gun is fired at B the interval between the flash and report observed at A is 2.96 s. Calculate the velocity of sound in air and velocity component of the wind in the direction BA. (O.C.)

6. Describe an outdoor experiment, employing echoes, for measuring the velocity of sound and show how the result is calculated.

How are ultrasonic waves used to measure the depth of the sea?

7. An observer carrying a metronome which makes a clicking sound at half-second intervals notices that echoes of the clicks from a wall 42 m away come midway between the clicks. Given that the velocity of sound lies between 300 and 400 m/s, calculate it. At what other greater distance from the wall could he hear the same effect? (O.C.)

8. A person stamps on a pavement which is bordered by an iron fence made of vertical rods at 15 cm intervals and hears a faint high-pitched note which lasts for a very short period. Explain how this occurs and calculate the frequency of the note. (Velocity of sound in air = 330 m/s.)

9. State the laws of reflection of sound and explain how an echo may be produced.

Two men stand facing each other, 200 m apart, on one side of a high wall and at the same perpendicular distance from it. When one fires a pistol the other hears a report 0.60 s after the flash and a second report 0.25 s after the first. Explain this and calculate:

(a) the velocity of sound in air;
(b) the perpendicular distance of the men from the wall. Draw a diagram of the positions of the men and the wall. (O.C.)

10. Explain the difference between longitudinal and transverse waves; state a practical example of each type.

Account for the fact that a distant lightning flash is seen before the thunder is heard. (S.)

11. A boy standing 100 m from the foot of a high wall claps his hands and the echo reaches him 0.5 s later. Calculate the velocity of sound in air using these observations.

(W.A.E.C.)

28. Musical sounds

Pitch and frequency

In the previous chapter it was mentioned that the *pitch* of a note, i.e., its position in the musical scale depends on the frequency of vibration of its source. This was demonstrated early in the nineteenth century by Felix Savart, who held a card against a rotating toothed wheel and showed that the pitch of the note emitted depends on the speed of rotation (Fig. 28.1). As the teeth strike the card it vibrates with a frequency equal to the number of teeth multiplied by the number of revolutions of the wheel per second.

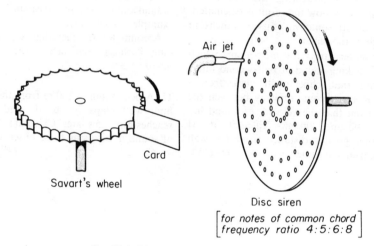

Savart's wheel

Card

Air jet

Disc siren
$$\left[\begin{array}{l}\text{for notes of common chord} \\ \text{frequency ratio } 4:5:6:8\end{array}\right]$$

Fig. 28.1. Pitch depends on frequency

If four wheels with teeth numbers in the ratio 4:5:6:8 are run at constant speed on the same shaft the notes given out are doh, me, soh, doh′. This well-known sequence of notes can be recognized whatever the constant speed of the shaft, showing that *the musical relation between notes depends on the ratio of their frequencies* rather than their actual frequencies.

Another device, the disc siren, is a rotating metal plate with holes drilled in concentric rings. When a jet of air is directed against the plate, puffs of air escape through the holes and produce a note of frequency equal to the product of the number of holes in a ring and the number of revolutions per second.

Music and noise

A sound of regular frequency is called a *tone* or musical note, and music is a combination of such sounds. Certain combinations of notes do, however, produce an effect of emotional tension, or *dissonance*, and this has played an important part in the vocabulary of music throughout the ages, in varying degrees.

Noise in relation to music is not easy to define but fairly general agreement may be expected with the definition that it is a sound or combination of sounds of *constantly varying* pitch, or *irregular* frequency.

Musical scales

As the art of music developed through the ages it came to be accepted that notes of certain frequency relationships gave a pleasing result when played together, while others produced a harsh effect. This experience led to the establishment of musical scales consisting of a series of notes whose pitch relationships enabled the maximum number of pleasing combinations to be obtained.

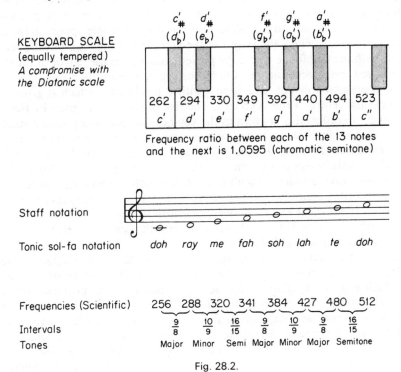

Fig. 28.2.

Music evolved along various lines in different parts of the world, and scales were adopted which differed both in the number as well as the pitch of the notes they contained. From the middle of the sixteenth to the middle of the nineteenth centuries European music, in particular, came to be based on the *diatonic scale*. This consists of eight notes which may be represented in various forms of notation (Fig. 28.2).

For scientific purposes, the diatonic scale has been standardized as a series of notes ranging from middle C (c′), 256 Hz to upper C (c″), 512 Hz. Most of the tuning-forks found in laboratories are in scientific pitch, but these are unsuitable for tuning musical instruments, as will be explained in due course.

At the outset it must be understood that, as far as music is concerned, the *actual* pitch of the notes has never been regarded as being so important as the *ratio* of the pitches of the various notes of the scale. Thus, during the eighteenth century it would be rare to find two organs whose middle C pipes had the same frequency, although, of course, the ratio of the pitches of the different pipes would be substantially the same for all instruments.

However, with the development of orchestral music, attempts were made at various times to establish a uniformity of pitch acceptable to musicians in all countries. In the year 1939 an international committee agreed that standard musical pitch should be based on a frequency of 440 Hz for a′.

Musical intervals

The ratio of the frequencies of two notes is called the musical interval between them. Certain recognized intervals have been given special names. Thus, a ratio of $9:8$ is called a *major tone*, $10:9$ a *minor tone* and $16:15$ a *semitone*.

Those who have studied music will be familiar with other intervals. For example, the interval between the top and bottom notes of the scale, $2:1$, is called an *octave*. The interval between c' and g', $3:2$, is a *fifth* and so on.

The problem of tuning a keyboard instrument

The lowest (or highest) note on the scale is called the *tonic* or *key-note*. Now it is an easy matter to make a change of key when singing, since the voice can be pitched to any frequency within its natural compass. Also the same principle applies in the case of stringed instruments, where the length of the vibrating strings can be adjusted by fingering; or with the trombone where the player manipulates the length of the slide. But with a keyed instrument such as a piano or organ a difficulty arises.

The scale represented in Fig. 28.2 begins with c' (middle C), which is given by a white key near the centre of the keyboard. The rest of the scale is given by the next seven white keys in order. If e' is taken as the key-note, an ascending scale cannot be played on a succession of white notes, as the intervals do not come in the correct order. For example, the interval between e' and the next white key f' is only a semitone, whereas the interval between the first and second notes of a diatonic scale must be a major tone. Similar difficulties arise with other intervals, and hence, in order to play a correct diatonic scale beginning with e', it would be necessary to have four extra keys. If sufficient extra keys were provided to enable diatonic scales to be played in all other possible keys we should end up with a piano which had so many keys that it would be impossible to play it.

Some compromise is therefore necessary. Space does not allow a detailed discussion of the various attempts which have been made in the past to overcome the difficulty. Suffice it to say that a satisfactory solution of the problem has been effected by substituting the *chromatic* or *equally tempered scale* for the diatonic scale. Five black keys, known as sharps (♯) or flats (♭) are added to each octave of white keys, making in all a range of thirteen notes each separated from its predecessor by an interval of $2^{\frac{1}{12}}$ or $1.0595:1$. This interval is called a *chromatic semitone*.

While these notes do not allow true diatonic scales to be played, the differences are so small that only those with extremely sensitive musical ears claim to be able to detect them. The actual frequencies employed in the equally tempered scale are shown in Fig. 28.2.

Intensity and loudness of sound

The intensity of a sound wave is defined as the rate of flow of energy per unit area perpendicular to the direction of the wave.

By mathematical treatment which is outside the scope of this book it may be shown that the intensity of a sound wave in air is proportional to: (*a*) the density of the air; (*b*) the square of the frequency; and (*c*) the square of the amplitude.

From the practical point of view we have to take the density of the air as we find it, and, therefore, *for a sound of given frequency the most important factor is the square of the amplitude*. Consequently, when the amplitude of vibration of a tuning fork or loudspeaker diaphragm is doubled the sound energy is not doubled but made *four times* as great.

Obviously, the *loudness* of a sound will depend on the intensity, but it does not follow that loudness is directly proportional to intensity. In the first instance, loudness depends on the varying pressure exerted on the eardrum by the incoming wave, and this will depend on the energy conveyed by the wave. But this is not the only factor involved. The ear varies in its sensitivity to sounds of different frequencies. Generally speaking, the ear is more sensitive to the higher frequencies.

Quality or timbre of a musical note

If a particular note on a scale is played on, say, a piano and a flute, it is easy to distinguish the tone of one instrument from that of the other. The two tones are said to differ in quality or *timbre*.

Generally speaking instruments do not give tones which are pure in the sense that they consist of single frequencies only. In the majority of cases a musical note consists of several different frequencies blended together. The strongest audible frequency present is called the *fundamental* and gives the note its characteristic pitch. The other frequencies are called *overtones*, and these determine the *quality of the sound*.

The notes of a trumpet possess a quality derived from the presence of strong overtones of high frequencies, while the flue pipes of an organ have a mellow tone. In the case of the latter, practically all the sound energy is centred in the fundamental frequency and the overtones are fewer in number and of smaller intensity.

Much experience and craftsmanship go into the design and construction of a piano with the aim of suppressing unwanted overtones and enhancing the desirable ones. The same can be said about other instruments, particularly the violin.

Beats

When two notes of nearly equal pitch are both sounded together a regular rise and fall occurs in the loudness of the tone heard. These alternations in loudness are called *beats*.

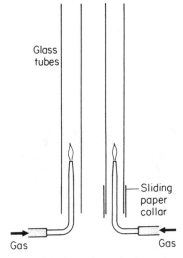

Fig. 28.3. Beats from singing tubes

This may be demonstrated with organ pipes, or singing tubes may be used instead. Two glass or metal tubes about a metre long and 4 or 5 cm in diameter are clamped in a vertical position and have small gas flames inside them about a quarter of the way up from the bottom. For this purpose, suitable burners may be made from drawn-out glass tubing, as shown in Fig. 28.3. After adjusting the flame height and position for the best results, the tubes will give out a loud, continuous note. Usually, the tubes are not exactly the same length, so their frequencies are slightly different and strong beats are heard.

By sliding a paper collar over the lower end of one tube its frequency may be altered as desired, with a corresponding change in the number of beats heard per second.

In the days when the majority of aircraft were powered by two or more piston-type engines a throbbing sound could often be heard owing to beats between the engine notes.

Piano tuners sometimes utilize beats for tuning a piano string to the pitch of a standard tuning fork. If the pitch of the string is not equal to that of the fork, beats will be heard when they are sounded together. The tension of the string is now altered until the beats become slower and finally disappear, showing that the two notes are in unison.

Two tuning forks of equal frequency may also be used to produce beats, if the frequency of one of them is reduced slightly by loading its prongs with tiny pieces of wax or plasticene. Let us suppose that two forks A and B of frequency 256 Hz, are sounded together, but that B has had its frequency reduced to 252 Hz in the manner described.

Consider an instant when both forks are exactly in step so that they are simultaneously producing compressions. Under these conditions they reinforce one another and produce a resultant sound of maximum loudness. After $\frac{1}{8}$ s A will have made 32 vibrations while B makes $31\frac{1}{2}$, and at this stage A will be producing a compression at exactly the same moment that B is producing a rarefaction. The resultant disturbance of the air will now be a minimum.

Again, after another $\frac{1}{8}$ s A will have completed 64 vibrations and B 63. Once more the two forks are in step with one another and the resultant air disturbance is a maximum. The loudness of the sound thus falls to a minimum and rises to a maximum every $\frac{1}{4}$ s or, in other words, there are 4 beats per second. Now 4 is equal to the difference in the two frequencies concerned, 256 and 252. In general, the number of beats per second given by two notes of nearly equal frequency f_1 and f_2 is given by $f_1 - f_2$.

Fig. 28.4 shows the resultant wave-form when two nearly equal waves combine to produce beats. The variation in the amplitude of the resultant wave indicates clearly how the beats occur.

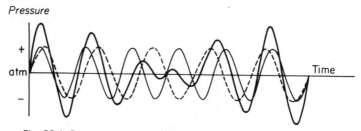

Fig. 28.4. Beat waveform from two notes of frequencies 5 and 6 Hz

QUESTIONS: 28

1. What do you understand by "frequency of a note"? A note has a frequency of 128; what is the frequency of a note two octaves higher?   (S.)

2. Define the terms frequency and amplitude as applied to sound and state how variations in these two factors affect the nature of a musical note.

Describe a simple experiment to illustrate the principle of the siren.

A rotating disc contains two sets of holes in the form of concentric circles. A jet of air is directed on the inner set and a note of frequency 256 is produced. If the number of holes in the set is 16, find the speed of revolution of the disc.

The speed is now reduced to three-quarters of its former value; the jet of air is directed to the outer set of holes, and the note has a frequency of 312. Calculate the number of holes in the outer circle.   (L.)

3. Describe a siren and how it may be used to determine the frequency of a tuning fork.

Calculate, in rev/min, the speed of rotation

of a disc siren for the note emitted to be in unison with an *E* tuning fork (320 Hz).

Unison between the two notes was determined by means of "beats". Explain this. (No. of holes in siren = 16.)   (*A.E.B.*)

4. Give reasons for the following:
(*a*) a noise differs from a musical note;
(*b*) musical notes differ in pitch, loudness, and quality.

Describe how you would determine the frequency of a note emitted by a tuning fork, showing how the result would be obtained from your observations.   (*L.*)

5. A wheel has 40 spokes and rotates at 8.5 rev/s. Calculate the frequency of the note obtained by holding a card lightly against the spokes as they rotate. What would you expect to hear if a tuning fork marked "341" were set in vibration near the card?   (*S.*)

6. Describe how you would demonstrate that the prongs of a tuning fork vibrate when it is emitting sound. State the meaning of the *frequency of the sound* produced by the fork.

Discuss the factors which determine
(*a*) the pitch,
(*b*) the quality,
(*c*) the loudness of a note which is produced by a tuning fork or other instrument and heard by an ear. Discuss also the difference between a musical note and a noise. Illustrate your answers with diagrams whenever possible.

A piano wire and a tuning fork produce different notes at the same time, and beats are heard. Explain this. Show how a piano wire can be tuned to emit a note of the same frequency as a vibrating tuning fork.   (*C.*)

7. Explain the terms *pitch* and *loudness* of a musical note. What factors determine the quality of a musical note?   (*W.A.E.C.*)

29. Strings and pipes

Stationary waves

A stationary wave is formed when two equal progressive waves are superposed on one another when travelling in opposite directions.

About the middle of the nineteenth century a German professor of physics called Franz Melde devised a method of demonstrating transverse stationary waves in a string. One end of a string passes over a pulley and is kept taut by attaching a weight, and the other end is fixed to the prong of a tuning fork. When the fork is bowed to set it in vibration a progressive wave travels along the string to the far end and is then reflected back. If the fork is kept in vibration the incident and reflected waves combine to form a *stationary* or *standing wave* and the string is seen to vibrate in a series of equal segments (Fig. 29.1). Instead of a tuning fork, some form of

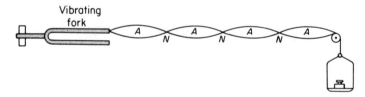

Fig. 29.1. Melde's experiment showing nodes and antinodes in a vibrating string

electric vibrator may be used for the experiment, and the string examined with a stroboscope (page 290). By this means it may be made to appear stationary or else to vibrate in slow motion.

The points marked N are called *nodes* and here the string remains at rest; in fact it makes no difference to the vibration of the string if it is touched with the edge of a thin card at any of the nodes. Between each pair of nodes the string vibrates with increasing amplitude towards the centre, where it reaches a maximum. The central points, A, of maximum amplitude are called *antinodes*.

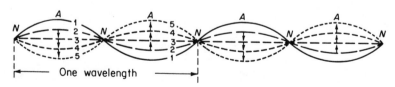

Fig. 29.2. Motion of string in which there is a stationary wave

Fig. 29.2 shows five stages covering half a complete vibration of the string. These diagrams should be compared with the progressive wave shown in Fig. 26.1, particularly with respect to the arrows indicating the movement of the string particles. Twice during a complete vibration the string is perfectly straight, and at these stages every particle simultaneously passes through its rest position. Furthermore, it will be seen from the diagrams that the distance between two successive nodes is equal to half a wavelength.

The sonometer

When experiments are being made to study the behaviour of vibrating strings a sonometer is used (Fig. 29.3). This consists of a long sounding board or box with a peg at one end and a pulley at the other. One end of the string or wire is attached to

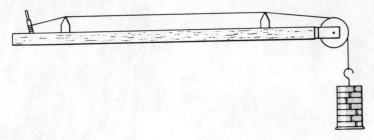

Fig. 29.3. The sonometer

the peg and the other end passes over the pulley. Iron weights are hung from the wire to vary the tension. In some types of sonometer, however, the tension is varied by a spring dynamometer or a lever arrangement. Two bridges are provided for the purpose of altering the effective vibrating length of the wire.

If the sonometer wire is gently plucked or bowed in the centre waves travel out to the two bridges and are then reflected back. A stationary wave is thus set up, and *in its simplest mode of vibration the wire vibrates in a single segment*. At the same time it gives out a note of definite pitch which is called the *fundamental*.

Overtones in a vibrating string

If the sonometer wire vibrates in more than one segment it gives out an overtone. This can be demonstrated in the following manner. The wire is lightly touched with a feather or strip of paper at its centre and is simultaneously bowed at a point midway between the centre and a bridge. The note emitted is the octave of the fundamental and the wire vibrates in two segments.

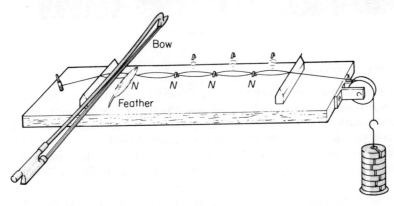

Fig. 29.4. Demonstrating nodes and antinodes with a sonometer

The experiment may be extended by placing small paper riders along the wire to coincide with node and antinode positions for a particular mode of vibration. On touching with a feather and bowing at appropriate points the wire is set in vibration. The riders jump off at the antinodes but remain in position at the nodes (Fig. 29.4). In the case of a violin or other stringed instrument the strings vibrate with a complex mode of this type which is superposed on the fundamental. Basically, this governs the characteristic tone of the instrument.

Fig. 29.5. Harp strings

Stringed instruments

Inspection of an instrument such as a harp or piano shows that it contains a number of stretched strings which vary in length and thickness. Harp strings (Fig. 29.5) are caused to vibrate by plucking them, while piano strings are struck by felt hammers. In contrast, the violin has only four strings of different thickness but the same length. The effective length of the strings is, however, altered by fingering. All stringed instruments are periodically tuned to correct pitch by adjusting the tensions of the strings.

In the following sections we shall discuss the factors which determine the pitch of the note emitted by a stretched string.

To investigate the relation between the frequency of a stretched string and its length ($f \propto 1/l$)

For this experiment several different tuning forks are required together with a sonometer fitted with a length of thin steel piano wire. The experiment is begun by adjusting the tension so that the length of the wire is *as long as possible when in unison with the lowest fork*. This will ensure that the lengths corresponding to the remaining forks will be as long as the dimensions of the sonometer will allow. The *percentage error* in measurement will thus be kept to a minimum.

Having tuned the wire to unison with the lowest fork, the tension is kept constant and the wire tuned to each of the other forks in turn by altering the length only.

Corresponding values of frequency and length are entered in a table as shown.

Frequency *f* (Hz)	Length *l* (cm)	$\frac{1}{l}$	*fl*

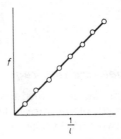

If $f \propto \dfrac{1}{l}$

then $f = \text{constant} \times \dfrac{1}{l}$

or $fl = \text{constant}$

The relationship is verified if the values of *fl* are found to be constant. Alternatively, the graph of *f* against $\dfrac{1}{l}$ should be a straight line through the origin.

Methods of tuning a sonometer to a tuning fork

Those who possess an average ear for music find it convenient to adjust the length or tension, as the case may be, until beats are heard when wire and fork are sounded together. A final adjustment is then made until the beats become slower and finally disappear when the wire is in unison with the fork. However, people who find it difficult to recognize pitch need not be afraid to tackle sonometer experiments, as a *visual* method of tuning can be used. A small paper rider is placed over the centre of the wire and the length or tension adjusted while the stem of the vibrating fork is held against one of the bridges. As soon as the natural frequency of the wire has become equal to that of the fork it begins to vibrate in sympathy with the fork and the paper rider is thrown off. This method employs the principle of *resonance*, which is discussed in more detail on pages 332–3.

Other factors which affect the frequency of a vibrating string

(*a*) *Tension*. The effect of tension can be investigated if the length of the string is kept constant while the tension is altered to bring the wire into unison with a number of standard forks in turn.

The results show that *the frequency is proportional to the square root of the tension T.*

(*b*) *Mass per unit length*. The thicker and heavier a string is, the lower is its frequency for a given length and tension. In musical instruments, however, merely using thicker strings of the same material is accompanied by stiffness leading to rapid decay in amplitude. To prevent this, the necessary loading without loss of flexibility is achieved by binding the strings with wire. This can be seen by looking inside a piano.

Experiment shows that *the frequency is inversely proportional to the square root of the mass per unit length, m.*

Summary. For a vibrating string,

$$(1)\, f \propto \frac{1}{l}$$

$$(2)\, f \propto \sqrt{T}$$

$$(3)\, f \propto \sqrt{\frac{1}{m}}$$

$\left. \right\}$ combining these results, $f \propto \dfrac{1}{l}\sqrt{\dfrac{T}{m}}$

To measure the frequency of a tuning fork by using a sonometer

The sonometer is tuned first to unison with a standard fork of known frequency and the length of the wire noted. Keeping the tension constant, it is now tuned to the unknown fork by altering the length only. Suppose the frequencies, in Hz, of the standard and the unknown forks are 320 and f and the corresponding wire lengths for unison are 82.0 and 61.5 cm respectively.

Since, at constant tension,

$$\text{frequency} \times \text{length} = \text{constant}$$
$$f \times 61.5 = 320 \times 82$$

$$f = \frac{320 \times 82}{61.5} = 426 \text{ Hz}$$

As a check, a second pair of lengths may be found using a different tension. Alternatively, the length may be kept constant and the tension varied, in which case f may be found by equating the two values of $\dfrac{frequency}{\sqrt{tension}}$ obtained.

Forced vibrations

When a vibrating tuning fork is held close to the ear the sound is heard quite loudly, but the loudness falls off rapidly as the fork is moved away. Since the prongs are of very small area, we may regard the fork as an approximately point source of energy, and allowing for *attenuation* (page 307), *the loudness of the sound is inversely proportional to the square of the distance from the source.*

In this respect a point source of sound may be compared with a point source of electromagnetic waves (page 306). Both obey an *inverse square law*.

If, however, the stem of the fork is pressed against a table top the sound can be heard clearly all over the room. Under these conditions the table is set into *forced vibration* and then acts as a large or extended source. It may thus be considered as equivalent to a large number of point sources, all of which contribute to the loudness of the sound at any point in the room.

It is also noticed that, although the sound is louder, it does not last so long. The amplitude of the table top is very much less than that of the fork, but its large area transfers energy to the air at a much greater rate than the small area of the prongs.

In the next section we shall consider the special case of forced vibration, in which the body set in vibration happens to have the same natural frequency as that of the body causing the vibration.

Resonance

Everyone knows that the best way to set a child's swing in motion is to give it small pushes in time with the natural period of the swing. This is an example of a general principle in physics which is called *resonance*.

Resonance is said to occur whenever a particular body or system is set in oscillation

**at its own natural frequency as a result of impulses received from some other system
which is vibrating with the same frequency.**

Under these conditions it is possible for a very large amplitude of vibration to be
set up. Cases have been known in the past where suspension bridges have been
damaged by the resonant vibrations caused by marching military columns. This can
happen if the rate of marching happens to bear a simple numerical relationship to
the natural frequency of oscillation of the bridge. Nowadays, in order to guard
against accidents, soldiers are always given the order to break step when crossing a
bridge.

When steel bodies were first introduced for motor-cars it was often noticed that a
drumming sound made its appearance at certain speeds. This was caused by resonance between the body panelling and the engine vibrations. It was soon discovered
that it could be prevented by coating the inside of the panels with a layer of plastic
material. This reduced the natural period of vibration of the panels and also damped
out the vibrations.

As well as mechanical examples of resonance, we find applications of the same
principle in other branches of physics. The action of tuning a radio set is to adjust
the value of the capacitance in a circuit until it has the same natural period of
oscillation for electricity as that of the incoming signal. The small alternating e.m.f.
set up in the aerial is then able to build up similar e.m.f. of large amplitude in the
tuned circuit.

The use of resonance in tuning a sonometer wire to have the same frequency as a
fork has already been mentioned on page 331. We shall now discuss the production
of resonance between a tuning fork and a column of air in a closed tube.

The resonance tube

Resonance in a closed tube is usually studied by means of the apparatus shown in
Fig. 29.6. Raising or lowering the water reservoir increases or decreases the effective

Tuning fork

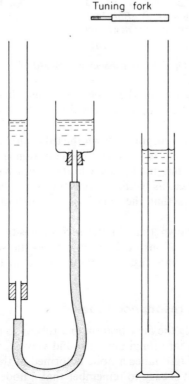

Fig. 29.6. Adjustable resonance tubes

length of the air column in the glass tube. The same figure also illustrates a simpler form of apparatus known as the resonance jar. A glass or metal tube stands in a tall jar full of water, and the length of the air column is varied by raising or lowering the tube.

Starting with a very short air column, a vibrating fork is held over the mouth of the tube, and the length of the column is then gradually increased. Strong resonance occurs when the column reaches a certain critical length. This is called the *first position of resonance*.

If the length of the column is now increased still further a *second position of resonance* is obtained when the column is approximately three times as long. Note that the resonance in this position is not so strong as in the first position.

Relation between length of air column and wavelength

Consider the first position of resonance and suppose the lower fork prong is in its extreme upper position. As the prong moves downward through half a vibration it sends a pulse of compression down the tube, which is reflected from the water surface and returns to the mouth of the tube (Fig. 29.7 (*a*)). If, at the moment the

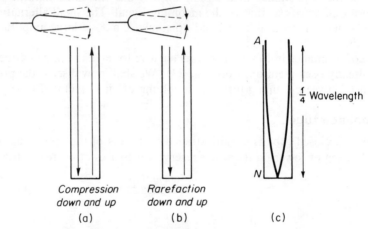

Fig. 29.7. Resonance in a closed tube

compression reaches the top of the tube, the prong has just reached its lowest position and is about to move upwards it will begin to send a rarefaction down the tube.

This rarefaction will, in turn, travel down the tube and be reflected back to the top in the time it takes the prong to complete the vibration (Fig. 29.7 (*b*)).

At this stage the fork has made one vibration and has sent out one complete wavelength of sound. Since, as we have seen, the sound has twice travelled up and down the tube, it follows that the length of the air column is one-quarter of a wavelength.

Similarly, it can be shown that, in the second position of resonance, the fork makes three complete vibrations in the time it takes the sound to travel four times the length of the tube. In this case the length of the air column is equal to three-quarters of a wavelength.

Stationary wave in a resonance tube

Another way of explaining the air vibration in a tube is to regard it as resulting from the formation of a stationary longitudinal sound wave.

In the first position of resonance a node is formed at the bottom of the tube and an antinode at the top. It is easy to remember that a node is formed at the bottom, since the air in contact with the bottom cannot move, whereas at the top it is free to move to and fro with maximum amplitude. In the second position of resonance an extra node and antinode are formed inside the tube.

It is useful to compare longitudinal stationary waves in a tube with transverse stationary waves in a string (page 328). In both cases **the nodes and antinodes are separated by a distance equal to one-quarter of a wavelength.**

Owing to the difficulty of representing longitudinal stationary waves diagrammatically it is customary to represent them symbolically by the use of transverse wave curves. This has been done in Fig. 29.7 (c).

To measure the velocity of sound in air by means of a resonance tube

We saw on page 315 that the velocity of sound is given by the formula

$$v = f\lambda$$

and hence if we can find the wavelength, λ, of the sound wave emitted by a standard tuning fork of frequency, f, we may easily calculate the velocity, v.

The resonance tube suggests a method of doing this. In the first position of resonance the length of the tube is $\lambda/4$, and therefore it would appear that λ can be found by multiplying the tube length by 4. However, there is a difficulty. The antinode at the top does not coincide exactly with the top of the tube, but projects slightly above it by an amount which is called the *end correction, c*.

Fortunately the difficulty may be overcome by measuring the length of the tube for the second position of resonance as well as the first (see Fig. 29.8). By subtracting one from the other, the end correction is eliminated and we obtain a correct value for half a wavelength.

Thus, if length of tube for 1st position of resonance $= l_1$
and   length of tube for 2nd position of resonance $= l_2$

then

$$\frac{\lambda}{4} = l_1 + c \quad \ldots \ldots \ldots \ldots \quad (1)$$

and

$$\frac{3\lambda}{4} = l_2 + c \quad \ldots \ldots \ldots \ldots \quad (2)$$

Subtracting (1) from (2),

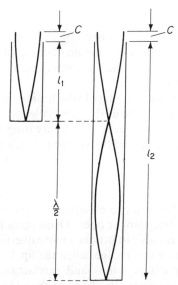

Fig. 29.8. First and second positions of resonance

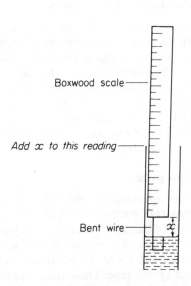

Fig. 29.9. Measuring the length of an air column

$$\frac{\lambda}{2} = l_2 - l_1$$

or

$$\lambda = 2(l_2 - l_1)$$

Substituting in the equation

$$v = f\lambda$$

we have

$$v = 2f(l_2 - l_1)$$

Either of the forms of apparatus shown in Fig. 29.6 are suitable for the above experiment. Fig. 29.9 illustrates a method for measuring the length of the air column which helps to overcome errors caused by the meniscus of the water. A bent wire is

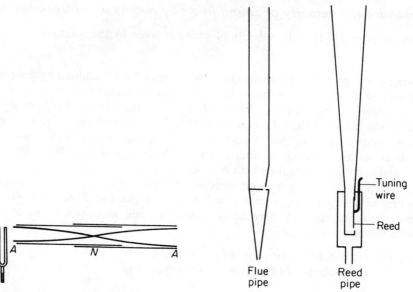

Fig. 29.10 Resonance in an open tube   Fig. 29.11. Organ pipes

inserted into a small hole drilled in the end of a boxwood scale. A reading of the top of the tube against the scale is taken when the tip of the bent wire is just level with the water surface. The distance, x, is afterwards measured and added to this reading in order to obtain the correct length l_1 or l_2.

Resonance in open pipes

Resonance may be obtained with a tuning fork and a pipe open at both ends, in which case there is an antinode at each end and a node in the middle. Bearing this in mind, we should expect the shortest length of open pipe which resonates to a given fork to be twice as long as the shortest closed pipe.

This assumption may be verified by the use of two tubes, one of which slides inside the other with a nice fit. Starting with this composite tube at its shortest length, the inner tube is gradually pulled out until resonance is obtained with a vibrating fork held near one end. The length is found to be double that of the shortest closed tube which resonates to the same fork (Fig. 29.10).

Organ pipes

Broadly speaking, the pipes of an organ are divided into two classes called *flue pipes* and *reeds* respectively. They may be either closed (stopped) or open. Open pipes have a more brilliant tone than stopped pipes, as they are able to produce more overtones.

In the case of a flue pipe a jet of air impinges on a sharp edge called the *lip*. Eddies are thus set up in the air-stream which causes compressions and rarefactions to travel up the pipe. These small impulses set up resonant vibrations in the pipe of a large amplitude, and hence a loud sound is given out. "Voicing" an organ pipe is a highly skilled craft, and consists of shaping the mouth and lip of the pipe so as to emphasize desirable overtones which enhance the beauty of the tone.

Reed pipes may be recognized by their conical shape. In these the air column is set in vibration by means of a strip of springy metal called the reed. The reeds are set in vibration by the air-stream. Tuning of the reeds is carried out by adjustment of a stiff wire, the end of which presses against the reed and controls its effective length (Fig. 29.11).

A large organ contains many hundreds of pipes, ranging in length from a few centimetres to over 9 metres in length. Certain ranks of pipes have two pipes for each note. One of each pair is de-tuned slightly so that beats are produced. The resultant tremolo effect is characteristic of those stops labelled "Vox humana" or "Vox angelica".

QUESTIONS: 29

1. What governs the pitch of a note emitted by a stretched string? How may the loudness be increased?

How do you account for the fact that two strings may be giving notes of the same pitch and loudness but of different quality?

Describe any observations you have made which confirms the fact that the velocity of sound is independent of its pitch.

(*C., part qn.*)

2. Describe an experiment to demonstrate that the prongs of a tuning fork vibrate when it emits sound and explain the increased loudness of the sound when the stem of the vibrating fork is held in contact with a table.

A tuning fork has a number 256 stamped on it. What does this signify? A sonometer wire is adjusted to be in unison with this fork. State exactly two different ways in which the wire could be adjusted so as to be in unison with a fork marked 384. (*L.*)

3. How does the frequency of vibration of a given string under constant tension and clamped at both ends depend upon the length? Describe an experiment to verify your statement.

Such a string has a length of 50 cm and emits a note of frequency 200 vibrations/second. What length of string under the same conditions would give a note in unison with a siren whose disc has 30 holes and is rotated at 10 revolutions/second? (*W.*)

4. When the wire of a sonometer is plucked it vibrates, emitting a note. State the effect on the fundamental frequency of this note, if

(i) the length of the wire were halved with no change in the tension;

(ii) the tension were made four times as large, with the original length unchanged;

(iii) the wire were replaced by one of mass per unit length four times as large, the length

and tension being the same as in the first experiment. (*A.E.B., 1982*)

5. (*a*) What is meant by "resonance"?

A steel wire is stretched between two points as shown in Fig. 29.12 and the tension in the wire is fixed such that, when the wire is plucked, the note which is produced is lower in pitch than that produced when a particular tuning fork is struck. Describe how you would tune the wire so that it has the same pitch as the fork.

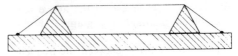

Fig. 29.12.

(*b*) A recording is made on a gramophone record turning at 33 revolutions/minute of a steady note with a frequency of 660 Hz.

(i) What is the wavelength of the sound in air if the velocity of sound in air is 330 m/s?

(ii) What is the speed of the record under the needle (stylus) when the needle is 0.10 m from the centre of the record?

(iii) What is the wavelength cut into the record groove at this radius? (*J.M.B.*)

6. Describe an experiment to determine the velocity of sound in the open air. Give reasons why your result may not be very accurate.

Transverse stationary waves are set up in a long thin wire using a suitable vibrator of frequency 50 Hz. The average distance between successive points where the wire is judged to be motionless is measured to be 47 cm. What name is given to these points? Calculate the velocity of transverse waves in the wire. (*S.*)

7. Explain:

(i) strings of different thickness are used on a stringed instrument such as a violin or a guitar;

(ii) the same note played on a violin and a flute sound different;

(iii) the strings of a stringed instrument are usually mounted on a hollow box of special shape. (C.)

8. (a) Illustrate the modes of vibration of the fundamental and the first two overtones when a musical note of fundamental frequency 256 Hz (cycles per second) is produced by air in a tube:

(i) open at both ends;

(ii) closed at one end.

Mark nodes and antinodes of *displacement* N and A respectively.

State the frequencies of each pair of overtones.

(b) Why is a musical note produced by an open tube or pipe more mellow than a note of the same fundamental frequency produced by a closed tube?

(c) While the tension of a vibrating string was kept constant, its length was varied in order to tune the string to a series of tuning forks. The results obtained were as follows:

Frequency of fork (Hz)	256	288	320	384	512
Length of string (cm)	78.1	69.5	62.5	52.1	39.1

By the appropriate use of the above readings, obtain a *straight line* graph and hence determine:

(i) the relationship between the frequency of vibration and the length of the stretched string;

(ii) the frequency of an unmarked fork which was in tune with 41.7 cm of the string. (A.E.B.)

9. Describe the resonance-tube method of measuring the wavelength in air of the sound emitted by a vibrating body. Why is the resonance tube so called?

If the shortest length of the tube for resonance is 0.12 m and the next resonant length is 0.37 m, what is the frequency of vibration? Take the speed of sound in air as 340 m/s.

Why would the result be in error if only one resonance position had been observed? (O.C.)

10. What is meant by resonance? Illustrate your answer by four examples from different branches of physics or from everyday life.

A tuning fork is struck and held over a jar full of water. The water is run out until the note given becomes very loud. Explain, with the aid of diagrams, exactly how the prongs of the fork and the air in the jar are vibrating.

How can this effect be used to determine the velocity of sound in air? (L.)

11. What are:

(i) *free vibrations*, and

(ii) *forced vibrations*? Hence explain *resonance*.

Two organ pipes A and B sound:

(i) their fundamental notes, and

(ii) their first overtones. Pipe A is closed at one end and pipe B open at both ends. Draw diagrams to show the mode of vibration in each of the four cases.

Describe an experiment to determine the relationship between the length of a pipe and its fundamental frequency of vibration. Sketch the graph you would draw to illustrate your results. (A.E.B.)

12. A string is set vibrating by plucking it:

(i) at the middle;

(ii) at a point one-quarter of the length from the end. Compare the frequencies of the first overtone in each case. Draw diagrams to illustrate your answer.

A column of air 26.25 cm long in a closed tube resonates to a sounding tuning fork. If the velocity of sound in air is 33 600 cm/s, what is the frequency of the fork? (S.)

13. Give two examples of resonance and explain briefly what is meant by *resonance* in an air column. (W.A.E.C.)

1000 MW coal fired power station at High Marnham, Dunham on Trent, Nottinghamshire

Electricity and magnetism

Electricity and magnetism

30. Magnets

As early as 600 B.C. the Greeks knew that a certain form of iron ore, now known as magnetite or lodestone, had the property of attracting small pieces of iron. Later, during the Middle Ages, crude navigational compasses were made by attaching pieces of lodestone to wooden splints floating on bowls of water. These splints always come to rest pointing in a N–S direction, and were the forerunners of the modern aircraft and ship compasses.

The word *lodestone* is derived from an old English word meaning *way*, and refers to the directional property of the stone mentioned above. Chemically, it consists of iron oxide having the formula Fe_3O_4. The word magnetism is derived from Magnesia, the place where magnetic iron ore was first discovered.

Magnetic poles

If a piece of lodestone is dipped into iron filings it is noticed that the filings cling in tufts, usually at two places in particular (Fig. 30.1). When the experiment is per-

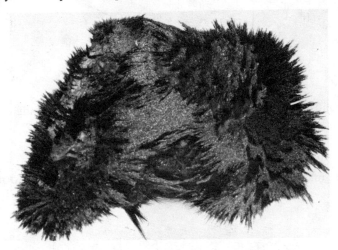

Fig. 30.1. Lodestone dipped into iron filings

formed with a bar magnet the filings are seen to cling in tufts near the ends. Few, if any, filings are attracted to the middle of the bar.

The places in a magnet where the resultant attractive force appears to be concentrated are called the poles.

Magnetic and non-magnetic substances

Apart from iron, the other common elements which are attracted strongly by a magnet are cobalt and nickel. These, together with certain strongly magnetic alloys are described as *ferromagnetic*. Substances such as copper, brass, wood, and glass are not attracted by a magnet, and are commonly described as non-magnetic. Nevertheless, experiments with very powerful magnets have shown that even the so-called non-magnetic substances have very feeble magnetic properties. We shall have something to say about these in the next chapter.

Suspending a magnet. Magnetic axis. Magnetic meridian

When a magnet is freely suspended so that it can swing in a horizontal plane it oscillates to and fro for a short time and then comes to rest in an approximate N–S direction. The magnet may be regarded as having a *magnetic axis* about which its magnetism is symmetrical, and it comes to rest with this axis in a vertical plane called the *magnetic meridian* (Fig. 30.2).

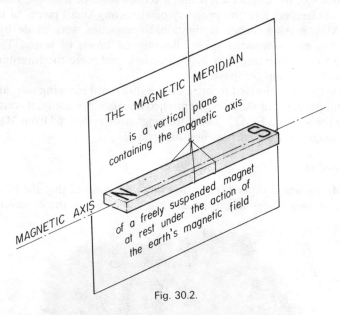

THE MAGNETIC MERIDIAN is a vertical plane containing the magnetic axis of a freely suspended magnet at rest under the action of the earth's magnetic field

MAGNETIC AXIS

Fig. 30.2.

The pole which points towards the north is called the north-seeking or simply the N pole; the other is called the south-seeking or S pole.

Action of one magnet on another

If the N pole of a magnet is brought near the N pole of a suspended magnet it is

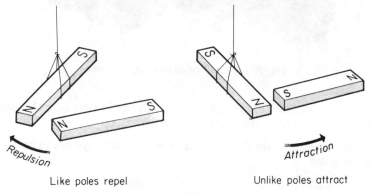

Repulsion

Attraction

Like poles repel

Unlike poles attract

Fig. 30.3.

noticed that repulsion occurs. Similarly, repulsion is observed between two S poles. On the other hand, a N and a S pole always attract one another (Fig. 30.3). These results may be summed up in the law,

like poles repel, unlike poles attract.

The above statement is sometimes called the *first law of magnetism.*

Test for polarity of a magnet

The polarity of any magnet may be tested by bringing both its poles, in turn, near to the known poles of a suspended magnet.

Repulsion will indicate *similar* polarity.

If attraction occurs, no firm conclusion can be drawn, since attraction would be obtained between either: (*a*) two unlike poles or (*b*) a pole and a piece of unmagnetized material.

Repulsion is, therefore, the only *sure* test for polarity.

To magnetize a steel bar by an electrical method

The best method of making magnets is to use the magnetic effect of an electric current. A cylindrical coil wound with 500 or more turns of insulated copper wire is connected in series with a 6 or 12 V electric battery and switch (Fig. 30.4). A coil of

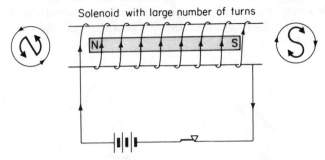

Fig. 30.4. Magnetization by an electric method

this type is called a *solenoid.* A steel bar is placed inside the coil and the current switched on and off. On removing and testing the bar it is found to be magnetized. It is unnecessary to leave the current on for any length of time, as the bar will not become magnetized any more strongly and the coil may be damaged through overheating.

The polarity of the magnet depends on the direction of flow of the current. **If, on looking at the end of the bar, the current is flowing in a clockwise direction, that end will be a S pole; if anticlockwise, it will be a N pole.** This may be easily be remembered by the symbols N and S as in Fig. 30.4. The magnetic effect of an electric current is discussed in further detail in chapter 36.

Commercially, C- and U-shaped magnets are made by linking them with one very thick copper conductor through which an enormous surge of current is passed for a fraction of a second (Fig. 30.5).

To magnetize a steel bar by the methods of single and divided touch

Before the magnetic effect of an electric current was discovered in the early nineteenth century magnets were made by stroking steel bars with a lodestone or with another magnet.

There are two ways in which this may be done, called the methods of *single* and *divided touch* respectively. In single touch a steel bar is stroked from end to end several times in the same direction with one pole of a magnet. Between successive strokes the pole is lifted high above the bar, otherwise the magnetism already induced in it will tend to be weakened (Fig. 30.6 (*a*)).

The disadvantage of the above method is that it produces magnets in which one pole is nearer the end of the bar than the other. It is better to use the method of

Fig. 30.5. Commercial production of magnets using an ignitron pulse-magnetizing apparatus

divided touch, in which the steel bar is stroked from the centre outwards with *unlike* poles of two magnets simultaneously (Fig. 30.6 (*b*)). In both of these methods it is to be noted that **the polarity produced at that end of the bar where the stroking finishes is of opposite kind to that of the stroking pole.** Steel knitting needles or pieces of clock spring may be magnetized by both methods and the polarity tested by obtaining repulsion with a magnetic needle. The magnets so made should also be dipped into iron filings, when the distribution of the filings will reveal the superiority of the method of divided touch.

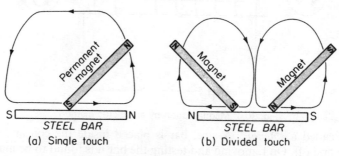

(a) Single touch   (b) Divided touch

Fig. 30.6. Older methods of making magnets

Another obsolete method of making magnets is described in a treatise on magnetism written by Dr. Gilbert in the reign of Elizabeth I. In this method red-hot steel bars are hammered while allowing them to cool when lying in a N–S direction. More is said about this on page 357.

Consequent poles

One should never assume, without prior test, that a bar magnet always has opposite *poles at its ends.* If a steel bar is magnetized by divided touch using two S poles we obtain a N pole at both ends of the bar and a double S pole in the centre. In this condition the bar is said to possess *consequent poles* (Fig. 30.7).

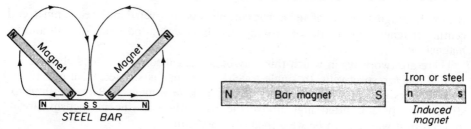

Fig. 30.7. Making a magnet with consequent poles

Fig. 30.8. Induced magnetism

Induced magnetism

When a piece of unmagnetized steel is placed either near to or in contact with a pole of a magnet and then removed it is found to be magnetized. This is called *induced magnetism* (Fig. 30.8).

Tests with a compass needle show that the induced pole nearest the magnet is of opposite sign to that of the inducing pole.

Induced magnetism can be used to form a "magnetic chain" as shown in Fig. 30.9.

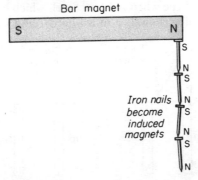

Fig. 30.9. Chain of nails experiment

Each nail added to the chain magnetizes the next one by induction, and attraction occurs between their adjacent unlike poles. Incidentally, the attraction between a magnet and an unmagnetized piece of material is always preceded by induction.

Demagnetization

The best way of demagnetizing a magnet is to place it inside a solenoid through which an alternating current is flowing. The current may be obtained from a 12 or 24 V mains transformer. While the current is still flowing, the magnet is withdrawn slowly to a distance of several metres from the solenoid in a W–E direction (Fig. 30.10). The alternating current takes the magnet through a series of ever-diminishing

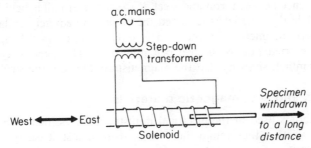

Fig. 30.10. Demagnetizing a magnet by use of alternating current

magnetic cycles, fifty times a second, until it is completely demagnetized. The magnet is held in an E–W direction so that it will not be left with some residual magnetism owing to induction in the earth's magnetic field (see page 357).

Used with care, the above method is useful for demagnetizing a watch. Watches should always be kept away from magnets, as the balance-spring, if made of steel, is liable to become magnetized. When this happens the watch no longer keeps good time.

Another method of destroying magnetism is to heat the magnet to redness and then to allow it to cool while it is lying in an E–W direction. This is not recommended as a practical method, since heat treatment will spoil the steel.

Finally, it should be noted that any vibration or rough treatment, such as dropping or hammering, particularly when the magnet is lying E–W, will cause weakening of the magnetism.

The principles underlying these last two methods will be understood after reading about the domain theory of magnetism in chapter 31.

To study the magnetic properties of iron and steel

It is important to distinguish between the magnetic properties of iron and steel. The term "soft" as applied to iron means reasonably pure iron. It is otherwise known as wrought iron, and is soft in the sense that it bends easily and can be readily hammered into any required shape when red hot. Steel, which consists of iron combined with a small percentage of carbon, is a much harder and stronger material.

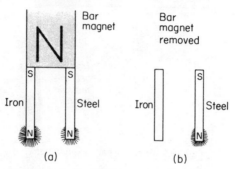

Fig. 30.11. Difference between iron and steel

A strip of soft iron and a strip of steel of the same dimensions, both initially unmagnetized, are placed side by side in contact with a pole of a magnet as shown in Fig. 30.11. Both strips become magnetized by induction, and on dipping their free ends into iron filings it is noticed that slightly more cling to the iron than to the steel. From this we conclude that the induced magnetism in the iron is slightly greater than that in the steel when both are subjected to the same magnetizing force.

If both strips are held firmly in the fingers while the magnet is removed it is noticed that practically all the filings fall from the iron, while few, if any, fall from the steel. The magnetic chain experiment illustrated in Fig. 30.9 may also be used to show the difference between iron and steel. If the topmost nail is held between finger and thumb while the magnet is removed the chain immediately collapses, showing that the induced magnetism in the iron is only temporary. When, however, the experiment is carried out using steel pen nibs or short pieces of clockspring, the chain remains intact, showing that the magnetism induced in the steel is permanent.

Uses of hard and soft magnetic materials

Magnetic materials used in the electrical industry are classified as *hard* or *soft* according as to whether they retain their magnetism or lose it easily. Both types are equally important.

Until the early years of the present century magnets were made of hard carbon steel (iron containing 1–1.5 per cent carbon). Subsequently it was found that the addition of small proportions of tungsten, chromium, and cobalt greatly improved the magnetic properties of the steel. Research along these lines has led to the discovery of special alloys for making powerful permanent magnets. Outstanding examples are Alcomax, Alnico, and Ticonal, which contain iron, nickel, cobalt, and aluminium in various proportions. Nowadays, magnets in a great range of types and sizes are used in the construction of electric motors, dynamos, and current-measuring instruments. Several magnets may be found in the ordinary household, for example, in the electricity meter, radio loudspeaker, and telephone earpiece.

In contrast with the above there are many types of electrical equipment in which rapid change or reversal of magnetism is required. Into this class come electric bells, relays, electromagnets, transformer cores and dynamo, and motor armatures. In the construction of these, soft magnetic alloys such as Mumetal (73 per cent nickel, 22 per cent iron, 5 per cent copper) and Stalloy (96 per cent iron, 4 per cent silicon) are used as well as soft iron.

Sintered materials

Of recent years, much research has gone into the development of *sintered* magnetic materials of which *Magnadur* is an example. Sintering is the name given to the conversion of powder into solid blocks by the application of heat and great pressure.

By using various metallic powders, either very soft or very hard magnetic materials may be made by this method. Many of the best permanent magnets for general use are of this type.

An interesting use for a strong permanent magnet is shown in Fig. 30.12.

(*For questions, see end of chapter 31*)

Fig. 30.12. *Early failure detection by permanent magnets.* This chip detector fitted to the scavenge side of the lubrication system of a gear-box gave timely warning of a bearing failure which might have led to expensive and possibly catastrophic damage to gears (see ball attached, from which the chips came). Inspection of the magnetic probe at regular intervals enables the engineer to keep reliable check on the equipment in his charge

31. Magnetic fields

The space surrounding a magnet in which magnetic force is exerted is called a magnetic field, and contains something which we call **magnetic flux**.

Magnetic flux is a vector quantity and, like all vectors, has magnitude and direction. It may be represented by magnetic field lines, and the pattern of these can be revealed by several methods which we shall now describe.

Direction of magnetic flux

Fig. 31.1 shows a bar magnet *NS* resting on the edge of a glass trough containing water, while a magnetized steel knitting needle *ns* is pushed through a large cork and floats on the water with its N pole uppermost. When the needle is held near the N pole of the magnet and then released it is repelled and travels towards the S pole along

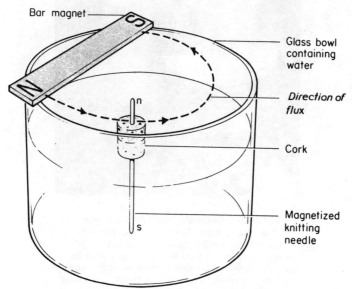

Bar magnet

Glass bowl containing water

Direction of flux

Cork

Magnetized knitting needle

Fig. 31.1. Direction of magnetic flux

a curved path which represents the direction of the magnetic flux. If the experiment is repeated with the S pole of the needle uppermost the needle travels in the reverse direction.

Clearly, the direction of travel depends on which pole of the needle is uppermost and one of these directions has to be chosen as a standard direction. It is conventional to choose the direction of the force on a N pole. Consequently, **the direction of the magnetic flux at any point is defined as the direction of the force on a N pole placed at the point.**

Magnetic flux patterns by the plotting-compass method

When a permanent record of a magnetic field is required the magnetic flux pattern may be traced out by means of a *plotting compass*. This consists of a very small magnetic needle pivoted between two glass discs in a brass case (Fig. 31.2). Fig. 31.3 shows how the compass is used. *NS* is a bar magnet placed on a sheet of white paper. Starting near one pole of the magnet, the position of the ends, *n, s* of the needle are

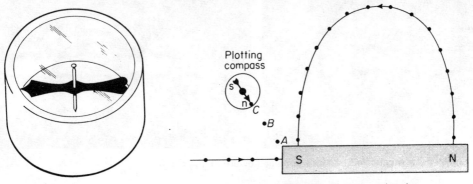

Fig. 31.2. Plotting compass

Fig. 31.3. Magnetic flux pattern by plotting compass method

marked by pencil dots *A, B*. The compass is now moved until the end *n* is exactly over the dot *B* and the new position of *s* is marked by a third dot *C*. This process is continued, and the series of dots obtained are joined, thus giving a magnetic field line which represents the direction of the magnetic flux. Other lines are plotted in the same way, and every line is labelled with an arrow to indicate its direction. Fig. 31.11 and 31.12 are examples of magnetic fields obtained by this method.

If magnetic flux patterns are plotted on a sheet of paper when no magnets are in the vicinity a series of parallel straight lines is obtained directed approximately from S to N geographically. This represents the earth's magnetic field in a horizontal plane (Fig. 31.4). The earth's magnetic field is described more fully later in the chapter.

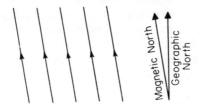

Fig. 31.4. Earth's magnetic flux in a horizontal plane

The main advantage of the plotting compass is that it is sensitive and can be used for plotting comparatively weak fields. It is unsuitable for fields in which the direction of the magnetic flux changes rapidly in a confined space, as, for example, in the neighbourhood of two magnets placed close together. Fields such as these are best investigated by the iron-filings method.

To plot magnetic flux patterns by the iron-filings method

The magnets whose field is to be studied are arranged beneath a sheet of stiff white paper, over which a thin even layer of iron filings is sprinkled from a caster. On tapping the paper gently with a pencil the filings form into chains along the magnetic field lines which reveal the flux pattern. Permanent records of these patterns may be obtained in two ways. The first method is to use waxed paper and to fix the filings by playing a bunsen flame quickly and evenly over the paper. The second and better method is to perform the experiment with photographic paper in a dark room. When a good pattern has been obtained the paper is exposed to light and then developed and fixed in the usual way.

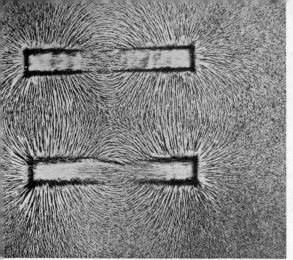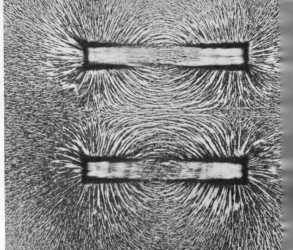

Fig. 31.5. Magnetic flux patterns obtained by use of iron filings

The principle of the method is as follows. Each filing becomes magnetized by induction. On tapping the paper the filings vibrate, and consequently are able to turn in the direction of the magnetic flux. The method fails with weak fields, which are unable to magnetize the filings sufficiently.

Fig. 31.5 and 31.18 show some typical magnetic fields which have been investigated by the iron-filings method.

The earth as a magnet

The idea that the earth is magnetized was first suggested towards the end of the sixteenth century by Dr. William Gilbert, who carried out experiments with spherical lodestones. Similarity between the magnetic field of a spherical lodestone and that of the earth led him to the conclusion that the earth was a magnet.

The origin of the earth's magnetism is still a matter of conjecture among scientists, but, broadly speaking, the earth behaves as though it contained a short bar magnet inclined at a small angle to its axis of rotation and with its S pole in the northern hemisphere (Fig. 31.6). The inclination of this supposed magnet to the earth's axis is inferred from the fact that a magnetic compass points towards the true north only at certain places on the earth's surface. Elsewhere it points either east or west of the true north.

Magnetic declination

In the previous chapter we explained that **the magnetic meridian at any place is a vertical plane containing the magnetic axis of a freely suspended magnet at rest under the action of the earth's field.**

The geographic meridian at a place is a plane containing the place and the earth's axis of rotation.

The angle between the magnetic and geographic meridians is called the magnetic declination (Fig. 31.7).

Inclination or dip

In 1576, Robert Norman, a compass-maker living at Wapping, made an experiment with a magnetic needle suspended at its centre of gravity. He found that it dipped with its N pole downwards and came to rest in the magnetic meridian with its axis making an angle of $71\frac{1}{2}°$ with the horizontal. This angle is called **the angle of dip or the inclination, and is defined as the angle between the direction of the earth's magnetic flux and the horizontal** (Fig. 31.10).

When pivoted at its centre of gravity the needle experiences no turning moment owing to the force of gravity on it. The only other turning forces acting are the magnetic forces on its poles, and these cause it to set along the true direction of the earth's magnetic flux. An ordinary compass needle shows no tendency to dip, since it is pivoted well above its centre of gravity. Mechanically, it is in stable equilibrium when horizontal.

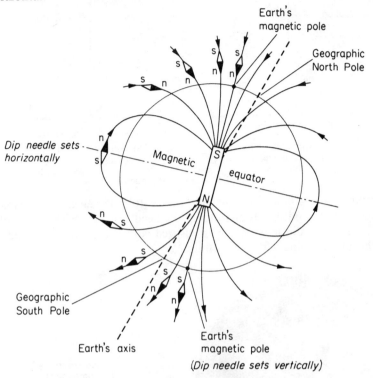

Fig. 31.6. The imaginary magnet inside the earth

Fig. 31.6 shows, in general, how the dip varies over the earth's surface. At the earth's magnetic poles the needle sets vertically, while along the magnetic equator it takes up a horizontal position. The N or S pole of the needle dips according as the needle is north or south of the magnetic equator.

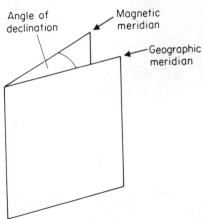

Fig. 31.7. Declination

Simple dip needle

The following experiment illustrates the principle used in measuring the angle of dip (Fig. 31.8).

An unmagnetized steel knitting needle is supported by a length of cotton and the position of the cotton adjusted until the needle balances horizontally. Under these conditions the needle will be supported approximately at its centre of gravity. With-

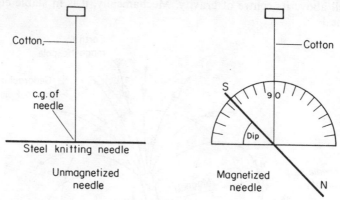

Fig. 31.8. A simple dip needle

out disturbing the position of attachment of the cotton, the needle is now magnetized by the solenoid method. On resuspending the needle, it dips with its N pole downwards. The angle of dip may now be measured with the aid of a protractor placed with its 90° mark in line with the cotton.

Fig. 31.9. The dip circle formerly used for measuring angle of inclination or dip. *Dip is now measured by electrical methods*

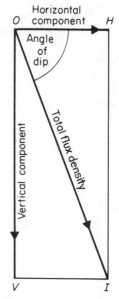

Fig. 31.10. The earth's magnetic flux density resolved into components

Owing to the difficulty of supporting the needle exactly at its centre of gravity, the above method cannot be relied upon for very accurate values of the dip. Better results may be obtained with a dip circle (Fig. 31.9), or by other methods outside our range of discussion.

Resolving the earth's magnetic field into components

We stated at the beginning of this chapter that magnetic flux is a vector quantity and so, like all vectors, it can be resolved into components by the parallelogram law.

When referring to the strength of a magnetic field we use the term **flux density**. The appropriateness of this expression becomes clear when we consider that the flux lines we have drawn by the methods described above are closer together in places where the magnetic field is stronger.

In England the earth's magnetic flux density has a certain magnitude and acts in a direction making an angle of about 70° (angle of dip) with the horizontal. We can therefore represent the earth's magnetic flux density to scale by a straight line OI at an angle of 70° to the horizontal (Fig. 31.10). If a rectangle of forces $OHIV$ is constructed about OI as diagonal the sides OH and OV will represent the horizontal and vertical components of the earth's flux density.

OI is said to represent the *total flux density of the earth's field*. Later in this chapter we shall describe an experiment in which the total flux density and the horizontal and vertical components respectively are used to magnetize a mumetal rod by induction (page 358).

By trigonometry, $\tan(\text{angle of dip}) = \dfrac{\text{vertical component}}{\text{horizontal component}} = \dfrac{OV}{OH}$

Changes in the earth's magnetism

A knowledge of the way in which the declination varies over the earth's surface is of importance to sea and air navigators who use magnetic compasses in order that they may calculate the correct compass course to steer for a given true course. Now the declination at any place is not constant, but changes slowly as the years go by. From time to time, therefore, fresh charts or maps of the world are prepared on which are drawn *isogonal lines* or lines joining all places having the same declination.

Our earliest record of the magnetic declination at London goes back to 1580, when its value was 11° 15′ east of north. After this it slowly decreased until the compass pointed true north at London in 1659. The declination then gradually increased in a westerly direction and reached a maximum of 24° 30′ west of north in 1820. From then onwards it began to decrease, and it has been estimated that the compass will once again point true north at London some time towards the end of the twenty-first century.

These changes in declination are accompanied by smaller changes in the inclination or dip. These changes are similar to those which would occur if a magnet inside the earth were slowly rotating about the earth's axis and taking 960 years to make a complete rotation.

To plot the magnetic flux pattern of a bar magnet in the earth's field

When a bar magnet is placed flat on a table the field round the magnet is a combination of the field of the magnet and the earth's horizontal field. We have already seen on page 349 that the earth's magnetic flux pattern alone in a horizontal plane consists of a series of parallel lines. The resultant pattern when a magnet's field is added to that of the earth depends on the direction in which the magnet is lying.

Fig. 31.11 shows the resultant flux pattern when a magnet is placed with its axis in the magnetic meridian and its S pole pointing north. This diagram was obtained by the plotting compass method (page 348).

In order to place the magnet correctly in the magnetic meridian, the following method is used. A plotting compass is placed with its centre over the edge of a sheet of paper and the paper rotated until the edge is parallel with the magnetic axis of the needle. This brings the edge of the paper into the magnetic meridian. The bar magnet is then placed on the paper with its axis parallel to the same edge and the magnetic flux pattern drawn with the aid of the plotting compass. (In order to prevent accidental movement of the paper during the course of the experiment, small pieces of self-adhesive tape may be used at the corners.)

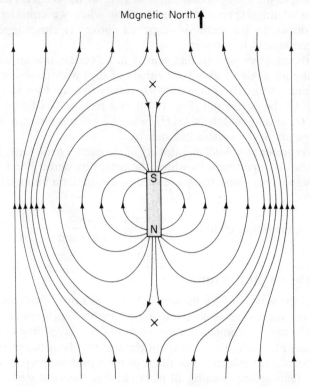

Magnetic North

Fig. 31.11. Horizontal magnetic flux pattern near a bar magnet with its axis in the magnetic meridian and its S pole pointing north

Neutral points

Along the axis of the magnet in Fig. 31.11 the earth's flux density and that of the magnet are in opposite directions. The earth's flux density remains constant in strength, while that due to the magnet is strong near the poles and weakens rapidly as the distance from the magnet increases. At the two points marked X in the diagram the horizontal component of the earth's flux density and the magnet's flux density are exactly equal and opposite and the resultant flux is zero. These points are called *neutral points*.

A neutral point is defined as a point at which the resultant magnetic flux density is zero.

Fig. 31.12 illustrates the resultant magnetic flux pattern when the magnet is placed in the magnetic meridian with its N pole pointing north. In this case neutral points are formed on the *magnetic equator* of the magnet.

Early theories of magnetism

During the eighteenth century, magnetism was attributed to the presence of "magnetic fluid". Electricity and heat were similarly regarded as fluids, but as time went on these fluid theories were found to be of very limited application, and so they were abandoned. An important advance was made about the middle of the nineteenth century by the introduction of *the molecular theory of magnetism* by Wilhelm Weber.

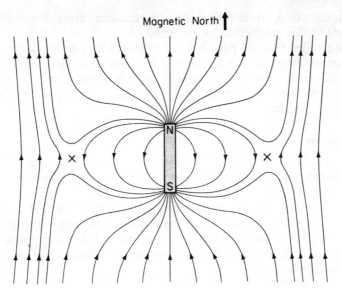

Fig. 31.12. Horizontal magnetic flux pattern near a bar magnet with its axis in the magnetic meridian and its N pole pointing north

Breaking a magnet

Weber considered that every molecule of a magnetic substance is itself a permanent magnet having two poles. He came to this conclusion by noticing that, when a magnet is broken in half, the two portions both have two poles, i.e., unlike poles have appeared on opposite sides of the break. No matter how many times a magnet is subdivided, each piece is found to possess two poles (Fig. 31.13). If this process could be carried out indefinitely, then, theoretically, single molecules would eventually be reached, and Weber thought it reasonable to suppose that these would also be magnets with two poles.

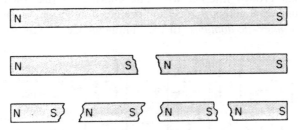

Fig. 31.13. Effect of breaking a bar magnet

The effect of breaking a magnet may be verified by taking a length of clockspring and magnetizing it by some convenient method. The magnetized clockspring is broken up with the aid of a pair of pliers and the polarity of the broken pieces tested by obtaining repulsion with a known pole of a pivoted compass needle (see page 342).

Modern views on magnetism

The molecular theory of magnetism was a considerable step forward but later there came an electrical explanation for the magnetism of atoms and this had wider applications.

We have already seen that magnets can be made by an electric current. Atoms contain particles of electricity called electrons which move in orbits about a central nucleus. These electrons behave like an electric current in a single turn of wire and so create a magnetic field. More importantly, the electrons are also spinning like tops and this adds further magnetism to the atom.

It may be recalled that, on page 342, we remarked that all substances show some kind of magnetic effect even if it is very small. The apparently non-magnetic substances are classified as either *diamagnetic* or *paramagnetic* respectively. We shall explain these terms very briefly before going on to deal with the strongly magnetic or *ferromagnetic* substances.

Diamagnetism

In many materials there are equal numbers of electrons spinning and orbiting in opposite directions so that, in the absence of some external magnetic field, their effects cancel out. If, however, a magnetic field is applied the electron orbits are very slightly disturbed by electromagnetic induction (page 484). This very slightly weakens the field inside the material giving a minute magnetic effect called *diamagnetism*. This name is derived from the fact that such substances set with their length perpendicular to a strong magnetic field instead of in line with it as do para- and ferromagnetic substances.

Paramagnetism

In other materials there are unbalanced electrons so that the individual atoms or molecules act like very tiny magnets. In the absence of an external magnetic field these molecular magnets are arranged at random, giving no resultant magnetic effect to the material as a whole, but if a magnetic field is applied the molecular magnets become partially aligned with it thus increasing its strength. This small effect is called *paramagnetism*.

Ferromagnetism. The domain theory of magnetism

In some materials, of which iron, steel, and certain alloys are outstanding examples, the atomic magnets or *dipoles* do not act independently as in paramagnetic substances but small groups interact with one another so that their magnetic axes spontaneously line up together in a certain preferred direction. Groups such as these are described as *magnetic domains*. In the "unmagnetized" state the resultant mag-

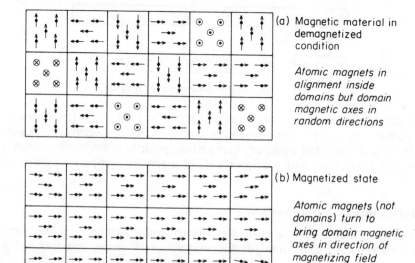

(a) Magnetic material in demagnetized condition

Atomic magnets in alignment inside domains but domain magnetic axes in random directions

(b) Magnetized state

Atomic magnets (not domains) turn to bring domain magnetic axes in direction of magnetizing field

Fig. 31.14. Domain theory of magnetism in ferromagnetic substances

netic axes of the domains point in all directions at random and so the bar as a whole shows no polarity.

This condition is illustrated in Fig. 31.14 (*a*) in which, for simplicity, we have drawn the domains as uniform cubes. In actual fact, however, the domains are not at all cubical in shape but can vary in size and their edges will lie in different directions from grain to grain throughout the material. The atomic magnets or dipoles have been represented by the symbol, —●—→ in which the arrow head signifies the N pole.

If the bar is placed in a gradually increasing magnetic field, e.g., inside a solenoid with steadily increasing current, then in some of the domains all the atomic dipoles may turn out of their preferred directions so that their axes all come into a new direction more closely related to that of the magnetizing field.

Eventually, if the magnetizing field is strong enough, the resultant magnetic axes of all the domains will be brought into the best possible alignment with the field and the material is said to be *magnetically saturated* (Fig. 31.14 (*b*)). We now have free atomic poles at the ends of the bar which will give rise to the poles of the magnet. Finally, when the magnetizing field is removed, the mutual repulsion between the free atomic poles at the ends of the bar will cause the domain axes to fan out slightly. This explains why the poles of a bar magnet are never situated exactly at its ends.

Magnetic phenomena explained by the domain theory

(1) *Demagnetization*
Anything which disturbs the dipoles in the domains and enables them to settle down back in their preferred directions will weaken or destroy the magnetism of the magnet as a whole. Mechanical vibration can have this effect, and so also will a rise in temperature, since this is accompanied by increased energy of vibration of the atoms.

(2) *Magnetic saturation*
It is a well-known fact that the strength of a magnet cannot be increased beyond a certain limit. This occurs when all the domain axes have been brought into alignment with the magnetizing field as explained in the previous section.

(3) *Breaking a magnet*
When a magnet is broken in two, the fracture exposes the ends of dipoles all pointing in the same direction, S poles on one side and N poles on the other (Fig. 31.13).

(4) *Magnetization by induction in the earth's magnetic field*
On page 344 we referred to an old method of making magnets by heating steel bars to redness and allowing them to cool in either a vertical or else a horizontal N–S direction. Heating causes the dipoles to vibrate and, on cooling, they turn out of their preferred directions in their domains and settle down in the direction of the applied field.

It is, however, better to place the bar in the magnetic meridian at an angle with the horizontal equal to the angle of dip. In the British Isles this is about 70°. In this position the bar becomes magnetized by the *total* magnetic flux density of the earth's field, which is greater than either the vertical or the horizontal component.

Instead of heating the bar, dipoles may be caused to vibrate mechanically by striking the bar with a hammer. This method has the advantage of not altering the crystalline structure of the steel. It is important to notice that the lower end of the bar becomes a N pole; this results from the fact that the dipoles turn so that their N poles point in the direction of the inducing magnetic field (Fig. 31.15). Iron may also be magnetized by this method, but in this case the induced magnetism is only temporary.

When *mumetal* (page 347) is used no hammering is required to assist the alignment of the dipoles. Mumetal becomes more strongly magnetized when in a magnetic field than either iron or steel. It is said to possess a greater *susceptibility*.

The effects described above may be studied by placing a mumetal rod in the magnetic meridian, first vertically, then horizontally, and finally at an angle with the horizontal equal to the magnetic dip. Tests should be carried out with a small compass needle to show that, in every case, the lower or north-pointing end of the rod has become a N pole and the opposite end a S pole. Furthermore, it should be verified that the polarity is reversed when the rod is turned end for end, showing that the dipoles have been completely turned round.

Magnetization of iron and steel used in building construction

Vertical steel pillars used in the construction of buildings are invariably found to be magnetized. This occurs owing to the induction by the vertical component of the earth's magnetic flux density. Similarly, steel joists lying horizontally in the magnetic meridian become magnetized by the horizontal component. In such cases the process of magnetization is assisted by the vibration set up by riveting during the course of construction.

North of the magnetic equator the lower end of a pillar is a N pole. South of the magnetic equator it is a S pole.

Joists lying E–W are not magnetized. In this position the earth's field tends to cause the dipoles to set at right angles to the length of the joist. This, however, cannot take place. If it did so, the large number of free atomic poles produced along the sides of the joist would set up powerful forces of mutual repulsion. The arrangement would thus be unstable.

To plot the magnetic flux pattern near the base of a magnetized iron pillar

If the compass needle method is used to plot the magnetic flux pattern on the floor near the base of a vertical magnetized iron pillar a map similar to that shown in Fig. 31.15 (*c*) is obtained. Note that a neutral point occurs south of the pillar at a point where the horizontal component of the earth's flux density and that due to the pillar are exactly equal and opposite.

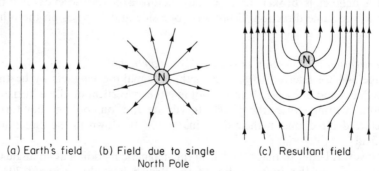

(a) Earth's field   (b) Field due to single North Pole   (c) Resultant field

Fig. 31.15. Magnetic flux pattern near base of a magnetized iron pillar

The pattern obtained may be explained by regarding the field as the resultant of the parallel straight lines of the earth's flux (Fig. 31.15 (*a*)) combined with the radial flux arising from the N pole at the base of the pillar (Fig. 31.15 (*b*)). We are, of course, assuming that the S pole at the top of the pillar is sufficiently far away to have a negligible effect.

For convenience, this experiment may be carried out with a long soft-iron rod which has been well hammered in a vertical position, so that it becomes magnetized by induction. The rod is then held vertically by a wooden stand with its lower end resting on a sheet of paper on the bench. If an iron rod is not available the same effect may be obtained by using a long ball-ended magnet, set up vertically with its N pole resting on the paper. The magnetic flux pattern is then plotted with a compass needle as described on page 348.

Magnetic shielding

The magnetic field near a soft-iron bar placed horizontally in a N–S direction is shown in Fig. 31.17. This map was obtained by the compass-needle method. The earth's magnetic flux appears to be drawn into the iron and concentrated through it. Iron is therefore said to be more permeable to magnetic flux than air is, or to possess a higher magnetic *permeability*.

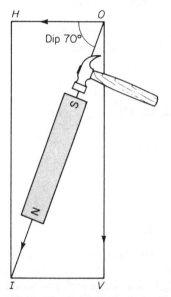

Fig. 31.16. Magnetization by induction in earth's magnetic field

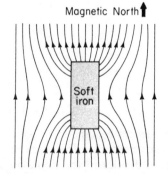

31.17. Soft iron in the earth's horizontal magnetic field

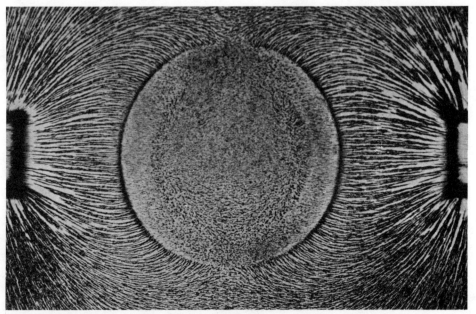

Fig. 31.18. Magnetic shielding by a soft-iron ring placed in a magnetic field

The behaviour of a soft-iron ring in a strong magnetic field is of particular importance (Fig. 31.18). In this case the magnetic field used is much stronger than the earth's field, and hence the iron-filings method may be used to map the field. Here, all the flux passes through the ring and no flux crosses the space inside. The inside of the ring is said to be shielded or *screened* from magnetic flux.

Magnetic shielding is put to practical use. For example, delicate measuring instruments which are liable to be affected by external magnetic fields can be protected by enclosing them in thick-walled soft-iron boxes.

Keepers

A bar magnet tends to become weaker with age, owing to *self-demagnetization*. This is caused by the poles at the ends of the magnet which tend to reverse the direction of the atomic dipoles inside it.

In order to prevent this, bar magnets are stored in pairs, with their opposite poles adjacent and with small pieces of soft iron, called *keepers*, placed across their ends.

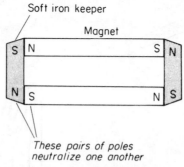

Fig. 31.19. Use of magnetic keepers

These keepers become strong induced magnets, and the opposite induced poles at their ends neutralize the poles of the bar magnets (Fig. 31.19). In other words the magnetic dipoles in the domains of both magnets and keepers form closed loops with no free poles. Consequently the demagnetizing effect disappears.

QUESTIONS: 30 and 31

1. Three steel knitting needles, exactly similar in appearance, are supplied to you. One is known to be magnetized with opposite poles at its ends, another has consequent poles in the middle and the third is unmagnetized. Describe how you would proceed to identify them if the only other apparatus available is a means for suspending them horizontally. (*L.*)

2. Describe, with the aid of diagrams, and explain how you would magnetize a steel knitting needle so as to obtain a S pole at a marked end of the needle:

(*a*) by using a permanent magnet;

(*b*) by using an electric current.

State, giving a reason, which method of magnetization you would expect to produce the stronger magnet and describe an experiment to test your conclusion.

Describe how a magnetized knitting needle can be effectively demagnetized. (*O.C.*)

3. Explain what is meant by:

(*a*) a magnetic field;

(*b*) a neutral point.

Draw diagrams of the magnetic fields due to:

(*a*) a bar magnet placed horizontally with its axis in the magnetic meridian and its N pole pointing south, and

(*b*) a bar of unmagnetized iron having the same shape and placed similarly in the earth's field.

Explain any one application of the effect illustrated by (*b*). (*L.*)

4. Describe how the magnetic flux pattern may be plotted in a horizontal plane in which a bar magnet rests in the earth's magnetic field. If the magnet is lying with its N pole pointing north, sketch the pattern you would expect to obtain, and mark the position of any neutral points.

Sketch also the magnetic flux pattern in a

horizontal plane through the S pole of a magnet placed vertically.

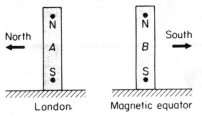

Fig. 31.20.

5. Fig. 31.20 shows two bar magnets, each standing on a bench. The one marked *A* is in London, *B* is at a place on the magnetic equator. The dots represent the positions of the poles of the magnets. Show on each diagram the direction of the force, due to the action of the earth's magnetic field, on each of the poles. (*C.*)

6. Explain what is meant by *magnetic induction* and give an example of it.

Describe an experiment to demonstrate the differences in the magnetic properties of iron and steel and state the conclusions to be derived from it.

Which of these metals would you use for:

(*a*) the core of an electromagnet;

(*b*) a compass needle; and

(*c*) magnetic screening? Give reasons for your answers. (*C.*)

7. Define the terms *declination, inclination* (dip) and show on a diagram how the earth's magnetic field may be resolved into two components.

Give a simple method for illustrating magnetic dip and explain any differences which might be observed according as the experiment is carried out in the northern or southern hemisphere.

8. A bar of iron is hammered for a short time at the latitude and longitude of Bristol in the plane of the earth's magnetic field with its longer axis successively:

(*a*) horizontal;

(*b*) vertical;

(*c*) inclined at approximately 60 to the horizontal, the end pointing north being lower than the other. Compare the magnetizing effect, giving reasons for your answer.

(*S.*)

9. Explain as fully as you can why:

(*a*) a pivoted compass needle free to rotate in a horizontal plane always sets in one particular direction at any given place;

(*b*) some iron filings on a piece of paper near a strong magnet will show a definite pattern after the paper has been tapped;

(*c*) vertical retort stand rods in laboratories are found to be magnetized, and the polarity of such rods in England is different from that found in Australia. (*W.*)

10. Draw a diagram to show how you would shield a small compass needle from the earth's magnetic field. Also, on your diagram, sketch the resultant magnetic flux pattern.

11. Give an account of a theory of magnetism which applies to iron and steel, and use it to explain the condition of:

(i) a magnetized, and

(ii) an unmagnetized bar of steel. Suggest a reason why it is not possible to increase the strength of a magnet indefinitely.

State any important differences between soft iron and steel and indicate which you would select for use as:

(iii) a magnetic keeper;

(iv) the core of an electromagnet;

(v) a compass needle.

Describe how you would make use of the earth's magnetic field in order to magnetize a bar of soft iron as strongly as possible. Show clearly the polarity produced at each end of the bar. (*L.*)

32. Static electricity

Everyone is familiar with the fact that if a pen made of certain plastic materials is rubbed on the coat-sleeve it will afterwards attract dust and small pieces of paper. The same effect is noticed when a mirror or window-pane is polished with a dry cloth in a very dry atmosphere. Dust and fluff from the cloth stick to the glass and are difficult to remove. Perspex, cellulose acetate and the vinyl compounds used for gramophone records also show the attraction, but to a more marked degree. The phenomenon is called *electric attraction*, and the rubbed materials are said to have become charged with *static electricity*. Knowledge of it goes back as far as the sixth century B.C., when the Greek philosopher, Thales, described the attractive properties of rubbed amber. The word electricity has, in fact, been derived from the Greek word "elektron", meaning amber.

Friction between certain textiles can also produce electrification. Robert Symmer first described this in the early eighteenth century. He noticed strong electrical attraction between a black and a white silk stocking when both had been put on and then withdrawn from the same leg. Anyone who has worn clothing made of Terylene or nylon knows that it is often strongly electrified especially when taken off at the end of a dry day.

Electrification by friction is sometimes associated with a crackling sound. This may be heard when very dry hair is combed with a vulcanite comb, or when an ebonite or alkathene rod is vigorously rubbed with fur. The crackling is caused by small electric sparks, which may be seen if the room is in darkness. Sparks from static electricity can be very dangerous when inflammable vapour is present. For example, instances have been known in hospital operating theatres where ether vapour has become ignited by a spark from a trolley used for transporting patients. If a trolley has wheels with insulating rubber tyres it can become charged as a result of friction between blankets and the plastic sheet on it. Nowadays, accidents are prevented by allowing a short length of chain to trail from the metal frame of the trolley. This conducts the electric charge away to earth, where it can do no harm.

In dry weather people alighting from cars and buses occasionally complain of a slight electric shock as their feet touch the ground. This has been attributed to electrification of the vehicle, either by friction where the exhaust gases leave the exhaust pipe or to friction between the person's clothing and the upholstery.

Electric repulsion

If an ebonite rod is charged by rubbing it with fur and then held just above a collection of small pith balls the balls will jump rapidly up and down for a short time between the rod and the bench. This is explained as follows. The balls are first of all attracted, become charged by contact and are then repelled. On striking the bench they lose their charge to the earth, and the action is repeated until the rod has lost most of its charge.

Positive and negative electricity

Electric repulsion was first described in 1672 by Otto von Guericke, who noticed that some feathers were attracted to a charged sulphur ball and then repelled from it. One hundred and fifty years later in France, Charles Du Fay discovered that charged bodies did not always repel each other, but that sometimes attraction took place. He came to the conclusion that there were two kinds of electricity. Charges of the same kind repel, while charges of opposite kinds attract one another.

To distinguish between the two kinds, Du Fay used the terms *vitreous* and *resinous* electricity. Vitreous (from the Latin *vitrum* = glass) electricity is obtained when glass is rubbed with silk, and resinous electricity is obtained when amber, sealing-wax, sulphur, shellac, and a host of other substances are rubbed with fur or flannel. Later on, these terms were found to be misleading, since, for example, ground glass gives resinous electricity and very highly polished ebonite gives vitreous electricity. Accordingly, Benjamin Franklin introduced the present-day terms *positive* and *negative* instead of vitreous and resinous respectively.

Traditionally, in experiments on static electricity the standard method for obtaining positive electricity is to rub glass with silk, and for a negative charge, to rub ebonite with fur or flannel. Nowadays, however, it may be considered preferable to use cellulose acetate for positive, and polythene (alkathene) for negative charge, since these are less affected by damp conditions. Also, a clean duster or a paper tissue may be used as a rubber instead of fur or silk.

Action between electric charges

A warm dry glass rod is rubbed with silk and placed in a wire stirrup suspended by a length of thread. On bringing near a second charged glass rod, repulsion occurs.

The experiment should now be performed using ebonite rods rubbed with fur. Repulsion again takes place.

Finally, the experiment is carried out with a charged glass rod and a charged ebonite rod. This time, attraction occurs.

These experiments verify a fundamental law, sometimes called the first law of electrostatics,

like charges repel, unlike charges attract.

Insulators and conductors

The first record of a serious approach towards the study of static electricity is to be found in one of the chapters of Dr. Gilbert's seventeenth-century treatise on magnetism, already mentioned on page 344. Here, Gilbert lists a large variety of substances which become electrified when rubbed. These *electrics*, as they were called, include precious stones and crystals besides amber, glass, resin, and sulphur. In contrast, he also gives a list of *non-electrics* or substances which cannot be electrified by rubbing. This list consists mostly of metals.

When current electricity was discovered, 200 years after the publication of Gilbert's book, it was found that an electric current would flow through a non-electric but not through an electric. Accordingly, these terms became obsolete. We now call an electric an *insulator* and a non-electric a *conductor*.

Incidentally, Gilbert mentions the importance of dryness in electrical experiments. Impure water is a conductor, and a film of moisture from condensation or moist hands on the surface of an insulator allows electricity to be conducted away to earth. *For successful results, all apparatus used in electrical experiments must be thoroughly dry.* Glass rods in particular are best warmed before use.

The gold-leaf electroscope

For the detection and testing of small electric charges, a gold-leaf electroscope is used. This instrument was invented towards the end of the eighteenth century by a Yorkshire clergyman named Abraham Bennet. Fig. 32.1 shows a common type of electroscope. It consists of a brass rod surmounted by a brass disc or cap and having at its lower end a small rectangular brass plate with a leaf of thin gold or aluminium attached. The leaf is protected from draughts by enclosing it in an earthed metal case with glass windows. The brass rod is supported by passing it through a plug of some good insulating material such as alkathene at the top of the case.

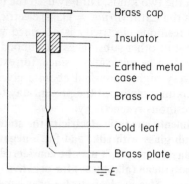

Fig. 32.1. Gold-leaf electroscope

The three horizontal parallel lines shown at *E* in Fig. 32.1 is the conventional symbol for an *earth connection*. The importance of having the case earthed will be understood after reading page 385.

Experiments with a gold-leaf electroscope

(1) *To detect the presence of charge on a body*

If a rod of some suitable material is charged by friction and then brought near to the cap of a gold-leaf electroscope the leaf is seen to diverge from the plate. A charge has been *induced* on the leaf and plate, and consequently repulsion occurs between them. On removing the charged rod, the leaf collapses, showing that the induced charge on the electroscope is only temporary.

Very small charges may be detected by this method.

(2) *To charge a gold-leaf electroscope by contact*

Generally speaking, it is not always easy to charge an electroscope by *contact* with a charged rod, but usually it can be done after a few attempts.

An ebonite rod is given a small charge by rubbing with fur, and is then *rolled* over the cap of an electroscope. The leaf will be seen to diverge, and then the rod is removed. If the leaf does not stay diverged the process is repeated until it does. We may now assume that the electroscope is charged with negative electricity by *conduction* from the ebonite rod.

If the cap of the electroscope is touched with the finger the charge flows to earth through the experimenter's body and the leaf collapses. This is called "earthing the electroscope". Before proceeding any further it must be pointed out that *charging by contact is not a good method and often gives a charge opposite to that expected.* It is better to use the method of *induction*, which is described later.

(3) *To test for the sign of the charge on a body*

Having charged the electroscope negatively as described above, the ebonite rod should be recharged and brought near to the cap. An increase in the leaf divergence is noted.

A glass rod rubbed with silk (positive charge) is now cautiously brought down towards the cap from a height of about 50 cm. This time, a decrease in divergence is noticed.

The electroscope is discharged by touching it with the finger and afterwards charged positively by contact, using a glass rod rubbed with silk. We shall now find that an increased divergence is caused by bringing a charged glass rod near the cap and a decreased divergence by a charged ebonite rod.

From these experiments we conclude that **an increase in divergence occurs when the charge on the electroscope and the test charge are of the same kind.**

We may not assume, however, that a decrease in divergence necessarily means that a charge of opposite kind has been brought near a charged electroscope. An uncharged body has the same effect. One's own hand is a convenient uncharged body to use for this purpose. It follows that the only sure test for the sign of charge on a body is to obtain an increase in divergence. The results of these experiments are summarized in the table shown.

Charge on electroscope	Charge brought near cap	Effect on leaf divergence
+	+	Increase
−	−	Increase
+	−	Decrease
−	+	Decrease
+ or −	Uncharged body	Decrease

(4) *To test the insulating properties of various materials*

The insulating or, conversely, the conducting property of a given substance may be tested by holding a sample of the substance in the hand and then bringing it into contact with the cap of a charged electroscope. If the substance is a good insulator there will be no leakage of charge through it and the leaf divergence will not alter. If, however, the leaf collapses instantly it shows that the substance is a good conductor.

Between these two extremes there are certain substances which produce a slow collapse of the leaf. These are classed as poor insulators or poor conductors. Examples of this type of material are paper, wool, cotton, and wood. Nevertheless, if these substances are dried thoroughly they become quite good insulators. This suggests that their ability to conduct electricity comes from their moisture content.

Among the good conductors are all the metals and carbon. The good insulators are sulphur, quartz, paraffin wax, polyvinyl chloride (P.V.C.), shellac, polythene, and silk.

The experiment should be carried out with as many different substances as possible and the results recorded in a table as below.

Good insulators	Good conductors	Partial insulators

Electrostatic induction

We saw earlier that a charged rod brought near to the cap of an electroscope causes the leaf to diverge from the plate, showing that a charge has been induced on both of them. The following experiment provides more information about the charges which are induced on an insulated conductor when a charged rod is brought near it (Fig. 32.2).

(a) Two insulated brass spheres A and B are placed together so that they touch one another and thus form, in effect, a single conductor.

(b) A *negatively* charged rod is now brought near to A. As a result, a positive charge is induced on A and a negative charge on B.

(c) Still keeping the charged rod in position, sphere B is moved a short distance from A.

(d) The charged rod is now removed and A and B are tested for charge.

The test is carried out as follows. Sphere A is brought near to the cap of a positively charged electroscope. An increase in divergence shows that it is positively

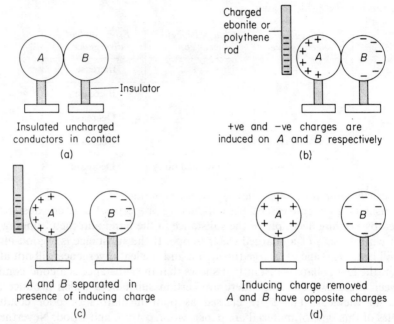

Fig. 32.2. Electrostatic induction

charged. Similarly, sphere B produces an increase in divergence when it is brought near to the cap of a negatively charged electroscope, thus showing it to be negatively charged.

If the whole experiment is carried out again using a *positively* charged rod as the inducing charge, the induced charges on A and B are reversed.

The process by which charges are induced on A and B will be explained after we have dealt with the electron theory of electricity. In the meantime, it is interesting to compare electrostatic induction with magnetic induction described on **page 345**. There it was seen that the induced pole on the end of the steel nearest to the inducing pole was of opposite polarity. In the case of electrostatic induction the induced charge on the end of the conductor nearest to the inducing charge is of opposite sign. There is, however, an important difference. While we are able to separate positive charges from negative charges, it is impossible to isolate magnetic poles.

Eighteenth-century ideas about electricity

At this stage we may well ask, "What is electricity?" Scientists are still unable to give a fully satisfactory answer, but since the end of the nineteenth century a great deal has been discovered about the nature of electricity and its relation to the atoms of which all matter is composed.

In the eighteenth century it was commonly believed that electricity was a kind of fluid. The American scientist, Benjamin Franklin, held the view that all substances in the so-called uncharged condition possessed a fixed standard quantity of electric fluid. A body was said to be positively charged when it had more than the standard quantity of electricity and negatively charged when it had less.

In England, Robert Symmer put forward another theory, namely, that there were two kinds of electric fluid, positive and negative. A body was positively or negatively charged according as it had an excess of one fluid or the other. An uncharged body possessed equal quantities of both fluids which neutralized each other. Both of these theories have long been abandoned in favour of the *electron theory*.

The electron theory. Atomic structure

Towards the end of the nineteenth century, Sir J. J. Thomson carried out some experiments with an electric discharge through a tube containing air at very low pressure. Following this investigation he came to the conclusion that negative electricity consists of tiny particles which came to be called electrons. During the succeeding years it became apparent that these negative electrons actually formed part of the atoms of which all substances are composed.

Previously, atoms were visualized as being like tiny solid billiard-balls, but investigations made during the present century have now given us an entirely different mind-picture of atomic structure.

We now believe that an atom has a central *nucleus* consisting of tightly packed particles called *neutrons* and *protons*, with electrons moving round it at various *energy levels*. The atoms of all elements are built on this pattern, but differ from one another by the number of protons and neutrons contained in the nucleus.

In the early days we used to think of the electrons as moving round the nucleus like planets round a sun but the mathematical theory relating to the energy, motion, and position of the electrons has developed to an extent that we can no longer accept this simple notion.

Pictures and models of atoms *can* be drawn but, unfortunately, they are too complicated to be discussed here. Nevertheless, we shall be saying something about the earlier and simpler ideas about atoms in chapter 45.

Electrons and protons

From the electrical point of view, the protons in the nucleus are important because they have a positive charge equal in magnitude to the charge on an electron. Furthermore, the number of protons in an atom is equal to the number of electrons, so that the atom as a whole is electrically neutral. Electricity, therefore, is not something entirely distinct from matter, but is actually part of the very substance of which the atoms of matter are made.

We shall see presently that it is possible for electrons to become detached from their atoms, and if this happens to a number of the atoms in a body the body as a whole will be left with a net positive charge. On the other hand, if a body gains extra electrons it acquires a net negative charge.

It must be clearly understood that a body cannot normally lose or gain extra protons: these remain inside the nuclei of the atoms and cannot move from one place to another. All the phenomena of static electricity are explained in terms of a transfer of electrons only.

Electrification by friction

When a glass rod is rubbed with a silk cloth some electrons from the glass attach themselves to the silk. Consequently, the glass becomes positively charged and the silk negatively charged. Likewise when ebonite is rubbed with fur electrons are transferred from fur to ebonite, thus making the ebonite negative and the fur positive. Why this happens, and why the electrons go from glass to silk and from fur to ebonite and not in the reverse direction, is not at present understood.

Electrons in insulators and conductors

The difference between an insulator and a conductor is that, in an insulator, the electrons are firmly bound to their atoms and will not move of their own accord, whereas in a conductor the electrons are able to move freely from one atom to another.

If an ebonite rod is held in the hand and rubbed with fur a charge of electrons is formed on its surface. These electrons cannot flow to earth through the hand, since they are unable to move through the insulating ebonite. When a brass rod is rubbed with fur it becomes charged with electrons in just the same way as the ebonite. However, the charge cannot be detected, since it is immediately conducted through the brass and the hand to earth. This may be prevented by mounting the brass rod on an insulating handle. The charge cannot now be conducted away, and its presence can be detected by bringing the rod near a gold-leaf electroscope. The charge can be tested and found to be negative by showing that there is an increase in divergence when the brass rod is brought near to the cap of a negatively charged electroscope.

Electrons and electrostatic induction

The electron theory gives a simple explanation of the manner in which opposite charges are induced on the two spherical conductors *A* and *B* in Fig. 32.2. Free electrons in the conductors are repelled to the remote side of *B* by the negative charge on the ebonite rod.

When the two conductors are separated the electrons are trapped on *B*, so that *B* now has a net negative charge. At the same time *A* is left with a positive charge, since it has lost the electrons which went on to *B*.

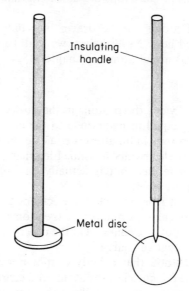

Insulating handle

Metal disc

Fig. 32.3. Proof planes

The proof plane

When testing a charge on a conductor it is often necessary to transfer a small sample of the charge to a gold-leaf electroscope. This is done by means of a proof plane, which consists of a small metal disc at the end of an insulating handle (Fig. 32.3). The proof plane is placed in contact with the surface of the conductor so that it becomes charged, and the charge is then transferred to the electroscope by touching the cap with the proof plane.

Fig. 32.4 illustrates the use of a proof plane to show the existence of induced charges on a conductor *AB* when a positively charged glass rod is brought near it.

The proof plane is first of all used to transfer some of the charge from the end *B* of the conductor to the electroscope. The sign of this charge is then tested by bringing a positively charged rod slowly down towards the cap of the electroscope from a good height above it. An increase in divergence of the leaf will show that the charge taken from *B* is positive.

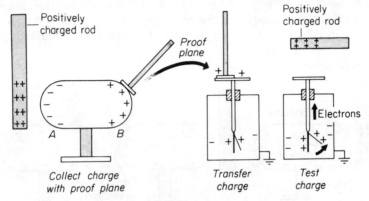

Fig. 32.4. Transferring electric charge

The electroscope and the proof plane are now both discharged by touching them with the finger and the experiment repeated using charge taken from the end *A* of the conductor. This time an increase in divergence is obtained when a negatively charged ebonite rod is brought near the cap, showing that the charge taken from *A* is negative.

The reason why the charged rods must be brought down towards the cap of the electroscope *from a good height above it* is explained on page 372.

Charging by induction

Electrostatic induction may be used to obtain an almost unlimited number of charges from a single inducing charge without any loss of the inducing charge. How this may be done is illustrated in Fig. 32.5.

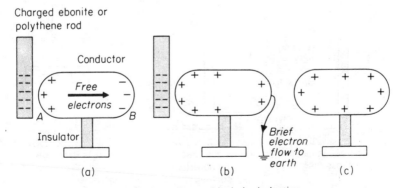

Fig. 32.5. Charging positively by induction

For a charge of given sign it is necessary to start with an inducing charge of the *opposite* kind. Thus, in Fig. 32.5 (*a*) a negatively charged rod has been brought near to the end *A* of a cylindrical insulated conductor *AB*. The shape of the conductor used does not matter: a cylindrical one is shown here merely for convenience in drawing. Some of the free electrons in the conductor are repelled towards *B* by the excess of electrons on the negative rod, thereby creating a negative induced charge at *B* and a positive induced charge at *A*. The conductor is now earthed by touching it with the finger, with the result that some of the free electrons are repelled to earth (Fig. 32.5 (*b*)). The finger is then removed from the conductor and afterwards the charged rod is taken away (Fig. 32.5 (*c*)). Tests with a proof plane and an electroscope can now be made to show that the conductor is positively charged all over its surface.

Fig. 32.6 shows the steps in the process of charging a conductor negatively by induction. In this case the inducing charge is a positively charged rod which attracts free electrons in the conductor towards the end *A*. When the conductor is earthed

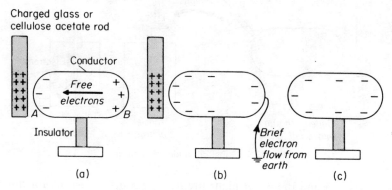

Fig. 32.6. Charging negatively by induction

more electons are attracted from earth into the conductor. Finally, after breaking the earth connection and removing the charged rod the conductor is left with a negative charge.

The electrophorus

In 1775, Alessandro Volta, who was Professor of Physics at Como in Italy, wrote a letter to Joseph Priestley in which he described his newly invented electrophorus. This instrument works by electrostatic induction and can be used to produce any number of positive charges from a single negative charge. It consists of a circular

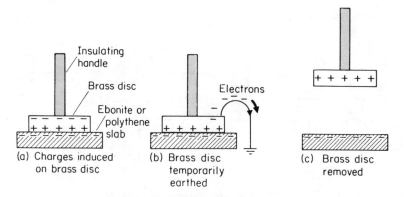

Fig. 32.7. Action of the electrophorus

slab of insulating material, together with a brass disc on an insulating handle. Volta used resin for making the slab, but polythene is now more commonly used.

The polythene slab is given a negative charge by flicking it with a piece of fur, and then the brass disc is placed on top of it. This results in positive and negative charges becoming induced respectively on the lower and upper surfaces of the brass disc (Fig. 32.7 (*a*)). It is important to notice that the disc does not become charged negatively by conduction, since it is never quite flat and makes contact with the slab only at three or four points.

The brass disc is now earthed momentarily by touching it with the finger so that the negative induced charge is repelled to earth (Fig. 32.7 (*b*)). In some instruments automatic earthing is effected by means of a brass stud passing through the slab.

On removing the disc from the slab it is found to be positively charged and a spark may be drawn from it (Fig. 32.7 (*c*)). The spark is usually hot enough to light a bunsen burner if the edge of the charged disc is presented to the burner with the gas turned on.

Source of the electric energy obtained from an electrophorus

Allowing for the inevitable slight leakage of charge from the slab, the brass disc may be charged very many times before the slab needs recharging. At first sight it might appear that these charges on the brass disc are being created from nothing. However, a closer examination of the problem shows that this is not so.

While the disc is still in contact with the slab, the positive charge on it possesses no available energy, since it is effectively neutralized by the negative charge on the slab. The disc will only release electric energy, e.g., by producing a spark, after the disc has been pulled away from the slab. In order to do this, mechanical work must be done against the force of attraction between the charges on the disc and slab, and this work becomes transferred to electric energy.

The positive charge on the disc thus acquires electric potential energy in much the same way as a stone acquires gravitational potential energy when it has been raised against the earth's attraction. Some of the electrophori made in the eighteenth century had discs over 2 m in diameter. Quite a lot of work had to be done in pulling these off their slabs before the charges could be made available.

Charging a gold-leaf electroscope by induction

When the method of charging a gold-leaf electroscope by contact was described earlier in this chapter it was pointed out that the method is unreliable and quite likely to give unexpected results. It is much better to charge an electroscope by induction.

Figs. 32.8 and 32.9 illustrate the process, which is identical with that described for the cylindrical conductor on page 369. A rod is used charged with electricity of opposite sign to that required on the electroscope. The rod is brought near the cap so that the leaf diverges by the amount desired, and the electroscope is then momentarily earthed by touching it with the finger. The rod is then removed, leaving the electroscope charged. Having studied the explanation of charging a conductor by induction on page 369, the reader should be able to give a similar explanation in the case of the electroscope.

Effect of a charged rod on a charged gold-leaf electroscope

On page 364 it was shown how a gold-leaf electroscope can be used to test the sign of the charge on a body. It will be remembered that a charge of the same kind as that on the electroscope causes an increase, while a charge of opposite kind causes a decrease in leaf divergence. We are now able to explain this in terms of the electron theory.

Fig. 32.10, which is self-explanatory, gives the reason for the increased divergence in the case where both charges are negative. The diagram and explanation for the positive case are left as an exercise.

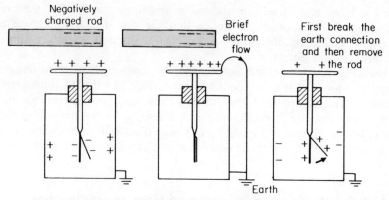

Fig. 32.8. Charging a gold-leaf electroscope positively by induction

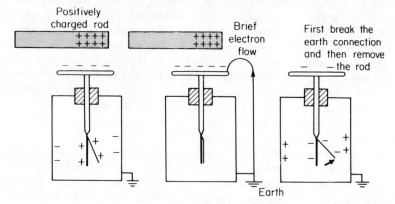

Fig. 32.9. Charging a gold-leaf electroscope negatively by induction

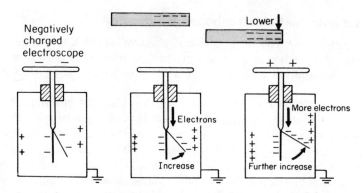

Fig. 32.10. Testing the sign of a charge

Need for care when using a gold-leaf electroscope

Fig. 32.11 shows what happens when a *strong* negative charge is brought near to the cap of a positively charged electroscope. Starting with the charge high above the cap, the leaf divergence decreases as the positive charge on the leaf and plate become partially neutralized by free electrons which are repelled downwards (Fig. 32.11 (*b*)). As the charged rod is gradually lowered, more electrons will be repelled to the leaf and plate, until eventually the former collapses completely, when its positive charges become exactly neutralized (Fig. 32.11 (*c*)). After this stage has been reached a further lowering of the charged rod will cause the leaf to diverge again, since the leaf and plate now acquire an excess of electrons (Fig. 32.11 (*d*)).

It is therefore obvious that a charged body must be brought *from a good height* slowly down towards a cap of a gold-leaf electroscope so that the initial decrease in leaf divergence (if any) will not be overlooked. Otherwise, if the charges are of opposite sign and the observer notices only the final increase in divergence he will

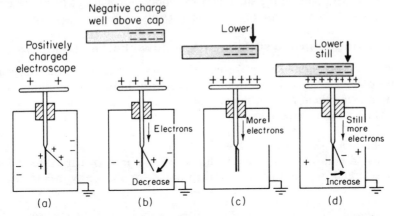

Fig. 32.11. If the test charge is brought too close to the cap a wrong conclusion might be drawn (compare with Fig. 32.10)

conclude wrongly that the test charge and the electroscope charge are of the same kind.

Distribution of charge over the surface of a conductor

A proof plane and gold-leaf electroscope may be used to investigate the distribution of charge over the surface of a conductor, by pressing the proof plane into contact with the surface at various places in turn and then transferring the charge to the electroscope (Fig. 32.12). The divergence of the leaf will give a rough measure of the amount of charge transferred, and hence some idea of the surface density of the charge.

Surface density is defined as the quantity of charge per unit area of surface of a conductor.

The experiment is usually carried out with a flat circular proof plane of the type shown on the left in Fig. 32.3. Strictly speaking, a set of proof planes should be used, *each having the same area but shaped so as to fit the curvature of the surface* at the various places chosen for test.

Fig. 32.12 also shows how charge is distributed over the surface of conductors of

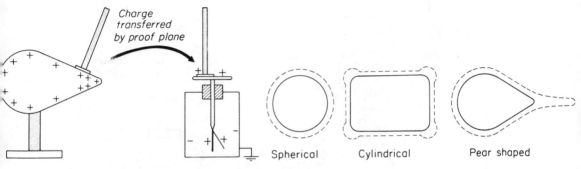

Fig. 32.12. Investigating surface distribution of a charge on conductors

different shapes. In these diagrams the distance of the dotted line from the surface is proportional to the surface density at any point. The most important fact shown by this experiment is that *charge is mostly concentrated at places where the surface is sharply curved.* This is particularly noticeable at the pointed end of the pear-shaped conductor.

The type of proof plane illustrated on the right in Fig. 32.3 is unsuitable for testing surface density. When in use, the short brass wire fixing the disc to the handle acts like a fairly sharp point on the surface of the conductor. Charge therefore tends to concentrate on this, giving rise to misleading results.

No charge on the inside surface of a hollow charged conductor

The electric conductors used in the experiments we have described are generally made of hollow brass or else of wood covered with tinfoil. No advantage is to be gained by making them of solid metal, since the charge resides only on the outside surface. The following experiments illustrate this fact.

(1) *The hollow charged conductor* (Fig. 32.13 (a))

An insulated hollow brass sphere with a small hole in it is charged by induction, and its inside and outside surfaces are then tested for charge by means of a proof

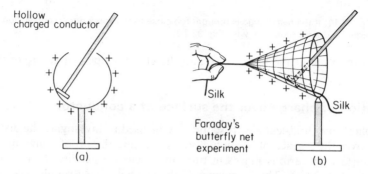

Fig. 32.13. No charge inside a hollow charged conductor

plane and electroscope. It is found that charge may readily be taken from the outside, but none at all from the inside.

(2) *Faraday's butterfly-net experiment* (Fig. 32.13 (b))

Michael Faraday used an insulated cotton net to act as a hollow conductor.

The net, to which an insulating silk string is attached, is given a charge by induction. Tests carried out with a proof plane and electroscope will now show that

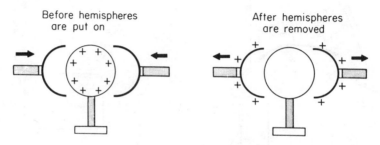

Fig. 32.14. Coulomb's hemispheres

charge may be taken from the outside of the net but not from the inside. The net is next turned inside out by pulling the silk string and again tested for charge. The results are the same as before, showing that the charge must have passed from one side of the net to the other when it was turned inside out.

(3) *Coulomb's hemispheres* * (Fig. 32.14)

Charles Coulomb demonstrated that charge always resides on the outside surface of a conductor with the aid of two hemispherical cups which fitted exactly round an insulated metal sphere.

The sphere is first charged, and afterwards the hemispheres are fitted over it while being held by insulating handles. On removing the hemispheres they are found to be charged, but no charge at all remains on the sphere. This shows that all the charge on the sphere must have passed to the outside of the hemispheres.

Discharging a charged rod. Ions in a flame

Under dry conditions a charged ebonite rod or polythene rod will retain its charge for a very long time. If the rod is to be discharged it may be done most effectively by passing it through the air above a bunsen flame.

In a flame the gas molecules are vibrating very rapidly. Frequent collisions occur, and some molecules have electrons knocked out of them. Molecules which have thus lost electrons are called *positive ions*. At the same time some of the free electrons may attach themselves to neutral molecules, and so form *negative ions*. When a charged rod is drawn through the ionized air above a flame ions of opposite sign are attracted to it, and the charge on the rod becomes neutralized.

Ions in the atmosphere

Ionization by the process described above is only one way of producing gaseous ions. Another method is to allow high-energy radiation to pass into the gas. Various experiments have shown that the atmosphere contains ions which have been produced by radiation from radioactive minerals in the earth's crust, by ultraviolet light from the sun and by the so-called cosmic radiation which enters the earth's atmosphere from outer space.

The presence of these ions in the atmosphere explains why a charged conductor will slowly lose its charge over a period of time even when mounted on a good insulator. The charge on the conductor is gradually neutralized by ions of opposite sign which are attracted towards it out of the surrounding air.

Electric machines

When a considerable quantity of static electricity is required an electric machine is used. The earliest machines which were constructed in the seventeenth and eighteenth centuries worked by friction. For example, Otto von Guericke used a rotating sulphur-ball to which a dry hand was applied to produce the friction. Later experimenters used glass spheres or discs which were rotated against pads made of wool or leather.

These friction machines were not very satisfactory, and after Volta's invention of the electrophorus, scientists of the nineteenth century turned their attention to *induction machines*. Very good machines of this type were made by the German workers Holtz, Voss, and Toepler, but the most efficient one was designed in 1883 by an Englishman named James Wimshurst, who was a consulting engineer to the Board of Trade. The Wimshurst machine, illustrated in Fig. 32.15 is often used in elementary science laboratories. The action of this machine is described in more advanced textbooks.

About 1930 Van de Graaff in America developed a highly efficient electrostatic generator which works on a different principle. Van de Graaff machines have been used a great deal in nuclear research (page 550). Simply Van de Graaff machines are available for elementary purposes but, normally, they provide charge of one kind only Fig. 32.16. The Wimshurst gives both kinds.

* Usually attributed to Biot.

Fig. 32.15. The Wimshurst machine. The two knobs give charge of opposite sign

Fig. 32.16. Hand-driven Van de Graaff generator for laboratory use

Action of highly charged points

When a sharp pointed wire is connected to the conducting knob of a Wimshurst or other electric machine, and the machine set in motion, the surface density of charge on the point becomes exceedingly high and a current of air known as an *electric wind* streams away from the point. This electric wind may be demonstrated by placing a candle flame a short distance away from the point. The flame is deflected by the draught, and may even be blown out (Fig. 32.17 (*a*)).

The electric wind may also be used to work a simple type of jet-propelled motor known as Hamilton's mill. This consists of several wires arranged as the spokes of a wheel on an insulated pivot and having their ends bent at right angles (Fig. 32.17 (*b*)). When connected to an electric machine an electric wind streams out from the ends of

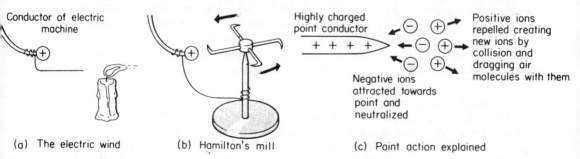

(a) The electric wind

(b) Hamilton's mill

(c) Point action explained

Fig. 32.17. Theory and effect of point action

the wires, and the resulting reaction on the wires causes the mill to rotate in the opposite direction. In these experiments the electric wind is caused by *point action*, and may be explained as follows (Fig. 32.17 (*c*)).

We have already seen that ordinary air contains a certain number of positive and negative ions. Therefore, in the neighbourhood of a highly charged point conductor ions of the same sign will be strongly repelled. When these fast-moving ions collide with air molecules they often knock electrons out of them, thus creating more ions. This process is cumulative, so that, in a very short time, an avalanche of ions will be moving rapidly away from this point. The surrounding air molecules get caught up in this stream and carried along with it, thus setting up the electric wind. At the same time ions of opposite sign are attracted towards the point, where they neutralize the charge on it. It has become customary to speak of charge as being "sprayed off" a point, but this is merely a figure of speech. It cannot be compared with, say, water sprayed out of a hosepipe.

To demonstrate the presence of both positive and negative ions in a flame

Before proceeding further with our discussion of point action we must draw the reader's attention to an effect which is observed when the wire from a conductor of an electric machine is placed *very close* to a candle flame (Fig. 32.18).

On starting up the machine the candle flame is drawn out in both directions. Negative ions in the flame are attracted towards the wire point while positive ions are repelled from it.

Note carefully the difference between this experiment and the previous one with a candle flame and a highly charged wire. In the former instance we placed the wire

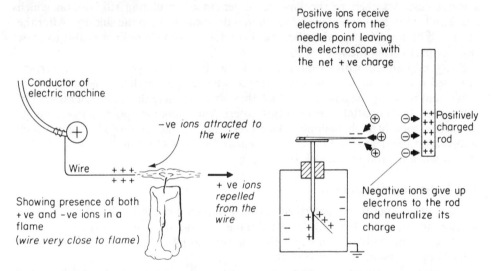

32.18. Positive and negative ions in a flame

Fig. 32.19. Charging an electroscope by point action

further away from the flame, so that the electric wind produced by point action on ions in the air between wire and flame, created a draught sufficiently strong to blow the whole flame away from the point.

Points as collectors of charge

A sharp point not only enables a conductor to lose charge but it can also act as a *collector* of charge. This aspect of point action can be shown by the following experiment.

A gold-leaf electroscope is provided with a sharp point by placing a needle or a piece of wire on the cap. When a positively charged rod is brought near to the point a divergence of the leaf rapidly builds up and the leaf will stay diverged when the rod is removed. On testing by the usual method the electroscope is found to be positively charged.

Fig. 32.19 explains what happens in this case. The positive charge on the rod acts inductively on the electroscope so that the point is negatively charged and the leaf positively charged. Point action then takes place, with the result that positive ions are attracted towards the point, receive electrons from it and become neutralized. The electroscope thus loses negative charge, i.e., it acquires a net positive charge.

At the same time negative ions are attracted towards the positively charged rod, where they give up electrons to it and neutralize its positive charge. The resultant effect, therefore, is a gain of positive charge by the electroscope and a loss of positive charge by the rod.

It is usual to describe this process by saying that charge has been "drawn off" the rod by the point, but once again this is merely a figure of speech, since positive charge does not actually pass into the electroscope any more than positive charge leaves the rod.

Atmospheric electricity. Franklin's experiments

About the middle of the eighteenth century the American scientist, Benjamin Franklin, came to the conclusion that lightning was a gigantic electric spark discharge occurring between two charged clouds or between a cloud and the earth. He had previously shown that it was possible to draw electricity from clouds by sending up a kite during a thunderstorm. The kite was anchored by a length of twine tied to a metal key. Attached to the key was a length of insulating silk ribbon which Franklin held in his hand while he stood in a doorway to keep the silk dry. After the twine had been wetted by rain it became a conductor, and Franklin was able to draw electric sparks from the key.

Being acquainted with point action, Franklin next suggested that a vertical insulated iron rod ought to become charged when a thundercloud passes over it.

Without realizing the danger to which they were exposing themselves, a number of experimenters successfully tried out this experiment, using long pointed iron rods set up on the tops of high buildings. However, enthusiasm for this kind of research waned after the death, in 1753 of the scientist Georg Richmann, who was killed in Russia when his rod was struck by lightning.

The lightning conductor

Based on the experiments described above, Franklin considered that it ought to be possible to protect a building from lightning damage by fixing, to the side of the building, a long pointed iron rod with its lower end buried in the earth. Tests were soon to prove the method successful.

In present-day practice a lightning conductor takes the form of a thick copper strip fixed to an outside wall, reaching above the highest part of the building and ending in several sharp spikes (Fig. 32.20). When a negatively charged thunder-cloud passes overhead it acts inductively on the conductor, charging the points positively and the earthed plate negatively. The negative charge on the plate is, of course, immediately dissipated into the surrounding earth. At the same time point action occurs at the spikes. Negative ions are attracted to the spikes and become discharged by giving up their electrons. These electrons then pass down the conductor and escape to earth. At the same time positive ions (see page 376), are repelled upwards from the spikes and spread out to form what is called a *space charge*. This positive space charge, however, has a negligible effect in neutralizing the negative charge on the cloud. The fact remains that because it projects above the highest part of a building a lightning conductor will accept any discharge which may occur and conduct it harmlessly to earth.

Without the protection of a lightning conductor the lightning usually strikes the highest point, generally a chimney, and the current passes to earth through the path of least resistance. Considerable heat is generated by the passage of the current, and masonry tends to split open through the sudden expansion of steam from the moisture contained in it. Sometimes the current has been known to find a path through soot inside a chimney, and to set it on fire.

Franklin was fortunate that his experiments were confined only to the inductive effect of a thundercloud on a vertical rod. Richman, less fortunate, received a lightning discharge as well.

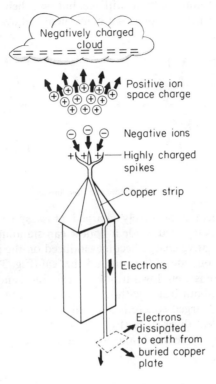

Fig. 32.20. Action of a lightning conductor

Electric fields

An electric charge sets up an electric field in the space surrounding it and an electric force is exerted on any charged body placed in the field. Electric fields may be represented by *electric field lines.*

An electric field line is a line drawn in an electric field such that its direction at any point gives the direction of the electric field at that point.

The direction of an electric field at any point is the direction of the force on a small positive charge placed at the point.

Michael Faraday went further than this in his idea of field lines and imagined them to have some kind of real existence. This does not mean that he thought of them as having a tangible existence like elastic cords or springs, but rather that they represent a state of strain in the electric field.

To these **electric field lines** he ascribed the following properties:

(1) **they begin and end on equal and opposite quantities of electric charge;**
(2) **they are in a state of tension which causes them to tend to shorten;**
(3) **they repel one another sideways.**

Faraday used this mind-picture of field lines to give an explanation of various electrical experiments. Thus, the attraction between two unlike charges is explained by the tension of the field lines joining them. Likewise, the repulsion between two like charges is caused by the sideways repulsion between their field lines (Fig. 32.21).

Faraday's ice-pail experiments

A particularly successful application of the properties of electric field lines has been made in the explanation of Faraday's *ice-pail experiments* on electrostatic induction. These experiments have nothing to do with ice, but owe their name simply to the fact that Faraday used an empty ice pail as a convenient hollow conductor.

Attraction   Repulsion

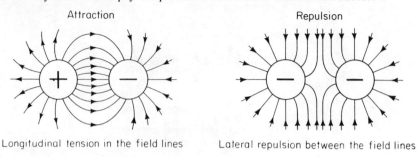

Longitudinal tension in the field lines   Lateral repulsion between the field lines

Fig. 32.21. Electric field lines

In the first experiment a positively charged brass sphere B, supported by an insulating silk thread, is lowered inside a hollow can standing on the cap of a gold-leaf electroscope. A negative charge becomes induced on the inside of the can, and a positive charge on the outside of the can and the leaf (Fig. 32.22 (a)).

So long as the sphere is well down inside the can and is not allowed to touch the can, it may be moved about inside without causing any alteration in the divergence of the leaf. When the charged sphere is allowed to touch the bottom of the can no change occurs in the leaf divergence (Fig. 32.22 (b)). On removing the sphere it is found to be completely discharged, and also there is no longer any charge on the

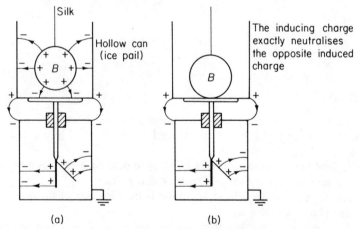

Fig. 32.22. Faraday's first ice-pail experiment

inside of the can. It is therefore concluded that the inducing charge on the sphere has exactly neutralized the opposite induced charge on the inside of the can, and hence the two are equal in magnitude.

This result may be explained in terms of the properties of electric field lines as follows. Before the sphere is put into the can its field lines spread out in all directions and end on the walls, ceiling, furniture and so on of the room. When well down inside the can, all the field lines terminate on the inside of the can. Since field lines begin and end on equal and opposite quantities of charge, it follows that a total charge, equal and opposite to that on the sphere, must be induced on the inside of the can. Finally, when the sphere is allowed to touch the can the field lines shrink towards the point of contact and the charges at their ends neutralize one another. In addition to the field lines inside the can there are other lines which arise from the induced positive charge on the outside of the can, the leaf and the plate. These will terminate on the earthed case of the electroscope, where they induce an equal negative charge. The tension in the lines between leaf and case causes the leaf to diverge.

The second ice-pail experiment is shown in Fig. 32.23. As in the previous experiment, the positively charged sphere *B* is lowered well down inside the can without touching, and the leaf divergence noted. The can and electroscope are now earthed momentarily by touching with the finger, with the result that the leaf collapses (Fig. 32.23 (*b*)). This occurs because the field lines between the earthed case of the electroscope and the outside of the can and the leaf have shrunk towards the earthing point and the charges at their ends have neutralized one another.

When the charged sphere is removed from the can the negative induced charge inside the can passes to the outside of the can and the leaf. Electric field lines are

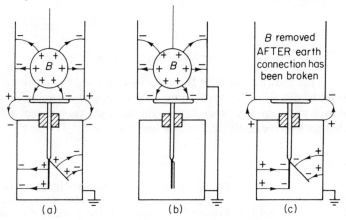

Fig. 32.23. Faraday's second ice-pail experiment

now set up between the leaf and the earthed case and the leaf diverges by exactly the same amount as it did previously (Fig. 32.23 (*c*)). This means that the negative induced charge must be equal in magnitude to the positive induced charge.

We finally conclude from these experiments that, **when all the electric field lines from an inducing charge terminate on a conductor, the two charges induced on the conductor are each equal in magnitude to the inducing charge.**

To show that equal and opposite charges are produced by friction

Whenever an ebonite rod is rubbed with fur the fur becomes charged as well as the ebonite, but with electricity of opposite sign. The same remarks apply to the case of glass rubbed with silk, or indeed to any pair of substances. Moreover, the two opposite charges produced by friction are equal in magnitude.

Normally it is not easy to show that fur becomes charged, since it is a partial conductor owing to the slight amount of moisture in it and so the charge leaks away through the hand. However, it is possible to show that the charge on the fur is equal and opposite to that on the rod if special precautions are taken, as in the following experiment (Fig. 32.24).

An ebonite rod is discharged completely by drawing it quickly through air above a gas flame and afterwards it is fitted with a small fur cap to which a thread of silk or other insulating material is attached (Fig. 32.24 (*a*)). The rod is then held in one hand while the cap is rotated several times round the rod by means of the silk thread. This causes both rod and cap to become charged by friction. The capped end of the rod is now placed inside a metal can standing on the cap of a gold-leaf electroscope, and it is observed that no divergence of the leaf occurs (Fig. 32.24 (*b*)). This implies that, either:

(i) there is no charge on either the rod or the fur cap; or
(ii) the rod and fur cap possess equal and opposite charges.

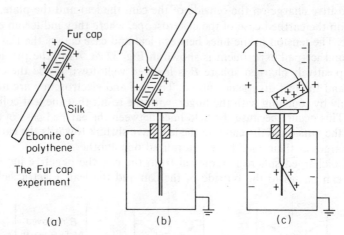

Fur cap

Silk

Ebonite or polythene

The Fur cap experiment

(a)   (b)   (c)

Fig. 32.24. Testing the equality of charges by friction

Condition (ii) is shown to hold in this case by removing the rod and leaving the fur cap inside the can (Fig. 32.24 (c)). Immediately, the leaf diverges, indicating that, when both rod and cap were together inside the can, they possessed equal and opposite charges which exactly neutralized one another.

QUESTIONS: 32

1. State what happens and interpret the results in the following:

(a) The cap of an uncharged electroscope is approached by a rubbed ebonite rod which is then removed.

(b) The rubbed ebonite rod is "rolled" on the cap and then removed.

After treatment of the electroscope as in (b):

(c) A rubbed glass rod approaches the cap.

(d) An earthed metal plate is held close to the cap. (S.)

2. State what happens under the following conditions:

(i) An ebonite rod is rubbed with fur.

(ii) The rod is held near a brass ball mounted on a glass stand.

(iii) The ball is touched momentarily with a finger while the rod is still held near.

(iv) The rod is removed.

How would you use a charged electroscope to arrange a number of substances in order of their conducting ability?

3. Describe the construction of the gold-leaf electroscope.

Given a gold-leaf electroscope, an ebonite rod and fur, explain how you would use them:

(a) to detect the presence of electrostatic charge on another body;

(b) to find the sign of that charge. (O.C.)

4. State and explain what happens when the

following are placed, in turn, near the plate of a positively charged electroscope:

(a) a polythene rod rubbed with cloth;

(b) a cellulose acetate rod rubbed with cloth;

(c) an uncharged metal rod.

5. Explain the term *electrostatic induction* and briefly describe and explain one practical application.

An uncharged metal box with a hole in the top stands on the top cap of a gold-leaf electroscope. Describe and explain the behaviour of the leaves when an insulated positively charged conductor is lowered into the box through the hole and:

(a) the box is temporarily earthed after which the metal conductor is withdrawn;

(b) the box is not earthed but the conductor is allowed to touch the inside of the box and is then withdrawn.

State the final sign of charge on the leaves in each case. (O.C.)

6. Sketch the electric field in Fig. 32.25. A is a positively charged spherical conductor and B is an earthed metal plate. (J.M.B.)

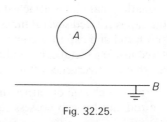

Fig. 32.25.

7. A polythene rod may be charged negatively by rubbing it with a cloth, but a brass rod held in the hand cannot be charged in this way.

(i) State clearly what happens when the polythene is being charged.

(ii) Explain why the brass cannot be charged by rubbing. (*J.M.B.*)

8. What is meant by *electrostatic induction*?

Given two equal metal spheres on insulating supports, an ebonite rod and a piece of fur, describe and explain how you would: (*a*) charge a sphere negatively by induction; (*b*) charge one sphere positively and the other negatively to the same extent.

In the last case, describe the experiments you would perform and any additional apparatus you require to prove the correctness of your procedure.

Illustrate your answers by diagrams.

(*O.C.*)

9. (*a*) Describe and explain what happens when a charged rod is brought near a pith ball suspended by a dry silk thread.

(*b*) Why has one end of a lightning conductor a sharp point? Describe a laboratory experiment to support your explanation.

(*J.M.B.*)

10. Give a labelled diagram of a gold-leaf electroscope. Explain how you would:

(*a*) charge it by induction;

(*b*) use it to show that the charge on a hollow conductor is situated on the outer surface.

You are given two insulated charged conductors of different shapes, an electroscope and a metal can of suitable size. Explain how you would determine without discharging the conductors whether they carry like or unlike charges, and which of them carries the greater charge. (*O.C.*)

11. A bunsen flame brought near to the cap of a charged electroscope causes the divergence of the leaf to become less. Explain this in terms of the movement of electric charges.

(*C.*)

12. Fig. 32.26 shows two metal spheres carrying equal positive charges hanging by insulating suspensions, and two metal cans supported by insulators.

Using the insulating suspensions the sphere in (i) is touched on to the outside of the can in (i) and then withdrawn; and the sphere in (ii) is touched on to the inside of the can in (ii) and then withdrawn.

. Show on the diagrams the resulting distribution of charges on the cans. In each case state whether or not any charge remains on the sphere. (*C.*)

13. Explain the nature of the charge on a gold-leaf electroscope if the divergence of the leaf is observed to decrease when an ebonite rod which has been rubbed with fur is brought towards the cap: Explain, also why, if the rod is brought still nearer to the cap, but without touching it, the divergence will first decrease to zero and then increase again.

14. Describe the gold-leaf electroscope, and show how it can be used to determine the sign of an electric charge.

If you were provided only with a rod bearing a negative charge, how would you charge the electroscope:

(*a*) negatively, and

(*b*) positively?

Describe an experiment which shows that, when charges are separated by friction (as, for example, when an ebonite rod is rubbed with fur), equal quantities of electricity of opposite signs are obtained. (*O.*)

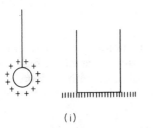

(i)

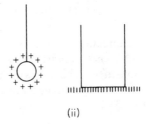

(ii)

Fig. 32.26.

33. Potential and capacitance

Potential difference

In the previous chapter we learned that when a negatively charged conductor is connected to earth, electrons (negative charges) flow from the conductor to earth. Similarly, in the case of a positively charged conductor there is a momentary flow of electrons from earth to the conductor. This movement of electricity takes place because there is a *difference of potential* between the conductors and the earth. We may therefore think of potential difference as being the electrical condition which

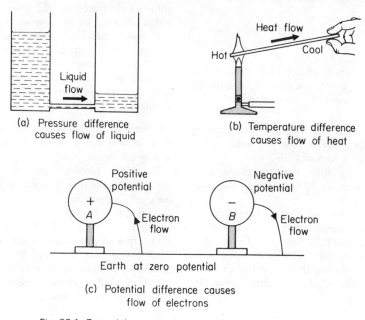

(a) Pressure difference causes flow of liquid

(b) Temperature difference causes flow of heat

(c) Potential difference causes flow of electrons

Fig. 33.1. Potential compared with temperature and pressure

governs the direction of flow of electricity from one point to another. In this sense, potential difference may be compared with temperature difference, which determines the direction in which heat will flow through a solid, or with pressure difference, which determines the direction in which water will flow between two communicating vessels (Fig. 33.1).

Earth at zero electric potential

When making measurements of electric potential it is necessary to choose a convenient zero level of potential, in just the same way that we choose the melting-point of ice as a suitable zero for temperature measurement.

In practice, we take the earth itself as being at zero potential. This is purely a matter of convenience. It does not mean that the earth has no electrical charge, any more than to say that pure melting ice is at 0 °C implies that ice contains no internal energy. Actually the earth has a negative charge, but this is so large that any charge given to or taken from it by man has a negligible effect.

Absolute zero of potential

On page 177 we explained the meaning of the absolute zero of temperature and showed how it was related to 0 °C. Similarly, in electricity we think of an absolute zero of potential. This is taken to be the potential of all points at infinity.

Potential gradient

In Fig. 33.1 the conductor *A* has a positive charge and a positive potential, while conductor *B* has a negative charge, and hence a negative potential. If *A* and *B* are connected to earth electrons will flow *from earth to A* and *from B to earth* until both *A* and *B* are at earth potential.

Since the potential of *A* is above earth potential, while that of *B* is below earth potential, it is clear that electrons tend to move from one place to another where the potential level is higher. In other words, **electrons tend to move up the potential gradient.** On the other hand, if positive charges were able to move they would tend to move down the potential gradient. In using the word gradient we are treating potential as a kind of electric level, just as we refer to the gradient on a road when it goes uphill or downhill.

In spite of what we have just said about the potentials of *A* and *B*, it must not be assumed that positive and negative charges always have positive and negative potentials respectively. A charged conductor not only has a potential due to its own charge but also a potential due to the presence of other charges which may be in its neighbourhood. A charge raises (or lowers, as the case may be) the potential of all points in its neighbourhood by an amount which decreases with distance from the conductor. It is therefore possible for a positively charged conductor to have a negative potential if another conductor with a large enough negative charge happens to be near by.

Potential and the gold-leaf electroscope

In the previous chapter we looked upon a gold-leaf electroscope as an instrument chiefly for the purpose of indicating the presence or sign of an electric charge, or for making a rough measurement of quantity of charge. Actually, **the leaf divergence of a gold-leaf electroscope indicates that there is a potential difference between the leaf and case.**

Fig. 33.2 (*a*) illustrates the normal use of an electroscope to test for the presence of charge. The positive charge held above the cap raises the potential of the cap and leaf, while the potential of the case remains at zero, since it is earth-connected. The resultant difference of potential between leaf and case causes the leaf to diverge.

In Fig. 33.2 (*b*) the electroscope has been placed upon a slab of paraffin wax so

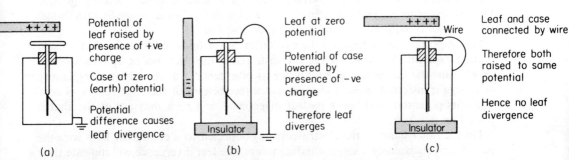

(a) Potential of leaf raised by presence of +ve charge / Case at zero (earth) potential / Potential difference causes leaf divergence

(b) Leaf at zero potential / Potential of case lowered by presence of −ve charge / Therefore leaf diverges / Insulator

(c) Leaf and case connected by wire / Therefore both raised to same potential / Hence no leaf divergence / Insulator

Fig. 33.2. The gold-leaf electroscope indicates potential difference between leaves and case

that the case is insulated while the cap and leaf are earth-connected. If, say, a negative charge is now brought *near the case*, the potential of the case is lowered while that of the leaf remains at zero. Once more a potential difference has been set up between leaf and case, and so the leaf diverges.

A third experiment is illustrated in Fig. 33.2 (c). This time the case is left standing on the wax slab and the cap and case are joined by a piece of copper wire. It is now impossible to cause the leaf to diverge, either by bringing near a charged rod or even by charging leaf and case by induction. Since leaf and case are connected together, they will always be at the same potential.

Potential over the surface of a charged conductor

Where static electricity is concerned, the potential is the same all over a charged conductor, no matter how the charge or charges may be distributed over it. Thus, in the experiment illustrated in Fig. 33.3 (a) the negatively charged rod produces a potential difference between the ends of the insulated conductor *AB*. Immediately, however, this potential difference causes some free electrons in the conductor to move towards the end *B*, with the result that the potential becomes equalized all over the conductor.

This fact may be tested by attaching a length of copper wire to the cap of an electroscope. The wire is then wound once or twice round a polythene rod to serve as an insulating handle, and the free end of the wire is moved all over the surface of the conductor. No matter where the wire touches the surface of the conductor, the leaf divergence remains the same, showing that **the potential is the same all over the surface**.

In the same manner it may be shown that the potential is everywhere the same over the surface of a charged pear-shaped conductor (Fig. 33.3 (b)). The reader

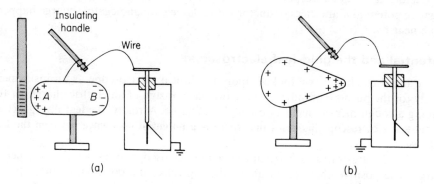

Fig. 33.3. Potential is the same all over the surface of a charged conductor

should compare this experiment very carefully with the experiment to investigate the distribution of charge over the surface of such a conductor, described on page 373. The two experiments make it clear that, *although the charge is unequally distributed, the potential is uniform over the surface.* (See also Fig. 33.4.)

It is important to understand the two different methods of using the electroscope in these experiments. When charge is transferred to the electroscope by means of the proof plane the leaf potential, and hence its divergence, will depend on the quantity of charge transferred. Since the capacitance (see below) of the electroscope is constant, its potential, and hence its leaf divergence, will be a measure of the charge placed on it.

On the other hand, if the electroscope is connected to a conductor by a wire the two effectively become a single conductor, and the leaf divergence will indicate their common potential.

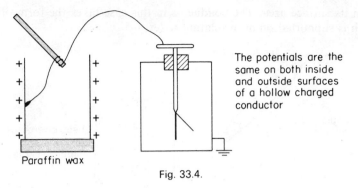

The potentials are the same on both inside and outside surfaces of a hollow charged conductor

Paraffin wax

Fig. 33.4.

Capacitance

If equal quantities of water are poured into vessels of different diameters the water will come up to different levels in each. Similarly, when equal charges are given to conductors of different sizes they will acquire different potentials. This may be demonstrated by standing two unequal metal cans on the caps of two identical electroscopes (Fig. 33.5 (a)). The cans are given equal charges of Q units from an

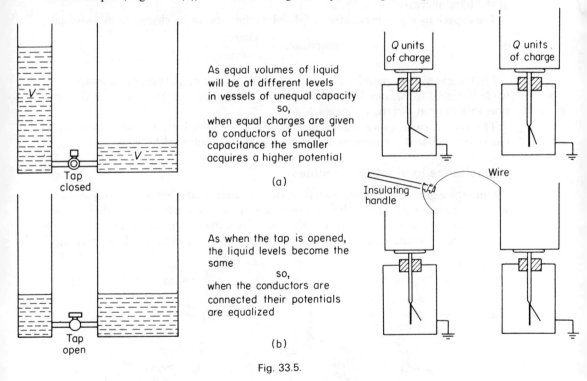

As equal volumes of liquid will be at different levels in vessels of unequal capacity
so,
when equal charges are given to conductors of unequal capacitance the smaller acquires a higher potential

(a)

As when the tap is opened, the liquid levels become the same
so,
when the conductors are connected their potentials are equalized

(b)

Tap closed

Tap open

Q units of charge

Q units of charge

Insulating handle

Wire

Fig. 33.5.

electrophorus disc. The charged disc is lowered inside a can until it touches the bottom. In this way the whole of the charge is given up to the can and goes to the outside. It will be noticed that the leaf divergence is greater for the small can, showing that it has acquired a higher potential than the large can. The large can is said to have a larger *capacitance*.

When the two cans are joined by a wire electricity flows from the small can to the large one until the potentials become equalized (Fig. 33.5 (b)). This may be compared with the equalization of water levels in two vessels which are connected through a pipe.

Fig. 33.6 illustrates an experiment to show that the capacitance of a metal plate depends on its surface area. The conductor in this case takes the form of a tinfoil blind which is supported on an insulating roller.

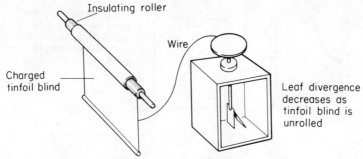

Fig. 33.6. Capacitance depends on area of a conductor

The blind is connected to an electroscope by means of a wire and is then given a charge. The leaf divergence indicates the potential of the blind. If the blind is now unrolled the leaf divergence, and hence the potential, is seen to decrease. Since the charge has remained unaltered, it follows that, as the area increases, the capacitance of the blind increases.

The capacitance of a conductor is defined as the ratio of its charge to its potential

or
$$\text{capacitance} = \frac{\text{charge}}{\text{potential}}$$

Electric charge (or quantity of electricity) is measured in units called **coulombs** (C), while potential is measured in **volts** (V). Both of these units are also used in connection with current electricity (pages 404–5).

The unit of capacitance, called the farad (F), is defined as the capacitance of a conductor such that a charge of 1 coulomb changes its potential by 1 volt.

Capacitance in electric machines

During the eighteenth century a large variety of electric machines were constructed, mostly consisting of glass globes or plates which were rotated against rubbing pads or even the human hand. These machines were capable of producing a succession of small sparks, but later it was discovered that much fatter sparks could be obtained if

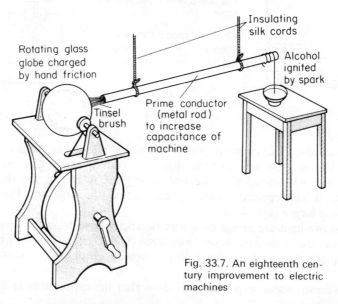

Fig. 33.7. An eighteenth century improvement to electric machines

the charge was collected on a long iron rod placed near the machine. A rod such as this was called a *prime conductor*. Its function was simply to provide extra capacitance so that a much larger charge could be collected at the same potential before sparking occurred (Fig. 33.7).

In 1746, the Dutch physicist, Pieter van Musschenbroek, was experimenting to see if electricity could be kept in a bottle of water, when he stumbled on a method for storing a large quantity of electricity in a very small space. Briefly, he had a wire hanging from the prime conductor of an electric machine and dipping into a bottle of water held in one hand. After working the machine for a short time he touched the wire with the other hand and immediately received a shock which "shook him like a thunderbolt".

This bottle later came to be known as the *Leyden jar*. In its modern form the Ley-

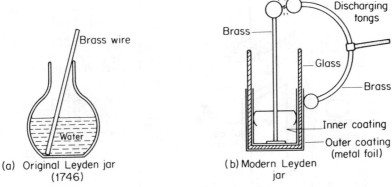

(a) Original Leyden jar
(1746)

(b) Modern Leyden
jar

Fig. 33.8.

den jar is simply a bottle with coatings of metal foil inside and out (Fig. 33.8). Bottles such as this were soon to replace prime conductors on electric machines. Two Leyden jars may be seen in the picture of the Wimshurst machine in Fig. 32.15.

The action of the jar will be understood after reading the next section.

The parallel-plate capacitor

A flat metal plate, A, is set up vertically on insulating legs and is connected to a gold-leaf electroscope by means of a wire. The plate is then given a positive charge by induction with a negatively charged ebonite rod. The divergence of the leaf indicates the potential of the plate (Fig. 33.9 (*a*)).

A second insulated plate B is now brought up slowly into a position parallel to A. When B is very close to A but not touching it, it will be noticed that the leaf divergence decreases very slightly. We conclude from this that the potential of A has been decreased by the presence of B, and hence its capacitance has increased slightly (Fig. 33.9 (*b*)).

This has been brought about as follows. The positive charge on A induces equal and opposite charges on opposite sides of B. These induced charges will respectively raise and lower the potential of all points in their neighbourhood and, in particular, they will affect the potential of plate A. As far as A is concerned, however, the negative induced charge will have the greater effect, since it is closer to A than the positive charge. The net result is that the potential of A is slightly reduced.

B is next earthed either by touching it with the finger or by connecting it to the nearest cold-water pipe (Fig. 33.9 (*c*)). Immediately the leaf shows a big decrease in divergence. This implies a big decrease in potential, and hence a big increase in the capacitance of A.

When B was earthed its positive induced charge disappeared. Previously this charge was helping to raise the potential of A. Now, only the negative induced charge remains and acts alone to cause a big decrease in the potential of A.

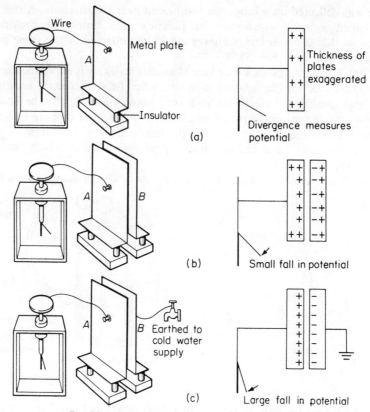

Fig. 33.9. Demonstrating the principle of a capacitor

The presence of the earthed plate *B* therefore results in a very large increase in the capacitance of *A*.

Any arrangement of two conductors such as these two metal plates, placed close together, is called a *capacitor*.

Factors affecting the capacitance of a parallel-plate capacitor

The three factors which affect the capacitance of a parallel-plate capacitor are the area of the plates, the distance apart of the plates and the nature of the insulating material or *dielectric* between them.

These factors may be investigated by the following experiments.

(1) *Area of plates*

The effect of altering the area of the plates may be demonstrated by setting up a charged insulated plate connected to an electroscope, alongside an earthed tinfoil roller blind, as illustrated in Fig. 33.10. The tinfoil is slowly unrolled and the leaf divergence noted.

As the blind is unrolled, the leaf divergence progressively decreases. An increase in the effective area of the plates is therefore seen to bring about a decrease in potential difference between the plates, and hence an increase in capacitance.

(2) *Distance apart of the plates*

A parallel-plate capacitor is set up as shown in Fig. 33.11 (*a*), one plate being earthed and the other connected to an electroscope and charged. The distance apart of the plates is now varied, and the effect of this on the leaf divergence noted. It is found that the closer the plates are together, the smaller is the divergence, and hence the lower the potential. It follows that the capacitance increases as the plates are moved closer together.

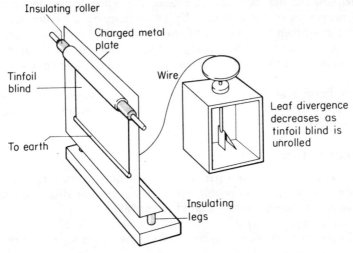

Fig. 33.10. Capacitance depends on effective area of plates

(3) *Dielectric between the plates*

The plates of the charged capacitor are placed a suitable fixed distance apart and slabs of various materials of equal thickness, e.g., polythene, glass, paraffin wax, etc., are placed in turn, between the plates (Fig. 33.11 (*b*)). In every case a decrease in the leaf divergence is noticed. As before, this indicates a decrease in potential, and hence an increase in capacitance.

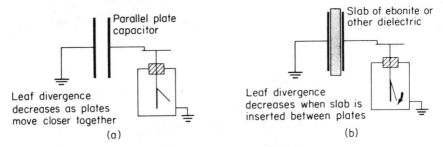

Fig. 33.11. Two other factors affecting capacitance of parallel plates (see also Fig. 33.10)

An insulating medium used between the plates of a capacitor is called a *dielectric*.

The relative permittivity or dielectric constant of a medium is equal to the ratio of the capacitance of a given capacitor with the medium as dielectric to the capacitance of the capacitor with a vacuum as the dielectric. Paraffin wax has a relative permittivity of about 2, while that of mica used in high-class radio capacitors is about 8.

The Leyden jar

We are now in a position to understand why Musschenbroek received so severe a shock from the Leyden jar described on page 389. His hand, holding the outside of the bottle, acted as an earthed conductor separated by a glass dielectric from the conducting water inside. The arrangement thus formed a capacitor. Later on, the use of water inside the jar was abandoned and the inside and outside of the jar were covered with lead foil to serve as the two conductors. The outer covering is automatically earthed when the bottle stands on a table. Fig. 33.8 (*b*) illustrates the construction of a modern form of Leyden jar.

Leyden jars were used a great deal in eighteenth-century electrical experiments.

During this period, electrical demonstrations became a popular form of entertainment. Rats and mice were killed and gunpowder and alcohol ignited by electric spark discharges. On one occasion in France the Abbé Nollet demonstrated the conducting properties of the human body by discharging a Leyden jar through a battalion of Guardsmen joined hand to hand. On a subsequent occasion the same experiment was repeated at a monastery in Paris. This time the human chain was formed by obedient monks. It is reported that they all leapt some distance into the air.

Nowadays, Leyden jars are rarely found in use outside elementary teaching laboratories, but they were used in the early days of wireless telegraphy in connection with morse transmitters.

Practical capacitors

Capacitors play a very important part today in various electrical circuits, especially radio circuits. For practical purposes, capacitance is measured in microfarads (μF). A **microfarad** is a millionth part of a farad (**page 388**).

Three common types of capacitor are illustrated in Fig. 33.12.

Fig. 33.12 (*a*) shows a variable capacitor of the kind used for tuning radio sets. It

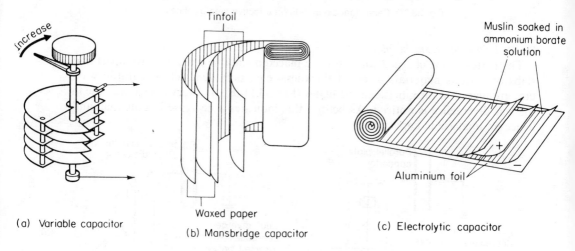

(a) Variable capacitor

Tinfoil

Waxed paper

(b) Mansbridge capacitor

Muslin soaked in ammonium borate solution

Aluminium foil

(c) Electrolytic capacitor

Fig. 33.12. Types of capacitor

consists of two sets of semicircular aluminium or brass plates separated by air. One set of plates is fixed and the other is rotated by a knob to alter the effective area of the plates.

Fig. 33.12 (*b*) illustrates the Mansbridge capacitor. It contains two long strips of tinfoil separated by thin waxed paper or polyester film. These are rolled up and sealed inside a metal box to prevent the entry of any moisture which would spoil the insulation.

Fig. 33.12 (*c*) shows the construction of the "dry" electrolytic capacitor. This takes the form of two sheets of aluminium foil separated by muslin soaked in a special solution of ammonium borate. These are rolled up and sealed in an insulating container. Wires attached to the foil strips are then connected to an electric battery. Electrolysis (**page 428**) takes place and a thin film of aluminium oxide forms on the positive foil. This film is highly insulating, and so the combination now forms a capacitor in which the oxide film acts as the dielectric. Owing to the extreme thinness of the film, very large capacitances which take up very little space may be obtained by this method. Obviously it is important to use this type of capacitor under conditions in which the oxide-covered foil never becomes negative with respect to the other foil.

QUESTIONS: 33

1. Draw and label a gold-leaf electroscope. What does such an instrument measure? (*S.*)

2. Define the *capacitance* of a capacitor.

Draw a clearly labelled diagram of a gold-leaf electroscope. Describe how you would use the electroscope to investigate the effect on the capacitance of a parallel-plate capacitor of changing:
 (i) the area of the plates;
 (ii) the distance between the plates;
 (iii) the dielectric between the plates.
State clearly the conclusions which you would expect to reach. (*A.E.B.*)

3. A capacitor is a device for storing electric charge.
 (i) What are the essential features in the construction of a capacitor?
 (ii) Explain how the charge is distributed when a capacitor is charged.
 (iii) Describe and explain what happens when the terminals of a charged capacitor are connected by a piece of copper wire.

In an experiment with a capacitor, the charge which was stored was measured for different values of charging potential difference. The results are tabulated below.

Charge stored/μC	7.5	30	60	75	90
Potential difference/V	1.0	4.0	8.0	10.0	12.0

 (i) Plot a graph of charge stored on the *y*-axis against potential difference on the *x*-axis.
 (ii) **Use the graph** to calculate the capacitance of the capacitor used in the experiment. (*J.M.B.*)

4. With the aid of diagrams, briefly describe two different types of capacitor.

Two small insulated spheres of unequal size possess equal charges. How would you:
 (*a*) show that the charges were equal;
 (*b*) compare the potentials of the spheres?

5. (i) The palm of a hand is held close to the plate of a charged gold-leaf electroscope and slowly moved across it. State and explain what happens.
 (ii) A gold-leaf electroscope stands on a block of paraffin wax and the plate is earthed. State and explain what happens when a rubbed ebonite rod is held close to the normal earthing terminal.
 (iii) How could you demonstrate that induced positive and negative charges are equal? (*S.*)

34. Electric cells

Towards the end of the eighteenth century, Luigi Galvani, Professor of Anatomy at Bologna University in Italy, published a book describing a series of investigations he had made on the subject of "animal electricity". In this book he described how a freshly dissected frog's leg could be thrown into muscular convulsions, simply by connecting the foot and the exposed nerves through a length of copper and iron wire (Fig. 34.1). It had been known for a long time that the muscles of dead animals could be caused to contract by means of an electric shock from a machine or Leyden jar, and therefore it was suggested that some source of electricity might be responsible for the contractions of the frog's leg.

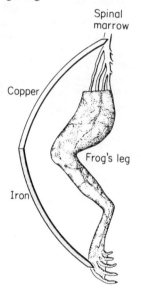

Fig. 34.1. Galvani's experiment with a frog's leg

Galvani himself was of the opinion that the muscle and nerve acted like a kind of charged Leyden jar and that discharge took place when they were joined by the copper and iron wires. His fellow countryman, Alessandro Volta, could not agree with this but believed that the two dissimilar metals were more important than the leg. Subsequently, Volta did some experiments to show that electricity was produced when two different metals were separated by various non-animal liquids and these practically settled the matter in his favour.

The voltaic pile

To support his views Volta constructed a pile of pairs of silver and zinc discs separated by discs of cardboard soaked in brine (Fig. 34.2 (*a*)). On touching the top and bottom discs simultaneously he received an electric shock. Further experiments showed that similar results could be obtained with other metals in conjunction with solutions of various salts and acids.

One disadvantage of the voltaic pile was that the weight of the discs squeezed out the liquid from the cardboard so that it ran down the outside and put the pile out of action. Volta solved this problem by making large numbers of small cells consisting of strips of zinc and either silver or copper dipping into dilute sulphuric acid. The zinc plate of one cell was then connected by a copper wire to the copper plate of the

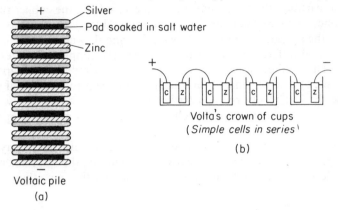

Fig. 34.2. Voltaic pile and simple cells

next, and so on, forming what Volta called "a crown of cups" (Fig. 34.2 (*b*)). Nowadays we should call it a *battery of simple cells*. With such a battery consisting of a hundred or more cells, a tingling sensation was felt when the free copper and zinc plates of the end cells were touched simultaneously by the fingers. Volta inferred from this that an *electric current* flowed from one plate to the other when they were connected through a conductor such as the human body or a metal wire.

The condensing electroscope

Volta was able to show that the copper plate of a simple cell was charged positively and the zinc plate negatively by means of the condensing electroscope. This was an instrument which he had designed for the purpose of detecting very small potential differences. It comprises a gold-leaf electroscope, the cap of which is a flat brass disc somewhat larger than usual, and on top of this is placed a similar disc which is connected to earth. The two discs are insulated from one another, either by a layer

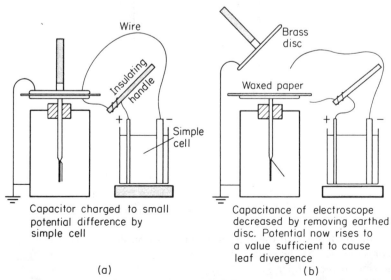

Fig. 34.3. Showing the presence of a small potential difference with a condensing electroscope

of shellac varnish on the upper disc or else by a thin sheet of waxed paper or polyester film placed between them. The arrangement forms a *parallel-plate capacitor and an electroscope combined* (Fig. 34.3).

The upper and lower discs of the condensing electroscope are connected respectively to the zinc and copper plates of a simple cell by lengths of copper wire. In order to avoid earthing the electroscope disc, the wire from the copper plate should be wrapped once or twice round an alkathene rod and the latter used as an insulating handle for the purpose of moving the wire. Having thus charged the capacitor to the small potential difference which exists between the simple cell plates, the wires are removed. The earthed disc is now taken away. This reduces the capacitance of the electroscope, and so the charge on it is able to raise it to a potential sufficiently high to cause the leaf to diverge. On testing with a glass rod rubbed with silk, an increased divergence shows that the charge obtained from the copper plate is positive.

If the experiment is repeated with the cell connections reversed it may be shown that the charge on the zinc plate is negative. Incidentally, this experiment confirms that current electricity and static electricity are one and the same thing in different forms.

Action of the simple cell

Although Volta found that various metals and salt or acid solutions could be used to make a simple cell, it is customary today to regard a simple cell as being composed of a plate of copper and a plate of zinc dipping into dilute sulphuric acid (Fig. 34.4).

When the copper and zinc plates of a simple cell are joined by a wire the zinc

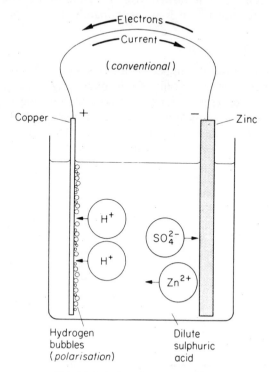

Fig. 34.4. Action of the simple cell

slowly begins to dissolve in the sulphuric acid, and bubbles of hydrogen are formed on the *copper* plate. At the same time a current of electrons drifts through the wire from the zinc to the copper.

This action may be explained as follows. Pure sulphuric acid has the chemical formula H_2SO_4, but when added to water the SO_4 group of atoms separates from the two hydrogen (H) atoms, taking two electrons with them, one from each hydrogen atom. The hydrogen atoms therefore have a net positive charge. In this electrically charged condition the atoms or groups of atoms are referred to as *ions*. The ionization of sulphuric acid in water can be represented by

$$H_2SO_4 = 2H^+ + SO_4{}^{2-}$$

As zinc atoms dissolve from the zinc plate they go into solution in the form of zinc ions, Zn^{2+}, each of which leaves two electrons behind on the plate. These electrons are the source of the electron current which goes through the wire from the zinc to the copper. We may think of the zinc ions as being attracted into the solution by the $SO_4{}^{2-}$ ions. From the chemical point of view this would be described by saying that the zinc dissolves in the sulphuric acid to produce zinc sulphate.

Normally, when zinc dissolves in acid, internal molecular energy is produced and the solution gets warm. In the simple cell the action of the acid on the zinc results in the production of electric energy instead.

At the same time as the zinc ions enter the solution from the zinc an equivalent number of hydrogen ions leave the solution and deposit themselves on the copper plate. Here they receive an electron from the copper, become neutral atoms and are liberated to form gas bubbles.

By losing electrons in this way the copper becomes positively charged, and this enables it to attract electrons from the zinc through the connecting wire. This movement of electrons through the wire is called an electric current.

Conventional direction of an electric current

When the simple cell was discovered in the year 1800 its action was not understood, and scientists had no way of finding out in which direction the current flowed. They therefore decided to adopt the convention of regarding the current as a flow of positive electricity from the copper to the zinc.

Although, as explained above, we now have good reason to believe that current is a flow of negative electrons from zinc to copper, yet we still retain the old convention. In all direct-current, (d.c.), circuit diagrams an arrow is therefore placed on the wires to indicate that the conventional current flows from the positive terminal of a cell towards the negative terminal.

Electromotive force

Any device such as a cell or dynamo which is able to drive an electric current round a circuit is said to possess *electromotive force* (e.m.f.). Electromotive force which is measured in volts is defined on page 406.

For the time being, however, we may regard the e.m.f. of a cell as being equal to the potential difference across its terminals when it is not producing current in a circuit. For practical purposes the e.m.f. of a cell is measured by connecting a voltmeter (page 412) to its terminals.

Faults of a simple cell

(1) *Polarization*

When a simple cell is in use it is found that the current quickly falls to a very small value. This defect results from the formation of a layer of *hydrogen bubbles on the copper plate* and is called *polarization*. The hydrogen layer weakens the current for two reasons. First, the hydrogen layer sets up a "back e.m.f." in the cell in opposition to that due to the copper and zinc; secondly, the gas partially insulates the plate, and hence increases the internal resistance of the cell (page 411).

(2) Local action

If the zinc used in the simple cell is of the impure commercial variety bubbles of hydrogen will be seen coming off the zinc. This is called local action, and must not be confused with polarization, which is the name given to the formation of hydrogen bubbles on the copper plate.

Local action is caused by the presence in the zinc of small impurities such as iron or carbon which set up tiny local cells at the zinc surface. Bubbles of hydrogen are given off from the impurity and the surrounding zinc slowly dissolves in the acid. This serves no useful purpose and merely wastes the zinc.

Fortunately, the trouble can easily be prevented by cleaning the zinc in sulphuric acid and then rubbing a small globule of mercury over the surface with a small piece of cotton wool. The mercury dissolves pure zinc out of the plate and forms a bright coating of zinc amalgam all over the surface. Local action will not now occur, since the amalgam covers up the impurities and prevents them from coming into contact with the acid.

To study the simple cell

A plate of copper and a plate of zinc, each fitted with a terminal, are dipped into a beaker containing very dilute sulphuric acid, and both plates are carefully examined. So long as the plates are not allowed to touch, no action will be observed at the copper plate. Owing to local action, however, hydrogen bubbles will generally be observed on the zinc. If this is the case the zinc should be taken out, amalgamated with mercury as described in the last paragraph and then returned to the cell.

When the terminals of the cell are joined by a length of copper wire hydrogen bubbles begin to form on the surface of the copper. The same thing happens if the plates are allowed to touch either outside or inside the acid.

The electromotive force of the cell may be measured by connecting a voltmeter to the cell terminals, but before taking a reading the copper plate should be wiped clear of hydrogen bubbles. The e.m.f. of a simple cell is slightly less than 1 V.

If a small pocket-lamp bulb is connected to the terminals of the cell its filament will glow dully for a few moments and then go out. Once more, hydrogen bubbles will be seen on the copper plate, and if a voltmeter reading is taken the terminal potential difference will be seen to have fallen to about 0.2 V.

The original e.m.f. of the cell may be restored by removing the hydrogen bubbles from the copper plate by vigorous brushing with a small paint brush, but it is much better to add a *depolarizing agent* to the cell. Potassium dichromate is suitable for this purpose, and it acts by *oxidizing the hydrogen to form water*. When a little potassium dichromate solution is added to the acid the hydrogen bubbles rapidly disappear and the original e.m.f. is restored. At the same time the lamp begins to glow continuously.

The Leclanché cell

Following the discovery of the simple cell, came the invention of a number of different cells in which polarization was either reduced or eliminated. All these have long since passed into history, but one which survived until comparatively recent times was the Leclanché cell, invented by Georges Leclanché about 1865. It has an e.m.f. of 1.5 V and came into use for working electric bells and telegraphs (Fig. 34.5).

Its positive pole is a carbon plate surrounded by a depolarizing mixture of powdered carbon and manganese(IV) oxide in a porous pot. This together with a zinc rod, is placed inside a glass jar containing a solution of ammonium chloride (sal ammoniac).

Polarization by the hydrogen atoms is prevented by the manganese(IV) oxide which oxidizes the hydrogen to form water. One disadvantage of this cell is that the depolarizing action of the manganese(IV) oxide is slow, so that if a large current is taken from the cell the manganese(IV) oxide cannot oxidize the hydrogen as fast as it is formed. Some polarization then takes place and the e.m.f. of the cell falls. If the cell is allowed to rest for a time the depolarizing action continues to completion and the original e.m.f. is restored.

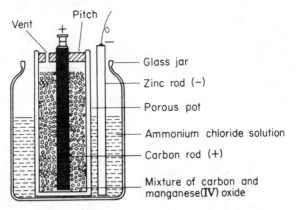

Fig. 34.5. Leclanché cell

For this reason, the Leclanché cell is unsuitable for maintaining a steady current over a long period. These cells were, however, found to be very satisfactory for working electric bells and other purposes where only an intermittent current is required. With the occasional addition of water to make up for loss by evaporation, Leclanché cells will run for very long periods without attention. They are rarely seen today: their place has been taken by dry cells.

The dry cell

The dry cell is a form of Leclanché cell in which the ammonium chloride solution is replaced with a jelly composed of starch, flour and ammonium chloride (Fig. 34.6). The positive element consists of a carbon rod surrounded by a core of a compressed

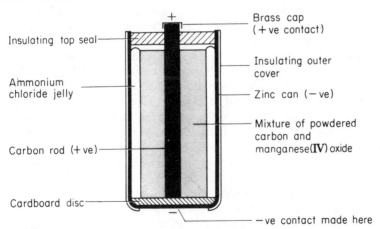

Fig. 34.6. Dry cell

mixture of manganese(IV) oxide and carbon. This is placed inside a zinc can and the space between filled with the ammonium chloride jelly. The zinc can also act as the negative element. The jelly is prevented from drying up by sealing the top of the cell either with pitch or else by means of a metal disc insulated from the can by a fibre disc. Cells of this type give a large current and have a much shorter recovery time than the "wet" type and are used for a great variety of purposes.

Owing to local action which cannot be entirely prevented, they slowly deteriorate when not in use. Modern cells, however, have a "shelf life" of months or even years if stored in a cool place.

Primary and secondary cells

The cells already described in this chapter are known as *primary cells*. In them the current is produced as a result of non-reversible chemical changes taking place between their various components. When all the zinc has been used up the cell cannot be restored to its original condition by passing a charging current through it in the reverse direction.

In contrast with primary cells there is another class of cells called *secondary cells*. These can be recharged after they have run down by passing a current through them from a dynamo or other source of current. They are also known as *storage cells* or *accumulators*, and the two most important types are the lead–acid cell and the nickel–cadmium alkaline cell. Lead cells are extensively used for ignition and lighting on motor cars. Their main advantage is that they have a very low internal resistance, and hence can give a large current with very little drop in terminal potential difference (page 413).

The lead–acid cell

Two plates of lead are immersed in a glass vessel containing dilute sulphuric acid, and a current of about 2 A is passed between them from a battery in series with a rheostat (page 406) and an ammeter. After a time the plate at which the current enters the cell turns a deep chocolate brown colour, owing to the formation of a layer of lead(IV) oxide (PbO_2) on its surface.

The cell is now disconnected from the circuit and connected to a small 2 V electric lamp. The lamp will glow for a short time as the cell sends a current through it in the reverse direction to that of the charging current.

In the year 1860, Gaston Planté, who was then twenty-six years of age, utilized this principle in the design of the first practical form of storage cell. He found that if the charging and discharging process was carried out a great number of times the thickness of the lead dioxide layer gradually increased and the cell was able to supply current for a much longer period.

In modern commercial practice the plates are made of grids of a lead–antimony alloy filled in with paste under hydraulic pressure. Red lead oxide (Pb_3O_4), is used for the positive plates and lead(II) oxide (PbO) for the negative plates. On their first charge such plates become converted to lead(IV) oxide (PbO_2) and lead respectively.

A low internal resistance which enables a lead cell to provide a very large current is achieved by having both sets of plates of as large a surface area as possible and placing them very close together (Fig. 34.7).

The current capacity of an accumulator is measured in ampere-hours (Ah). Thus, a battery rated at 80 Ah will give a current of 8 A for 10 hours. It also might be expected to give a current of 1 A for 80 hours or 20 A for 4 hours. This, however, can be taken only as an approximation, since the capacity is generally very much greater for small currents than when it is discharged rapidly. Commercially, accumulators are rated at the "10 hour rate", i.e., the ampere-hour capacity if discharged completely in a period of 10 hours.

As a lead cell discharges, both sets of plates slowly change to lead sulphate and the acid becomes more dilute. This causes the density of the acid to fall, so that the state of the cell may be ascertained by testing the acid by means of a bulb hydrometer (Fig. 12.7). When fully charged, the e.m.f. of a lead cell is 2.2 V and the relative density of the acid 1.25. The cell is regarded as fully discharged when the relative density of the acid has fallen to 1.15. These figures vary somewhat with different makes of cell.

Fig. 34.7. *Interior construction of a high power output lead cell.* Note the tubular construction of the positive plates which provides maximum surface area for large current output

Care of lead cells

The level of the sulphuric acid in a lead cell should be inspected regularly and any loss from evaporation made up with distilled water only. Acid should never be added except in rare cases where spillage has occurred. Lead cells must be charged regularly, using the maker's recommended charging current, and not left standing for any length of time in a discharged condition. When not in use, they should be given a "topping up" charge at least once a month. When fully charged, hydrogen is freely evolved from the negative plates and oxygen from the positive. This is called "gassing". Needless to say, cells on charge should never be examined in the light from a naked flame, as a mixture of hydrogen and oxygen is dangerously explosive.

Over-discharging and "shorting", i.e., connecting a wire directly across the terminals are very detrimental. Such treatment causes swelling and buckling of the plates, with the result that the active material becomes loosened and falls to the bottom as a sludge. Under these conditions, and particularly if the cell is left in a discharged condition, the lead sulphate in the plates changes to a white crystalline

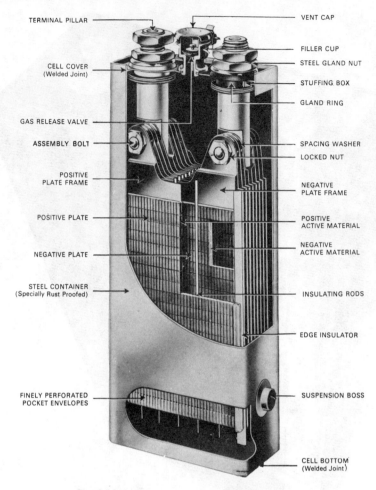

TERMINAL PILLAR

VENT CAP

FILLER CUP

CELL COVER
(Welded Joint)

STEEL GLAND NUT

STUFFING BOX

GLAND RING

GAS RELEASE VALVE

ASSEMBLY BOLT

SPACING WASHER

LOCKED NUT

POSITIVE
PLATE FRAME

NEGATIVE
PLATE FRAME

POSITIVE PLATE

POSITIVE
ACTIVE MATERIAL

NEGATIVE
ACTIVE MATERIAL

NEGATIVE PLATE

STEEL CONTAINER
(Specially Rust Proofed)

INSULATING RODS

EDGE INSULATOR

FINELY PERFORATED
POCKET ENVELOPES

SUSPENSION BOSS

CELL BOTTOM
(Welded Joint)

Fig. 34.8. Interior construction of an alkaline cell

form which cannot be reconverted into lead(IV) oxide and lead respectively. Once this has occurred the cell is said to be "sulphated", and for this there is no remedy except to renew the plates.

Alkaline batteries

In the year 1900, Thomas Edison in America and Valdemar Jungner in Sweden both invented storage cells using a solution of caustic potash as the electrolyte. Edison used iron for the negative plate and Jungner used cadmium. Both men used nickel hydroxide for the positive plate.

Cadmium has certain advantages over iron, and present-day cells are mostly of this type, although it is found that the addition of a little iron to the cadmium improves the efficiency of the cells. The active material is in powder form and is enclosed in perforated pockets in specially constructed steel plates. The container is made of nickel-plated steel (Fig. 34.8).

Although more expensive, alkaline cells have a very much longer life than lead cells and also possess a number of other advantages. Large currents, such as would ruin a lead cell, may be freely taken from alkaline cells without ill effect, and they may be left for months in a discharged condition without harm.

They are therefore very suitable for installation in railway rolling-stock and yachts for lighting and other purposes, as they require no special maintenance when out of use for extended periods. Many large ships, hospitals and public buildings are provided with alkaline-battery installations for emergency lighting.

One disadvantage is that the e.m.f. of an alkaline cell is only about 1.25 V, and this tends to fall continuously on discharge. Five alkaline cells in series give an e.m.f. about equal to that of three lead cells.

QUESTIONS: 34

1. Give an account of the action of a simple cell.

Explain the terms *polarization* and *local action*. Show how these defects are overcome in one practical form of cell.

2. Draw a labelled diagram of a dry Leclanché cell and explain the function of the various components. (*J.M.B.*)

3. Name two advantages which a lead accumulator has over a dry cell.

4. State the components of a fully charged lead accumulator and the changes which occur as a result of discharge. Give, with reasons, three precautions which should be taken to maintain the efficiency of such an accumulator. (*S.*)

5. Two accumulators differ only in that one, *P*, has twice the linear dimensions of the other, *Q*. What is the ratio of:

(i) their storage capacities (i.e., the energy they can store), and

(ii) their internal resistances? (*O.C.*)

6. (*a*) Give a labelled diagram showing the structure of a lead–acid accumulator. State clearly the chemical composition of the two plates when the accumulator is:

(i) in the charged condition;

(ii) in the discharged condition. What change takes place in the electrolyte when the accumulator is being charged? (No chemical equations need be given.)

(*b*) Give two precautions regarding the care of a lead–acid accumulator which is used only occasionally and two precautions to be observed while the accumulator is being charged. For what purpose are cells of the lead–acid type still in very common use? (*A.E.B.*)

35. Current, electromotive force and resistance

In this chapter we shall deal with the principles and definitions underlying the measurement of current electricity.

A simple electric circuit may consist of a cell connected by copper wires to one or more *resistors* or other components to be described later. The cell provides an *electromotive force* which sets up *potential differences* across the various circuit components and drives the *current* through them. These components themselves offer varying degrees of *resistance* to the flow of the current.

Thus, in any electric circuit there are three things which have to be measured:

(1) current, measured in *amperes* (A);
(2) electromotive force and potential difference, both measured in *volts* (V);
(3) resistance, measured in *ohms* (Ω).

Unit of electric current. The ampere

The ampere is the current which, if flowing in two straight parallel wires of infinite length, placed 1 metre apart in a vacuum, will produce on each of the wires a force 2×10^{-7} newton per metre length.

The manner in which the current sets up this force is explained on page 426.

As far as the practical measurement of current is concerned we do not pass the current through two very long wires and measure the force between them as it is not possible to do this with any reasonable degree of accuracy. Instead, we apply the definition to calculate the force between two circular coils of wire in an instrument called a *current balance*. By measuring the force between the coils, which can be done very accurately, we are then able to determine the value of the current flowing through them in amperes.

Current balances are large and expensive pieces of apparatus and are used only in standardizing laboratories such as the National Physical Laboratory (Fig. 35.1). The *ammeters* (page 444) used in ordinary laboratories for measuring current have been calibrated so as to agree with the readings of a standard current balance.

The sub-units of current commonly used are the milliampere (1 mA = 0.001 or 10^{-3} A) and the microampere (1μA = 0.000 001 or 10^{-6} A).

Unit of quantity of electricity. The coulomb

An electric current in a wire is a drift of electrons and the quantity of electricity which passes any point in a circuit will depend on the strength of the current and the time for which it flows. The unit of quantity of electricity is called the *coulomb*.

A coulomb is the quantity of electricity which passes any point in a circuit in 1 second when a steady current of 1 ampere is flowing.

It follows that the total quantity of electricity Q, in coulombs, which passes any

Fig. 35.1. Current balance used at the National Physical Laboratory, Teddington, to determine the ampere in terms of the fundamental units of length, mass and time

point in an electric circuit is given by multiplying the current I, in amperes, by the time t, in seconds,

thus,   coulombs $=$ amperes $\times$ seconds

or $Q = It$

Unit of potential difference. The volt

The meaning of electric potential and potential difference (p.d.) was discussed in chapter 33. When an electric current flows through a wire a potential difference is said to exist between the ends of the wire. The potential difference between two points is defined as the work done, in joules, when one coulomb of electricity moves from one point to the other. The unit of potential difference is called the *volt* and is defined as follows:

Two points are at a potential difference of 1 volt if 1 joule of work is done per coulomb of electricity passing from one point to the other.

Electromotive force

The electromotive force of a cell is measured in volts and may be regarded as the sum total of the potential differences which it can produce across all the various components of a circuit in which it is connected including the potential difference required to drive the current through the cell itself.

The e.m.f. of a cell in volts is therefore defined as the total work done in joules per coulomb of electricity conveyed in a circuit in which the cell is connected.

Resistance

An experiment to compare the electric conducting powers of various substances was described on page 365. There it was seen that different materials could be classified as good or poor conductors and insulators.

As far as current electricity is concerned, we usually think in terms of the ability of a substance to *resist* the flow of electricity through it. A good conductor is therefore said to have a low resistance and a poor conductor a high resistance.

With regard to insulation, materials which are good insulators for static electricity are also good insulators for current electricity.

We shall see later (page 457) that the resistance of a wire depends on its dimensions and the material from which it is made. For a wire of given dimensions, silver offers least resistance to the current. Silver is too expensive for normal use, and therefore copper, which comes next on the list, is used for electric cables and for connecting wires in electric circuits.

When a high resistance is required, e.g., to reduce the current in a circuit, special alloys are used. The most common of these are constantan (otherwise called *contra* or *eureka*), manganin and nichrome.*

Constantan is used for general purposes, while manganin is used for making high-quality standard resistors. Nichrome is the alloy from which the elements of electric fires are made, since it resists oxidation when red hot. Constantan and manganin wires respectively have a resistance of about twenty-five times, and nichrome about sixty times, that of a copper wire of the same dimensions.

Rheostats

The current flowing in a circuit may be increased or decreased by changing the length of resistance wire inserted into it. This is done by means of a *rheostat*. Rheostats are obtainable in a variety of different patterns, but the type most com-

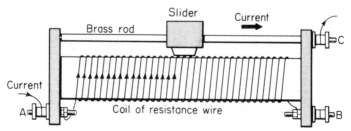

Fig. 35.2. Wire-wound rheostat

monly used in laboratories is illustrated in Fig. 35.2. It consists of a coil of constantan wire wound on a slate or enamelled iron former and provided with a sliding contact on a brass rod. Current enters the coil through a terminal at one end and flows along the coil until it reaches the sliding contact. From here it flows along the brass rod, and thence through the circuit. The resistance thus introduced into the circuit will depend on the position of the slider along the coil.

* Constantan (60% copper, 40% nickel), Manganin (84% copper, 12% manganese, 4% nickel), Nichrome (80% nickel, 20% chromium).

Ohm's law

In the year 1826, Georg Simon Ohm, a teacher of physics at Cologne, published a book containing details of some experiments he had made to investigate the relationship between the current passing through a wire and the potential difference between the ends of the wire. As a result of these experiments, he arrived at the following law.

The current passing through a wire at constant temperature is proportional to the potential difference between its ends.

A conductor for which this relationship is true is said to obey Ohm's law.

This law may also be expressed as

$$\frac{\text{potential difference}}{\text{current}} = \text{constant}$$

For a given potential difference, a high resistance will pass a small current and a low resistance a large current. Thus, the value of the constant in the above equation, which is high when the current value is small and low when the current is large, can be used as a measure of the *resistance* of the wire.

We may therefore write

$$\frac{\text{potential difference}}{\text{current}} = \text{resistance}$$

In other words, **the resistance of a conductor is the ratio of the potential difference across it to the current flowing through it.** It must be appreciated that this relationship defines the resistance of a conductor and applies whether Ohm's law is obeyed or not.

We have already chosen the units of potential difference and current, namely, the volt and the ampere. The above definition enables us to define the unit of electric resistance. This is called the **ohm** and is defined as follows:

The ohm (Ω) is the resistance of a conductor such that, when a potential difference of 1 volt is applied to its ends a current of 1 ampere flows through it.

It follows that,

$$\frac{\textbf{volts}}{\textbf{amperes}} = \textbf{ohms}$$

or, in symbols

$$\frac{V}{I} = R$$

whence also

$$I = \frac{V}{R}$$

and

$$V = IR$$

Any one of these expressions may be derived algebraically from any other. The last one is usually the easiest to memorize.

Limitations of Ohm's law

Ohm's work was carried out using metal conductors and the law also applies to certain other materials. We have mentioned the need for constant temperature, but if the law is to be obeyed then *all* physical conditions must remain constant. For example, the resistance of some conductors will alter if they are bent or placed under tension or if put at right angles to a strong magnetic field.

Ohm's law does not apply to semiconductors or rectifiers (chapter 44), nor to the conduction of electricity through gases.

Effect of temperature on resistance. Semiconductors

The resistance of a pure metal increases with temperature but the resistance of certain other conducting materials, e.g., carbon decreases with temperature.

Various metal alloys, notably manganin and constantan which are used for making standard *resistors*, show very small resistance changes with temperature under normal laboratory conditions.

Certain other substances notably germanium, silicon, and selenium are classed as *semiconductors*. The resistance of these substances *decreases* with rising temperature, especially when they contain certain impurities. In single-crystal form with controlled amounts of impurity such as arsenic or indium they are used in the manufacture of solid-state diodes and the transistors used in radio and computer circuits and for many other purposes. See also Chapter 45.

A semiconducting device known as a *thermistor* and which consists of a mixture of oxides of manganese and nickel has a resistance which can decrease by a factor of many hundreds from room temperature to 100 °C. Thermistors can be used as sensitive thermometers and also as devices to compensate for the increase in resistance with rising temperature of other components in a circuit.

Resistors

Radio and television receivers contain large numbers of resistors with resistances of anything from a few ohms to millions of ohms. Some are wire wound, while others are made from carbon or graphite. They are supplied with either wire ends or terminals for connection purposes.

The special standard resistors designed for laboratory purposes are described in chapter 40.

Electric circuit. Use of ammeters and voltmeters

Fig. 35.3 illustrates some of the symbols used in electric circuit diagrams.

When the terminals of a cell are joined to the ends of a resistor an electric circuit is formed. If several resistors are connected end to end in the circuit so as to form a

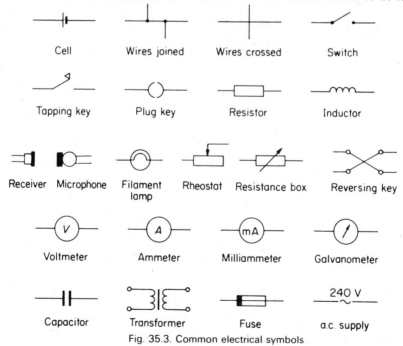

Fig. 35.3. Common electrical symbols

continuous path they are said to be *in series* and the same current flows through each. The current may be measured with an ammeter which is placed in series with the resistors (Fig. 35.4). An ammeter has a low resistance, so that it introduces as little extra resistance as possible into the circuit.

A voltmeter is used to measure the potential difference between the ends of a resistor, and is therefore connected across (i.e., in parallel with) the resistor. Voltmeters have a high resistance, and the current they take is usually negligible.

The construction and action of ammeters and voltmeters is dealt with in chapter 39.

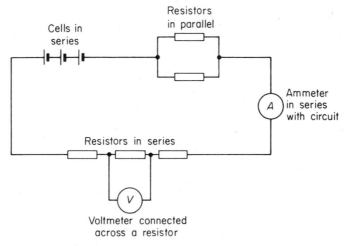

Fig. 35.4. Use of symbols in an electric circuit

Resistors in series

A number of resistors R_1, R_2, R_3, in ohms, are said to be connected in series if they are connected end to end consecutively so that the same current I, in amperes, flows through each (Fig. 35.5).

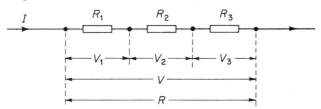

Fig. 35.5. Resistors in series

If R is the combined resistance and V, in volts, is the total potential difference across the resistors, then

$$V = IR$$

But
$$V = \text{sum of the individual p.d.s. across } R_1, R_2 \text{ and } R_3$$
$$= V_1 + V_2 + V_3$$
$$= IR_1 + IR_2 + IR_3$$

therefore
$$IR = IR_1 + IR_2 + IR_3$$

hence, dividing throughout by I

$$R = R_1 + R_2 + R_3$$

The same argument may be applied to any number of resistors.

Resistors in parallel

Resistors are said to be *in parallel* when they are placed side by side and their corresponding ends joined together (Fig. 35.6). The same potential difference will thus be applied to each, but they will share the main current in the circuit.

We will suppose that the main current I divides into I_1, I_2, and I_3 through the resistors R_1, R_2, and R_3 respectively and that the common potential difference across them is V.

If R is the combined resistance we may write

$$I = \frac{V}{R}$$

The total current,

$$I = I_1 + I_2 + I_3$$

$$= \frac{V}{R_1} + \frac{V}{R_2} + \frac{V}{R_3}$$

therefore

$$\frac{V}{R} = \frac{V}{R_1} + \frac{V}{R_2} + \frac{V}{R_3}$$

Dividing throughout by V

$$\frac{1}{R} = \frac{1}{R_1} + \frac{1}{R_2} + \frac{1}{R_3}$$

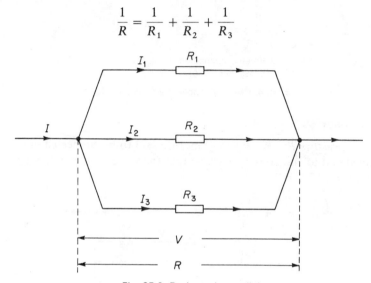

Fig. 35.6. Resistors in parallel

Special case of two resistors in parallel

The formula obtained above may be extended for any number of resistors. In practice, however, it is rare to find more than two resistors in parallel, and therefore we shall examine this case in more detail.

For two resistors R_1 and R_2 in parallel

$$\frac{1}{R} = \frac{1}{R_1} + \frac{1}{R_2}$$

Putting the right-hand side over the same common denominator,

$$\frac{1}{R} = \frac{R_2 + R_1}{R_1 R_2}$$

Inverting both sides,

$$R = \frac{R_1 R_2}{R_1 + R_2}$$

Hence, for *two* resistors in parallel,

$$\text{combined resistance} = \frac{\text{product of resistances}}{\text{sum of resistances}}$$

It is useful to learn this formula by heart.

Practical importance of the internal resistance of a cell

Cells of one particular kind, for example, dry cells, may be obtained in several different sizes, but so long as they are all made of exactly the same materials and have the same strength of electrolyte they will have identical e.m.f.s.

The strength of the current which is obtained from a cell depends not only on its e.m.f. but also on the *internal resistance*. In order to obtain a large current, the internal resistance must be low. In the case of a lead cell this means that the plates must be of large area and placed close together. Also the strength of the electrolyte must be such that its resistance is as low as possible.

According to its size and construction, the internal resistance of a dry cell varies from about 0.5 to 1.0 Ω, and the e.m.f. is about 1.5 V. If, therefore, the terminals of a dry cell were shorted by a piece of thick copper wire of negligible resistance the maximum current obtainable would be of the order of 3 to 0.5 A. This would be quite safe, but on the other hand, it would be dangerous to connect a piece of copper wire across the terminals of a 2V lead cell which had an internal resistance of about 0.01 Ω. The current in this case could possibly be of the order of 200 A. Harm would be done to the cell, and the wire would be liable to vaporize explosively, and burn anyone handling it.

The electric starter motor of a car requires a very high current to operate it. Consequently, car batteries are made of cells having many thin plates separated by as small a distance as possible. Eight dry cells in series have the same e.m.f. as a 12V car battery, but they would be quite useless for working the starter owing to their high internal resistance.

Arrangement of cells

A group of cells connected together is called a *battery*. Normally, cells are connected in series, the positive of one being connected to the negative of the next and so on (Fig. 35.7). On occasions, however, they may be connected in parallel, i.e., all the positives connected together and the same with the negatives (Fig. 35.8).

Fig. 35.7. Cells in series   Fig. 35.8. Cells in parallel

When maximum current is required from a given number of cells the arrangement used will depend on the external circuit resistance. Generally speaking, series connection is used when the circuit resistance is high compared with that of the cells and parallel connection when it is low. This principle is illustrated in the worked example on page 414.

When cells are in series the total e.m.f. of the battery is equal to the sum of the separate e.m.f.s, and the internal resistance is equal to the sum of the separate internal resistances of the cells. When cells of equal e.m.f. and internal resistance are connected in parallel the resultant e.m.f. is the same as that of one cell only and the internal resistance of the battery is calculated from the formula for resistors in parallel.

One advantage of connecting cells in parallel is that there is less drain on the cells, since they share the total current, whereas with series connection the same main current is supplied by each cell.

Cells should never be left connected in parallel when not in use, for if the e.m.f. of one is slightly greater than that of another current will circulate in the battery itself and the cells become exhausted. This cannot happen when they are in series.

"Lost volts" when a cell is producing current in a circuit

For practical purposes, the e.m.f. of a cell may be measured to a very close approximation by taking the reading of a *high-resistance* voltmeter connected directly across the cell terminals when the cell is not connected to anything else.

The reason why this value is only approximate will be explained at the end of this section, but for the time being we shall ignore the slight error.

Let us suppose that a voltmeter connected to the terminals of a dry cell of internal resistance 2.0 Ω gives a reading of 1.5 V. This is the e.m.f. of the cell (Fig. 35.9 (a)).

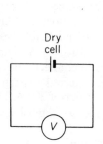

(a) High resistance voltmeter reads
e.m.f. 1.5 volts
(negligible current flows)

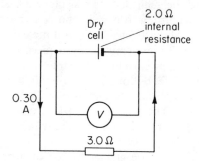

(b) Voltmeter reads only 0.90 volts.
0.60 *Lost voltage* drives current
through internal resistance

Fig. 35.9.

A 3.0 Ω resistor is now connected to the terminals of the cell so that the cell drives a current through it. It is noticed that the voltmeter reading has dropped to 0.90 V (Fig. 35.9 (b)). The cell now appears to have "lost" $(1.5 - 0.90) = 0.60$ V.

This may be explained as follows.

The current flowing in the circuit is given by

$$I = \frac{\text{e.m.f.}}{\text{total resistance}} = \frac{E}{R + B}$$

where E = e.m.f. of cell
       R = external circuit resistance
       B = internal resistance of cell

$$= \frac{1.5}{3.0 + 2.0} = 0.30 \text{ A}$$

Now the p.d. required to drive this current of 0.30 A through the external 3.0 Ω. resistor is given by

$$\begin{aligned} V &= IR \\ &= 0.30 \times 3.0 \\ &= 0.90 \text{ V} \end{aligned}$$

which is, of course, the same as the voltmeter reading.

The voltmeter is connected to the cell terminals, but if it were connected to the 3.0 Ω resistor terminals instead it would make no difference to the reading. Owing to the fact that the connecting wires from cell to resistor are of negligible resistance, the p.d. across them is also negligible, and hence the cell terminal p.d. is equal to the p.d. across the resistor.

The p.d. required to drive the current through the cell itself is given by

$$\begin{aligned} \text{current} \times \text{internal resistance} &= 0.30 \times 2.0 \\ &= 0.60 \text{ V} \end{aligned}$$

which is equal to the "lost volts" $(E - V)$ of the cell.

To sum up, the reading of a voltmeter connected to the terminals of a cell on open-circuit may be assumed equal to the e.m.f. E of the cell.

When the cell is sending current through an external resistor the voltmeter gives the terminal p.d. V which is the p.d. required to send the current through the external resistor.

The terminal p.d. is always less than the e.m.f., and the difference between them, or "lost volts", represents the p.d. required to send the current through the internal resistance of the cell. *This cannot be read directly from a voltmeter, but can be obtained only by subtracting the terminal p.d. from the e.m.f.*

We are now able to explain why a voltmeter gives only a close approximation for the e.m.f. of a cell. Even a very high-resistance voltmeter *must take some current*, and hence a small part of the e.m.f. will be "lost" in driving this current through the cell. If, however, the voltmeter resistance is very high compared with the cell resistance the current taken will be very small, and consequently the "lost volts" in this case will be negligible.

For the measurement of e.m.f. to a high degree of accuracy a potentiometer is used (page 459).

To measure the internal resistance of a cell

A high-resistance voltmeter is connected to the cell terminals and the reading noted.

Cell e.m.f. = E
Internal res. = B

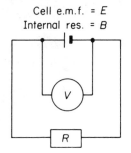

Fig. 35.10. Measurement of internal resistance

This will give the e.m.f. of the cell. A 1 Ω resistor is now connected to the cell terminals and the voltmeter reading taken again (Fig. 35.10). Further readings are taken using 2, 3, 4, etc., ohms, and the results are recorded in the table shown.

E.M.F. of cell $= E =$ volts		
Resistance R (ohms)	*Voltmeter reading* V (volts)	*Internal resistance* $B = \dfrac{R(E - V)}{V}$ (ohms)
1 2 3 etc.		
Mean value of internal resistance		ohms

The internal resistance of the cell is calculated from the readings as follows:

Let E = e.m.f. of the cell in volts
  V = p.d. across cell terminals when cell is producing current in a resistance R
  B = internal resistance of cell in ohms

Since this is a simple series circuit, the current is the same through both cell and the external resistor. The current through the high-resistance voltmeter is negligible. Thus,

p.d. driving current through the external resistor $R = V$

p.d. driving current through the cell $=$ "Lost voltage" $= (E - V)$

hence the current is given by
$$I = \frac{V}{R}$$

and also by
$$I = \frac{(E - V)}{B}$$

Therefore
$$\frac{V}{R} = \frac{(E - V)}{B}$$

cross multiplying
$$VB = R(E - V)$$

whence
$$B = \frac{R(E - V)}{V}$$

Note. Consistent results must not be expected if the cell polarizes during the experiment. The method is unsuitable for cells of very low internal resistance such as lead cells.

Worked examples

1. *Two cells each having an e.m.f. of 1.5 V and an internal resistance of 2 Ω are connected:* (a) *in series, and* (b) *in parallel. Find the current in each case when the cells are connected to a 1 Ω resistor.*

  If the 1 Ω resistor is substituted by an 11 Ω resistor, calculate the new current in both cases.

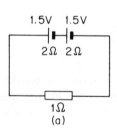

Fig. 35.11.

The first step in all electrical problems is to draw a circuit diagram (Fig. 35.11). The unit symbols V, A, and Ω are used to denote volts, amperes, and ohms respectively.

Fig. 35.11 (*a*) total e.m.f. of 2 cells in series $= 1.5 + 1.5 = 3$ V
  Total resistance in circuit $= 2 + 2 + 1 = 5 \,\Omega$

$$\text{Current} = \frac{\text{e.m.f.}}{\text{total resistance}} \qquad = \frac{3}{5} = \underline{0.6 \text{ A}}$$

Fig. 35.11 (*b*) e.m.f. of 2 cells in parallel $= 1.5$ V
  Resistance of 2 cells in parallel $= R$ given by,

$$\frac{1}{R} = \frac{1}{2} + \frac{1}{2} = 1$$

hence
  Total resistance in circuit $\qquad R = 1 \,\Omega$
$\qquad\qquad\qquad\qquad\qquad\qquad\qquad = 1 + 1 = 2 \,\Omega$

$$\text{Current} = \frac{\text{e.m.f.}}{\text{total resistance}} \qquad = \frac{1.5}{2} = \underline{0.75 \text{ A}}$$

Last part

(a) The total resistance is now $2 + 2 + 11 = 15\ \Omega$, and

$$\text{Current} = \frac{3}{15} = \underline{0.2\ \text{A}}$$

(b) Total resistance $= 1 + 11 = 12\ \Omega$

$$\text{Current} = \frac{1.5}{12} = \underline{0.125\ \text{A}}$$

Answer. (a) 0.6 A, 0.2 A

(b) 0.75 A, 0.125 A

Notice that, in the case of the 1 Ω resistor, the larger current is obtained with the cells in parallel. In the case of the 11 Ω resistor, however, it is better to have the cells in series.

2. *A cell supplies a current of 0.6 A through a 2 Ω coil and a current of 0.2 A through a 7 Ω coil. Calculate the e.m.f. and internal resistance of the cell.*

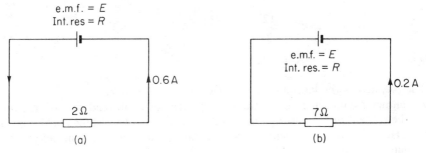

e.m.f. = E
Int. res = R

0.6 A

2 Ω

(a)

e.m.f. = E
Int. res. = R

0.2 A

7 Ω

(b)

Fig. 35.12.

In this problem we have to find two unknown quantities, namely, the e.m.f. and internal resistance of the cell. It is therefore necessary to obtain two equations which can be solved simultaneously.

Let E = e.m.f. of the cell

R = internal resistance of cell

Then since   e.m.f. = current × total resistance

in the first case $E = 0.6(2 + R)$.   .   .   .   (1) (Fig. 35.12 (a))
and in the second case $E = 0.2(7 + R)$.   .   .   (2) (Fig. 35.12 (b))

Equating the right-hand sides of these two equations,

$$0.6(2 + R) = 0.2(7 + R)$$
$$1.2 + 0.6\,R = 1.4 + 0.2\,R$$
$$0.4\,R = 0.2$$
$$R = \frac{0.2}{0.4} = \underline{0.5\ \Omega}$$

Substituting for R in equation (1)

$$E = 0.6(2 + 0.5)$$
$$= \underline{1.5\ \text{V}}$$

Answer. E.M.F. of cell = 1.5 V

Internal resistance of cell = 0.5 Ω

3. *Six cells each having an e.m.f. of 2 V and an internal resistance of 0.1 Ω are connected in series with an ammeter of negligible resistance, a 1.4 Ω resistor and a metal-filament lamp. The ammeter reads 3 A. Find: (a) the resistance of the lamp; (b) the p.d. across the lamp. What reading would you expect from a high-resistance voltmeter connected across the battery terminals?*

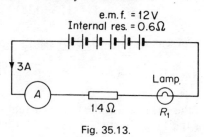

Fig. 35.13.

Let the lamp resistance = R_1 (Fig. 35.13)
Total circuit resistance = $(6 \times 0.1) + 1.4 + R_1 = (2 + R_1)$
The total e.m.f. $= (6 \times 2) = 12$ V

Applying the formula $R = \dfrac{E}{I}$ to the whole circuit, we have

$$2 + R_1 = \frac{12}{3}$$

whence $R_1 = 2\,\Omega$

The p.d. across the lamp (from $V = IR_1$) = $3 \times 2 = 6$ V.

A voltmeter connected across the battery terminals will read the p.d. required to drive the current of 3 A through the total external circuit resistance of $(1.4 + 2)$, i.e., 3.4 Ω. Hence (from p.d. = current × resistance), voltmeter reading across battery terminals = $3 \times 3.4 = 10.2$ V.

> Answer. Lamp resistance   = 2 Ω
> P.d. across lamp   = 6 V
> P.d. across battery = 10.2 V

4. *Fig. 35.14 shows a 12 V battery of internal resistance 0.6 Ω connected to three resistors A, B, and C. Find the current in each resistor.*

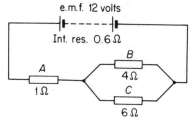

Fig. 35.14.

The combined resistance R, in ohms, of B and C is given by

$$\frac{1}{R} = \frac{1}{4} + \frac{1}{6} = \frac{6 + 4}{24} = \frac{10}{24}$$

whence $R = \dfrac{24}{10} = 2.4\,\Omega$

Total circuit resistance = $0.6 + 1 + 2.4 = 4\,\Omega$

$$\text{Main current} = \frac{\text{e.m.f.}}{\text{total resistance}} = \frac{12}{4} = 3 \text{ A}$$

The p.d. across both B and C (current × combined resistance) = 3 × 2.4 = 7.2 V

$$\text{Current through } B = \frac{\text{p.d.}}{\text{resistance}} = \frac{7.2}{4} = 1.8 \text{ A}$$

$$\text{Current through } C = 3 - 1.8 = 1.2 \text{ A}$$

$$\text{(alternatively, Current through } C = \frac{\text{p.d.}}{\text{resistance}} = \frac{7.2}{6} = 1.2 \text{ A)}$$

Answer. Current through $A = 3$ A
through $B = 1.8$ A
through $C = 1.2$ A

QUESTIONS: 35

1. What is meant by the electromotive force of a cell?

A battery consists of three accumulators in series, each having an e.m.f. of 2 V. A second battery consists of four dry cells also in series, each having an e.m.f. of 1.5 V. What is the e.m.f. of each battery? Why could you get a bigger current from the battery of accumulators? (S.)

2. State Ohm's law and define resistance.

Derive an expression for the resistance of two conductors connected:
(a) in series;
(b) in parallel.

What values of resistance could you obtain if you were supplied with three coils each of resistance 1 Ω? (O.C.)

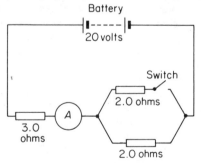

Battery

20 volts

Switch

2.0 ohms

A

3.0 ohms

2.0 ohms

Fig. 35.15.

3. In the circuit shown, (Fig. 35.15), find the reading of the ammeter A when the switch is:
(i) open;
(ii) closed. (Assume that the battery resistance is negligible.) (C.)

4. A cell has an e.m.f. of 1.5 V and an internal resistance of 1 Ω, and is connected to two resistances of 2 and 3 Ω in series. Find the current flowing and the potential difference across the ends of each resistance. (S.)

5. Two cells each having an e.m.f. of 1.5 V and an internal resistance of 1 Ω are con-

nected to a resistance of 4 Ω. What is the current in this resistance if the cells are connected in parallel? (J.M.B.)

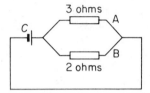

3 ohms

A

C

B

2 ohms

Fig. 35.16.

6. Fig. 35.16 shows a cell C supplying current to resistors A and B. The resistance of each is shown on the diagram. The current in A is 1.2 A. Find:
(a) the current in B;
(b) the current in C. (C.)

7. When a cell having an e.m.f. of 1.50 V is giving a current of 0.50 A, a high-resistance voltmeter across its terminals reads 1.20 V. Explain why the voltmeter reading is less than 1.5 V and find what the voltmeter will read when the cell is giving 0.60 A. (O.)

8. One 4 ohm and two 2 ohm resistors are available. **All three** are to be connected together in two different arrangements such that the total (resultant) resistance is
(i) less than 2 ohm,
(ii) more than 4 ohm but less than 8 ohm.
Draw a diagram of each arrangement and calculate the total (resultant) resistance in each case. (J.M.B.)

9. A cell can supply a current of 1.2 A through two 2 Ω resistors connected in parallel. When they are connected in series the value of the current is 0.4 A. Calculate the e.m.f. and the internal resistance of the cell. (O.C.)

10. Two identical cells have each an e.m.f. of E in volts and an internal resistance of R in

ohms and are connected in series to an external resistor of 2 Ω.

(a) Find in terms of E and R:

(i) the combined e.m.f.;

(ii) the combined internal resistance;

(iii) the main current;

(iv) the p.d. across the 2 Ω resistor.

(b) Find these four values when the cells are in parallel.

(c) If the p.d. across the 2 Ω resistor in (a) is 0.8 V and in (b) is 1.0 V find the values for E and R. (*S.*)

11. A cell is joined in series with a resistance of 2 Ω, and a current of 0.25 A flows through it. When a second resistance of 2 Ω is connected in parallel with the first the current through the cell increases to 0.3 A. What is:

(a) the e.m.f.;

(b) the internal resistance of the cell?

(*J.M.B.*)

12. Explain the meaning of the terms *electromotive force* of a cell, *internal resistance* of a cell.

Would you expect two identical cells in parallel to drive more current through a resistor than one cell does?

Why do two identical cells in series drive more current through a resistor than one does, and why do they not double the current?

A cell of 6.0 V e.m.f. and negligible internal resistance is connected to a resistor and drives a current of 3.0 A through it. Another cell of e.m.f. 1.5 V is inserted in the circuit in series with the first one. The current remains at 3.0 A. What is the internal resistance of the second cell? (*O.C.*)

13. Fig. 35.17 shows two identical high resistance voltmeters connected across two identical d.c. power sources. The left hand voltmeter reads 6.0 V but the right hand voltmeter reads 4.0 V.

(i) Suggest why there is a difference between the readings of the two voltmeters.

(ii) What would be the reading on the voltmeter if the 10 Ω resistor were replaced by a 15 Ω resistor? (*A.E.B., 1982*)

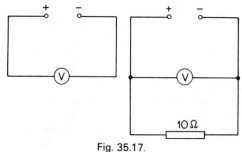

Fig. 35.17.

14. Describe with the aid of a circuit diagram how you would measure the internal resistance of a cell. Explain your calculation.

A cell has an e.m.f. of E in volts and an internal resistance of r in ohms. When resistances of 2 Ω and 5 Ω are connected in turn across its terminals currents of 0.5 A and 0.25 A are passed. Calculate the values of E and r. What is the p.d. across the cell when the 2 Ω resistance is connected between its terminals? (*A.E.B.*)

15. When a particular cell is on open circuit the p.d. between its terminals is 1.5 V. When a 10 Ω resistance is connected between the terminals the p.d. falls to 1.0 V and when the 10 Ω resistance is replaced by a resistance R the p.d. becomes 0.5 V. What is:

(a) the internal resistance of the cell;

(b) the value of R? (*J.M.B.*)

16. A battery of four cells in series, each having an e.m.f. of 1.1 V and an internal resistance of 2 Ω is to be used a charge a small 2 V accumulator of negligible internal resistance. What is the charging current?

(*O.*)

17. Define *potential difference, electric current,* and *resistance.*

"The electromotive force acting in a circuit, in volts, is numerically equal to the energy in joules supplied by the source of e.m.f. in causing unit electric charge, 1 coulomb, to move once around the circuit." Show that this statement is in agreement with the expression $E = i(R + r)$, relating the e.m.f. E of a battery with the current i it causes to flow in a circuit of external resistance R, the internal resistance of the battery being r.

In an experiment to determine E and r for a storage battery, the following readings are made of the current i for different values of R.

R (Ω)	0.70	2.50	5.50
i (A)	5.0	2.0	1.0

Calculate the value of l/i for each value of i. Plot l/i against R and hence or otherwise find the values of r and E. (*O.C.*)

18. (a) Draw TWO clear-circuit diagrams to show how a 6 V battery may be used to operate efficiently:

(i) 3 torch bulbs each marked "2 V 0.04 A", and

(ii) 3 headlamp bulbs each marked "6 V 24 W".

In each case calculate the current supplied by the battery and the total effective resistance of the bulbs. (*S.*)

36. Magnetic effect of an electric current

In the year 1819 Hans Christian Oersted, Professor of Physics at Copenhagen, discovered that a wire conveying a current was able to deflect a pivoted magnetic needle. It happened during the course of a lecture which he was giving on the simple cell. Having shown that a wire connected to a cell had no effect when placed at right angles to the needle, Oersted then placed the wire parallel with the needle. Immediately, to his surprise and that of his students, the needle deflected out of the meridian.

On further investigation, Oersted found that the direction of the deflection depended on the direction of the current and also on whether the wire was above or

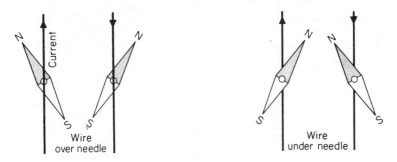

Fig. 36.1. Oersted's experiment

below the needle (Fig. 36.1). Later, André Ampère, the French mathematician and physicist gave a rule, called the *swimming rule*, for predicting the direction of deflection of the needle.

Ampère's swimming rule

Let the observer imagine himself to be swimming along the wire in the direction of the current and facing the needle, then the N pole of the needle will be deflected towards his left hand.

Schweigger's galvanometer

The newly found relationship between electricity and magnetism attracted a good deal of attention, and soon the German scientist, Johann Schweigger, applied the discovery to the construction of a simple *galvanometer* or current-indicating instrument.

Schweigger's instrument is shown in Fig. 36.2. It consists of a pivoted magnetic needle placed at the centre of a number of turns of insulated wire. When an electric current passes through the coil each turn produces its own effect on the needle, so that quite small currents can be detected.

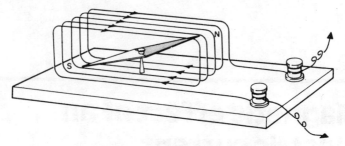

Fig. 36.2. Schweigger's simple galvanometer

Magnetic flux pattern due to a current in a long straight wire

Since an electric current has a magnetic effect, we should expect it to be surrounded by magnetic flux. The following experiments show that this is so.

A large rectangular coil consisting of about twenty turns of insulated copper wire is set up vertically with one of its vertical sides passing through a hole in the centre of a piece of cardboard supported horizontally (Fig. 36.3). A current of about 3 A

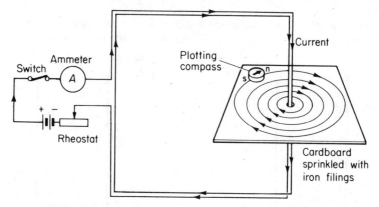

Fig. 36.3. Magnetic flux pattern due to current in a straight wire

may be passed through the coil from a 6 V battery in series with a rheostat, an ammeter and a switch. The reason for having a coil of twenty turns instead of one is that a single wire would require a current of 60 A to produce the same effect. This would need not only a very large battery but the wire would also have to be very thick indeed, or it would become exceedingly hot when the current was passing.

A fine layer of iron filings is then sprinkled on the cardboard; the current is switched on and the card tapped gently. The filings set in a series of concentric circles about the wire as centre. A small plotting compass placed on the card indicates the direction of the flux. If the current is reversed by changing over the battery connections the compass needle will swing round and point in the opposite direction, but the pattern of the flux remains unaltered.

James Clerk Maxwell, a great mathematical physicist of the nineteenth century, gave a rule relating the direction of the magnetic flux round a wire to the direction of the current flowing through it. This is known as the "screw rule", and is stated as follows:

Maxwell's screw rule

Imagine a corkscrew being screwed along the wire in the direction of the current. The direction of rotation of the screw gives the direction of the magnetic flux.

Many people favour a somewhat simpler rule, called the *right-hand grip rule*.

Right-hand grip rule

Imagine the wire to be grasped in the right hand with the thumb pointing along the wire in the direction of the current. The direction of the fingers will give the direction of the magnetic flux (Fig. 36.4).

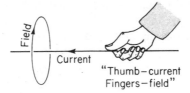

"Thumb–current
Fingers–field"

Fig. 36.4. Right-hand grip rule

Magnetic flux due to a solenoid

When an electric current is passed through a solenoid or long cylindrical coil the resultant magnetic flux is very similar to that of a bar magnet. One end of the solenoid acts like a N pole and the other a S pole (Fig. 36.5 (*b*)).

The experiment is carried out with the aid of a coil wound like a helical spring on a rectangular piece of hardboard. Two other pieces of hardboard are fixed to the sides of the first to form a flat table on which the iron filings are sprinkled (Fig. 36.5

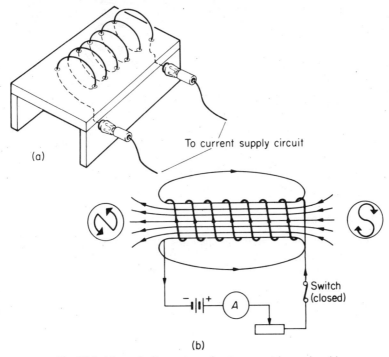

Fig. 36.5. Magnetic flux pattern due to current in a solenoid

(*a*)). The coil is connected in series with a battery, ammeter, rheostat and switch. The current should be adjusted to at least 20 A, as it is not possible to multiply the effect as in the last experiment by having a number of wires close together. A heavy current such as this should not be left on any longer than is necessary to enable the filings to set in the direction of the magnetic flux.

Rule for the polarity of a coil carrying a current

When viewing one end of the coil, it will be of N polarity if the current is flowing in an aNticlockwise direction, and of S polarity if the current is flowing in a clockwiSe direction.

Fig. 36.5 (*b*) illustrates an easy way of remembering this rule, by attaching arrows to the letters N and S.

Magnetic flux due to a short circular coil

When a current is passed through a short circular coil the pattern of the magnetic flux is shown in Fig. 36.6 (*b*). The coil used for this experiment consists of twenty or more turns closely wound together and fixed to a thin hardboard table (Fig. 36.6 (*a*)). Current is passed through the coil and the magnetic flux pattern demonstrated with

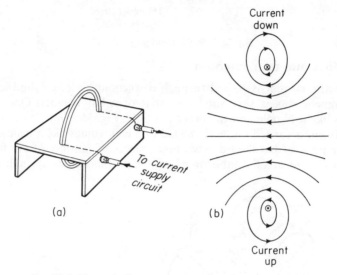

Fig. 36.6. Magnetic flux pattern due to current in a short coil

iron filings as in the previous experiment. The coil may be regarded as having a N face and a S face, and the same rule may be applied as already given in the case of the solenoid.

One important feature should be noted. The magnetic flux at the centre of the coil is straight and is perpendicular to the plane of the coil.

The electromagnet

The solenoid method for magnetizing steel described on page 343 was first used for making magnetic needles by Ampère in 1820. Five years later a Lancashireman named William Sturgeon used soft iron instead of steel and found that the iron was strongly magnetized only when the current was flowing. Sturgeon's electromagnet

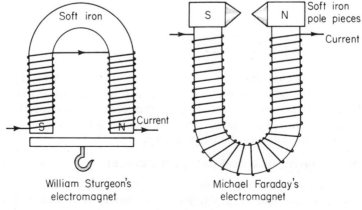

Fig. 36.7. Nineteenth-century electromagnets

was made of a bar of soft iron bent into a U-shape and having coils of insulated copper wire wound in opposite directions on the two arms. Michael Faraday in England and Joseph Henry in America made similar electromagnets, which were provided with soft-iron pole pieces for the purpose of concentrating the magnetic flux in a small gap (Fig. 36.7). Fig. 36.8 shows a modern electromagnet designed for general experimental purposes.

Lifting magnets

Fig. 36.9 illustrates a powerful double electromagnet used in industry for lifting and transporting steel plates, girders, scrap iron, and so on. When the load has been carried to the desired position it is released by switching off the current. Fig. 36.10 shows a lifting magnet in section. It is made of magnetically soft iron, with a coil of

Fig. 36.8. A Newport electromagnet designed for general research purposes

insulated copper strip wound in an annular recess inside it. The inside and outside portions of the face of the magnet are of opposite polarity, and hence greater lifting power is obtained.

Fig. 36.9. *Lifting a steel slab with electromagnets*. The picture shows two 1 metre Witton Kramer magnets being used to turn steel slabs at the Abbey Works of the Steel Company of Wales

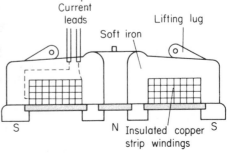

Fig. 36.10. Section of lifting magnet

The electric bell

The electric bell consists of two solenoids wound in opposite directions on two soft-iron cores joined by a soft-iron yoke (Fig. 36.11). One end of the windings is connected to a terminal T_1, and the other to a metal bracket which supports a spring-mounted soft-iron armature. The armature carries a light spring, to which is soldered a small silver disc to act as a contact. The latter presses against the end of a silver-tipped contact screw from which a wire is taken to a second terminal T_2. The electrical circuit is completed through a battery and push switch connected to the terminals T_1 and T_2.

When the switch is pressed current flows through the circuit and the cores become magnetized. The resultant attraction of the armature separates the contacts and breaks the circuit. The magnetism in the cores then disappears and to armature is returned by the spring to its original position. Contact is now remade and the action repeated. Consequently, the armature vibrates and a hammer attached to it strikes the gong.

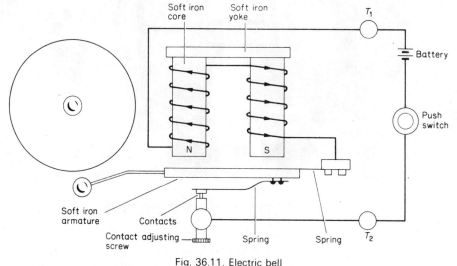

Fig. 36.11. Electric bell

The solenoid switch (magnetic relay)

The current required by the starter motor of a car is very high, and this requires very thick cable and a switch with heavy contacts.

For convenience, a solenoid switch is used, the operation of which will be understood from Fig. 36.12. The starter switch, taking only a small current, is usually incorporated in the key-operated ignition switch.

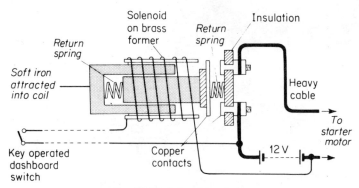

Fig. 36.12. Solenoid switch

Magnetic separators

An electromagnetic separator, used for the purpose of removing *tramp iron* from crushed copper ore, is shown in Fig. 36.13. A miscellaneous collection of iron objects, including nuts and bolts, etc., finds its way into the ore, and must be removed before smelting. This is done by passing the ore on a conveyor belt past a powerful electromagnet. The illustration shows a battery of separators of this type installed at a copper mine.

Fig. 36.13. Battery of electromagnets installed at a copper mine for the purpose of removing tramp iron from the copper ore as it slowly passes the magnets on conveyor belts

The telephone receiver (earpiece)

The telephone receiver contains a U-shaped magnet formed by placing a short permanent bar magnet across the ends of two soft-iron bars (Fig. 36.14). This is placed so that it exerts a pull on a springy magnetic alloy diaphragm. Two solenoids are wound in opposite directions on the soft-iron bars.

When a person speaks into the microphone at the other end of the line a varying electric current is set up having the same frequency as the sound waves. A similar

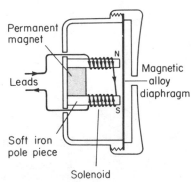

Fig. 36.14. Telephone receiver

electric current is caused to pass through the solenoids in the earpiece. This alters the strength of the magnetic flux in the U-shaped magnet and produces a corresponding variation in the pull of the diaphragm. The latter therefore vibrates and reproduces a copy of the sound waves which entered the microphone.

Resultant magnetic flux in a horizontal plane due to the earth and the current in a long vertical wire

The iron-filings method described on page 420 to investigate the magnetic flux due to a current in a long straight wire gives information only about the strong field very close to the wire. In order to study the field further away from the wire, where its strength is comparable with the earth's weak field, it is necessary to use the more sensitive compass-needle method. Fig. 36.15 shows the pattern of the magnetic flux obtained. East of the wire, the earth's horizontal field and the field due to the current are in opposite directions. A neutral point X is formed where these two fields are equal and opposite.

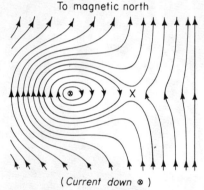

(*Current down* ⊗)

Fig. 36.15. Resultant magnetic flux pattern in horizontal plane due to current in vertical straight wire in earth's magnetic field

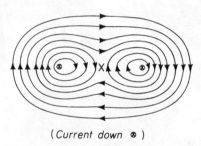

(*Current down* ⊗)

Fig. 36.16 Magnetic flux pattern in plane at right angles to two parallel currents in the same direction (longitudinal tension in the magnetic field lines results in attraction between the wires carrying the currents)

Magnetic flux due to current in two vertical parallel wires

If two parallel wires are arranged side by side vertically and carry currents *in the same direction* the magnetic flux in a horizontal plane close to the wires is as shown in Fig. 36.16. A neutral point is formed between the wires, its exact position depending on the relative strengths of the currents. Note that some of the magnetic flux surrounds both wires.

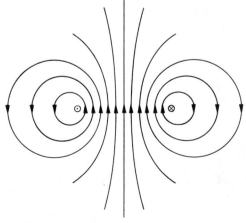

Current up ⊙   Current down ⊗

Fig. 36.17. Magnetic flux pattern in plane at right angles to two parallel currents in opposite directions (lateral repulsion between magnetic field lines results in repulsion between wires carrying the currents)

Michael Faraday visualized a magnetic field as containing magnetic *field lines* which possessed longitudinal tension and lateral (= sideways) repulsion. In this sense they may be compared with the electric field lines described on page 380.

The tension in the lines sets up a force which pulls the two wires inwards. On the other hand, two parallel wires carrying currents in opposite directions repel one another since the magnetic field lines between them are in the same direction and hence show lateral repulsion (Fig. 36.17).

QUESTIONS: 36

1. Draw diagrams, with clear indication of current and flux directions to show the magnetic flux patterns associated with an electric current flowing in:

(*a*) a plane circular coil;

(*b*) a solenoid.

2. Draw a clearly labelled diagram of an electromagnet.

Describe, with the aid of a circuit diagram, how you would investigate the relation between the lifting strength of such a magnet and the current passing. Indicate briefly, preferably on a sketch graph, the results you would expect.

State briefly TWO uses of electromagnets.
(*S.*)

3. Describe an experiment to investigate the magnetic effect of a current flowing in a long straight wire. Draw a diagram showing the nature of the field produced.

Sketch the field due to the same current flowing through two long straight parallel wires:

(*a*) when the directions of the current are the same;

(*b*) when they are opposite. Neglect the earth's magnetic field. (*O.C.*)

4. Draw a diagram of a trembler bell showing the polarity of the electromagnet and the direction of the current. Show on the same or on another diagram how the bell may be rung from two places using only one battery.

A bell takes a current of 0.25 A from a battery of two cells in series. Each cell has an e.m.f. of 1.5 V and an internal resistance of 1 Ω. What is the effective resistance of the bell? What current would the bell take if the cells were arranged in parallel? (*J.M.B.*)

5. Describe how an electric current may be used:

(*a*) to magnetize a steel bar;

(*b*) to demagnetize it.

Draw a diagram of the magnetic field produced by a current flowing in a long straight wire in a plane at right angles to the wire. State a rule which gives the relation between the direction of the current and that of the field.

A long vertical wire carrying a current passes through a horizontal bench. Give a diagram of the resultant magnetic field on the surface of the bench around the wire due to the current and the earth. Mark the positions of any neutral points formed in this field.
(*L.*)

6. Describe and explain the action of a telephone receiver. (*L.*)

7. (*a*) Two long vertical wires carry electric currents:

(i) in the same direction;

(ii) in opposite directions. In each case sketch a plan of the lines of magnetic flux around the wires, ignoring the effect of the Earth's magnetic field. Show on your diagrams the directions of the currents, the flux lines, and the forces acting on the wires.

(*b*) Draw a labelled diagram showing the essential features of an electrical relay and describe how the relay works.

Name one situation in which such a relay could be used and explain why an ordinary type of switch is not suitable. (*J.M.B.*)

8. Sketch the form of magnetic flux pattern due to a current flowing in a solenoid. Indicate on your diagram the direction of the current flow and of the lines of force (field lines) due to the current.

A solenoid has at its centre a small compass needle. The axis of the solenoid is horizontal and at right angles to the magnetic axis of the needle when there is no current flowing.

When a certain current is passed through the coil the needle is deflected through 45°. When the current is reversed the needle moves to a position 45° on the other side of the N–S line.

Explain these observations and, if the strength of the earth's field is 1 unit, deduce the strength of the magnetic field at the centre of the solenoid, due to the current.

Would the deflection

(*a*) increase,

(*b*) decrease,

(*c*) remain the same, if the needle were moved to one end of the solenoid while remaining on the axis? (*O.C.*)

37. Electrolysis

One of the earliest industrial applications of the electric current was the coating of base metals with a layer of a more expensive metal. This is called *electroplating*, and it serves a double purpose. It protects the base metal from atmospheric corrosion and also gives it a more attractive appearance. Silver plating has long been used to enhance the appearance of household cutlery and other articles made from nickel alloys. At one time, nickel plating was very popular, but nowadays it has largely been replaced by chromium and cadmium. Mainly by reason of its freedom from corrosion, gold plating has a wide application for the contacts of electronic components used in computers and elsewhere.

The discovery of electrolysis

As soon as the news of Volta's invention of the simple cell reached London a professor of anatomy named Sir Anthony Carlisle constructed a voltaic pile consisting of silver half-crowns, zinc discs and cardboard soaked in brine. While experimenting with this he decided to place a drop of water on the top disc for the purpose of improving the electrical contact made with the connecting wire. He was surprised to notice that bubbles of gas rose from the wire when it was allowed merely to dip into the water drop without touching the disc. Sir Anthony had a friend named William Nicholson, who had also constructed a voltaic pile, and the two decided to work together in making a further investigation of this new phenomenon, now called *electrolysis*.

In one experiment they connected the pile to two platinum wires dipping into a shallow vessel containing water to which a few drops of sulphuric acid had been added. Clouds of bubbles arose from each wire and were collected in separate bottles. When the gases were tested it was found that the wire connected to the zinc ($-$ve) disc gave off hydrogen, while that connected to the silver disc ($+$ve) gave oxygen.

It was already known that water consisted of hydrogen and oxygen in chemical combination. This had been demonstrated by several chemists, who, at various times, exploded mixtures of hydrogen and oxygen and obtained the formation of a dew which proved to be water. Now, for the first time, Nicholson and Carlisle had shown that water could be decomposed into its elements by an electric current. It was a discovery of first-class importance. The obvious inference was that it might be possible to decompose other substances by this means and hence discover their chemical composition. This was done with striking success a few years later by Sir Humphry Davy. He isolated the metals sodium and potassium by passing an electric current through molten sodium hydroxide and potassium hydroxide respectively.

Terms used in connection with electrolysis

The work of Nicholson, Carlisle, and Davy created the need for several new technical terms, and this task was undertaken by Michael Faraday. Faraday asked the advice of his friend Dr. Whewell, the classical scholar, and the words they coined are still in use today.

The process by which a substance is decomposed by the passage of an electric current is called **electrolysis.** A substance which conducts current and undergoes decomposition is called an **electrolyte.** The same word is also commonly used to refer to a solution of an electrolyte in water when, strictly speaking, the term **electrolytic solution** should be used.

The two wires or plates at which the current enters and leaves the electrolyte are called **electrodes.** The electrode at which the current enters the electrolyte is called the **anode** and that by which it leaves, the **cathode.**

The apparatus consisting of vessel, electrolyte, electrodes, and so on in which the electrolysis is carried out is called an **electrolytic cell** or **voltameter.** This latter term must not be confused with *voltmeter.*

The Hofmann voltameter

A convenient form of apparatus for collecting the gases liberated during the electrolysis of various liquids was designed by the German chemist Hofmann (Fig. 37.1).

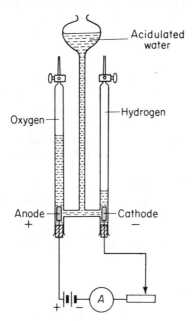

Fig. 37.1. Hofmann's voltameter

It consists of two graduated tubes fitted with taps and provided with platinum electrodes. The tubes are connected near the bottom by a short cross tube which has an upright tube and reservoir.

In order to study the electrolysis of water, the taps are opened, and water to which a few drops of sulphuric acid have been added is poured in until the graduated tubes are both full. The taps are then closed and current is passed through the voltameter from a 6 V battery in series with a rheostat. The current is adjusted to give a steady stream of bubbles from the electrodes. When sufficient gas has collected it will be noticed that the volume of hydrogen from the cathode is approximately twice that of the oxygen from the anode.

On opening the tap above the cathode and applying a light, the issuing gas is seen to burn with a pale-blue colour which is characteristic of hydrogen. Also, droplets of water are formed on the outside of a test-tube of cold water held in the flame. These are the usual tests for hydrogen.

When the tap above the anode is opened a glowing splint bursts into flame if held in the stream of escaping gas. This is a standard test for oxygen.

Ionic theory of electrolysis

The use which Sir Humphry Davy had made of electrolysis as a new method of chemical analysis attracted a great deal of attention in the scientific world, and a number of different ideas were put forward to explain it. The most important suggestion was made by the German physicist Rudolf Clausius, who considered that the current was conveyed through an electrolyte by charged particles which are now called *ions*. This theory was later developed by the Swedish scientist Svante Arrhenius.

Ions have already been mentioned on page 375, where it was explained that an ion is an atom or a group of atoms possessing either a positive or a negative electric charge.

We shall now explain what is believed to occur when an electrolyte dissolves in water. Take copper sulphate as an example. This is a substance consisting of equal numbers of copper ions, Cu^{2+}, and sulphate ions, SO_4^{2-}, arranged in a definite pattern so that the whole is electrically neutral. A copper ion is a copper atom which has lost two electrons, and hence has a net positive charge. A sulphate ion is an SO_4 group which has acquired the two electrons lost by the copper atom and is therefore negatively charged. When copper sulphate is dissolved in water the ions dissociate or separate and are able to move freely in the electrolytic solution; in fact, the word *ion* means "wanderer".

Other substances ionize in a similar manner. Sulphuric acid, H_2SO_4, dissociates into two hydrogen ions, H^+ (H minus one electron), and one SO_4^{2-} ion (SO_4 plus two electrons). Silver nitrate, $AgNO_3$, consists of positive silver ions, Ag^+, and negative nitrate ions, NO_3^-, and these separate when in solution. Water ionises only very slightly into hydrogen ions, H^+, and hydroxyl ions, OH^-. Generally speaking, all acids, bases and salts are ionized in solution, but not organic compounds such as sugar, alcohol, and naphthalene.

When two electrodes are dipped into an electrolyte and connected to a battery positive ions (**cations**) are attracted towards the negative cathode and negative ions (**anions**) towards the positive anode. This migration of ions in both directions is the process by which an electric current flows through the electrolyte.

In the following sections we shall describe the action which occurs at the electrodes when a current is passed through various electrolytes.

Electrolysis of acidified water

Earlier in this chapter we described the electrolysis of acidified water in a Hofmann voltameter. In this case, only the water is decomposed. The small quantity of acid added simply provides a vast number of extra ions to make the solution more highly conducting.

Water acidified with sulphuric acid contains hydrogen ions, H^+, and hydroxyl ions, OH^-, produced by dissociation of the water, together with hydrogen ions, H^+, and sulphate ions, SO_4^{2-}, from the sulphuric acid.

When the current is passing H^+ ions migrate towards the cathode. Here they receive electrons from it and become neutral hydrogen atoms. The H atoms then combine in pairs to form molecules of hydrogen, H_2, which are liberated in the form of gas bubbles. An ion which has received an electron from the electrode in this manner is said to be *discharged*.

The action at the cathode may be summarized by the following equations, in which e represents an electron:

$$H^+ + e = H$$
$$H + H = H_2$$

While the H^+ ions are moving towards the cathode, the OH^- and SO_4^{2-} ions move towards the anode. Here only the OH^- ions lose electrons to the anode and are discharged. The SO_4^{2-} ions remain in solution. This is called the *preferential discharge* of ions. When thus liberated, pairs of OH groups combine to form water, H_2O, and oxygen atoms, O. Finally, the oxygen atoms combine in pairs to form oxygen molecules, O_2, which are set free in the form of gas bubbles.

The action at the anode may therefore be symbolized as follows:

$$OH^- - e = OH$$
$$OH + OH = H_2O + O$$
$$O + O = O_2$$

Electrolysis of copper sulphate solution using platinum or carbon electrodes

Copper sulphate solution contains Cu^{2+} ions and SO_4^{2-} ions together with H^+ and OH^- ions from the water.

During electrolysis, Cu^{2+} and H^+ ions migrate to the cathode, but only the Cu^{2+} ions are discharged. The H^+ ions remain in solution. As each Cu^{2+} ion is discharged it becomes a neutral Cu atom and is deposited on the cathode. After a time, therefore, the cathode becomes covered with a reddish layer of pure copper.

At the anode the action is identical with that which occurs in the case of acidified water described in the previous section. OH^- ions are discharged and combine in pairs to give water and oxygen gas.

If the current is passed through the electrolyte for a sufficiently long time all the copper ions are removed from the solution and deposited on the cathode. The electrolyte therefore loses its blue colour and becomes dilute sulphuric acid. When this stage has been reached the action becomes simply the electrolysis of dilute sulphuric acid. Hydrogen is liberated at the cathode and oxygen at the anode.

Electrolysis of copper sulphate solution using copper electrodes

When copper electrodes are used the action at the cathode is exactly the same as with platinum or carbon electrodes. Cu^{2+} ions are discharged and deposited on the cathode.

At the anode, however, copper ions go into solution in preference to the discharge of either OH^- or SO_4^{2-} ions.

As the electrolysis continues, the cathode increases in thickness while the anode slowly dissolves away. The concentration of copper sulphate in the electrolyte remains constant, and the action ceases when the anode is completely dissolved away.

Applications of electrolysis

Some of the commercial applications of electrolysis were mentioned at the beginning of this chapter. Industrially, electrolysis also plays an important part in the refining of copper (Fig. 37.2). Large plates of crude copper enclosed in canvas sacks are made to form the anodes in vats containing copper sulphate solution. The cathodes are thin sheets of pure copper (Fig. 37.3). When current is passed through a cell, copper is dissolved out of the anode and deposited on the cathode. Impurities in the anode are left behind in the canvas sack and are later removed and disposed of. Copper purified in this way is called electrolytic copper. By reason of its exceptional purity it has a very low electric resistance, and this makes it suitable for the manufacture of electric cables.

Before it was discovered that aluminium could be extracted from its ore, bauxite, by an electrolytic process this highly useful metal was little more than an expensive chemical curiosity. From bauxite an oxide of aluminium called alumina is extracted (Al_2O_3). This does not dissolve in water, but dissociates when dissolved in the

Fig. 37.2. An electrolytic copper refining plant

Fig. 37.3. Loading up with copper starting-sheets in the electrolytic refining cells

molten mineral cryolite (Na_3AlF_6). The process is carried out in specially designed electrolytic furnaces (Fig. 37.4). These furnaces have a bed made of carbon, which acts as the cathode. The anodes are thick carbon rods which dip into a mixture of alumina and cryolite. A very high current is passed which causes the mixture to melt. Electrolysis then occurs, with the result that aluminium is liberated at the cathode and forms a molten layer on the bed of the furnace. From time to time the aluminium is run off and fresh alumina and cryolite added.

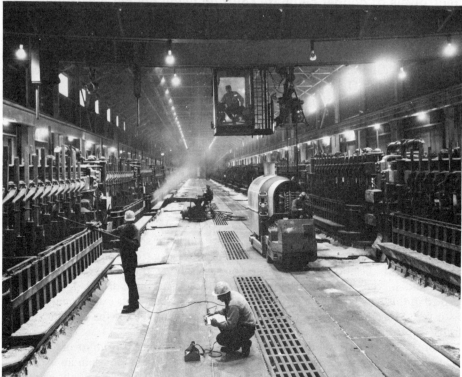

Fig. 37.4. Potline of electrolytic reduction cells for the production of pure aluminium

QUESTIONS 37

1. Explain the terms *electrolyte*, *anode*, *cathode*, *ion*, *anion*, *cation*.

Describe and explain what happens when an electric current is passed through dilute sulphuric acid and using platinum electrodes.

2. Use the ionic theory of electrolysis to explain what happens when an electric current passes through a copper(II) sulphate solution using:

(*a*) carbon electrodes,

(*b*) copper electrodes.

38. The principle of the electric motor

After Oersted's discovery of the magnetic effect of an electric current the possibility of producing an electric motor occurred to a number of scientists. In particular, William Wollaston and Michael Faraday discussed the problem together at the Royal Institution in London. Wollaston unsuccessfully tried some experiments and then lost interest. Faraday, however, continued his investigations during 1821, and by the end of the year was able to show how continuous rotation could be obtained with a magnet and an electric current.

Faraday's rotation experiment

A modern reproduction of Faraday's final experiment is shown in Fig. 38.1. A thin cylindrical magnet is placed in a glass tube *A*, containing mercury and held down at

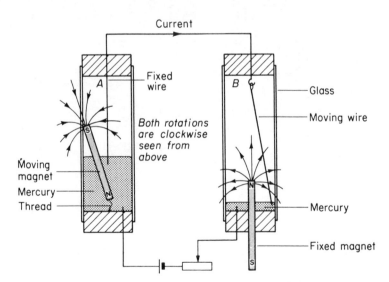

Fig. 38.1. Faraday's magnet and current rotations

its lower end by a short thread fixed by sealing-wax. A straight copper wire dips into the mercury and passes up the centre of the tube and through a cork. Here it is bent over at right angles and continued through a cork at the top of a second tube *B*. The end of the wire is bent into the form of a hook, and from it is suspended a second vertical wire which dips into a shallow pool of mercury. Tube *B* also contains a fixed cylindrical magnet passing through the cork at the bottom and projecting above the surface of the mercury. Wires are passed through the lower corks on both sides to make contact with the mercury, and a circuit is completed by connecting these wires to a battery and rheostat.

When current flows through the wires the magnet in *A* revolves round the wire and the suspended wire in *B* revolves round the magnet. Faraday recognized that these experiments illustrated that to every action there is an equal and opposite reaction (*Newton's Third Law of Motion*, page 44). Since a current exerts a force on a magnet, it follows that the magnet exerts an equal and opposite force on the current. Either the magnet or the current will move, depending on which of the two is fixed.

The kicking wire experiment. Force on a conductor in a magnetic field

The experiment illustrated in Fig. 38.2 has been designed to show how the direction of the force on a wire carrying a current is related to the direction of the magnetic field in which the wire is situated.

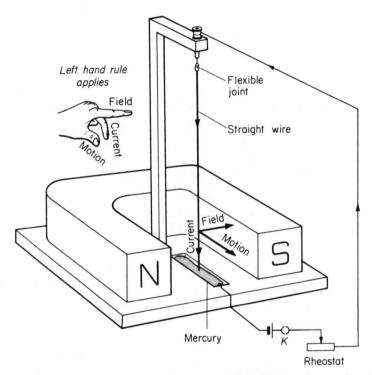

Fig. 38.2. Force on a conductor in a magnetic field (kicking wire experiment)

A straight wire with a flexible connection at its upper end hangs between the poles of a U-shaped magnet and dips into a small pool of mercury in a depression cut in a piece of wood. A circuit is provided so that the current may be adjusted to the minimum value required to make the experiment work. Too large a current results in over-heating of the wires.

On closing the key *K*, the current flows downwards and the hanging wire swings forwards. This causes it to leave the mercury and break the circuit. The wire falls back, remakes contact with the mercury, and the action is repeated. If the battery connections are reversed so that the current flows up the wire the direction of the force on it will be reversed, and it will now swing backwards out of the mercury. The direction of the force on the wire may likewise be reversed by turning the magnet over so that the direction of the magnetic field is reversed.

It will be noticed in this experiment that, initially, *the current, the magnetic field, and the direction of the force on the wire are all three mutually at right angles.* Professor J. A. Fleming gave a rule for relating the direction of motion of the wire to the directions of the current and magnetic field.

Fleming's left-hand rule (motor rule)

Place the forefinger, second finger, and thumb of the left hand mutually at right angles. Then, if the Forefinger points in the direction of the Field and the seCond finger in the direction of the Current, the thuMb will point in the direction of the Motion (Fig. 38.3).

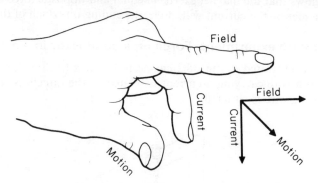

Fig. 38.3. Fleming's left-hand rule (motor rule)

Force on a conductor explained by properties of magnetic field lines

Fleming's left-hand rule merely tells us the direction in which a current carrying wire moves in a magnetic field: *it does not explain* the motion. An explanation can, however, be given which is based on the properties of magnetic field lines. Faraday's concept of magnetic field lines was mentioned on page 427. Properties ascribed to them are:

1. They tend to contract (longitudinal tension).
2. They tend to repel one another sideways (lateral repulsion).

If we take a horizontal section through the vertical wire of Fig. 38.2 where it passes through the magnetic field, and there plot the magnetic field, the pattern shown in Fig. 38.4 (c) is obtained. This is the resultant force obtained when the parallel lines

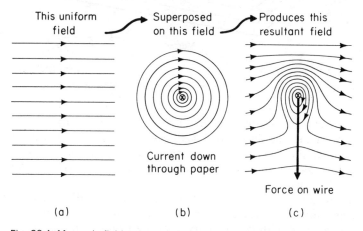

Fig. 38.4. Magnetic field pattern due to current in a straight wire at right angles to a uniform field

between the poles of the magnet are combined with the concentric circular lines due to the current in the wire. It will be noticed that the lines are closer together on one side of the wire than on the other, and thus they will exert a resultant sideways or *lateral* push on the wire. Also, the tendency of the lines to shorten, and hence to straighten out, will likewise cause them to exert a force on the wire. In a sense we may liken the action to that of an elastic catapult.

Barlow's wheel

Peter Barlow utilized the force on a conductor in a magnetic field in order to produce continuous rotation. His apparatus is illustrated in Fig. 38.5. It consists of a star-shaped wheel made of copper which is able to rotate in a vertical plane with its points dipping into a pool of mercury. A wire is taken from the mercury to the terminal of a cell and a circuit completed from the other terminal of the cell, through a switch and rheostat and thence to the axle of the wheel. A magnetic field is set up at right angles to the wheel spokes by two bar magnets placed in line, one on either side of the wheel, and with their opposite poles adjacent.

When the switch is closed current flows down the spoke dipping into the mercury, and hence a force is exerted on it causing it to move forward. The next spoke then dips into the mercury, and the same thing happens. This process is repeated, with the result that the wheel rotates continuously.

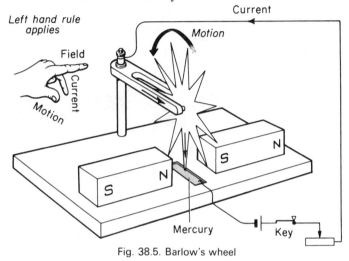

Fig. 38.5. Barlow's wheel

Use of mercury in the laboratory

In the experiment just described the sparking which occurs when the spokes of the wheel leave the mercury hastens the production of mercury vapour, which is poisonous. So, the experiment should be done in a fume chamber or under an extraction hood. In all experiments involving the use of mercury care should be taken to avoid spillage. Where possible, experiments should be done on trays, so that if any mercury is spilt it may be more easily recovered.

The simple electric motor

Fig. 38.6 illustrates the construction of a simple direct current (d.c.) electric motor. It consists of a rectangular coil of wire mounted on a spindle so that it can rotate between the curved pole pieces of a U-shaped permanent magnet. The two ends of the coil are soldered respectively to the two halves of a copper split ring or *commutator*.

Two carbon *brushes* are caused to press lightly against the commutator by means of springs, and when these are connected in circuit with a battery and rheostat the coil rotates.

Suppose the coil is in the horizontal position when the current is first switched on. Current will flow through the coil in the direction shown, and by applying Fleming's left-hand rule it will be seen that the side *ab* of the coil experiences an upward force and the side *cd* a downward force. These two forces form a *couple* which causes the coil to rotate in a clockwise direction until it reaches the vertical position. In this position the brushes touch the space between the two halves of the commutator and the current is cut off. The coil does not come to rest, since its momentum carries it

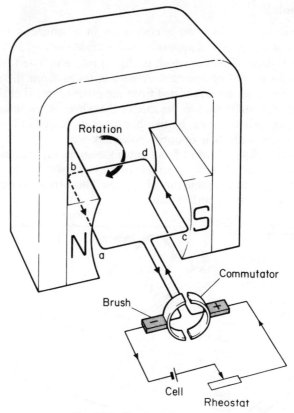

Fig. 38.6. Simple electric motor

past the vertical, and when this has occurred the two commutator halves automatically change contact from one brush to the other. This reverses the current through the coil, and consequently also reverses the directions of the forces on the sides of the coil. Side *ab* is now on the right-hand side with a downward force on it, and side *cd* on the left-hand side with an upward force. The coil thus continues to rotate in a clockwise direction for so long as the current is passing.

The magnetic field pattern in the simple electric motor is shown in Fig. 38.7. The tendency of the lines to shorten and also to repel one another sideways gives rise to equal and opposite forces on the two sides of the coil thus producing a turning couple.

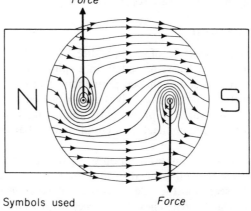

Symbols used

⊙ Point of arrow (current towards observer)

⊗ Tail of arrow (current away from observer)

Fig. 38.7. Magnetic field pattern in a simple electric motor

Practical d.c. electric motors

The simple electric motor as described above is not very powerful or efficient, but it can be improved by increasing the number of turns in the rotating coil and also winding them on a soft-iron armature. If there are n turns of wire instead of one turn the force on the side of the coil will be n times as great. In addition, the iron armature becomes magnetized and greatly increases the power of the motor by adding its magnetic flux to that of the coil. This method of construction is often found in toy electric motors.

In commercial practice, electric motors have cylindrical armatures built up from soft-iron discs. These are slotted and a number of coils are wound in the slots. Each coil has its own pair of segments on a multi-segment commutator (Fig. 38.8). This

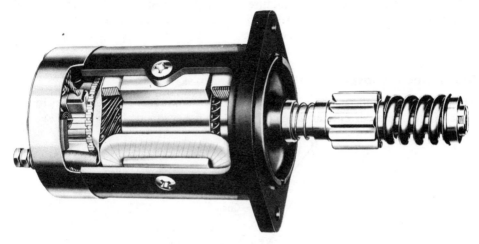

Fig. 38.8. *A car starter-motor*. This cut-away picture clearly shows the multi-segment commutator on the left. Notice also that the field coils are wound from copper strip. These motors necessarily take a very large current, as they work at low voltage and have to produce a big starting torque

design gives increased power coupled with smooth running. Usually, the magnetic field in which the armature runs is produced by an electromagnet, and the coils wound round this are called the *field* coils. Sometimes the field coils are connected in series with the armature coil (series wound) and sometimes in parallel with it (shunt wound).

The particular method of winding used depends on the use to which the motor is put. Series-wound motors have a high torque or turning force at low speeds, and are therefore used where heavy loads have to be moved from rest, as in the case of electric trains. On the other hand, shunt-wound motors have the advantage of running at constant speed under varying loads. This characteristic renders them very suitable for driving power tools, such as lathes and drills, where steady speeds are desirable.

Efficiency of an electric motor

If a practical electric motor is regarded as a machine for transferring electric energy to mechanical energy its efficiency will depend on its design. In this connection, the factors leading to energy losses are the same as those involved in the design of a transformer described in chapter 43. The only additional energy loss in a motor comes from the work done against friction in the bearings and commutator.

Back e.m.f. in an electric motor

When an electric motor is running it also acts as a *dynamo* (page 496) and so sets up a *back e.m.f.* E_b in opposition to the e.m.f. E applied to the motor to drive it.

The resultant current I, in amperes, flowing through the motor windings of resistance R, in ohms, is therefore given by

$$I = \left(\frac{E - E_b}{R}\right) \quad \left(\text{instead of } \frac{E}{R}\right)$$

The work done by the current against the back e.m.f. becomes transferred to useful work done by the motor.

The work done by the resultant terminal p.d. in driving the current through the motor is simply wasted in the various energy losses mentioned in the previous section. Hence, for maximum useful power output coupled with minimum ultimate wastage in the form of heat, E_b should be as nearly equal to E as possible.

If, for any reason, the motor armature is brought to rest while the current is still turned on, the back e.m.f. becomes zero and all the energy of the current is now converted into internal energy which raises the temperature of the windings. This is to be avoided as it might easily cause the motor to burn out.

Moving-coil loudspeaker

The moving-coil loudspeaker used in radio receivers and record players works by the force exerted on a current-carrying coil situated in a magnetic field (Fig. 38.9).

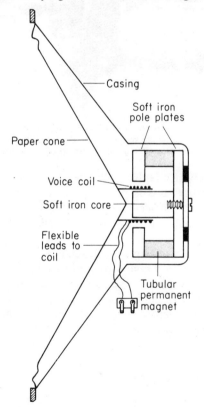

Fig. 38.9. Moving-coil loudspeaker

Varying electric currents which correspond with the sound to be reproduced are passed through a short cylindrical *voice coil* which is free to move in the radial magnetic field set up by a *pot magnet*. This magnet is constructed by fitting a soft-iron core and pole plates to a powerful tubular permanent magnet, as shown. Its radial magnetic field cuts the turns of the coil at right angles, and consequently, as the current through the coil varies, it will move to and fro in accordance with Fleming's left-hand rule.

The coil is attached to a cone made of specially treated paper or other material which moves with the coil and sets the surrounding air in vibration.

QUESTIONS: 38

1. Give a diagram showing the lines of magnetic force due to a current in a straight conductor, in a plane perpendicular to the conductor.

Describe an experiment to show that a current-carrying conductor which is perpendicular to a magnetic field experiences a mechanical force. Show clearly the directions of the current, the magnetic field and the force.

Describe a simple electric motor and explain why a commutator is necessary to obtain continuous rotation. (*L.*)

2. Draw a diagram of a simple d.c. electric motor. Mark in clearly the direction of the supply current and the direction of rotation of the armature. Explain the action of the motor.

Explain what is meant by back e.m.f. in a motor.

Give FOUR reasons why the efficiency of an electric motor is always less than 100 per cent. (*A.E.B.*)

3. Fig. 38.10 represents the section of a wire between the opposite poles of two bar mag-

Fig. 38.10.

nets. Draw the lines of force between the poles when a current flows down the wire (into the paper). (*J.M.B.*)

4. Draw a labelled diagram of a moving-coil loudspeaker and explain in detail how it works. (*A.E.B.*)

5. A 240 V vacuum-cleaner motor takes a current of 0.6 A. What is its efficiency if the useful mechanical power output is 72 W? How is most of the energy being wasted?

39. Galvanometers, ammeters and voltmeters

The simple galvanometer or current-indicating instrument designed by Schweigger has already been described on page 419. Instruments of the moving-magnet type with various modifications to make them more sensitive were in general use during the nineteenth century. They did, however, have certain disadvantages. Not only were they affected by stray magnetic fields set up by near-by electric motors and power cables, but also they always had to be set up in one particular direction owing to the fact that the earth's field acted as the control field. For these reasons they became obsolete and have been replaced by moving-coil galvanometers.

The idea of making a galvanometer in which the magnet was fixed and the coil moved was first put into practice by William Sturgeon in 1836. Later the same principle was adopted by Lord Kelvin as a means of detecting small electric currents sent through the first submarine telegraph cables.

The moving-coil galvanometer

The construction of a moving-coil galvanometer, calibrated for use as a milliammeter, is shown in Fig. 39.1.

A rectangular coil carrying a pointer is pivoted on jewelled bearings and free to move in the annular (= ring-like) space between a soft-iron cylinder and the cylindrical pole faces of a strong magnet.

Current is led into and out of the coil by two phosphor-bronze hairsprings wound in opposite directions to compensate for thermal expansion. These hairsprings also provide the control couple.

When a current flows in the coil the resultant magnetic field in the annular space is as shown in Fig. 39.2. This sets up forces on the sides of the coil as already explained in the case of a simple electric motor (page 438). Consequently, in accordance with Fleming's left-hand rule, equal and opposite parallel forces act respectively on the two vertical sides of the coil. **These two forces together form a deflecting couple which causes the coil to rotate until the deflecting couple is just balanced by the control couple set up by the hairsprings.**

The function of the soft-iron cylinder is to concentrate the magnetic flux radially in the annular space. Thus for all positions of the coil the magnetic flux density is constant and in the plane of the coil, and hence the force on the sides of the coil will be proportional to the current. This, coupled with the fact that the controlling hairsprings obey Hooke's law (page 145) ensures that the deflection will be proportional to the current, and so gives the instrument a linear or evenly divided scale.

The coil itself is wound on a light aluminium former. The function of this is to damp the movement of the coil and so make it *dead beat*. That is to say, the pointer swings out to a deflection and comes to rest immediately instead of oscillating to and fro. This is a case of *eddy current* damping and will be understood after reading chapter 43. Briefly, as soon as the coil starts to move, the aluminium former cuts magnetic flux. This sets up an induced current which opposes its motion.

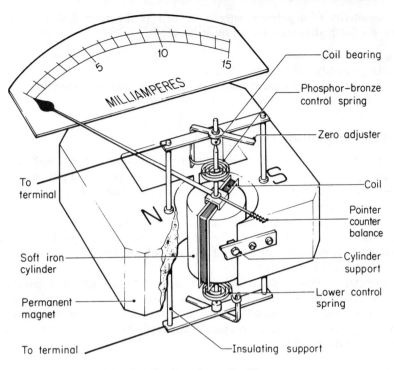

Fig. 39.1. Details of moving-coil milliammeter

Pivoted moving-coil instruments are made in various ranges from those with a full-scale deflection (f.s.d) of a few microamperes up to 25 or more milliamperes. Some have a centre zero or offset zero for use in *null-deflection* experiments with Wheatstone bridges and potentiometers (chapter 40).

Sensitivity of a moving-coil galvanometer

The deflecting couple on the coil is equal to the force on a vertical side multiplied by the width of the coil (see page 64). For a given current, the force on a vertical side will be proportional to its length. Hence the deflecting couple will be proportional to (length × width) which is equal to the area of the coil.

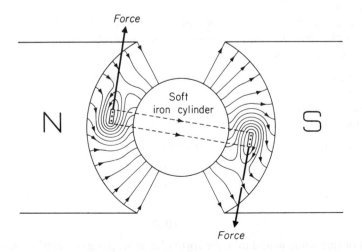

Fig. 39.2. Deflecting couple on coil of moving-coil galvanometer

The sensitivity of a galvanometer is defined as the scale deflection per micro-ampere. So, for high sensitivity we must have:

(1) a coil of large area;
(2) a large number of turns in the coil;
(3) a special alloy permanent magnet which gives high magnetic flux;
(4) weak hair springs to give a small control couple.

Ammeters and voltmeters

Ammeters and voltmeters are electric measuring instruments fitted with scales which have been graduated respectively to read current in amperes and potential difference in volts. The correct method of connecting an ammeter and a voltmeter in a circuit is shown on page 409.

An ammeter *is always placed in series* with the resistance or other circuit components through which the current is to be measured. Ammeters should therefore have a *low resistance* compared with that of the rest of the circuit, so that they do not introduce unwanted resistance.

A voltmeter *is always placed in parallel* with the resistance or apparatus across which the potential difference has to be measured. Voltmeters ought therefore to have a *high resistance* compared with the resistance across which the voltage is to be measured, so that they take a comparatively negligible current, and so disturb the circuit as little as possible.

Construction of an ammeter. Use of shunts

Moving-coil instruments are not made to take currents of more than a few milli-amperes. If designed for larger currents the coil would have to be wound with much thicker wire, and the resultant instrument would be both clumsy and expensive. Fortunately there is an easy way out of the difficulty. When currents of several amperes have to be measured a low resistance which by-passes the greater part of the current is placed in parallel with a milliammeter or a microammeter, and then only a small known fraction of the total current passes through the meter itself. A resistance used in this way is called a *shunt*.

Suppose, for example, a milliammeter of resistance 5 Ω and full-scale deflection 15 mA is to be used for the purpose of measuring currents to 1.5 A. It would be necessary to use a shunt which passed $(1.5 - 0.015) = 1.485$ A while the meter carried only 0.015 A (Fig. 39.3). We shall now show how to calculate the value of the shunt required.

Let the resistance of the shunt be R. Since the shunt and the meter are in parallel, *there will be the same potential difference V across each.* We can use this fact to obtain two equations from which R may be calculated.

Remembering that we may write:

$$\text{p.d.} = \text{current} \times \text{resistance}$$

for the milliammeter
$$V = 0.015 \times 5$$
and for the shunt,
$$V = 1.485 \times R$$

equating the right-hand side of these equations,

$$1.485 \times R = 0.015 \times 5$$

whence
$$R = \frac{0.015 \times 5}{1.485}$$

$$= 0.0505 \ \Omega$$

A shunt of this value would take the form of a short piece of fairly thick manganin wire or strip carefully adjusted to have the exact resistance required. Manganin is chosen for shunts, since its resistance does not alter when its temperature is raised by the passage of current.

When the shunt is in use the scale readings of the instrument are divided by 10 and called amperes instead of milliamperes.

An ordinary ammeter with a range of 1.5 A will, of course, have its shunt connected up inside the case of the instrument.

Milliammeter used as a millivoltmeter

The milliammeter described on the preceding page has a full-scale deflection of 15 mA and a resistance of 5 Ω. It follows that the potential difference across it when it is giving a full-scale reading is given by

$$\text{p.d.} = \text{current} \times \text{resistance}$$
$$= 0.015 \times 5$$
$$= 0.075 \text{ V} \quad \text{or} \quad 75 \text{ millivolts (mV)}$$

This milliammeter can therefore be used as a millivoltmeter provided scale readings are multiplied by 5 and called millivolts. Sometimes two scales are engraved on the same instrument, one in milliamperes and the other in millivolts.

Use of a multiplier to convert a milliammeter into a voltmeter

We have just seen that, provided we know its resistance, a milliammeter may be used as a millivoltmeter. Larger potential differences may be measured by placing a high resistance or *multiplier* in series with the milliammeter (Fig. 39.4).

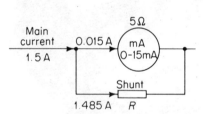

Fig. 39.3. A shunt increases the range of a moving-coil milliammeter

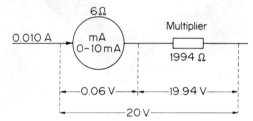

Fig. 39.4. a multiplier (series resistor) enables a milliammeter to be used as a voltmeter

Suppose we wish to measure potential differences up to 20 V using a milliammeter of resistance 6 Ω and full-scale deflection of 10 mA.

At full-scale deflection the p.d. across the milliammeter is given by

$$\text{p.d.} = \text{current} \times \text{resistance}$$
$$= 0.010 \times 6$$
$$= 0.06 \text{ V}$$

Since the total p.d. across the instrument and multiplier together has to be 20 V, it follows that the p.d. across the multiplier alone is 20 − 0.06 = 19.94 V.

Let the resistance of the multiplier be *R*. The same current, 0.01 A, passes through both instrument and multiplier. Hence

$$R = \frac{\text{p.d.}}{\text{current}}$$

$$= \frac{19.94}{0.01}$$

$$= 1994 \text{ Ω}$$

When used with a multiplier of this value the readings of the milliammeter are multiplied by 2 and called volts.

Unlike the shunt used in an ammeter, the multiplier used in a voltmeter has a high resistance, and is therefore made of a fairly long length of double silk-covered manganin wire wound on a bobbin.

The Multimeter

Fig. 39.5 shows a galvanometer known as a *universal indicator*. Basically, it is a microammeter with a range of 0–100 μA. Its resistance in 1000 Ω, so it can also be used as a voltmeter with a full-scale deflection of

$$E = IR$$
$$= (100 \times 10^{-6}) \times 1000 \text{ V}$$
$$= 100 \times 10^{-3} \text{ V}$$
$$= 100 \text{ mV}$$

In the picture, however, it is shown fitted with a multiplier which extends its range from 0 to 25 V. Multipliers such as this are provided with a series of ranges up to 500 V,

Fig. 39.5. *Offset zero universal indicator* showing 25 V multiplier attached to provide the necessary series resistance. An appropriate transparent scale (see left) is placed over the indicator window according to the range of current or voltage in use

and they have plugs which fit into sockets on the instrument's own terminals. The instrument scale is not numbered, but each multiplier has an appropriately numbered transparent slide (shown in Fig. 39.5) which fits into the window recess, thus removing the necessity to multiply scale readings by a factor such as we described in the previous section.

For the measurement of current a similar set of plug-in shunts are available in ranges up to 10 A.

The **Avometer** shown in Fig. 39.6 is a multimeter with built-in shunts and multipliers, and the desired range is obtained by the rotary switches seen on the front of the meter. The instrument shown in the picture is specially designed for use with a.c. (alternating current) as well as d.c. (direct current).

Moving-iron instruments

Apart from the moving-coil instruments, there is another important class of meters known as moving-iron instruments. These are of two types, one based on repulsion and the other on attraction.

Fig. 39.6. The Avometer

The *repulsion moving-iron instrument* consists of a coil inside which are two soft-iron bars, one fixed and the other attached to a pivoted pointer. A hairspring is attached to the pointer spindle to provide the control couple (Fig. 39.7 (*a*)).

When current is passed through the coil both soft-iron bars become magnetized with like poles adjacent. Repulsion therefore takes place and the pointer deflects until the turning moment due to the repulsion is just balanced by the opposing torsional couple in the spring.

With an instrument of this type the pointer oscillates for some time before settling.

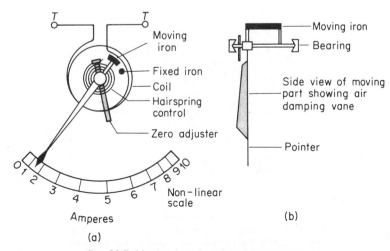

Fig. 39.7. Moving-iron (repulsion) instrument

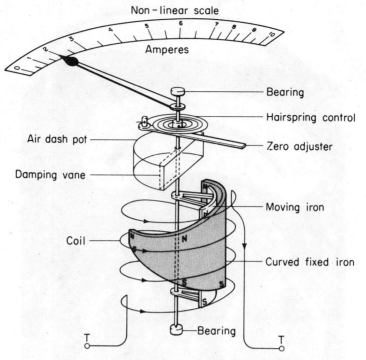

Fig. 39.8. Moving-iron ammeter (improved pattern)

down to a steady deflection, unless steps are taken to prevent it. One method is to fix a light aluminium vane to the moving part. Air resistance to this slows down the oscillation (Fig. 39.7 (*b*)). The damping effect is made more efficient if the vane is able to move inside a small box or *air dash-pot* (Fig. 39.8).

One disadvantage of the moving-iron instrument is that its scale is not linear or evenly divided. This arises from the fact that the force of repulsion between the irons is approximately proportional to the square of the current. The graduations tend to be crowded together at the beginning of the scale, and therefore low readings are apt

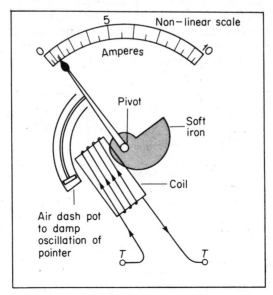

Fig. 39.9. Moving-iron (attraction) instrument. (The control couple is obtained from a hairspring attached to the pointer—not shown in the diagram)

to be unreliable. Nevertheless, by making the fixed iron curved and tapering as shown in Fig. 39.8 the scale is considerably improved.

In the *attraction moving iron* instrument a specially shaped piece of soft iron with a pointer attached is pivoted near to the open end of a short solenoid (Fig. 39.9).

When current is passed through the solenoid it becomes magnetized and the soft iron is drawn into it with a force which depends approximately on the square of the current. The control couple is provided by a hairspring. The scale is unevenly divided like that of the repulsion instrument.

Both types of moving-iron instrument can by the addition of suitable shunts or multipliers be used either as ammeters or voltmeters.

Use of moving-iron instruments for measuring a.c.

The main advantage of moving-iron instruments is that they can be used with alternating current (a.c.) as well as direct current (d.c.). Direct current flows in one direction only, while alternating current flows backwards and forwards many times a second. For example, a.c. from the electric supply mains has a frequency of 50 Hz, i.e., it oscillates at 50 cycles per second.

When a.c. is passed through a moving-iron repulsion instrument both irons will have their polarity reversed when the current reverses. They therefore continue to repel one another, and so a deflection is obtained whatever the direction of the current. On the other hand if a.c. were passed through a moving-coil instrument the direction of the couple on the coil would change 50 times a second. The average couple is therefore zero and no deflection occurs.

QUESTIONS: 39

1. A moving-coil galvanometer has a resistance of 40 Ω and gives a full-scale deflection of 2 mA.

(*a*) What is the potential difference across its terminals when this current is flowing?

(*b*) How can the galvanometer be converted into a voltmeter?   (*J.M.B.*)

2. A student is required to measure currents up to 1.00 A. She is given a reel of wire of resistance 0.5 Ω/m and a moving coil galvanometer of resistance 10 Ω which has a full scale deflection for a current of 0.01 A. Explain how the student could adapt the galvanometer in order to measure the required range of currents.   (*A.E.B.*)

3. Describe a moving-coil galvanometer and explain the principle on which it works. What features of the design will make such a galvanometer sensitive?

A galvanometer of resistance 20 Ω is to be provided with a shunt such that only one-tenth of the whole current in a circuit shall pass through the galvanometer. What is the resistance of the shunt?   (*W.*)

4. An ammeter gives its full-scale reading for a current of 0.1 A and its resistance is 0.5 Ω. Explain how you would adapt it:

(*a*) to give a full-scale of 2 A;

(*b*) for use as a voltmeter to read up to 100 V.   (*O.C.*)

5. Describe the construction of a moving-coil galvanometer. Explain the principles of its action.

Explain how a sensitive moving-coil galvanometer may be converted to serve:

(*a*) as an ammeter, and

(*b*) as a voltmeter.

A moving-coil meter which gives full-scale deflection with 0.005 A is converted to a voltmeter reading up to 5 V, using an external 975 Ω resistance. What is the resistance of the meter?   (*O.*)

6. A moving-coil galvanometer (milliammeter) has a resistance of 5 ohms and will give a full-scale deflection when a current of 0.015 A flows through it.

Calculate

(i) the potential difference across the meter when a current of 0.015 A flows through it,

(ii) the value of the resistance which would convert the meter into an ammeter reading up to 3 A, showing how the resistance would be connected to the galvanometer,

(iii) the value of the resistance which would convert the meter into a voltmeter reading up to 15 V, showing how the resistance would be connected to the galvanometer.   (*J.M.B.*)

7. A steady current in a circuit is read by an ammeter. A piece of wire of resistance 1.2 Ω connected across the terminals of the ammeter causes the reading to change from 2.0

to 1.5 A. Calculate the resistance of the ammeter pointing out any assumption you need to get an answer. (*O.C.*)

8. Describe a moving-iron ammeter and explain its action, pointing out why iron is preferred to steel. Explain why this instrument can be used to measure both alternating and direct current. (*L.*)

9. Draw and label a diagram to show the essential features of a moving-coil galvanometer. Explain the features which are responsible for:

   (i) the proportionality between deflection and current;

   (ii) producing a high deflection from a small current.

   Briefly explain why such an instrument does not show a deflection when connected to a suitable circuit in which alternating current from a mains supply is flowing.

   A moving-coil milliammeter gives its full-scale deflection for a current of 1 mA. The potential difference between its terminals is then 100 mV. What resistor placed in series with the meter will limit the current flow to 1 mA when 10 V is applied across the combination? (*W.*)

10. A galvanometer of internal resistance 100 Ω gives a full-scale deflection for a current of 10 mA. Calculate the values of the resistances necessary to convert the galvanometer to:

   (i) a voltmeter reading to 5 V,

   (ii) an ammeter reading to 10 A.
   (*W.A.E.C.*)

11. Give the meaning of the term *couple*. Explain the way in which a current in a moving-coil ammeter can cause a deflecting couple to act on the coil. What gives rise to a couple which opposes the deflecting couple and results in a steady deflection for a steady current?

   Give TWO reasons why in a good-quality moving-coil ammeter the deflection is proportional to current.

   Describe how you think the needle of a centre-zero ammeter would behave if the direction through the ammeter of a current giving, say, a half-scale deflection were reversed at a gradually increasing rate starting at 1 Hz and finally reaching 100 Hz. (*O.C.*)

40. Measurement of resistance and potential difference

For laboratory use, standard resistors are obtainable either singly or in boxes giving various ranges. A good standard resistor should not alter in value as its gets older, and should not be affected by changes in temperature. The wires used for standard resistors are therefore made from special alloys, the two generally used being *manganin* and *constantan* (see page 406). Both of these increase in resistance only by about one-hundred-thousandth part per kelvin rise in temperature. In particular, the resistance of manganin is not affected by age.

In order to avoid temperature rise, only comparatively small currents should be passed through resistance boxes. In this respect they differ from rheostats (page 406), which are used for the purpose of controlling fairly large currents in a circuit. Rheostats are designed to run fairly warm, and hence are wound on an open former so that air cooling can take place.

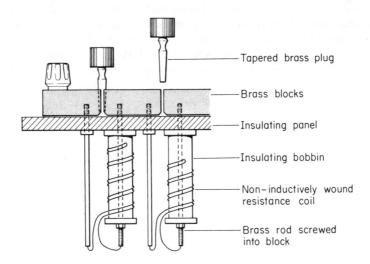

Fig. 40.1. Construction of standard resistance box

Fig. 40.1 shows the details of construction of a plug resistance box. The ends of the resistance coils are connected respectively to adjacent pairs of brass blocks, which may be shorted by inserting a brass plug. When the box is in use any particular resistance may be brought into the circuit by taking out the appropriate plug. Fig. 40.2 show a different type of standard resistance box in common use. Each resistance coil is wound in the form of a loop on an insulating bobbin. This makes it equivalent to two equal coils wound in opposite directions, and hence there will be no resultant magnetic field when current flows through it. Winding of this type is said to be *non-inductive*. Its importance will be explained in chapter 43.

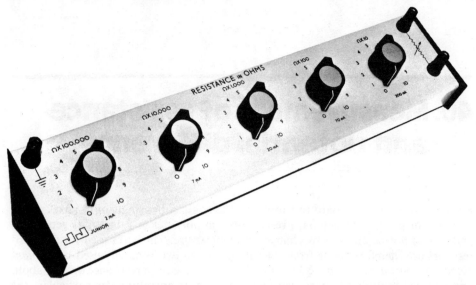

Fig. 40.2. *A decade resistance box* with a range of 1 111 100 Ω in steps of 10 Ω. The resistance coils are brought into circuit by the rotary switches shown. The symbol —◆◆— on the right-hand side is now replaced by —▱—

To measure resistance by the substitution method

The unknown resistor R_1, in ohms, to be measured, is connected in series with a cell and a galvanometer or milliammeter. A rheostat is connected across the latter to serve as a shunt (Fig. 40.3). The rheostat is then adjusted until the galvanometer gives a suitable deflection, which is noted. The unknown resistor is now taken out of the circuit and replaced by a resistance box. Without altering the rheostat, the value of the resistance R, in the resistance box is altered until the galvanometer gives the same reading as before. The value of the box resistance R will now be equal to R_1.

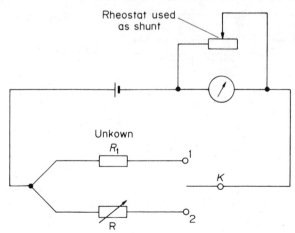

Fig. 40.3. Measuring resistance by substitution method

Criticism of the substitution method

The method of substitution is suitable only for measuring resistances of 100 Ω or more. There are two reasons for this. First, it is not usual to employ resistance boxes containing fractions of an ohm, and therefore the resistance can be found only to the nearest ohm. Secondly, if the resistance of the rheostat and galvanometer is large compared with that of the unknown resistor the circuit is *insensitive*. This means that a change of several ohms or more in the box makes so little difference to the total circuit resistance that the change in the galvanometer deflection is too small to be detected.

Satisfactory results can be obtained by this method only if the combined resistance of the rheostat and galvanometer are both small compared with the unknown resistance.

To measure resistance by the ammeter–voltmeter method

For this method we require a voltmeter of very high resistance and an ammeter of very low resistance.

The unknown resistor, R, is connected in series with an ammeter, a rheostat, a plug key or other switch, and one or more cells. The voltmeter is connected across the ends of R (Fig. 40.4 (a)). *As a safety precaution to avoid damage to the ammeter*

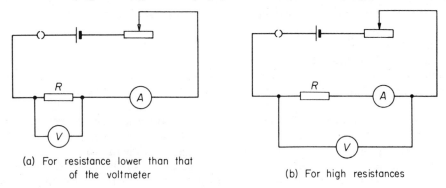

(a) For resistance lower than that of the voltmeter

(b) For high resistances

Fig. 40.4. Measuring resistance by ammeter-voltmeter method

through overload, the rheostat should be set to maximum resistance, before closing the circuit. The rheostat is then altered to give a series of suitable ammeter and voltmeter readings.

Now, $$\text{resistance} = \frac{\text{p.d.}}{\text{current}}$$

or $$R = \frac{V}{I} = \frac{\text{voltmeter reading}}{\text{ammeter reading}}$$

The results may be entered in a table as below and a mean value for R calculated

Voltmeter V (volts)	Ammeter I (amperes)	Resistance $R = \dfrac{V}{I}$ (ohms)
	Mean value of R	ohms

Alternative treatment of results

The equation $R = \dfrac{V}{I}$ may be written as

$$V = RI$$

Remembering that V and I are variables and R is a constant, we recognize this equation as being of the form,

$$y = mx$$

In mathematics we learn that if a graph of y against x is plotted from values which satisfy this equation its gradient is equal to m. Hence if a graph is plotted of V against I its gradient will give the values of R. (The term "gradient" is explained on page 25.)

Errors in the ammeter–voltmeter method

In the circuit just described (Fig. 40.4 (*a*)) the voltmeter gives the true p.d. across R, but the ammeter reads the current through R plus the current going through the voltmeter. If, however, the voltmeter resistance is very high compared with R the current through the voltmeter will be comparatively small, and hence the error made in calculating R will be negligible.

This circuit would, of course, give absurd results if the unknown resistance were of the same order of magnitude or greater than that of the voltmeter. Under these conditions the true current through R would only be approximately half that given by the ammeter. Consequently, the value calculated for R would be about half what it should be.

Fortunately the difficulty may be easily overcome. *When high resistances are being measured the voltmeter is connected across R and the ammeter together* as in Fig. 40.4 (*b*). The ammeter now gives the true current through R, while the voltmeter gives the p.d. across R plus the p.d. across the ammeter. If, as is usually the case, the resistance of the ammeter is very small compared with R the p.d. across it will likewise be small, and hence the error made in calculating R will be negligible.

Finally, it should be noted that in all experiments to measure resistance the current should be kept low to avoid undue temperature rise, which would increase the resistance (see page 407). For high resistances in particular it would be wise to use a milliameter instead of an ammeter and limit the current accordingly.

Resistance of an electric lamp and other conductors

The ammeter–voltmeter method is an excellent one for measuring the resistance of an electric lamp when run at different voltages. The circuit shown in Fig. 40.4 (*a*) is used in which the single cell is replaced by any convenient 6 V d.c. source and the resistor R changed for a 6 V electric lamp. The results show that the resistance of the lamp filament increases with temperature. A mean value should therefore *not* be worked out. Instead, a graph of resistance against voltage, or else voltage against current, is more instructive.

Fig. 40.5 shows typical voltage–current curves for various conductors. Look again at what we said earlier about the gradient of such curves, and also page 26.

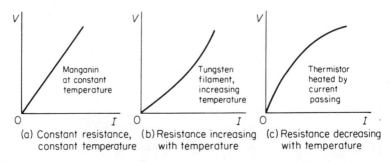

(a) Constant resistance, constant temperature

(b) Resistance increasing with temperature

(c) Resistance decreasing with temperature

Fig. 40.5. Voltage-current characteristics

The Wheatstone bridge

One of the most accurate methods of measuring resistance was devised in 1843 by Charles Wheatstone, the first Professor of Physics at King's College, London. It is now known as the Wheatstone-bridge method.

Suppose that four resistors, R, S, P, and Q in ohms are connected so as to form the four arms ab, bc, ad and dc of a bridge or network (Fig. 40.6 (a)). A 1.5 V dry cell is connected across ac and a sensitive centre-zero galvanometer across bd. The battery and galvanometer arms are provided with tapping keys, to which we shall refer as the battery* and galvanometer keys respectively. For the time being we shall suppose that R is an unknown resistance and that S, P, and Q are standard resistance boxes.

S, P, and Q are given values roughly equal to that of R and the battery key is closed. If the galvanometer key is also closed the galvanometer will usually give a deflection in one direction or the other. It is now possible, by suitably altering the values of the resistances S, P, and Q, to "balance" the bridge, i.e., to obtain no current through the galvanometer when first the battery key and then the galvanometer key is closed.

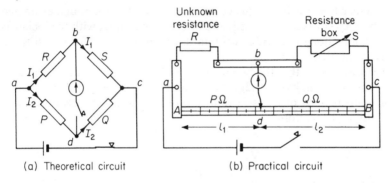

(a) Theoretical circuit   (b) Practical circuit

Fig. 40.6. The metre bridge

Suppose the current entering the network at a divides up into I_1 through R and I_2 through P. If no current flows through the galvanometer the current through S and Q must also be equal to I_1 and I_2 respectively. Also, since no current flows through the galvanometer, the potentials of b and d must be equal. Hence, for a balance, or *null deflection* of the galvanometer,

$$\text{p.d. across } R = \text{p.d. across } P$$

and
$$\text{p.d. across } S = \text{p.d. across } Q$$

or, since
$$\text{p.d.} = \text{current} \times \text{resistance}$$
$$I_1 R = I_2 P$$

and
$$I_1 S = I_2 Q$$

Dividing,
$$\frac{I_1 R}{I_1 S} = \frac{I_2 P}{I_2 Q}$$

whence
$$\frac{R}{S} = \frac{P}{Q}$$

i.e., the unknown resistance
$$R = S \times \frac{P}{Q}$$

Earlier on we described S, P, and Q as being three standard resistance boxes. However, since only the *ratio* of P to Q is required for calculating R, Wheatstone

---

* Although only a single cell is generally used, it is usual to speak of the battery key and not the cell key in this circuit.

found it more convenient to replace *P* and *Q* by a uniform resistance wire which could be divided into two parts by a movable contact. He used this principle in the *metre bridge* form of the Wheatstone bridge.

The metre bridge

The metre bridge consists of a straight uniform resistance wire *AB*, 1 m long, stretched over a boxwood scale graduated in millimetres and mounted on a board (Fig. 40.6 (*b*)). The ends of the wire are soldered or clamped to two stout copper or brass strips *a* and *c*, and a third and longer strip *b* is screwed to the board parallel to the wire. Terminals are provided on the strips for making the necessary connections, and a movable contact or jockey enables contact to be made at any point along the wire.

   The theoretical circuit shown in the diagram should be compared with the practical circuit. The unknown resistance *R* is connected in the gap between strips *a* and *b*, and a standard resistance box *S* in the gap between *b* and *c*. A sensitive centre zero galvanometer is inserted between *b* and the contact on the wire at *d*. A cell and a tapping key are connected across *ac*.

   The resistance box *S* is first adjusted to a value equal to the rough estimated value of *R*. The battery key is then pressed, and afterwards contact with the jockey is made at various points along the bridge wire until a point is found for which the galvanometer gives null deflection.

   We have already seen that the two resistances *P* and *Q* in the theoretical circuit are formed by the two lengths of bridge wire l_1 and l_2 in the practical circuit. Since the wire is of uniform area of cross-section, the resistance will be proportional to the length of wire and therefore

$$\frac{P}{Q} = \frac{l_1}{l_2}$$

Hence instead of using

$$R = S \times \frac{P}{Q}$$

we write

$$R = S \times \frac{l_1}{l_2}$$

Precautions when using a metre bridge
(1) The value of the known resistance *S* should be chosen so that the balance point comes somewhere on the "middle third" of the wire, i.e. the balance point should come between about 30 and 70 cm. It is bad practice to have a balance point near one end of the wire. For example, suppose the length l_1 was 1 cm. Assuming that the balance point can be read to within half a millimetre on the scale, the percentage error on the length of 1 cm would be $\frac{0.05}{1.0} \times 100 = 5$ per cent. On a length of 50 cm the percentage error would only be $\frac{0.05}{50.0} \times 100 = 0.1$ per cent. Secondly, most metre bridges require an *end correction* which may be expressed in millimetres of bridge wire. This arises from the solder or contact resistance where the wire joins the copper end strips. On short lengths of wire the end correction will be a very high percentage of the measured length, and hence the calculated value of *R* will be unreliable. On the other hand, if l_1 or l_2 are greater than about 30 cm, the percentage error will be very much smaller.

   (2) The battery key should always be pressed before the galvanometer contact is made on the bridge wire. This is to ensure that a steady current is flowing in the circuit before attempting to find the balance point. Otherwise a counter e.m.f. due to *self-induction* may be set up somewhere in the circuit which would render it impossible to find the exact balance point. Self-induction is explained in chapter 43.

(3) When the first balance has been obtained the resistors R and S should be interchanged and a second pair of values of l_1 and l_2 obtained. This compensates for errors due to the position of the millimetre scale with respect to the wire and, to some extent, for errors due to slight lack of uniformity in the wire.

It should be noted that interchanging the resistors does not eliminate the end errors mentioned earlier. A method of correcting for these will be found in more advanced textbooks.

Factors affecting the resistance of a wire at constant temperature

When resistors are being made, short lengths of thick wire are used for the low resistances and long lengths of thin wire for the high resistances. Besides length and thickness the material of which the wire is composed is another important factor to be considered when deciding the length and gauge of wire for a particular resistance.

Suppose, for instance, that the length of a given wire is doubled. This doubles the resistance, since twice the length of wire is equivalent to two equal resistances in series. If the length of wire is increased five times the resistance likewise becomes five times its previous value, and so on. The resistance of a wire is therefore proportional to its length, l, or in symbols,

$$R \propto l \quad . \quad . \quad . \quad . \quad . \quad . \quad . \quad . \quad . \quad . \quad (1)$$

With regard to the thickness or area of cross-section of the wire, thick wires may be regarded as equivalent to a number of thinner wires of equal area joined in parallel. Doubling the area will therefore halve the resistance and so on. In other words, the resistance of a wire is inversely proportional to its area of cross-section, A, or

$$R \propto \frac{1}{A} \quad . \quad . \quad . \quad . \quad . \quad . \quad . \quad . \quad . \quad (2)$$

Resistivity

By measuring the resistance of:

(*a*) various lengths of wire of the same thickness and made of a given material;
(*b*) several wires of equal length but different areas of cross-section also of the same material;

we may verify the relations (1) and (2) above.

Thus, by experiment $\qquad\qquad R \propto l$

and $\qquad\qquad\qquad\qquad\qquad R \propto \dfrac{1}{A}$

combining these two results, we obtain

$$R \propto l \times \frac{1}{A}$$

This relation may be turned into an equation by putting in a constant.

Thus, $\qquad\qquad\qquad\qquad R = \dfrac{\rho l}{A}$

where ρ is a constant called the *resistivity* of the material of the wire.
If, in this equation, we write $l = 1$ m and $A = 1$ m^2

$$R = \rho \times \frac{1 \text{ m}}{1 \text{ m}^2} = \frac{\rho}{\text{metre}}$$

whence $\qquad\qquad\qquad\qquad \rho = R$ ohm metre

From this expression we see that **the resistivity of a material is expressed in ohm metre units and is numerically equal to the resistance of a conductor made of the material of length 1 metre and area of cross-section 1 metre²**.

We talked in terms of ohms and metres in the foregoing discussion from which we see that the **SI unit of resistivity is the ohm metre (Ω m)**.

The sub-unit, ohm centimetre (Ω cm) is often used especially when dealing with wires of small diameter.

To measure the resistivity of constantan

A metre-bridge circuit, in which the unknown resistance consists of a length of constantan wire, is connected up as in Fig. 40.6 (*b*). It is convenient to use from 0.75 to 1 m of wire for this purpose, and the experimenter must use his own ingenuity to ensure that the length of wire between the terminals is measured as accurately as possible. One way of doing this is to begin by clamping one end of the wire under the terminal on *a*. The zero end of a metre rule is then placed at the point where the wire leaves the terminal. The wire, which has previously been freed from bends and kinks, is now laid along the rule and the required length, *l*, indicated by the thumb-nail when the wire is held between finger and thumb. Without allowing the finger and thumb to slip, the wire is placed in position under the terminal on *b* and the screw securely tightened.

The value of the resistance in the box, *S*, is adjusted so that a balance point is obtained somewhere near the middle of the bridge wire. The lengths l_1 and l_2 of the bridge wire segments are noted and a second pair of values obtained when the constantan wire and the resistance box *S* are interchanged. The mean value of l_1/l_2 is used to find *R* from the expression

$$R = S \times \frac{l_1}{l_2}$$

The next part of the experiment is to find the mean diameter, *d* in metres, of the wire. With the aid of a micrometer screw gauge, the diameter of the wire is measured at ten or more different places along the wire. At each place chosen, the diameter is noted in two directions at right angles, in case the wire is not truly circular. The mean diameter of the wire is found from the readings and used to calculate the area of cross-section *A* of the wire in metres².

Thus,
$$A = \pi \times (\text{radius})^2$$

$$= \pi \times \left(\frac{\text{diameter}}{2}\right)^2$$

or, in m²,
$$A = \frac{\pi d^2}{4}$$

Calculation of the resistivity

$$R = \frac{\rho l}{A} \text{ (page 470)}$$

therefore, in Ω m,
$$\rho = \frac{AR}{l}$$

hence, substituting for *A* and *R*, the value of ρ in Ω m

is given by
$$\rho = \frac{\pi d^2}{4l} R$$

Note that, since we are working in SI units, we expressed all our length measurements in metres before substitution in the equation.

The potentiometer

The potentiometer is a piece of apparatus for comparing potential differences. The simple potentiometer used in elementary work consists of a length of one or more metres of uniform resistance wire mounted on a board between thick brass or copper strips of negligible resistance. A millimetre scale is fixed to the board for measuring the distance from one end of the wire to the position of a sliding contact or jockey.

When the potentiometer is in use a 2 V lead cell or a battery of two alkaline cells in series is connected across the wire so that a steady current flows in it. Since the wire is uniform, there will thus be a steady fall in potential along it. In circuit diagrams the symbol for a potentiometer is a straight line AB to represent the wire together with a lead cell E connected directly across it (Fig. 40.7).

To compare the e.m.f.s of two cells by using a potentiometer

In the circuit diagram of Fig. 40.7 the points D and F may be considered as the two terminals of a voltmeter. D is connected directly to A, and F is connected to a sliding contact on the potentiometer wire via a sensitive centre-zero galvanometer. As it is easy to damage the galvanometer by too large a current if contact is made on the wire too far from the balance point, a safety resistor of 1000 Ω or more is placed in series with the galvanometer to protect it by limiting the current. Once the balance point has been located approximately, the safety resistor is shorted out by closing the key K, and the final adjustment made with the galvanometer at maximum sensitivity.

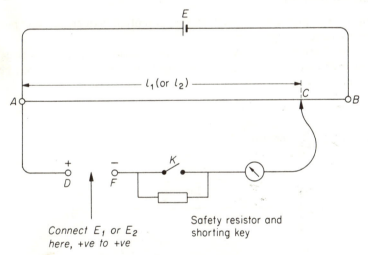

Connect E_1 or E_2 here, +ve to +ve

Safety resistor and shorting key

Fig. 40.7. Comparing e.m.f.s. by potentiometer

With the circuit as shown in Fig. 40.7 there will be a potential difference between D and F equal to that between A and C on the potentiometer wire. This potential difference can be varied by altering the position of C, and will be proportional to the length AC. In the diagram D will be positive relative to F, since the positive terminal of the cell E is connected to A.

Suppose we wish to compare the e.m.f.s of two cells. The two points D and F are connected to the terminals of the first cell E_1 (Fig. 40.7) just as the leads from a voltmeter would be connected across the cell, care being taken to connect D to the positive terminal of the cell. On making contact with the jockey on the wire at C current will, in general, flow through the galvanometer in one direction or the other, depending on whether the e.m.f. of E_1 is greater or less than the potential difference between D and F. The position of C is now adjusted until no current is indicated by the galvanometer. This will occur when the potential difference between D and F

(and therefore between A and C) is equal to the e.m.f. of the cell E_1. Note that the p.d. across the terminals of E_1 is equal to its e.m.f. because the cell is delivering no current. We now note the length AC, say l_1.

The experiment is repeated using the second cell E_2 and the corresponding length l_2 obtained. Since the wire is uniform, the p.d. between any two points on the wire will be directly proportional to the length of wire between them.

Therefore, since l_1 and l_2 are the lengths corresponding to the e.m.f.s of E_1 and E_2 respectively

We have,
$$\frac{\text{e.m.f. of } E_1}{\text{e.m.f. of } E_2} = \frac{l_1}{l_2}$$

When the experiment is carried out as described above the result is simply the ratio of the e.m.f. of the two cells concerned. However, a special cell called a *Weston cadmium standard cell*, which has an e.m.f. of 1.02 V, can be used to *standardize* a potentiometer so that it can be used to measure e.m.f.

Thus if the balance length for a Weston cell is l_2 in cm, then in volts,

$$E_1 = \frac{1.2}{l_2} \times l_1$$

In this way, the actual value of the e.m.f. E_1 may be calculated in terms of that of a Weston cell.

The potential divider

The potentiometer principle is used in the potential divider for the purpose of providing a variable voltage supply.

For example, if the terminals A and B of the resistor in Fig. 35.2 on page 406 are connected to a 12 volt battery, then any voltage from zero to 12 volts may be tapped off from A and C by moving the slider from A to B.

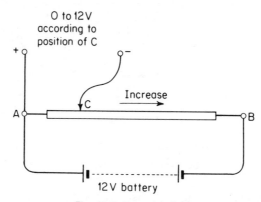

Fig. 40.8. Potential divider

The arrangement is shown in Fig. 40.8. To suit the requirements of the particular circuit it supplies, a potential divider must be chosen for its maximum current output without undue temperature rise. This maximum current value is usually marked on the slider and will depend on the thickness of the wire from which the instrument is made.

A potential divider is used in the circuit shown in Fig. 44.19 (*a*), on page 520.

QUESTIONS: 40

1. Draw a circuit to show the voltmeter–ammeter method of measuring resistance. What would be the effect on the reading of the ammeter if the voltmeter had a resistance comparable with that of the resistance? State briefly why. (*S.*)

2. You are supplied with an accumulator ammeter, voltmeter, rheostat, key, copper wire, and an unknown resistance X. Describe with the aid of a circuit diagram, how you could use this apparatus to determine the resistance X.

Explain why the method does not lead to an accurate result.

A cell having an e.m.f. of 1.5 V and an internal resistance of 2.0 Ω is connected in series with an ammeter of resistance 0.5 Ω and a 5.0 Ω resistance. What will be:

(i) the ammeter reading, and
(ii) the potential difference across the terminals of the cell? (*J.M.B.*)

3. Draw a circuit diagram to show how you would use the substitution method to measure the resistance of a coil of wire believed to be in the region of 150 Ω. Label each of the components, indicating appropriate values of e.m.f., resistance or range as applicable.

Why is this method generally unsuitable for the measurement of comparatively low resistances?

4. Draw a diagram of the circuit you would use to measure the potential difference (V) across a coil of wire for various currents (i) passing through it. Sketch a graph to show how V is related to i.

How would you use the results to find the resistance of the coil? Why should the current in such an experiment be reasonably small?

What further measurements would be necessary if the resistivity of the coil material was to be found? State also the required formula. (*S.*)

5. (*a*) Describe, with the help of a circuit diagram, how you would measure the resistance of a resistor using an ammeter, a voltmeter, and any other equipment you require. Suggest one error which could affect the accuracy of your result and state the way in which it would affect the result.

(*b*) Two electric-light bulbs, both marked 0.3 A, 4.5 V are connected (A) in parallel, (B) in series, across a 4.5 V battery of negligible internal resistance. Assume that the resistance of the filament does not change. In each case A and B:

(i) state what might be seen;
(ii) calculate the currents through each bulb, and
(iii) calculate the current supplied by the battery. (*J.M.B.*)

6. (i) A wire 0.40 m long and of diameter 0.60 mm has a resistance of 1.5 Ω. What is the resistivity of the material of which it is made?

(ii) A voltmeter of 1000 ohm resistance reads 200 V when connected across a battery of 60 Ω internal resistance. What is the e.m.f. of the battery? (*O.C.*)

7. What length of resistance wire of diameter 0.6 mm and resistivity 1.1×10^{-6} Ω m, would you cut from a reel in order to make a 44 Ω resistor?

8. A six volt battery is used, together with a potential divider, to provide a variable voltage first across a 3 ohm resistor alone, then across a torch bulb alone.

(*a*) Sketch one circuit which shows

(i) the arrangement of the potential divider and
(ii) how the current through either the resistor or the bulb and the potential difference across it could be measured.

(*b*) Sketch two graphs, one for each component, showing how the current varies with the potential difference. (*J.M.B.*)

41. Electric light, energy and power

Electricity plays so large a part in the modern world that it would be difficult to imagine life without it. Energy generated by hydro-electric and nuclear power stations is transmitted cheaply to big towns and industrial areas and its distribution in country districts eases the labours of farmers and dairy workers. The advantage of electric energy is the ease with which it may be transferred to light, heat, and other forms of energy. Electric light is taken for granted almost everywhere, and in most homes radio and television are regarded as necessities. People appreciate the cleanliness and convenience of electric heating and cooking, and their daily toil is lightened by electric washing machines, polishers and vacuum cleaners. The well-equipped house may have more than a dozen items of electrical equipment, ranging from the refrigerator to the electric razor. Underground electric railways have helped to solve the passenger transport problem in many of the big cities of the world, and the replacement of steam by electric power in factories makes for a cleaner atmosphere.

Fig. 41.1. *Applications of the electric arc*

(*a*) Developed for use in lighthouses, this high-pressure xenon arc lamp provides a luminous intensity of 1 million candelas, yet it is only about 30 cm long. (The candela is a unit of luminous intensity approximately equal to that of a candle)

(*b*) Steel melting furnace of 80 tonnes capacity. The trodes have just been withdrawn and the roof swung prior to top charging

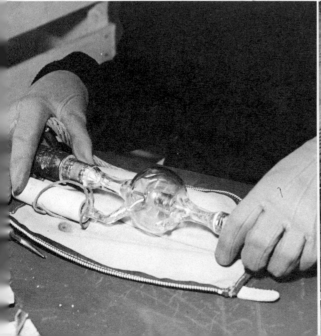
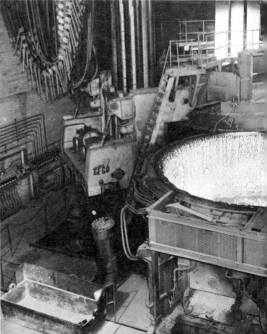

Fig. 41.2. A twin automatic flash butt arc welder in operation

The electric arc

Early in the nineteenth century Sir Humphry Davy showed that a brilliant source of light could be obtained by connecting a pair of carbon rods to a battery of e.m.f. greater than 40 V. This source consisted of an *arc* or incandescent flame which was struck by bringing the ends of the rods together and drawing them slightly apart.

During the second half of the century considerable use was made of this form of illumination, and inventors showed much ingenuity in designing mechanism to strike the arc and advance the tips of the rods as they slowly burnt away. The cumbersome arc lamps of those days were expensive to run and maintain, but they were a great advance over gas lamps. Although after 1880 they were gradually replaced by filament lamps, they lingered well into the twentieth century for lighting streets, factories and theatres.

Nowadays, arc lamps are restricted to such purposes as search-lights and projection lamps, where an intensely concentrated source is required. Fig. 41.1 shows a high power arc lamp designed for lighthouse lanterns. It has metal electrodes surrounded by xenon gas.

The temperature reached in the electric arc is in the region of 3700 °C. This is well above the melting points of metals, and therefore the main applications of the arc are in electric furnaces (Fig. 41.1), and welding equipment (Fig. 41.2). Two types of arc furnaces are used. The larger ones employ carbon arcs and are used for melting special steels. The smaller furnaces are used in research work. These have a tungsten rod which acts as one electrode, while the metal to be melted forms the other electrode.

In welding, a metal rod is connected to one terminal of the supply and the two pieces of metal to be joined are connected to the other. The arc is struck by touching the joint with the end of the welding rod. The heat generated melts the rod and the two components are fused together.

Filament lamps

The electric filament lamps with which we are familiar today had their origin in the late 1850s. During those years experiments were made with lamps consisting of short carbon rods contained in an evacuated glass globe. When current was passed the rods became white hot and gave out light, but at the same time the absence of air prevented the rods from burning away. The trouble with these early *glow lamps*, as they were called, was that the rods were very liable to break without warning. This prompted one inventor to take out a patent for a lamp with five rods. Each time a rod broke another one automatically came into operation until all were used up.

Eventually, Swan in England and Edison in America made greatly improved filaments using carbonized bamboo and cotton respectively. These had a useful life of at least 800 hours, and after 1880 filament lamps became a practical commercial proposition.

At the beginning of the twentieth century it was found that a much brighter light could be obtained from the same consumption of electric energy if tungsten filaments were used. Tungsten is a metal with the very high melting point of 3400 °C, and therefore can be run at a temperature of 2000 °C or more, compared with 1300 °C for carbon. The early filament lamps were vacuum filled, but most lamps used today are gas filled. A little nitrogen or argon is introduced into the bulb at low pressure. This helps to prevent evaporation of metal from the filament and enables the lamps to be run at a much higher temperature. The higher the temperature, the greater is the proportion of electric energy transferred to light.

The old carbon lamps transferred very little of the energy into visible radiation and most of it into internal energy which was lost as heat. Tungsten filament vacuum-filled lamps are twice as efficient from this point of view while the *coiled-coil* lamps first introduced in 1934 give up to three times as much light as carbon lamps for the same consumption of electric energy.

A disadvantage of gas filling is that convection currents are set up which cool the filament, but by constructing the filament in the form of a coiled coil it occupies a much smaller space, and therefore the rate of loss of heat by convection is greatly reduced.

Improvements go on all the time. For example, tungsten tends to evaporate and condense on the inside of the bulb thereby darkening it. In projector lamps this is eliminated by introducing a little iodine which forms tungsten iodide from the tungsten vapour and this remains as vapour when the lamp is in use.

Discharge lamps

If some gas at low pressure is contained in a tube fitted with metal electrodes at each end the gas glows with a characteristic colour when a high voltage is applied to the electrodes. The electric field set up inside the tube causes ions present in the gas to move with high speeds. As a result of collisions which occur between ions and gas molecules, light is emitted of a colour which depends on the nature of the gas. Neon, for example, gives the familiar bright orange-red seen in advertising signs.

Following the successful commercial introduction of these coloured gas tubes in the 1920s, manufacturers turned their attention to the development of discharge tubes for general illumination. One of the results of their research is to be seen in the orange sodium vapour lamps and the blue-green mercury vapour lamps used for street lighting. Although these emit coloured light, they give five times more luminous energy per watt than the best filament lamps, and so are cheaper to run.

Fluorescent lighting

Besides giving out coloured light, a mercury vapour discharge tube also emits *ultraviolet* light. When ultraviolet light, which is itself invisible, falls on certain minerals they glow brilliantly with various colours. This phenomenon is called *fluorescence*. Accordingly, the inside of a mercury discharge tube may be coated with a mixture of various powders which give out either a white or tinted light. Some of these powders contain beryllium compounds which are highly poisonous if they enter a cut in the skin. The greatest care should therefore be taken when handling a broken tube. Owing to the fact that they require special starting equipment, fluorescent lamps are initially more expensive than filament lamps. On the other hand, they are roughly three times as efficient.

Heating by electricity

The heating element in an ordinary radiant electric fire is a length of resistance wire which becomes raised to a temperature of about 900 °C when current is passed through it. The wire is supported on a fireclay rod or bar or coiled inside a fused silica tube. The alloy from which the wire is made is a mixture of nickel and chromium which resists oxidation in air when red hot.

In the convector type of heater the element is raised to a temperature of only about 450 °C. This is popularly known as *black heat*, and the elements in this case are not designed to radiate but only to warm the air, which circulates through the heater by either free or forced convection.

In another method of space heating current is passed through iron wires embedded in the concrete of floors and the plaster of walls and ceiling. Storage heaters use cheap electricity during the night and store up internal energy in fireclay blocks which subsequently is given out as heat.

Many domestic electric appliances contain heating elements. The kettle, laundry iron, toaster, and electric blanket are examples which readily come to mind, and the reader will be able to think of at least a half a dozen others. Whenever electricity is used for raising the temperature of water or other liquids the element is well insulated and enclosed in a metal tube or sheath. *It is important not to let the element come into direct contact with water, otherwise the latter will become live and therefore dangerous.*

The elements of laundry irons are made of strip instead of wire, so that they are as flat as possible and also present a large surface for conducting away heat into the sole of the iron. Elements of this type are wound on mica and sandwiched between two thin sheets of the same material. Mica is a mineral which can be readily cleaved into thin sheets or laminae. It is not only highly insulating but can also withstand high temperatures.

Energy in an electric circuit

Before studying the rest of this chapter the reader should revise definitions, given earlier in the book, of the following quantities: *joule, watt, ampere, coulomb, volt,* and *ohm*.

When a potential difference is applied to the ends of a conductor some of the electrons inside it are set in motion by the electric forces. Work is therefore done and the electrons acquire energy. The moving electrons form an electric current, and the energy of this current appears in various forms according to the type of circuit of which the conductor forms a part. Thus, radio transmitters transfer the energy of rapidly oscillating electric currents to energy in the form of electromagnetic waves. The function of a radio or television receiver is to transfer wave energy back to electric current energy, and thence into energy of sound and light. An electric motor is designed to transfer as much of the electric energy as possible to mechanical energy of rotation. In the element of an electric fire the energy of the current is transferred to internal energy which is then given out in the form of heat.

In the present chapter we shall be concerned mainly with cases such as the last mentioned, where the current flows through a simple resistor and all the electric energy is eventually transferred to internal energy and thence to heat.

Calculation of the work done by an electric current

From the definition of the volt (page 405), it follows that if a p.d. of 1 volt is applied to the ends of a conductor and 1 coulomb of electricity passes through it the work done is 1 joule.

Hence if the p.d. applied is V in volts and the quantity of electricity which passes is Q in coulombs the work done is VQ in joules.

But $\qquad\qquad\qquad Q =$ current in amperes $\times$ time in seconds

or $\qquad\qquad\qquad\qquad Q = It$ coulombs

therefore work done in joules $\quad = VQ \qquad\qquad = VIt$ . . . . (1)

Two other expressions for the work done may be obtained by using the equation based on Ohm's law and substituting

either $\qquad\qquad\qquad\qquad V = IR$

or $\qquad\qquad\qquad\qquad I = \dfrac{V}{R}$ in the expression (1) above

Thus, $\qquad$ work done in joules $= VIt = IR \times It = I^2Rt$ . . . . (2)

and also $\quad$ work done in joules $= VIt = V \times \dfrac{Vt}{R} = \dfrac{V^2t}{R}$ . . . . (3)

The work done becomes transferred to internal molecular energy in the conductor, accompanied by a rise in temperature. Subsequently this energy may be given out in the form of heat. We thus have three alternative expressions for the heat produced in terms of V, I, and R.

Electric power

If a piece of electrical equipment is examined it will usually be found to have a label or engraved plate giving the working *voltage* and the power consumption in *watts*. For example, an electric lamp may be marked 240 V 60 W, or a fire 240 V 3 kW. The abbreviation kW stands for kilowatt (1 kW = 1000 W).

By definition (page 82) a power of 1 watt is a rate of working of 1 joule per second, and therefore

$$\text{power in watts} = \text{rate of working in J/s}$$

The three expressions, VIt, I^2Rt and $\dfrac{V^2t}{R}$ (as above) represent the work in joules done by the current in time t in seconds. Hence the wattage or power in J/s is found by dividing each expression by t.

Thus, power in watts $= \dfrac{\text{work in joules}}{\text{time in seconds}}$

$$= \frac{VIt}{t} = VI \ (= \text{volts} \times \text{amperes})$$

$$= \frac{I^2Rt}{t} = I^2R \, (= \text{amperes}^2 \times \text{ohms})$$

$$= \frac{V^2t}{Rt} = \frac{V^2}{R} \left(= \frac{\text{volts}^2}{\text{ohms}}\right)$$

Electricians use these expressions frequently. It is advisable to memorize the first one in words.

i.e., $\qquad\qquad\qquad\qquad$ **watts = volts $\times$ amperes**

The other two expressions may then be derived from this by substituting for either volts or amperes, using the formula

$$\text{volts} = \text{amperes} \times \text{ohms}$$

Example. Find (a) *the current taken,* (b) *the resistance of the filament of a lamp rated at* 240 V 60 W.

$$watts = volts \times amperes$$

therefore
$$amperes = \frac{watts}{volts}$$

hence
$$current\ taken = \frac{60}{240} = \underline{0.25\ A}$$

also
$$ohms = \frac{volts}{amperes}$$

hence
$$resistance\ of\ filament = \frac{240}{0.25} = \underline{960\ \Omega}.$$

Board of Trade unit of electric energy. The kilowatt hour

If an electricity meter is inspected it will be found to have the abbreviation kWh inscribed on it. This stands for *kilowatt hour*, which is the commercial unit of electric energy.

As its name implies, **the kilowatt hour is the energy supplied by a rate of working of 1000 watts for 1 hour**

or
$$1\ kilowatt\ hour = 1000\ watt\ hours$$
$$= 1000\ joules\ per\ second\ for\ 1\ hour$$
$$= 1000 \times 60 \times 60\ joules$$
$$= 3\ 600\ 000\ joules$$
$$= 3.6\ megajoules\ (MJ)$$

When working out problems on the consumption of electric energy it is unnecessary to convert to joules. It will be found more convenient to work in *watt hours*, as the following example illustrates.

Example. Find the cost of running five 60 W *lamps and four* 100 W *lamps for* 8 h *if electric energy costs* 5.0p *per unit.*

Total power consumption $= (5 \times 60) + (4 \times 100) = 700$ watts

Time $= 8$ hours

Therefore energy consumed $= 700 \times 8$
$= 5600$ watt hours
$= \frac{5600}{1000}$ kilowatt hours
$= 5.6\ kWh$

Cost $= 5.6 \times 5.0 = 28p.$

Domestic electric installation

The cable bringing the mains electricity supply into a house contains two wires, one of which is "live" and the other "neutral". The neutral wire is earthed at the local transformer substation, so it is at earth potential. At some convenient place inside the house the service cable enters a sealed box, where the live wire is joined to the Electricity Board's fuse. A fuse is a device containing a short length of thin wire which melts and breaks the circuit if the current exceeds a safe value.

The supply is controlled at the intake by a *Consumer's Service Unit*. This consists of one large main double-pole switch combined with a single-pole fuse or an earth-leakage contact breaker for each of the lighting circuits, cooker circuit, water heater circuit, and ring circuit (Fig. 41.3).

In modern domestic installations the power sockets are tapped off a *ring circuit*. This is a cable which is taken through the various rooms in the house and has both pairs of conductors connected to the one fuse. There is, therefore, a double path for the current to any particular circuit, which effectively doubles the capacity of the cable. A system such as this is used in conjunction with a special type of fused plug, rated to carry 13 A (Fig. 41.5).

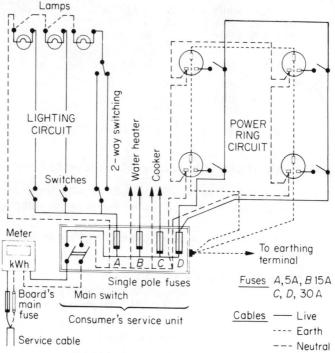

Fig. 41.3. Domestic electric installation

It will be seen from the circuit diagram that all light and power switches and fuses are placed in the live side of the supply. If they were in the neutral side the ceiling roses and power sockets would still remain live when the switches were in the off-position. It would, therefore, be possible to get a shock by touching the element of an electric fire even when it was cold and the current had been switched off.

With regard to the wiring, most authorities favour the use of BSRI (butyl and silicone rubber insulated) or PVC (polyvinyl chloride) sheathed cable enclosed in metal or PVC high-impact conduit. However, it is sometimes more convenient to use rubber or PVC sheathed cable for surface wiring in older houses. Cable may be buried under the plaster but it is advisable to provide mechanical protection if there is likelihood of accidental damage.

Fuses

The Electricity Board's fuse is usually of the cartridge type, in which the fuse element is enclosed in a tubular container filled with quartz sand. The ordinary house fuses generally consist of short lengths of tinned copper wire or cartridge fuses fitted into carriers made of porcelain or other insulating material. The special fused plug used in ring circuits contains a small cartridge fuse with a capacity of 3, 5, 10, or 13 A to protect the flexible cable connected to the plug.

Replacing a fuse

When a fuse has blown the first step is to turn off the main switch before looking for the cause of the trouble. Sometimes a fuse melts simply because the wire is very old and has become weakened by oxidation, or it can melt from overloading of the circuit. Frequently, fusing is caused by a short-circuit in flex where the insulation has become worn and frayed or by insulation failure to a component part of an appliance. If the cause cannot be traced or if a defect has occurred in the house wiring it is best to seek the advice of an electrician. Obviously the fault must be rectified before fitting a new fuse.

The method of replacing a small cartridge fuse needs no special description: the old fuse cartridge is removed and a new one slipped into the clips provided. Certain precautions must be observed when replacing fuse wire in the ordinary type of carrier. First, all traces of the old wire must be removed. Secondly, new fuse wire must be chosen of the correct current capacity, and finally, the fuse wire must not be over-tightened so as to stretch and weaken it. Those who use odd pieces of any kind of wire, paper clips, etc., instead of proper fuse wire, are asking for trouble.

Very rarely the Board's sealed fuse may melt. Should this occur, it is necessary to inform the Board. Their sealed fuse must not be touched or interfered with in any way.

Power circuits

Besides the live and neutral wires, all correctly installed power circuits are provided with a third continuous and covered wire which has been earthed by a good electrical joint to an earthing terminal provided by the Electricity Board. This wire is used for earthing the metal casing of any apparatus, e.g., an electric fire, and is a safeguard to prevent anyone from receiving a shock should the casing become live. Such a danger would occur if the insulation on the live flex lead became worn and allowed the live wire to come into contact with the framework of the fire. If this happened in the case of a properly earthed fire a fuse would immediately blow and the current be cut off.

In England it is conventional to wire three-pin sockets so that when the earthing socket is at the top the live socket is on the right and the neutral on the left. However, it cannot be taken for granted that this convention has been observed in

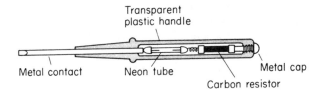

Transparent plastic handle

Metal contact   Neon tube   Metal cap

Carbon resistor

Fig. 41.4. Live mains lead indicator

each case. It is best to make sure by using the indicating device illustrated in Fig. 41.4. This can take the form of a small probe with a hollow insulating handle containing a tiny neon discharge tube. One electrode of the neon tube is in contact with the probe and the other is connected through a high carbon resistance to a metal cap on the handle. When the probe is inserted into the live socket leakage of current to earth takes place through the user's body and the neon tube glows. Owing to the very high resistance, the current is negligible, and hence there is no danger of shock. These indicators are often made in the form of a small screwdriver. *It is exceedingly dangerous to insert an ordinary screwdriver into the live socket.*

Wiring a three-pin plug

When electric apparatus is fitted with a three-core flexible lead the insulation on the three wires is distinctively coloured so that the correct connections can easily be made to the plug. The old British Standard convention was red for live, black for neutral, and green for earth.

The present International convention is BROWN for live, LIGHT BLUE for neutral, and GREEN or GREEN/YELLOW for earth. These colours have the advantage, in normal circumstances, of being easily distinguished by colour blind people.

Wiring must be checked to make sure that the earth lead goes to the metal framework and the live lead to the switch if one is provided.

A three-pin plug of reliable make will have its pins marked *L*, *N*, and *E*, standing for live, neutral and earth respectively (Fig. 41.5). The requisite amount of insulation is removed from each of the wires either by the aid of a penknife of a pair of wire strippers. When this is being done care must be taken not to cut or nick any of the wire strands, or else they will break off when the wire is bent. The wire strands are twisted gently together with a pair of pliers and the wire is then bent in a *clockwise* direction round the terminal stud or screw. There should be a brass washer between the wire and the terminal screw head or nut to prevent drag on the wire when

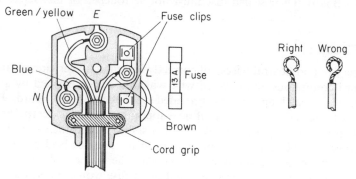

Fig. 41.5. Wiring a universal 3-pin 13 A mains plug

tightening up. When the job is finished the insulation on the wires should extend right up to the terminals. If the plug is provided with a cord grip this should be tightened just sufficiently to grip the cord gently. Overtightening serves no useful purpose, and may result in the insulation giving way under the pressure.

These plugs have the added advantage that an appropriate fuse may be chosen, e.g., 13 A for an electric fire, 2 A for a clock and so on.

QUESTIONS: 41

1. What do the following units measure:
   (a) the volt;
   (b) the coulomb;
   (c) the watt;
   (d) the kilowatt hour?   (*J.M.B.*)

2. An electric lamp uses energy at the rate of 48 W on a 12-V supply. Calculate:
   (a) the current in the lamp;
   (b) the resistance of the lamp while in use.
                                       (*C.*)

3. (a) An electric lamp is marked 250 W, 230 V. What does this mean?
   (b) In what time would this lamp use one Board of Trade unit when connected to 230V mains?   (*J.M.B.*)

4. A lamp is marked 12 V, 24 W. How many joules does it consume in an hour and what is the current it passes?   (*S.*)

5. An electric kettle is rated "2 kW, 240 V" and when filled with cold water takes 5 min to boil. Calculate:
   (a) the resistance of the element when the kettle is in use;
   (b) the average weekly cost of using the kettle, assuming that it is filled six times each

day with cold water which is then boiled, and 1 kWh costs 5p. (Neglect all heat losses.)
                                       (*A.E.B.*)

6. Define *the volt*. From your definition, derive an expression for the rate of production of heat in a wire of resistance R (ohms) carrying a current of 1 ampere.
   A 250 V, 2 kW electric fire has two elements in parallel rated at 750 W and 1250 W respectively. Calculate for each element:
   (a) the current taken, and
   (b) the resistance. Assuming that the resistances remain constant, find the power dissipated by the whole fire if the voltage of the mains supply drops to 200.   (*O.*)

7. State and explain a formula for the rate of production of heat in an electric circuit.
   Given two equal resistance coils and an accumulator of negligible internal resistance, compare the rate of production of heat of the whole system when the coils are connected first in series and then in parallel with the accumulator.   (*J.M.B.*)

8. How long will it take a 240 V, 3000 W

electric immersion heater to raise the temperature of 150 litres of water in a well-lagged copper tank of mass 20 kg from 15 °C to 70°C? Find the cost at 5p per kWh. (Specific heat capacities: water, 4200 J/kgK; copper, 390 J/kgK.)

9. Describe with the aid of a circuit diagram how you would measure the power and resistance of a 12 V electric filament lamp using a voltmeter and ammeter as well as other basic apparatus.

Explain why the current when the lamp is first switched on is different from when the lamp is operating.

Draw a further circuit diagram to show how a lamp may be switched on and off at two independent switch positions.   (*S.*)

10. An electric fire operated from the 240 V a.c. mains supply is rated at 1.5 kW. When operating at rated power, calculate

(i) the current used by the fire,

(ii) the energy given out by the fire in 10 hours,

(iii) the cost of running the fire for 10 hours, if the cost per kWh is 5p.   (*A.E.B.*, 1982)

11. Water in an electric kettle connected to a 240 V supply took 6 min to reach its boiling point. How long would it have taken if the supply had been one of 210 V?   (*L.*)

12. A 240 V washing-machine has a motor rated at 250 W and a heating element at 3 kW. Find the annual cost for electricity of running it for 1.5 h per week assuming that, during this time the motor is on all the time and the heater element only two-thirds of the time. (Electricity costs 5p per kWh.)

13. A coil of resistance wire is immersed in liquid in a calorimeter of a total heat capacity 950 J/K. If the temperature rises from 9 °C to 29 °C in 5 min when a steady current of 4 A is passed, find:

(*a*) the resistance of the coil;

(*b*) the p.d. across it.

14. 84 tonnes (t) of coal per hour are used in steam boilers which supply a turbo-alternator having a power output of 200 000 kilowatts (kW). If the heat value of the coal is 22 000 kilojoules per kilogram (kJ/kg), calculate the efficiency of the installation.

15. A 3 kW electric fire is designed to operate from a 240 V supply. Calculate the resistance of the fire. The fire is connected to the supply by long leads of resistance 0.8 Ω. Assuming their resistance to remain unaltered, determine:

(i) the current in the leads, and

(ii) the power dissipated in the leads.

Explain why a single pole switch (cutting only one wire) should be placed so that it cuts the live wire.   (*L.*)

16. Describe with the aid of a diagram the structure of an electric filament lamp. Give reasons why the filament of some lamps consists of a closely wound coil of wire and the bulb is filled with an inert gas.

Write a short account of the ways in which the energy supplied to a filament lamp is dissipated when the lamp is in use.   (*C.*)

17. (*a*) State the internationally accepted colours for the live (line), neutral, and earth leads of a 3-core flex. Draw a labelled diagram of a correctly connected fused mains plug.

(*b*) Explain what is meant by a *ring-main* and state the advantages of this method of domestic wiring.

Illustrate your answer by a circuit diagram for **two** outlets only.

(*c*) Select the most appropriate value of fuse for: (i) a 240 V, 1 kW heater, and (ii) a 240 V, 3 kW heater from 3 A, 5 A, 10 A, 13 A, 15 A.   (*A.E.B.*)

18. An electric bulb is marked 40 W, 220 V. What do the markings mean? Another bulb is marked 40 W, 110 V. Calculate the ratio of their resistances.   (*W.A.E.C.*)

42. Electrical calorimetry

The term *calorimetry* means the science of heat measurement and is a word which first came into use in the eighteenth century at a time when heat was still being thought of as a fluid called *caloric*. Similarly, any piece of apparatus with which heat measurements may be made is called a *calorimeter*.

In the previous chapter we explained how the work done by an electric current when it passes through a simple resistor may be measured in joules, and transferred, first to internal energy and thence to heat. So we are now in a position to deal with some elementary methods, based on electric heating, for the measurement of specific heat capacities and specific latent heats. But before proceeding further with this chapter the reader will find it useful to revise the heat definitions and calculations given in chapters 18 and 19.

To measure the specific heat capacity of a metal (solid block method)

This method is suitable for a metal which is a good thermal conductor, e.g., copper or aluminium.

A cylindrical block of the metal is drilled with two holes, one to receive an electric immersion heater and the other a thermometer. A little oil may be used in the holes to ensure good thermal contact. Heat losses may be reduced by standing the block on a slab of expanded polystyrene and covering its top with the same material. The lagging is completed by wrapping its sides with several thicknesses of 2 mm thick expanded polystyrene. The material sold in rolls for house wall insulation is suitable for this purpose and, when cut to size, may be fastened with adhesive tape (Fig. 42.1).

The heater circuit is connected as shown but before inserting the heater into the block the rheostat should be adjusted for a suitable current (2 to 4 A may be suitable with a 12 V supply). The current is switched off and the heater allowed to cool before insertion in the block.

The temperature of the block is noted. The current is switched on and simultaneously the time is read from a seconds clock (alternatively a stopclock may be used). Readings of the voltmeter and ammeter are taken and the current kept constant by adjustment to the rheostat as necessary.

After the temperature has risen by about 10 degrees, the time is noted and simultaneously the current is switched off. The final steady reading of the thermometer is taken, remembering that a slight delay may be necessary to ensure thermal equilibrium between heater and block. The block alone is weighed when cool, either before or after the experiment, and the readings recorded as shown.

Readings		*(Units)*
Mass of metal block	$= m$	(g)
Initial temperature	$= \theta_1$	(°C)
Final temperature	$= \theta_2$	(°C)
Current	$= I$	(A)
P.d. across heater coil	$= V$	(V)
Starting time	$= t_1$	(s)
Finishing time	$= t_2$	(s)

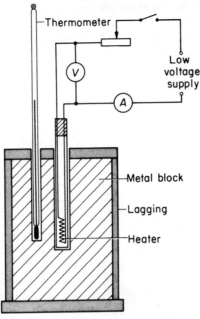

Fig. 42.1. Simple solid block calorimeter

Calculations

Time current is passed	$= (t_2 - t_1) = t$	(s)
Energy supplied by heater	$= VIt$	(J)
Rise in temperature of metal	$= (\theta_2 - \theta_1) = \theta$	(K)
Let specific heat capacity of metal	$= c$	(J/g K)

If we assume that,

energy received by block = energy supplied by heater

then, $$mc\theta = VIt$$

or, specific heat capacity of metal $= c = \dfrac{VIt}{m\theta}$ (J/g K)

$$= 1000 \,\dfrac{VIt}{m\theta} \quad \text{(J/kg K)}$$

Note.

(1) The small heat capacities of the heater and thermometer have been neglected.

(2) The SI unit of specific heat capacity is the J/kg K, although some scientific workers find the sub-unit, J/g K, more convenient.

To measure the specific heat capacity of a liquid

A copper calorimeter together with a copper stirrer is weighed, first empty, and then when about three-quarters full of liquid. A heating coil is placed in the liquid and is supported by a cover which has holes to take the stirrer and a thermometer. The calorimeter is highly polished and is supported on a poor heat conductor inside a polished outer jacket to reduce heat losses from radiation, conduction, and convection. The coil is connected in series with a d.c. supply, an ammeter, a rheostat and a plug key or switch. A voltmeter is connected directly across the coil terminals (Fig. 42.2).

The voltage of the d.c. supply used will depend on the resistance of the heating coil, but the current should be adjusted to give a temperature rise of about 10 K in not more than 5 minutes. It is first necessary to carry out a preliminary test for the purpose of adjusting the current to the chosen value. The current is then switched off and the liquid well stirred.

Having taken the initial temperature of the liquid, the current is now switched on again, and simultaneously a stopclock is started. The voltmeter reading is also noted

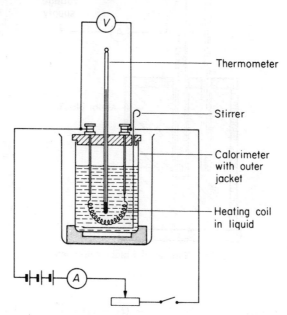

Thermometer

Stirrer

Calorimeter with outer jacket

Heating coil in liquid

Fig. 42.2. Specific heat capacity of a liquid (electrical method)

and a steady current maintained for 5 minutes. Meanwhile, the liquid is stirred continuously. At the end of the 5 minutes the current is switched off and the final steady temperature of the liquid is noted. The results may be recorded, as shown, symbols being replaced by actual readings.

Readings   (*Units*)

Mass of calorimeter and stirrer	$= m_1$	(g)
Mass of calorimeter, stirrer and liquid	$= m_2$	(g)
Mass of liquid $\quad = (m_2 - m_1) = m$		(g)
Initial temperature of liquid	$= \theta_1$	(°C)
Final temperature of liquid	$= \theta_2$	(°C)
Rise in temperature $\quad = (\theta_2 - \theta_1) = \theta$		(K)
Current	$= I$	(A)
P.d. across heater coil	$= V$	(V)
Time	$= t$	(s)

Assume, specific heat capacity of copper $= 0.40$ J/g K

Calculation

Energy supplied by heater	$= VIt$	(J)
Let specific heat capacity of liquid	$= c$	(J/g K)
Heat received by liquid	$= mc\theta$	(J)
Heat received by calorimeter and stirrer	$= m_1 \times 0.40 \times \theta$	(J)

Hence, $mc\theta + 0.40 m_1 \theta = VIt$

or,

specific heat capacity of liquid $= c = \dfrac{VIt - 0.40 m_1 \theta}{m\theta}$ (J/g K)

Note. Owing to the fairly long time of heating suggested for the above experiment, heat losses are inevitable. If, however, the liquid is first cooled to about 5 K below room temperature and the heating continued until it is about 5 K above, we may assume that the heat gain from the atmosphere during the first half of the time will be compensated by that lost to the atmosphere during the second half.

It is of interest to note that this simple cooling correction was first suggested by Count Rumford back in the eighteenth century.

The last two experiments are based on a steady rate of supply of heat to the substances concerned. We shall now describe a different procedure known as the method of mixtures.

To measure the specific heat capacity by the method of mixtures

This method is suitable for substances not readily available in block form or which are not such good heat conductors as copper or aluminium. A copper calorimeter similar to that used in the previous experiment, and containing a fixed mass of

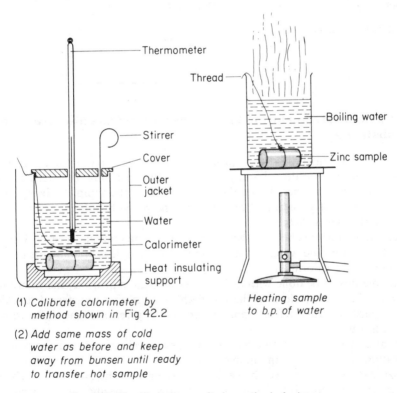

(1) *Calibrate calorimeter by method shown in Fig 42.2*

(2) *Add same mass of cold water as before and keep away from bunsen until ready to transfer hot sample*

Heating sample to b.p. of water

Fig. 42.3. Specific heat capacity by method of mixtures

water, is calibrated electrically so that we can subsequently use it as a heat energy measuring device of known total heat capacity (Fig. 42.3).

Calibration of calorimeter

The calorimeter is about two-thirds filled with cold water. For our present purpose we shall not need to know either the mass of the calorimeter or the water so long as we always use the same mass of water. This is therefore best done by adding water with a pipette so that we can easily and quickly get the same mass of water in later experiments.

The experiment is carried out exactly as described in the last section except that the results are treated differently.

Readings

Mass of calorimeter + water in grams = m (for reference only in subsequent experiments)

Initial temperature	$= \theta_1$	(°C)
Final temperature	$= \theta_2$	(°C)
Rise in temperature	$= (\theta_2 - \theta_1) = \theta$	(K)
Current	$= I$	(A)
P.d. across heater coil	$= V$	(V)
Time	$= t$	(s)

Calculation

Energy supplied by heater = VIt (J)

Let total heat capacity of calorimeter
+ water = C (J/K)

Heat energy received by calorimeter
+ water = $C\theta$ (J)

Hence, $C\theta = VIt$

or, total heat capacity of calorimeter

$$+ \text{ water} = C_, = \frac{VIt}{\theta}$$

$$= \underline{\hspace{4cm}} \quad \text{J/K}$$

Use of a calibrated calorimeter to measure the specific heat capacity of a substance

A sample of the material whose specific heat capacity is required, e.g., a piece of zinc, is suspended by a thread and placed in a beaker or mug of water kept boiling gently by a bunsen burner. While the temperature of the sample is being raised to 100 °C, the previously calibrated calorimeter is prepared by adding exactly the same mass of cold water as before.

When the sample has been in the boiling water long enough to reach 100 °C, the temperature of the cold water is noted, and the hot sample is lifted by the thread and transferred to the calorimeter.

To enable this to be done with as little loss of heat as possible, the calorimeter in its jacket is held fairly close to the hot bath and the zinc transferred quickly, but without splashing. Just before transfer, the zinc is given a slight shake to remove adhering hot water.

The calorimeter cover is put on, the mixture is well stirred and the final steady temperature read from the thermometer.

The zinc sample may be weighed either before or after the experiment. The results of an experiment are given below to show how the specific heat capacity is calculated.

From previous calibration experiment using 100 g of water in the calorimeter

Heat capacity of calorimeter + water	= 440 J/K
Mass of zinc	= 150 g

Temperatures

Hot zinc	= 100 °C
Cold water	= 14.2 °C
Mixture	= 24.2 °C
Change in temp. of zinc	= (100 − 24.2) = 75.8 K
Change in temp. of cal. + water	= (24.2 − 14.2) = 10.0 K

Calculations

Heat given out by zinc	= mass × sp. ht. cap. × temp. change
	= 150 × c × 75.8
Heat received by cal. + water	= heat capacity × temp. change
	= 440 × 10.0

Hence,

$$150 \times c \times 75.8 = 440 \times 10.0$$

$$c = \frac{440 \times 10.0}{150 \times 75.8}$$

$$= 0.39 \text{ J/g K}$$

Note.

(1) The result is affected, of course, by errors inherent in the experiment, such as the inevitable small loss of heat in transferring the hot zinc to the calorimeter together with losses from the calorimeter itself.

(2) The method is of interest in that, broadly speaking, it illustrates the principle of calibration of the *fuel calorimeter*. In order to find the *heat of combustion* of various fuels, a small weighed sample is placed in a platinum crucible contained in a stainless steel "bomb" containing oxygen under high pressure. The bomb is immersed in water in a calorimeter and the fuel ignited by an electrically heated platinum wire.

The calorimeter, water, and bomb without fuel are calibrated electrically by a separate experiment to find their total heat capacity, C in J/K. Having thus measured C, the heat of combustion of the fuel sample is calculated from $C\theta$ where θ is the measured rise in temperature in K.

To measure the specific latent heat of steam (approximate method)

A large calorimeter provided with a lid is weighed, first empty, and then when three-parts full of water. An electric immersion heater is placed in the water and connected in circuit as shown in Fig. 42.4.

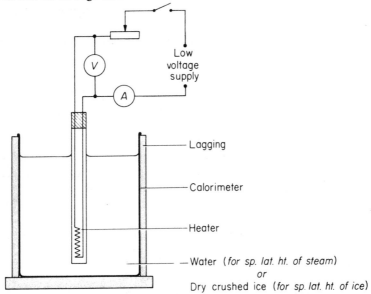

Fig. 42.4. Specific latent heat (approximate method)

Loss of heat may be reduced by lagging the calorimeter with expanded polystyrene in the manner as described for the solid block experiment on p. 472.

The lid is removed, and the current switched on and adjusted to a suitable value. As soon as the water starts to boil, a stopclock is started. The current is kept constant by the rheostat, and switched off 15 minutes later. The immersion heater is then removed and the calorimeter immediately covered with its lid to prevent further loss of vapour.

Having removed the polystyrene jacket the calorimeter is cooled by immersion in cold water and then dried with a cloth and reweighed. The various readings to be taken are detailed below.

Readings		*(Units)*
Mass of calorimeter, lid and water	$= m_1$	(g)
Current	$= I$	(A)
P.d. across heater coil	$= V$	(V)
Time of boiling	$= t$	(s)
Mass of calorimeter, lid and water after		
boiling	$= m_2$	(g)

therefore,   mass of water boiled away $= m_1 - m_2 = m$

Calculation

Energy supplied by heater in vaporizing		
water	$= VIt$	(J)
Let specific latent heat of steam	$= l$	(J/g)
Energy absorbed by steam at constant		
temperature	$= ml$	(J)

Hence, $ml = VIt$

i.e., specific latent heat of steam $= l = \dfrac{VIt}{m}$ (J/g)

or, converting to SI units $l = 1000\,\dfrac{VIt}{m}$ (J/kg)

Comment

In the above experiment, although we took precautions to reduce heat loss and prevent loss of vapour, we could not avoid error owing to loss of vapour during the period when the water was coming to the boil.

We shall now describe a more accurate experiment in which these particular errors are largely eliminated.

To measure the specific latent heat of steam (more accurate method)

The apparatus for this experiment consists of an inverted vacuum flask containing distilled water, the level of which can be observed and maintained as required by means of a funnel and side tube (Fig. 42.5).

The water inside is boiled by a heater coil supplied with a current of I in amperes at a p.d. of V in volts. After the water has been boiling for some time, a steady temperature state is reached and, under these conditions the rate of supply of heat energy, VI in watts (i.e., joules per second) is equal to the rate of absorption of latent heat into the vapour.

The vapour passes down the central glass tube projecting up into the boiler and is condensed back to water as it passes through a straight tube surrounded by a cold water jacket.

When steady temperature conditions have been reached, a dry weighed beaker is placed under the outlet tube and the time for about 20 g of water to collect is measured. The beaker and water are then weighed to find the mass of water condensed.

Readings		*(Units)*
Current	$= I$	(A)
P.d. across heater	$= V$	(V)
Time	$= t$	(s)

Mass of empty beaker $= m_1$ (g)
Mass of empty beaker + water $= m_2$ (g)
Mass of water evaporated $= (m_2 - m_1) = m$ (g)

Calculations
Energy supplied by heater $= VIt$ (J)
Latent heat absorbed by vapour $= ml$ (J)
(where l = specific latent heat of steam in J/g)
Hence, $ml = VIt$

or, $l = \dfrac{VIt}{m}$ (J/g)

or, in SI units $l = 1000\,\dfrac{VIt}{m}$ (J/kg)

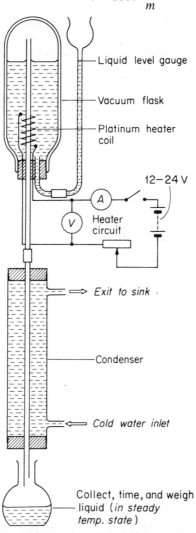

Liquid level gauge

Vacuum flask

Platinum heater coil

12–24 V

A

V   Heater circuit

Exit to sink

Condenser

Cold water inlet

Collect, time, and weigh liquid (*in steady temp. state*)

Fig. 42.5. Specific latent heat of steam (more accurate method)

Comment

This is an example of a steady temperature state, continuous flow experiment, in which errors due to heat and vapour loss during the warming up period and heat loss during the boiling period have been almost entirely eliminated.

The heat capacity of the apparatus itself does not enter into the calculation, and the use of a vacuum vessel reduces heat losses during the steady temperature state to a minimum.

To measure the specific latent heat of ice by direct heating

A calorimeter lagged with sheet expanded polystyrene, as described for the approximate determination of the specific latent heat of steam, is used in this experiment also. This is preferable to the use of a polished calorimeter in an outerjacket owing to the fact that water vapour from the air is liable to condense on the surface and give up unwanted latent heat.

The circuit shown in Fig. 42.4 is connected up, and the calorimeter is then weighed empty. After being lagged in the manner described it is filled with small pieces of melting ice which have been previously dried on a cloth.

The immersion heater is then inserted into the ice. Some forms of heater may be themselves used as stirrers; otherwise a separate stirrer is essential. A stopclock is started and simultaneously the current is switched on. Stirring is kept up continuously, and more dried ice added as the water level sinks. As soon as all the ice has just melted the current is switched off and the time noted. Finally, the jacket is removed and the calorimeter is weighed.

Readings *(Units)*

Mass of empty calorimeter	$= m_1$	(g)
Current	$= I$	(A)
P.d. across heater	$= V$	(V)
Time	$= t$	(s)
Mass of calorimeter + melted ice	$= m_2$	(g)
Mass of ice used $= (m_2 - m_1)$	$= m$	(g)

Calculation

Energy supplied by heater	$= VIt$	(J)
Let specific latent heat of fusion of ice	$= l$	(J/g)
Heat absorbed by ice at 0 °C in melting to form water at 0 °C	$= ml$	(J)

Then,
$$ml = VIt$$

therefore
$$l = \frac{VIt}{m} \qquad \text{(J/g)}$$

or, in SI units
$$l = 1000\,\frac{VIt}{m} \qquad \text{(J/kg)}$$

To measure the specific latent heat of ice by the method of mixtures

The first step is to measure the specific heat capacities of copper and water by the methods described earlier in this chapter. For our present purpose we may assume the following values, using the SI sub-unit J/g K.

Specific heat capacity of water $= 4.2$ J/g K
Specific heat capacity of copper $= 0.40$ J/g K

In this experiment we shall use a polished copper calorimeter in a polished copper jacket similar to that shown in Fig. 42.3. The copper calorimeter and stirrer are first weighed empty, and then with the calorimeter about two-thirds full of water warmed to a temperature about 6 K above room temperature.

A thermometer is placed in the water, and some small pieces of ice are carefully dried on a cloth or filter paper. The water is stirred, the temperature noted, and immediately the dry ice is added. Stirring continuously, sufficient ice is added to bring the final steady temperature down to about 6 K below room temperature. This final steady temperature is noted and the calorimeter and its contents weighed.

The following is a typical set of results obtained for this experiment.

Weighings
Calorimeter and stirrer	$= 55.0$ g
Calorimeter, stirrer and water	$= 148.0$ g
Calorimeter, stirrer, water and ice	$= 161.4$ g

Temperatures
Warm water	$= 22.2\ ^\circ$C
Final steady temperature of mixture	$= 10.0\ ^\circ$C
Fall in temperature $= (22.2 - 10)$	$= 12.2$ K

Calculations
Mass of warm water	$= 93.0$ g
Mass of ice	$= 13.4$ g
Let specific latent heat of ice	$= l$ (in J/g)
The specific heat capacity of water	$= 4.2$ J/g K
The specific heat capacity of copper	$= 0.40$ J/g K

In this experiment the heat given out by the calorimeter, stirrer and water in cooling from the initial to the final temperature is equal to the heat received by the ice. Now the ice may be regarded as receiving *two* lots of heat. First, the latent heat necessary to melt it to form water at 0 °C; and secondly, the heat to raise the now melted ice from 0 °C to the final temperature 10.0 °C. Thus:

Heat given out by calorimeter and stirrer	$= 55 \times 0.40 \times 12.2$ J
Heat given out by warm water	$= 93 \times 4.2 \times 12.2$ J
Heat received by ice to melt it to water at 0 °C	$= 13.4 \times l$
Heat received by melted ice to raise it from 0 °C to 10.0 °C	$= 13.4 \times 4.2 \times 10.0$ J

Equating the heat received to the heat given out,
$$(13.4 \times l) + (13.4 \times 4.2 \times 10.0) = (55 \times 0.40 \times 12.2) + (93 \times 4.2 \times 12.2)$$
whence $\qquad\qquad\qquad l = 330$ J/g
or, in SI units $\qquad\qquad l = 330 \times 1000$ J/kg
$$= 3.3 \times 10^5 \text{ J/kg}$$

It will be noted that Rumford's cooling correction, mentioned earlier, had been applied in this experiment. It is important not to allow the temperature to go more than 5 or 6 K below room temperature otherwise water vapour in the atmosphere may condense to form dew on the outside of the calorimeter and give up latent heat to it.

QUESTIONS: 42

(Note. Additional numerical examples will be found at the ends of chapters 18 and 19)

1. A manufacturer designs an immersion heater which has a power output of 120 W. The heater is used to raise the temperature of 2 kg of a liquid from 15 °C to 35 °C in 10 minutes. Assuming that 25% of the energy supplied by the heater is lost from the liquid to the surroundings, calculate
 (i) the energy supplied by the heater in 10 minutes,
 (ii) the specific heat capacity of the liquid.
The immersion heater is made of insulated wire of resistance 12 Ω/m. If the immersion heater is designed for use with the 240 V main supply, calculate
 (iii) the current through the immersion heater,
 (iv) the resistance of the immersion heater,
 (v) the length of wire required.
$\qquad\qquad\qquad$ (A.E.B., 1982)
2. Explain the meaning of the term *specific heat capacity* of a substance.
 Describe how you would determine the

specific heat capacity of aluminium. You may, if you wish, take the specific heat capacity of water as known.

An electric kettle, having a 1.5 kW heating element, contains 925 g of water. With the heater on, the rate of rise of temperature of the water when its temperature is 60 °C is 20.2 K per minute. After the heater is switched off the temperature falls and the rate of fall, again at 60 °C, is 1.4 K per minute.

What would be the rate of rise of temperature at 60 °C if there were no heat loss?

Calculate a value for the specific heat capacity of water, given that the heat capacity of the kettle including the heater is the same as that of 75 g of water. (*O.C.*)

3. What is meant by the statement that the specific heat capacity of water is 4200 J/kg K?

Describe how you would attempt to verify this statement by experiment.

An electric kettle which produces energy at a rate of 2250 W contains 0.80 kg of water. It takes 150 seconds to heat the water from 10 to 100 °C.

Calculate:

(i) the heat produced by the kettle in this time;

(ii) the heat taken in by the water in this time.

Suggest *two* reasons for the difference between these two quantities.

If the heater continues to supply energy for a further 100 s it is found that 0.100 kg of water is converted to steam.

Calculate a value for the *specific latent heat of steam.* (*C.*)

4. With the aid of a labelled diagram describe an experiment, based on a steady rate of supply of heat, for measuring the specific heat capacity of paraffin oil. Show how the result is calculated and explain the procedure you would adopt to effect a simple cooling correction.

5. Find the time taken by a 500 W heater to raise the temperature of 50 kg of material, of specific heat capacity 960 J/kg K, from 18 to 38 °C. Assume that all the heat from the heater is given to the material. (*C.*)

6. A 50 W heating coil is totally immersed in 100 g of water contained in an insulated flask of negligible heat capacity.

(i) If the temperature of the water is 20 °C when the heater is switched on, how long would it take for the water to boil?

(ii) After the water has been boiling for 15 minutes it is found that the mass of water in the flask has decreased to 80 g. Assuming no external heat losses, calculate a value for the specific latent heat of vaporisation of water.

(Assume that the specific heat capacity of water = 4200 J/kg K.) (*J.M.B.*)

7. A tin can contains water at 290 K and is heated at a constant rate. It is observed that the water reaches boiling point after 2 min and after a further 12 min it has completely boiled away. Calculate the specific latent heat of steam and state why your value is only approximate. (Specific heat capacity of water = 4200 J kg^{-1} K^{-1}.)* (*S.*)

8. Some water initially at 15 °C, contained in an aluminium can, is heated at a steady rate and comes to the boil in 8 min; 13 min later, one-quarter of the water has boiled away. Calculate the specific latent heat of steam. Compare the value obtained with that given on page 205 and explain any discrepancy. (Sp. ht. cap. of water = 4.2 J/g K.)

9. A well-lagged electric steam boiler has a heat capacity of 1800 J/K and is fitted with a safety valve which allows steam to escape when the temperature rises to 106 °C.

A litre of water at 26 °C is placed in the boiler and steam begins to blow from the valve 6 min after switching on the heater. After a further 14 min, half of the water has boiled away. Assuming that heat losses are negligible, find the specific latent heat of steam at 106 °C. (Mean specific heat capacity of water between 26 °C and 106 °C = 4.2 J/g K.)

10. 600 g of lead is melted in a crucible and it is found that a 42 W electric immersion heater will keep it in the liquid state at its melting-point. If the heater is switched off, the temperature of the lead begins to fall 5 min later. Calculate the specific latent heat of lead.

11. 1 t of iron is heated from 40 °C in an electric furnace until it is all melted at 1540 °C, and it is found that 1664 MJ of electric energy are required. If the mean specific heat capacity of iron over the temperature range involved is 590 J/kg °C, and assuming that 30 per cent of the energy is wasted, calculate the specific latent heat of iron.

12. Describe fully a method for the determination of the specific latent heat of fusion of ice indicating any precautions taken to achieve an accurate result. Show how the result is obtained from your observations. (*L.*)

13. A solid of mass 2 kg receives heat at the constant rate of 1 × 10⁵ J per minute. Its

---

* See footnote page 169.

temperature T (in K) during the first 30 min of the experiment was as follows:

Time t (minutes)	0	1	2	6	9	13	18	23	28	29	30
Temperature T (K)	230	250	270	270	270	270	310	350	390	390	390

Sketch a graph of temperature against time.

Calculate those of (i) to (vii) below which can be determined from your sketch and the information given. If a property cannot be determined, simply state this as your answer.

Indicate which of your numerical answers you would expect to be inaccurate due to heat losses during the experiment, stating whether your answer would be too high or too low.

(i) The specific heat capacity of the substance in the solid state.

(ii) The specific heat capacity of the substance in the liquid state.

(iii) The specific heat capacity of the substance in the vapour state.

(iv) The melting point of the solid.

(v) The boiling point of the liquid.

(vi) The specific latent heat of fusion or melting.

(vii) The specific latent heat of vaporisation or boiling. (*A.E.B.*)

14. 200 g of a liquid at 21 °C was heated to 51 °C by a current of 5 A at 6 V for 5 minutes. Neglecting heat losses, calculate the specific heat capacity of the liquid.

If the current was changed to 2.5 A and all other factors remained the same, calculate the temperature rise. (*W.A.E.C.*)

43. Electromagnetic induction

In 1791, the year in which Galvani performed his famous experiment with frog's legs, a son was born to a blacksmith in Newington, London. This was Michael Faraday, destined to become one of the greatest scientists of the nineteenth century. When he reached the age of fourteen his father apprenticed him to a bookbinder, and in this occupation he found plenty of opportunity for reading. It was not long before young Faraday developed a liking for physics and chemistry and became ambitious to take up a career in science. One day a customer who was aware of Faraday's interest in science presented him with tickets for a course of lectures to be given by Sir Humphry Davy at the Royal Institution in Albemarle Street. The Royal Institution had been founded by Count Rumford in 1800 for the advancement of applied science, and in the early years of the century Sir Humphry Davy had become famous there, both for his brilliant scientific researches and his skill as a lecturer.

Faraday attended the lectures with much enthusiasm, and afterwards wrote out his notes and bound them in leather. He then sent the notes to Sir Humphry Davy, together with a letter earnestly expressing his desire to become a scientist and asking if he could be employed at the Royal Institution. At first Davy was inclined to advise Faraday to stay as a bookbinder. However, he was so impressed with Faraday's sincerity that when a vacancy for a laboratory assistant occurred some months later he offered it to him.

Faraday immediately accepted the offer, and so started on a career which ultimately led to his appointment as director of the Institution. We cannot here go into details of all Faraday's scientific researches during the lifetime he spent at the Institution, but there are several biographies available which tell the fascinating story of this great scientist's life and achievements.

Electricity from magnetism

Ever since Oersted's discovery, in 1819, that magnetism could be produced by an electric current scientists had been concerned with the problem of how to bring about the reverse effect and to produce electricity from magnetism. Faraday, in particular, had always been interested in this subject, but it was not until 1831 that he found time to tackle it seriously. His notebooks, which are preserved at the Royal Institution, show that in August of that year he began a series of experiments in which he demonstrated the principle of the *dynamo* or electric generator. It is the purpose of this chapter to discuss one or two of the more important of these experiments.

Faraday's experiments on electromagnetic induction

In an entry in his notebook dated 17 October 1831, Faraday described how he made a cylindrical coil of copper wire wound round a paper tube. In those days one could not purchase insulated wire, and so the coil turns had to be separated from one another by means of string wound on the tube at the same time as the wire. Successive layers of turns were separated by strips of calico. When the coil was finished its ends were connected to a sensitive galvanometer. Faraday then plunged a magnet into the coil and noticed that the galvanometer needle gave a momentary deflection, showing that a current had been induced in the coil. On removal of the magnet from the coil the galvanometer gave another deflection, this time in the opposite direction (Fig. 43.1). This effect is called *electromagnetic induction*. No current was induced simply by allowing the magnet to remain at rest inside or outside the coil.

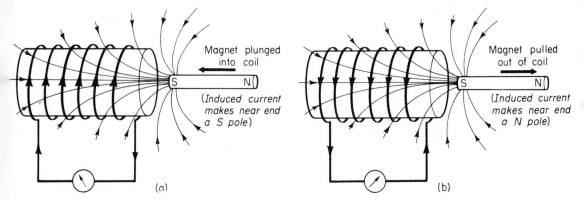

Fig. 43.1. Electromagnetic induction

After much thought Faraday came to the conclusion that current was induced in the circuit as a result of the wire being *cut by magnetic flux* when the magnet moved. Another way of describing this phenomenon is to say that an **electromotive force is induced whenever there is a change in the magnetic flux linked with the coil.**

Continuing his investigations with other coils and magnets, Faraday found that the strength of the induced current depends on:

(1) *the number of turns in the coil;*
(2) *the strength of the magnet;*
(3) *the speed with which the magnet is plunged into the coil.*

In another experiment Faraday wound a coil on a bar of soft iron and connected the ends of the coil to a galvanometer (Fig. 43.2). The opposite poles of a pair of bar

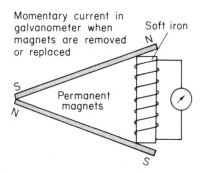

Fig. 43.2. One of Faraday's experiments on electromagnetic induction

magnets were then brought into contact with the ends of the iron bar. This produced a large concentration of magnetic flux through the coil, and the galvanometer gave a kick, showing that a momentary current had been induced. On removing the magnets a current was induced in the coil in the opposite direction.

The above experiment will also work without iron inside the coil, but in this case the induced current is much weaker. The effect of the iron is to increase the magnetic flux through the coil, and consequently the induced electromotive force is greater.

Using the earth's magnetic flux to induce a current

During the course of his investigations it occurred to Faraday that it ought to be possible to use the earth's magnetic flux to induce a current in a coil. He accordingly placed an unmagnetized bar of soft iron inside a coil which was pointing in the direction of a dipping needle (Fig. 43.3). On reversing the coil end for end, a momentary deflection was obtained in a galvanometer connected to the ends of the coil.

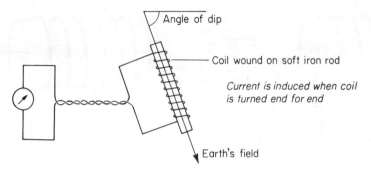

Fig. 43.3. Electromagnetic induction by the earth's magnetic field

In this experiment magnetic flux was set up through the iron and coil. When the coil was reversed the direction of the flux through it was reversed, and hence an e.m.f. was induced.

The results of the experiments which have been described are summed up in the following law.

Faraday's law of electromagnetic induction

Whenever there is a change in the magnetic flux linked with a circuit an electromotive force is induced, the strength of which is proportional to the rate of change of the flux linked with the circuit.

After publication of the details of Faraday's work on the production of electricity from magnetism, scientists in other countries repeated the experiments for themselves and continued the investigations.

In Russia, Henry Lenz made observations of the direction of the induced current in a number of experiments, and in 1834 he was able to state another law relating to electromagnetic induction.

Lenz's law of electromagnetic induction

The direction of the induced current is always such as to oppose the change producing it.

Fig. 43.1 illustrates the application of this law in the case of a magnet plunged into a coil. When the S pole of the magnet is plunged into the coil the induced current must flow in such a direction as to give S polarity to the end of the coil facing the magnet. Thus, since like poles repel, the motion of the magnet is opposed by the induced current. On the other hand, when the S pole of the magnet is pulled away from the coil the direction of the induced current is reversed. The near end of the coil becomes a N pole and, since unlike poles attract, the motion of the magnet is again opposed.

Looked at from another point of view, Lenz's law is simply an application of the law of conservation of energy. Energy is expended when the induced current flows, and the source of this energy is the work done when the magnet is moved. Hence, if work is to be done when the magnet moves with respect to the coil, it must experience an opposing force.

Direction of the induced current in a straight wire

In one of his numerous experiments Faraday showed that a current was induced in a straight wire when it was moved at right angles to the magnetic flux near a bar magnet. Many years later Fleming chose this experiment to illustrate a simple rule relating the direction of motion of a wire, the direction of the magnetic flux and the direction of the induced current.

First of all, a centre-zero galvanometer is connected in series with a cell and a suitable high resistance and the direction of movement of the pointer noted when a

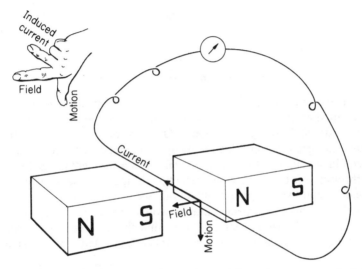

Fig. 43.4. Current induced in a wire moving at right angles to a magnetic field

small current is passed in a known direction. This knowledge will afterwards enable the instrument to be used for testing current direction.

The galvanometer is now connected to the ends of a straight wire placed at right angles to the magnetic flux between two opposite magnetic poles (Fig. 43.4). If the wire is moved downwards the galvanometer indicates that an induced current flows in the direction shown in the diagram. When the wire is moved upwards the induced current is reversed.

It occurred to Fleming that the thumb and first two fingers of the right hand could be used to predict the direction of the induced current, so he accordingly stated the following rule (Fig. 43.5):

Fleming's right-hand rule (dynamo rule)

Extend the thumb, forefinger, and second finger of the right hand mutually at right angles. If the forefinger points in the direction of the field and the thumb in the direction of the motion, then the second finger will point in the direction of the induced current.

Aid to memory: thuMb – Motion, foreFinger – Field, seCond finger – Current.

Mutual induction. Faraday's iron ring experiment

As well as using a magnet to induce a current in a coil, Faraday also employed a second coil carrying a current. The manner in which he did this is shown in Fig. 43.6.

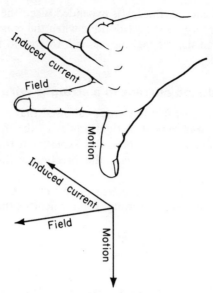

Fig. 43.5. Fleming's right-hand rule (dynamo rule)

A primary coil P and a secondary coil S are wound on opposite sides of an iron ring about 15 cm in diameter and 2 cm thick. Coil P is connected to a battery and tapping key, and coil S to a galvanometer. On pressing the key an electric current flows through the primary coil. This builds up a magnetic flux through the iron ring and the secondary coil, and the galvanometer gives a momentary deflection. When the circuit through P is broken the magnetic flux disappears and a current is induced in the secondary coil in the opposite direction. See also Fig. 43.7.

As a variation of this experiment, Faraday wound two coils side by side on a piece of wood and obtained a similar effect, except that the induced current was weaker. When iron is present the magnetic flux becomes concentrated in it. This ensures that

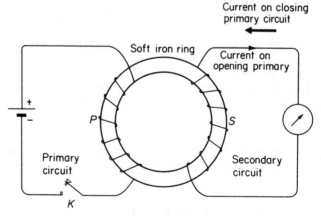

Fig. 43.6. Faraday's iron ring experiment

practically all of the primary flux becomes linked with the secondary, with a consequent increase in the secondary e.m.f.

Any arrangement of two coils such that a current is induced in one when the current is changed in the other is said to possess *mutual inductance*.

Application of Lenz's law to the iron ring experiment

Lenz's law may be applied to the iron ring experiment in order to predict the direction of flow of the induced current in the secondary coil on either the make or

Fig. 43.7. *The original Faraday ring.* One of the treasures of the Royal Institution

the break of the primary current. Thus, in Fig. 43.6 when the primary circuit is closed a magnetic flux will be set up in the iron ring in a clockwise direction. Consequently, the induced current in the secondary will flow in such a direction as to set up a flux in an anticlockwise direction. Likewise, when the primary circuit is broken the flux in the ring will die away. The induced current in the secondary tries to prevent this by setting up a flux in the same, i.e., clockwise, direction.

By applying one of the rules given on pages 420–1, the reader should check that the induced current directions in Fig. 43.6 have been correctly indicated.

Self-induction

When the coil of a large electromagnet is connected in series with a 6 V battery and a tapping key a small arc occurs at the switch contacts when the circuit is broken. This is caused by the large self-induced e.m.f. in the coil as a result of the disappearance of its own magnetic flux. The effect is called *self-induction*. In the case of a coil with a very large number of turns the self-induced e.m.f. will be much greater than that of the battery.

The high e.m.f. of self-induction may be demonstrated by connecting a neon discharge lamp across the terminals of an electromagnet which has already been wired up to a battery and key (Fig. 43.8). A neon lamp requires at least 180 V to cause it to glow, yet it flashes every time the current in the coil is switched off.

What has happened here is that some of the electric energy from the battery became transferred to magnetic energy in the field of the electromagnet. When the circuit was broken the magnetic field disappeared, feeding its energy back into the circuit in the form of electric energy.

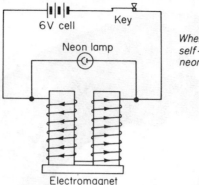

When the circuit is broken the high self–induced e.m.f. causes the neon to flash

Fig. 43.8. Demonstrating self-induction

Non-inductive resistance coils

In certain experiments self-induction in resistance coils is a disadvantage, and therefore the coils are wound as shown in Fig. 40.1. By this method of winding the coils have zero resultant magnetic field, and consequently no self-induced counter e.m.f. is set up when the current through them changes.

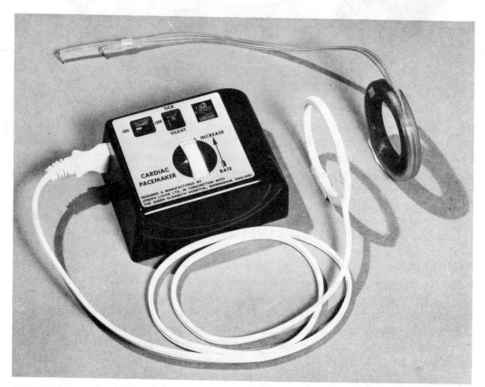

Fig. 43.9. An external heart pacemaker of the 1960s which transferred electric pulses to the heart by electromagnetic induction. The white primary coil was attached, by adhesive tape, on the outside of the chest immediately above a plastic covered secondary coil which had been surgically implanted inside the chest wall. When the primary was supplied with current impulses from the small electronic generator, it induced pulses in the secondary which were conveyed to the heart muscles by the electrodes seen at the top.

Nowadays, miniaturization of components and the development of materials resistant to body fluids, have resulted in much smaller units. Powered by very long-life batteries, these units are completely implanted in the chest where they apply a direct stimulus to the heart

Applications of electromagnetic induction

Faraday's work on electromagnetic induction in 1831 aroused great interest the world over, and men quickly saw the possibility of putting the new discoveries to practical use. Within a dozen years dynamos had been designed and built and were being used for industrial purposes. These machines were used at first mainly to provide the current required in electroplating works. Later, during Faraday's lifetime, dynamos were installed for running the arc lamps in lighthouses. The first electric power stations for public supply were built soon after 1880, but unfortunately Faraday did not live to see them. In the remainder of this chapter we shall deal with some of the applications of electromagnetic induction, including the construction and action of simple dynamos.

Before proceeding to the next section, take a look at Fig. 43.9.

Eddy currents. Waltenhofen's pendulums

If something is done to cause a change in the magnetic flux through a piece of metal induced currents, called *eddy currents*, will circulate inside it. The flux change may be brought about either by moving the metal in a constant magnetic field or else by subjecting the metal to a varying magnetic field.

Waltenhofen showed how to demonstrate the effect of eddy currents induced in a piece of copper moving in a magnetic field. His apparatus consists of a pendulum having a curved strip of copper at its lower end and arranged so that it hangs between the poles of an electromagnet (Fig. 43.10 (a)). So long as no magnetic flux passes through the copper, the pendulum swings freely, but as soon as the electromagnet is switched on the oscillations of the pendulum are heavily damped and die away rapidly. This may be explained as follows. As the pendulum swings to and fro it cuts magnetic flux, and consequently an e.m.f. is induced in it. This induced e.m.f. sets up eddy currents inside the copper, which, in accordance with Lenz's law, set up magnetic fields which oppose the motion of the pendulum.

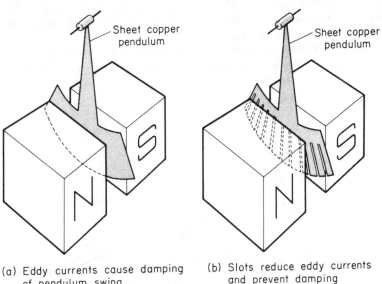

(a) Eddy currents cause damping of pendulum swing

(b) Slots reduce eddy currents and prevent damping

Fig. 43.10. Eddy current damping (Waltenhofen's pendulums)

When this pendulum is replaced by a similar one having slots cut in the copper strip the oscillations are much less damped whether the magnetic field is present or not (Fig. 43.10 (b)). In this case, although an e.m.f. is still induced in the metal, eddy currents cannot flow across the gaps.

The induction furnace

Eddy currents are put to good use in the induction furnace (Fig. 43.11). Basically, this is a crucible containing metal to be melted, which is surrounded by a coil made of copper tubing. When a very high frequency alternating electric current is passed

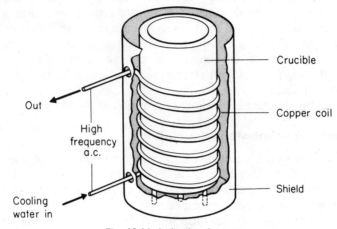

Fig. 43.11. Induction furnace

through the coil the rapidly changing magnetic flux through the metal induces eddy currents in it. The internal energy produced causes the metal to melt.

Cold water is passed through the tubular coil to keep it cool.

The speedometer

If a copper disc is rotated so that it cuts the magnetic flux of a magnet pivoted above it eddy currents are set up in the disc, which, in accordance with Lenz's law, oppose the relative motion of magnet and disc. Consequently, the magnet tends to rotate and follow the motion of the disc. This experiment was first carried out by the French physicist, Dominique Arago.

The same principle is applied in one type of motor car speedometer, in which a rotating magnet, driven by a flexible cable, sets up eddy currents in a pivoted

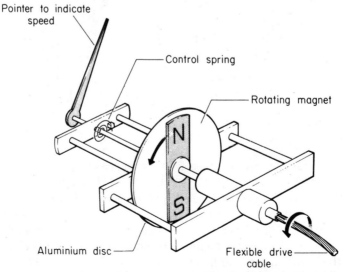

Fig. 43.12. Speedometer movement based on Arago's rotation effect

aluminium disc (Fig. 43.12). The magnetic reaction of the eddy currents sets up a couple on the disc which varies with the speed of the car. The disc therefore rotates until the electromagnetic couple is just balanced by an opposing couple set up by a hairspring. A pointer attached to the disc spindle indicates the speed on a suitably calibrated scale.

A further application of eddy currents is illustrated in Fig. 43.23 at the end of this chapter.

The induction coil

In the year 1851, a German instrument maker named Heinrich Ruhmkorff showed how an electric spark 5 cm long could be produced by using a battery of only a few volts e.m.f. The apparatus he used is called an induction coil or Ruhmkorff spark coil (Fig. 43.13). It has a primary coil of comparatively few turns of thick wire wound on a core consisting of a bundle of soft-iron wires. On top of this is wound a

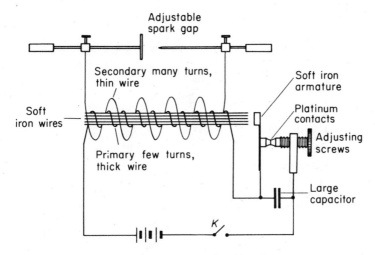

Fig. 43.13. Induction (spark) coil

secondary coil having many thousands of turns of thin wire. The ends of the secondary are connected to an adjustable spark gap.

The primary circuit, which may be designed for use with either a 6 or a 12V battery, is provided with a make-and-break device similar to that used on an electric bell, but of heavier construction. When the switch K is closed current flows through the primary coil and magnetizes the iron core. Attraction of the soft-iron armature then occurs and causes the platinum contacts to separate and break the circuit. When this happens the core becomes demagnetized and the armature is pulled back again by the spring on which it is mounted.

After this the whole process is repeated. As a result, the armature vibrates to and fro, causing a rapid make and break of the primary current. Each time the primary current is broken the magnetic flux through the core collapses suddenly and an e.m.f. is induced in the secondary. Because the secondary has such a large number of turns, the induced e.m.f. is very high indeed, and is sufficient to cause a spark to jump across the gap.

Owing to the self-induction of the primary, a small arc tends to form between the make-and-break contacts every time they open, and this makes them wear away. This is prevented by connecting a large capacitor across the contacts. The use of insulated soft-iron wires in the core reduces eddy currents and consequent energy loss (see also under transformer on page 498).

Uses of the induction coil

Ruhmkorff's coil of 1851 gave a 5 cm spark. At the Paris Exhibition of 1881 a coil was shown producing a spark 60 cm long. During the nineteenth century these coils played a part in the investigation of high voltages and the study of electric discharge through gases. In the early years of the present century they were used in morse radio transmitters and also for operating X-ray tubes. For many years the main importance of the induction-coil principle has been its application to the coil-ignition system of motor vehicles. This is being displaced by electronic systems.

Coil ignition

When a petrol engine is running a mixture of petrol vapour and air is drawn into each cylinder in turn, compressed, and then ignited by an electric spark. This spark is produced by an induction coil and is timed to occur at the right moment by means of a contact-breaker operated by a rotating cam (Fig. 43.14). In a four-cylinder engine the cam has four projections. As the cam rotates these projections press against a

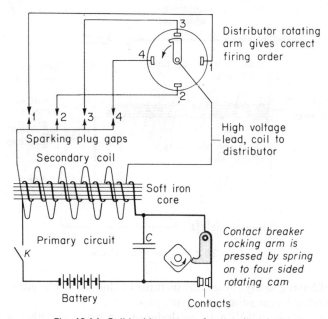

Fig. 43.14. Coil ignition system for petrol engine

pivoted rocking arm, causing contacts to open and break the primary circuit. Each time this happens a current pulse at high voltage is generated in the secondary coil. This pulse is conveyed through a well insulated cable to a rotating distributor arm and thence connected to the correct sparking plug at the right moment. A sparking plug has a central insulated electrode, and the spark takes place between this and another electrode fixed to the metal body of the plug. The current finds a return path to the secondary coil through the engine and the framework of the vehicle.

The simple alternating current generator

Fig. 43.15 illustrates the construction of a simple form of a.c. dynamo for the continuous production of electric current by electromagnetic induction. The machine has a rectangular coil of wire which is rotated in the magnetic field between the poles of a U-shaped permanent magnet. This magnet is called the *field magnet*. The ends of the coil are connected to two *slip rings* mounted on the coil spindle. Current may be obtained from the coil through two *carbon brushes*, which are made to press lightly against the slip rings.

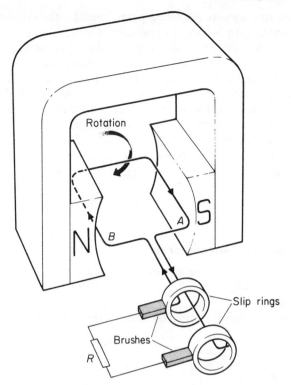

Fig. 43.15. Simple alternating current generator (a.c. dynamo)

As the coil rotates its sides cut the magnetic flux, and therefore a current is induced in it. Application of Fleming's right-hand rule indicates that the current flows from back to front along side *A* and from front to back along side *B*. When viewed from above the current therefore flows in a clockwise direction round the coil.

Fig. 43.16 shows graphically how the electromotive force generated in the coil varies

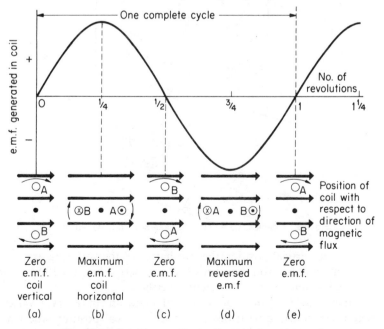

Fig. 43.16. E.m.f from a simple alternator (a.c. dynamo)

over one complete rotation. As a starting-point, consider the instant that the coil is in the vertical position with its side *A* uppermost (Fig. 43.16 (*a*)). In this position the sides of the coil are moving *along* the magnetic flux. No cutting of flux is taking place, and therefore the e.m.f. generated at this instant is zero. During the first quarter of a rotation the e.m.f. increases from zero to a maximum or *peak value* when the coil is in the horizontal position (Fig. 43.16 (*b*)). After this the e.m.f. decreases again during the second quarter of the rotation, and once more becomes zero when the coil is in the vertical position with side *B* uppermost (Fig. 43.16 (*c*)). During the second half of the rotation the e.m.f. generated follows the same pattern as that in the first half, except that the direction of the e.m.f. is reversed. This reversal occurs, since the direction of motion of the sides sides *A* and *B* across the magnetic flux is now reversed (Fig. 43.16 (*d*)).

If this alternating e.m.f. is applied to an external resistance *R* an alternating current (a.c.) will flow through it.

The simple direct current generator

If the slip rings and brushes of the simple a.c. dynamo are replaced by a single *split ring* with two diametrically opposed brushes the machine becomes converted into a simple d.c. dynamo.

Fig. 43.17 shows how this is done. The brushes are arranged so that when a coil is

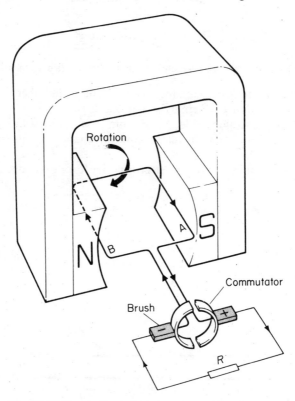

Fig. 43.17. Simple direct current generator (d.c. dynamo)

passing through the vertical position the two halves of the split ring are just on the point of changing contact from one brush to another. When used in this way the split ring is called a *commutator* or current reverser.

Fig. 43.18 shows how the e.m.f. obtained from a machine of this type varies during one complete cycle. Starting with the coil in the vertical position, the e.m.f. generated during the first half rotation rises from zero to a maximum and then falls to zero again (Fig. 43.18 (*a*), (*b*), (*c*)). As far as the external resistance *R* is concerned, the right-hand brush is positive and the left-hand brush negative. Now as the coil passes through the vertical position the two halves of the commutator change contact from one brush to the other (Fig. 43.18 (*c*)). Hence, although during the second half rotation the current is reversed in the coil itself, the changeover between brushes and commutator halves ensures that the right-hand brush remains positive and the left-hand brush negative. The e.m.f. of the brushes during the second half rotation of the coil is thus identical with that during the first half.

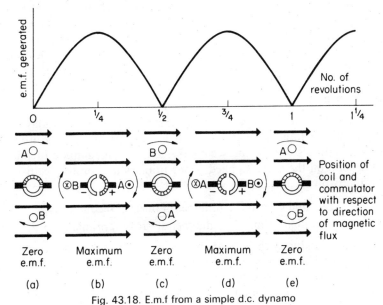

Fig. 43.18. E.m.f from a simple d.c. dynamo

D.c. dynamo compared with an electric motor

Comparison of the d.c. dynamo described above with the electric motor on page 437 will show that the two machines are identical in construction. If a simple d.c. dynamo is connected to a battery it will run as a motor. Conversely, if a simple electric motor is made to rotate it will behave as a dynamo and deliver current at the brushes.

When, therefore, an electric motor is running it acts as a dynamo and so produces an e.m.f. in opposition to that applied to it. This, is called a *back e.m.f.* (See also page 439.)

How to increase the e.m.f. obtained from a simple dynamo

Anything which increases the rate of cutting of magnetic flux in a dynamo will increase the electromotive force obtained from it. There are several ways in which this may be done:

(1) by increasing the number of turns in the coil;
(2) by winding the coil on a soft-iron armature so as to increase the magnetic flux through the coil;
(3) by increasing the speed of rotation; and
(4) by making the field magnet as strong as possible.

Practical dynamos

One of the disadvantages of the simple dynamo as described above is that its e.m.f. is reduced to zero every half rotation. In practice, this difficulty is overcome by having a number of coils wound in slots cut in an iron cylinder called the *armature*. Each coil has its own pair of segments in a multi-segment commutator similar to the commutator of the electric motor in Fig. 38.8. The e.m.f. obtained from a dynamo of this sort is fairly steady and merely has a slight ripple. The iron armature of a dynamo is built up of discs insulated from one another either by varnish or oxide coatings. This form of construction prevents large eddy currents from being set up as the armature rotates in the magnetic field.

Quite often an electromagnet, rather than a permanent magnet, provides the magnetic flux. The coils of this electromagnet are called the *field coils*, and the dynamo itself supplies the necessary current to them. These field coils may be either in series or in parallel with the armature coils. Dynamos are thus described as being *series wound* or *shunt wound*. Occasionally there are two field coils, one in series and the other in parallel with the armature coils. This is called *compound winding*.

The transformer

If an alternating current is passed through the primary coil of Faraday's iron ring (page 488) an alternating magnetic flux will be set up through the iron and will induce an alternating e.m.f. in the secondary coil. The magnitude of this induced e.m.f. will depend on the e.m.f. applied to the primary and on the relative numbers of turns in the two coils. It may be shown that

$$\frac{\text{secondary e.m.f.}}{\text{primary e.m.f.}} = \frac{\text{number of turns in secondary}}{\text{number of turns in primary}}$$

When a mutual inductance is used in this way it is called a *transformer*. The output e.m.f. will be larger or smaller than the input e.m.f. according as the number of turns in the secondary is larger or smaller than the number of turns in the primary. We therefore speak of either *step-up* or *step-down* transformers.

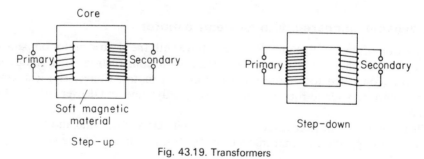

Fig. 43.19. Transformers

Fig. 43.19 shows the construction of one type of transformer. The primary and secondary coils are wound on a laminated core of magnetically soft material (silicon iron or stalloy). See also Fig. 43.20.

Energy losses in a transformer

A transformer is designed so that as little energy as possible is wasted inside it thus ensuring that its efficiency is as high as possible. Important features are:

(1) low-resistance copper coils so that internal energy (I^2R) losses in the windings are small;
(2) laminated core to reduce eddy current losses;
(3) core made of soft magnetic material to reduce the energy required to bring about magnetic reversals;
(4) efficient core design to ensure that all the primary flux is linked with the secondary.

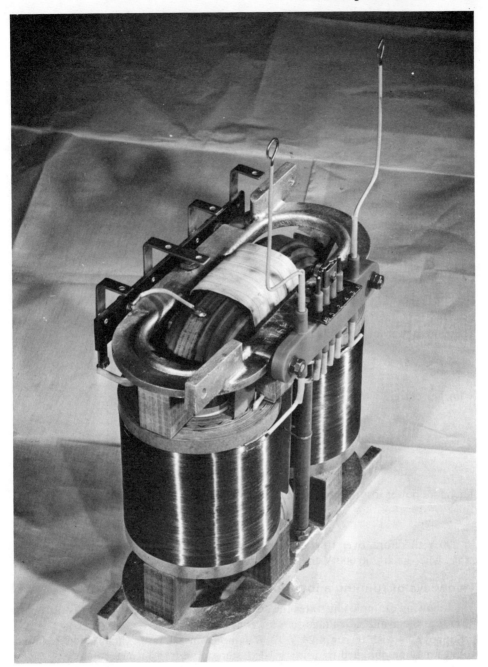

Fig. 43.20. Interior construction of a small power transformer

In a well-designed transformer the above losses will be very small, and therefore we may assume

$$\text{secondary power output} = \text{primary power input}$$

Advantage of a.c. over d.c. power transmission

One of the main advantages of a.c. is that it can be easily and cheaply changed from one voltage to another by a transformer with very little loss of energy.

For this reason, electric power generally is conveyed by a.c., as it can be transformed to very high voltage and transmitted over long distances with minimum power loss. This makes it possible to produce electricity where water power or coal are easily obtainable or from conveniently sited nuclear generating stations and to convey it all over the country by high-voltage overhead power lines (the *Grid*).

Electricity is generated in the power stations at 11 000 to 33 000 V and then stepped up to 400 000 V (400 kV) by transformers. It is fed into the grid at this voltage and subsequently stepped down in successive stages at substations in the neighbourhood of towns and other areas where the energy is to be consumed.

The following calculations show that a much smaller power loss occurs in the cables when energy is transmitted at high voltage.

Example. Find the power wasted as internal energy in the cable when 10 kW is transmitted through a cable of resistance 0.5 Ω (a) at 200 V; (b) at 200 000 V.

(*a*) The current in this case is given by amperes $= \dfrac{\text{watts}}{\text{volts}}$

$$= \frac{10\,000}{200}$$

$$= 50\,\text{A}$$

Therefore power loss in cable in watts $= I^2R$

$$= 50^2 \times 0.5\,\text{W}$$
$$= 1250\,\text{W} = 1.25\,\text{kW}$$

(*b*) The current which passes at the higher voltage $= \dfrac{\text{watts}}{\text{volts}}$

$$= \frac{10\,000}{200\,000}$$

$$= 0.05\,\text{A}$$

Therefore power loss in cable in watts $= I^2R$

$$= 0.05^2 \times 0.5\,\text{W}$$
$$= 0.001\,25\,\text{W}$$

At 200 V therefore, over 10 per cent of the energy is wasted merely in warming the cable, whereas at 200 000 V the energy losses are negligible.

Two ways of running a low-voltage lamp from the mains

The following example illustrates the advantage of a.c. over d.c. when it is necessary to reduce the available voltage to a lower value.

Suppose we wish to run a 24 V 12 W lamp from 240 V a.c. mains. The required voltage may be obtained by using a 10 : 1 step-down transformer.

The current taken by the lamp may be calculated from the formula,

$$\text{watts} = \text{volts} \times \text{amperes}$$

Therefore, $\text{current} = \dfrac{\text{watts}}{\text{volts}} = \dfrac{12}{24} = 0.5\,\text{A}$

Assuming no energy loss in the transformer,

$$\text{power input} = \text{power output}$$

or, $240 \times \text{primary current} = 12\,\text{W}$

i.e., transformer primary current $= \dfrac{12}{240} = 0.05\,\text{A}$

$$= \text{current taken from mains}$$

Let us now consider how this lamp could be run from 240 V d.c. mains. In this case it would be necessary to connect a limiting resistor in series with the lamp, the value of which is calculated as follows:

The p.d. across the lamp must be 24 V. Hence the p.d. across the limiting resistor must be (240 − 24) = 216 V.

Since lamp and resistor are in series, the same current, 0.5 A, passes through both.

The value of the resistor required is $\left(\text{from } R = \dfrac{V}{I}\right)$ equal to $\dfrac{216}{0.5}$ or 432 Ω.

The mains current is the same as that through the lamp, namely, 0.5 A, and so the total power output from the mains is given by

$$\text{watts} = \text{volts} \times \text{amperes} = 240 \times 0.5 = 120 \text{ W}$$

Of this, the lamp uses only 12 W. The remaining 108 W is wasted raising the temperature of the limiting resistor.

The choke

The current in a d.c. circuit is controlled by means of a rheostat, but as the above example shows, this method is very wasteful of energy.

The current in an a.c. circuit can also be controlled by a rheostat, but for some purposes it is better to use a *choke*. This consists of a low resistance copper coil wound on a laminated iron core. When a.c. passes through the choke it sets up a counter e.m.f. owing to self-induction. This counter e.m.f. impedes the current without leading to waste of energy as occurs in the case of a resistance. The property of a choke to impede current in this way is called *inductive reactance* in order to distinguish it from ordinary electric resistance.

Telephone communication

The action of a telephone receiver or ear-piece has already been described on page 425.

The transmitter or *microphone* of a telephone instrument consists essentially of two polished carbon discs, placed a short distance apart, the space between being filled with small carbon granules (Fig. 43.21). One of the discs is fixed and the other

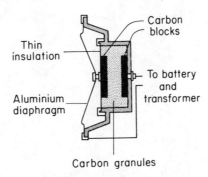

Fig. 43.21. Carbon microphone

connected to a metal diaphragm. When a person speaks into the microphone the diaphragm is set in motion by the sound waves, and so the carbon granules are subjected to variable compression. Consequently the electric resistance between the carbon blocks varies in a corresponding manner.

A steady current is passed through the microphone by connecting it in series with a battery and the primary of a step-up transformer. Thus, when sound waves enter the microphone the changes in resistance will cause the current in the primary circuit

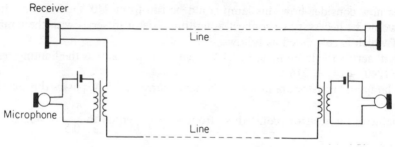

Fig. 43.22. Simple 2-way telephone circuit

to vary. A high-voltage a.c. will therefore be set up in the secondary of the transformer with the same frequency as that of the original sound waves entering the microphone. This a.c. is transmitted along lines to the receiver at the other end, where the electric energy is transferred back to sound energy.

Fig. 43.22 illustrates a simplified circuit to show how messages can be sent and received in both ways. It would, of course, be possible to transmit messages without

Fig. 43.23. Invisible cracks may become fractures and endanger life. This crack detector probe has a coil carrying an alternating current which sets up eddy currents in the rail over which it is passed. A crack will reduce eddy currents and consequently less energy will be transferred from the probe coil by e.m. induction. This leads to a rise in the voltage across the coil which is indicated on a meter, thereby giving warning of the presence of a crack. (Compare the corresponding case in direct current where the terminal voltage across a cell varies according to the current taken from it)

using a transformer, but only over short distances. For long distances the line resistance would be so great that the small changes of resistance in the microphone would have a negligible effect on current changes and the signals would be too weak to be heard.

The transformer thus serves a double purpose. It not only enables the microphone circuit resistance to be kept low so that the sound waves produce large current changes, but also it steps up the transmission voltage so that the energy loss in the connecting lines is reduced to a minimum.

Moving-coil microphone

The moving-coil microphone (Fig. 43.24) has a diaphragm with a light coil attached. The coil is situated in the strong magnetic field of a cylindrical pot magnet. Sound waves cause the diaphragm to vibrate, and as the coil moves back and forth in the magnetic field an alternating e.m.f. is induced in it. This is fed into an amplifier by a light flexible lead.

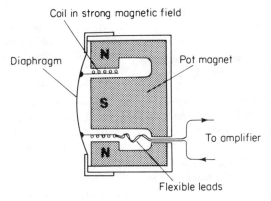

Fig. 43.24. Moving-coil microphone

It is of interest to compare the construction of the moving-coil microphone with that of the moving-coil loudspeaker in Fig. 38.9. In one, energy is transferred from electric current into mechanical energy of vibration in a cone and thence to sound waves. In the other the reverse occurs. Energy in sound waves is transferred to mechanical energy in a vibrating diaphragm and thence to electric energy.

QUESTIONS: 43

1. Explain the terms *electromagnetic induction* and *induced electromotive force*. Describe THREE experiments to show how an induced e.m.f. can be produced and the factors on which its direction and magnitude depend.

Briefly describe and explain one practical application of electromagnetic induction, illustrating your answer by a drawing.
(*O.C.*)

2. Describe how you can demonstrate by experiment the transfer of mechanical energy into electrical energy by electromagnetic induction and the variations you would make to test any two of the factors on which the magnitude of the induced e.m.f. depends.

Draw a labelled diagram of a transformer suitable for transforming from 240 V a.c. to 12 V a.c. Give details of the materials used in its structure and reasons for their choice.

Explain what is meant by the statement that the efficiency of a transformer is less than 100 per cent and give one reason why this must be so. (*L.*)

3. State the laws which determine the magnitude and direction of a current produced by electromagnetic induction. Describe a simple experiment to verify ONE of the laws.

A plane coil of wire is rotated about a diameter, which is perpendicular to a uniform magnetic field. Sketch a graph of the induced e.m.f. against the angle between the coil and the field. Indicate clearly the values of the angle on the *x*-axis. What is the effect on the e.m.f. of using a coil of:

(*a*) greater number of turns;

(*b*) greater area? (*O.C.*)

4. With the aid of labelled diagrams explain the action of any two of the following:

(*a*) an ignition coil;

(*b*) a primary cell;

(*c*) a simple a.c. generator. (*J.M.B.*)

5. Explain why a small potential difference may be measured between two railway lines along which a train (non-electric) is passing.

Discuss how the magnitude and sign of this p.d. depends on:

(*a*) the speed of the train;

(*b*) the direction relative to the Magnetic North;

(*c*) whether the train is travelling towards or away from the observer;

(*d*) the latitude? (*O.C.*)

6. Describe a simple a.c. generator and how it is modified by a commutator to provide d.c.

What are the advantages to be gained by using a.c. at high voltage when electrical power is to be transmitted over a long distance? (*C.*)

7. What features in the design of a practical form of dynamo determine the magnitude of the e.m.f. which it generates?

Describe, with diagrams, two methods, one involving a large loss of energy and the other a much smaller loss, by which a 240 V a.c. dynamo could be used to light a 12 V lamp. Explain for each method how the energy is lost. (*J.M.B.*)

8. Draw a section through a simple a.c. transformer and explain how the laws of electromagnetic induction are applied in its construction and operation.

100 kW of power are being supplied to a factory through wires of resistance of 0.1 Ω. What power is lost in the leads if the voltage at the factory end of the wires is:

(*a*) 230,

(*b*) 10 000? (*O.C.*)

9. What are *eddy currents*? Describe one application of their use.

Give an account, with a diagram, of the structure and mode of action of a transformer which supplies 12 V when connected to 240 V mains. If this transformer takes 1.1 A from the mains when used to light ten 12 V 24 W lamps in parallel, find:

(*a*) its efficiency;

(*b*) the cost of using it for ten hours at 3p per kWh.

10. (*a*) (i) Draw a labelled diagram showing the essential features of a voltage step-up transformer.

(ii) State the law of electromagnetic induction on which the working of the transformer depends.

(iii) Use the law stated in answer to (ii) to explain how the transformer works.

(*b*) When a 240 V electrical supply is connected to the primary winding of a transformer, a current of 50 mA flows in the circuit. The secondary winding is connected to a 5 ohm resistor in which a current of 1.5 A flows.

Calculate

(i) the power supplied to the transformer,

(ii) the power dissipated in the 5 ohm resistor and

(iii) the efficiency of the transformer.

(*J.M.B.*)

11. Draw a labelled diagram to show the structure of a transformer suitable for operating a 12 V lamp from the normal mains supply. State *two* reasons why transformers are not 100 per cent efficient. (*S.*)

12. A transformer is used on the 240 V a.c. supply to deliver 9.0 A at 80 V to a heating coil. If 10 per cent of the energy taken from the supply is dissipated in the transformer itself, what is the current in the primary winding? (*O.C.*)

13. A step-up transformer is designed to operate from a 20 V supply and deliver energy at 250 V. If the transformer is 90 per cent efficient, determine the current in its primary winding when the output terminals are connected to a 250 V 100 W lamp.

(*L.*)

14. (*a*) Describe how you would demonstrate:

(i) the production of an e.m.f. in a circuit when the magnetic flux linking with it changes;

(ii) the factors upon which the magnitude and direction of the e.m.f. depend.

(*b*) An electric heater has a resistance of *R* (in ohms). At what rate will it produce heat when connected to a supply of *V* (in volts)?

A generator with a power output of 20 kW at 4 kV distributes power to a workshop along cables having a total resistance of 16 Ω.

Calculate:

(i) the current in the cables;

(ii) the power loss in the cables, and

(iii) the potential drop between the ends of the cables.

How could the power supplied to the workshop be increased? (*A.E.B.*)

15. Explain the construction and action of a transformer which will operate from a 240 V

a.c. supply to give an output of 8 V for ringing a door bell.

(i) If the primary coil has 4000 turns, about how many turns will there be in the secondary?

(ii) Why are the coils wound on an iron core?

(iii) Why is the core made of sheets of iron instead of being in one solid piece?

(iv) What is the primary current if a current of 2.4 A flows through the bell and the and the efficiency of the transformer is 100 per cent? (*W.*)

16. Two coils *X* and *Y*, in the circuits shown in Fig. 43.25, are placed end to end. Describe the explain what happens to the reading of the galvanometer *G* when the switch *S* is:

(*a*) first closed;
(*b*) kept closed;
(*c*) opened again.

State, giving reasons, one feature which the galvanometer should possess in order to be effective in this circuit.

Suggest and explain three possible ways of

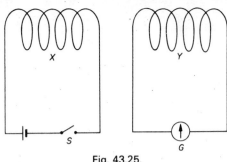

Fig. 43.25.

altering the apparatus so as to make the deflection of the galvanometer larger.

How could the supply connected to *X* and the meter connected to *Y* be changed so that the meter gave a steady deflection? (*L.*)

17. Explain with the aid of a labelled diagram how a moving coil microphone works.

(*A.E.B.*, 1982)

18. Draw a fully labelled diagram of an a.c. generator and state how the device is modified to produce a direct current. (*W.A.E.C.*)

Proton synchrotron at CERN, Geneva.
A view of the one hundred units of the 3 400 tonne ring magnet

Electronics, atomic and nuclear physics

44. Electrons in gases and vacuo

It was known to the scientists of the eighteenth century that the succession of sparks from an electric machine changed to a quiet luminous discharge when the pressure of the surrounding air was reduced. One of the investigators of this phenomenon was a Fellow of the Royal Society named William Watson, who, in 1750, produced a luminous electric discharge between two brass rods at the opposite ends of an evacuated glass tube, about a metre long. Also Lord Charles Cavendish, father of the eccentric Henry Cavendish, showed that a similar discharge could be sent through the Toricellian vacuum at the top of a barometer tube.

About the middle of the nineteenth century an expert glass-blower named Geissler obtained some remarkably beautiful coloured discharges by using other gases besides air, contained in skilfully shaped tubes in a variety of patterns. Sometimes these tubes were rotated like Catherine-wheels and their colours were further enhanced by the use of glass containing various minerals which fluoresced brilliantly while the discharge was passing.

Discharge tube phenomena

In order to study the electric discharge in air at low pressure an extra high tension (e.h.t.) unit is required. This is used to apply a potential difference of 1000 V or more between two electrodes at opposite ends of a glass tube 15 or more cm long. The

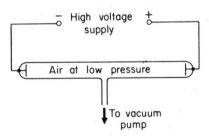

Fig. 44.1. Electric discharge tube

tube is connected to a vacuum pump through a side tube (Fig. 44.1). The appearance of the discharge at various stages while the air is being pumped out of the tube is shown in Fig. 44.2.

Shortly after starting the pump a brush discharge appears at the electrodes, but by the time the pressure is down to about 20 mmHg the electrodes are joined by one or more wavy violet streamers. On further reduction of pressure the streamers broaden out into a deep salmon-pink discharge which practically fills the space between the electrodes. At 5 mmHg pressure a dark region, called the *Faraday dark space*, appears near the cathode and divides the column into two parts, the *pink positive column* ending in the anode glow and a *blue negative glow* in the neighbourhood of the cathode. As the tube is evacuated still further the positive column shrinks towards the anode and begins to break up into striations. The Faraday dark space (F.D.S.) and the negative glow increase in length, and a second dark region, the *Crookes dark space* (C.D.S.), appears near the cathode.

It is important to note that the length of the dark spaces depends only on the pressure and not on the length of the tube. The remainder of the tube, whatever its length, is filled with pink positive column.

By the time the pressure has reached 0.01 mmHg, the pink positive column and the negative glow have disappeared and the Crookes dark space extends to fill the whole of the tube. At this stage also the walls of the tube show a green *fluorescence*.

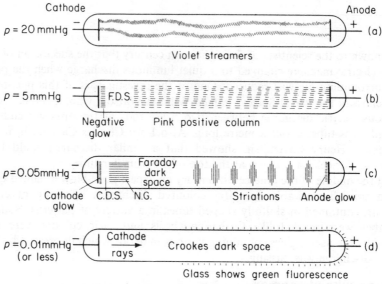

Fig. 44.2. Electric discharge through air at low pressures

It may be noted in passing that the type of discharge shown in Fig. 44.2 (*b*) is used in the familiar advertising sign tubes. In these the colour obtained depends on the nature of the gas and the material of the glass. The orange-red tubes contain neon, while a green colour is obtained by the use of mercury vapour in uranium glass.

Crookes dark space

During the last decade of the nineteenth century a number of investigators began to seek an explanation of the discharge phenomena. Interest became centred on what happened inside the Crookes dark space, and various tubes were made which were evacuated to such a pressure that the whole of the space between the electrodes was occupied by this space. Small pieces of certain minerals placed in these tubes were seen to fluoresce brilliantly with characteristic colours. We have already mentioned the green fluorescence of the glass tube itself.

The Maltese cross tube. Cathode rays

In 1859, Julius Plücker of Bonn University made a special tube, the anode of which was a Maltese cross made of aluminium (Fig. 44.3). This cast a sharp shadow on the end of the tube, suggesting that some kind of radiation was proceeding in straight lines from the cathode towards the anode. At the time the nature of this radiation was open to question, but general opinion held that it consisted of streams of particles shot out from the cathode with high velocity. Sir William Crookes described these particles as "radiant matter". Sir J. J. Thomson referred to them as "negative corpuscles". Nowadays we use the term *cathode rays*, an expression which was originally suggested by the German physicist Eugen Goldstein.

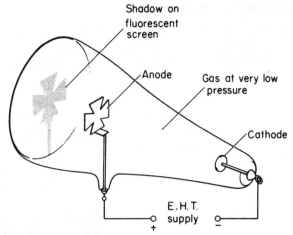

Fig. 44.3. Cathode rays travel in straight lines (Maltese cross tube)

Nature of cathode rays

During the last years of the nineteenth century experimental evidence gradually accumulated which demonstrated beyond doubt that cathode rays consist of streams of particles of negative electricity now known as *electrons*.

Cathode rays deflected by a magnetic field

The path of cathode rays can be rendered visible by allowing them to impinge upon a screen coated with zinc sulphide and placed at a small angle to the line joining the electrodes of the discharge tube (Fig. 44.4). If a bar magnet is placed at right angles

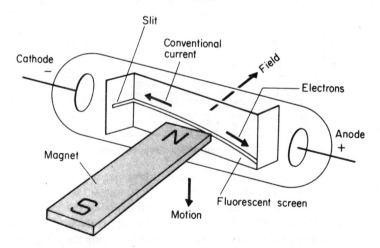

Fig. 44.4. Cathode rays are deflected by a magnetic field

to the tube with its N pole against the glass as shown in the diagram the rays are seen to bend downwards. If the rays are a stream of negative electrons moving from left to right they are equivalent to a positive or conventional electric current in the reverse direction. It is clear, therefore, that if this assumption is true the rays behave like an ordinary electric current and obey Fleming's left-hand rule (page 436).

Incidentally, this experiment was a strong argument against the notion that cathode rays were of the same nature as light waves, since no one had ever been able to show that light could be deflected by a magnetic field.

Cathode rays convey negative charge

Professor Jean Perrin of the University of Paris designed a tube to demonstrate that cathode rays convey negative charge (Fig. 44.5). The anode of this tube is an aluminium tube open at both ends, inside which is an insulated metal cylinder, called a

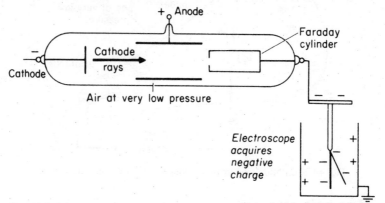

Fig. 44.5. Perrin's experiment shows that cathode rays convey negative charge

Faraday cylinder, having a small opening at one end. Cathode rays shot out from the cathode enter the Faraday cylinder and build up a charge on it as indicated by the divergence of a gold-leaf electroscope connected to the cylinder. On testing the electroscope in the usual way, with a charged ebonite rod, it is found to be negatively charged.

Energy conveyed by cathode rays

One of Sir William Crookes's tubes contained a small paddle wheel running on rails inside it and having mica vanes (Fig. 44.6). The two electrodes are arranged in such a

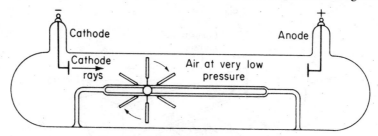

Fig. 44.6. Paddle-wheel discharge tube

way that cathode rays impinge only on vanes above the axis. On passing the discharge the paddle wheel rotates towards the anode. It appears to be pushed along by the cathode rays. Actually, this force is not directly due to the momentum of the electrons but is really a secondary effect. When the electrons strike the vanes their energy becomes transferred to internal energy, thereby raising the temperature of the vanes' surface. Gas molecules in the tube receive heat energy when they come into contact with the warm surface and rebound with high velocity, thus imparting an equal and opposite momentum to the vanes. The force of reaction thus set up causes the paddle wheel to rotate.

More is said about what goes on inside an electric discharge tube on page 516, followed by a description of the more recent *hot-cathode* types of discharge tube.

Work of Sir J. J. Thomson

Between 1894 and 1900 a series of experiments was carried out by Sir J. J. Thomson in the Cavendish Laboratory at Cambridge to determine the mass and charge of the electrons as well as the speed with which they travelled between the electrodes. He found that the electrons were all alike and always possessed the same properties whatever the nature of the gas in the tube or the material of the electrodes.

He first of all designed a special tube in which cathode rays were deflected by both a magnetic and an electric field. From this he was able to calculate their velocity (about 29 000 km/s) and also the ratio of the charge to the mass (e/m_e) for an electron. In a later experiment he obtained a rough value for the electronic charge (e). From the the results of these two experiments he was able to calculate the mass (m_e) of an electron and found it to be approximately $\frac{1}{2000}$ of the mass of a hydrogen atom. Subsequently a more accurate determination of the electronic charge (e) was made by the American physicist Robert Millikan.

Specific charge

The ratio $\dfrac{e}{m_e}$ mentioned in the previous paragraph is called the *specific charge* of an electron. Its units are $\dfrac{\text{coulombs}}{\text{kilograms}}$ (C/kg). Note the use of the word *specific* which, as we have seen before in this book, refers to a property of unit mass of something.

The discovery of X-rays

In December 1895 the German physicist Wilhelm Röntgen noticed that some barium platinocyanide crystals glowed brightly in the neighbourhood of a working cathode ray tube even when the tube was covered up. He also noticed that some wrapped photographic plates left near the tube had become fogged. It appeared that some kind of invisible radiation proceeded in straight lines from the tube and was able to penetrate substances opaque to ordinary light. He subsequently found that this radiation, which he called *X-rays*, was coming from the fluorescent glass wall of the tube at the place where it was struck by the cathode rays. The rays were otherwise known as *Röntgen rays*, in honour of their discoverer.

Röntgen found that if a sheet of cardboard covered with barium platinocyanide was placed in the path of the X-rays it became luminous. On placing his hand between tube and screen a faint shadow of the hand was formed in which the bones were clearly visible, since they were more opaque to the X-rays than flesh.

It was found that the newly discovered X-rays were not deflected by magnetic or electric fields, from which it was concluded that they could not consist of electrically charged particles. Röntgen assumed that the X-rays were some kind of longitudinal wave motion, but we now know them to be of the same nature as light waves but with a very much shorter wavelength.

The gas-filled X-ray tube

Later investigation showed that X-rays were produced whenever fast moving electrons were stopped on impact with a target. This led to the development of tubes containing air at very low pressure in which a beam of cathode rays was focused from a concave cathode on to a tungsten disc called the *anticathode* (Fig. 44.7). The anticathode is set into a block at the end of a copper tube which serves the purpose of conducting away as heat, the internal energy transferred by the impact of the electrons.

The Coolidge hot-cathode tube

For reasons which cannot be discussed here the wavelength of the X-rays produced by the early gas tubes was difficult to control. Modern tubes are based on a method devised by W. D. Coolidge in 1916 (Fig. 44.8). In this tube the cathode takes the

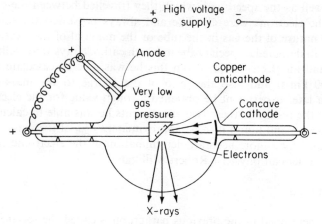

Fig. 44.7. Gas-filled X-ray tube

form of a spiral of tungsten wire which, when heated by an electric current, gives out a stream of electrons by a process called *thermionic emission* (page 516). The spiral is surrounded by a short metal cylinder which serves to focus the cathode rays on to the anticathode.

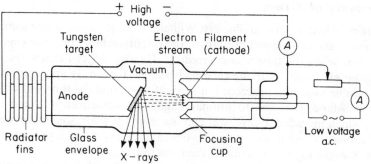

Fig. 44.8. Coolidge X-ray tube

The electrons are speeded up in the usual way, by applying a high p.d. between the cathode and anticathode. The higher the voltage used, the shorter is the wavelength of the X-rays and the greater is their penetrating power. Very short wavelength X-rays are described as *hard*, as distinct from the less penetrating *soft* rays of longer wavelength which are produced at lower voltages. The intensity of the rays is governed by the filament current, which controls the temperature of the tungsten filament, and hence the rate of emission of electrons.

Only part of the energy conveyed by the fast-moving electrons is transferred to X-radiation. The rest becomes transferred to internal energy in the target when the electrons are brought to rest on impact. This energy, in turn, is conducted away as heat through the copper anode to the cooling fins outside.

Applications of X-rays

The X-ray photographs or radiographs used in medical practice are made by allowing X-rays to pass through parts of the body and on to a photographic film.

X-rays are used in hospitals also in the treatment of malignant growths, as it is found that cancer cells can be destroyed by this means. Nevertheless, great care is always taken to avoid unwanted doses of this radiation, as these have a harmful effect on normal cells also which often does not become apparent until some years afterwards. Before this danger was realized many of the early experimenters became very ill after working with X-rays, and a number lost their lives. Nowadays X-ray tubes are always surrounded by lead shields to absorb stray radiation.

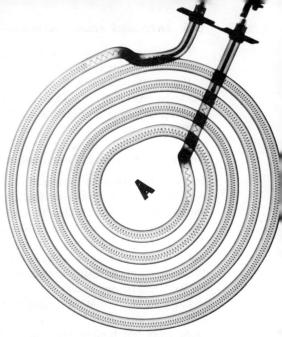

(a) 2 million volt X-ray equipment being used to examine the seam weld of a fusion-welded pressure vessel. The inset picture is a reproduction of the radiograph showing a defective joint

(b) X-ray photograph of a boiling plate taken at the Electricity Council's Appliance Testing Laboratories at Leatherhead, Surrey. The picture reveals that the element spiral is wound uniformly, thus avoiding the development of dangerous "hot-spots" which would lead to local high temperature rise and early breakdown. All kinds of electrical equipment is tested at the laboratories to the advantage of manufacturer and consumer alike

Fig. 44.9. X-radiography

Fig. 44.10. *St. Sebastian by Francia*. X-ray photograph superimposed, showing original head overpainted

In industry X-ray photographs are used to reveal hidden flaws in metal castings and welded joints (Fig. 44.9).

Fig. 44.10 shows how X-radiography is used to detect alterations which have been made to works of art. The technique owes its success to the fact that pigments containing lead compounds absorb X-rays more strongly than those which contain lighter elements.

The Edison effect. Thermionic emission

In 1883, the American scientist and inventor, Thomas Edison, was experimenting with an incandescent electric filament lamp which also contained a small metal plate supported by a wire sealed through the glass. When a battery and galvanometer were connected between the plate and filament he found that a small current flowed through the galvanometer if the plate was positive with respect to the filament, but not if it was negative. This was called the *Edison effect*, and remained nothing more than a scientific curiosity for many years until an explanation was forthcoming from O. W. Richardson.

Early in the twentieth century Richardson, who was for a time Professor of Physics at Princeton University, carried out investigations of the emission of electrons from hot bodies. He explained the Edison effect by assuming that electrons evaporate from the hot filament in much the same way as vapour molecules leave a hot liquid. Thus, in a very short time the region surrounding the hot filament becomes occupied by a *space charge* of electrons. Eventually, a state of equilibrium is reached in which the rate at which the electrons leave the filament is balanced by the rate at which they return to it. This was happening inside the bulb of Edison's lamp. When the plate was positive it attracted electrons and a current flowed in the plate circuit, but if the battery was reversed so as to make the plate negative the electrons were repelled and current ceased to flow. The process by which electrons leave the hot filament is called *thermionic emission*.

What happens inside an electric discharge tube?

We shall now return to the electric discharge tubes described earlier in this chapter. The processes which go on inside these tubes are complex and difficult to investigate, so that physicists hold different views regarding explanations of the various dark spaces, striations and so on.

However, all agree that ions play the most important part in the conduction of electricity through gases. The meaning of the term gaseous ion was explained on page 375. In a discharge tube at low pressure the ions which are always present are accelerated by the electric field between the electrodes; positive ions towards the cathode and free electrons and negative ions towards the anode. Collisions between electrons and atoms occur resulting in the production of further ions.

It is believed that the impact of positive ions on the cathode causes electrons to be knocked out of the metal of the cathode itself, and these constitute the cathode rays which stream towards the anode at high velocity.

The various discharge tubes discussed earlier in this chapter played a very important part in our knowledge and understanding of the nature and properties of electrons. But, in order to obtain a copious stream of electrons in tubes such as the low gas pressure Maltese cross tube of Fig. 44.3, very high voltages are required and this can cause undesirable X-radiation when the electrons strike the walls of the tube.

The danger can be avoided by the use of evacuated tubes which have a *thermionic electron gun*. It is therefore preferable to use this type of tube for investigating the properties of electrons since they produce a good beam of cathode rays at lower accelerating voltages, with the added advantage of reduced danger from X-radiation.

Experimental thermionic electron tubes

Fig. 44.11 shows a thermionic tube designed to show the deflection of moving electrons in an electric field.

Its electron gun consists of a cylindrical cathode with a hot metal filament inside to provide a supply of electrons by thermionic emission. The electrons are ac-

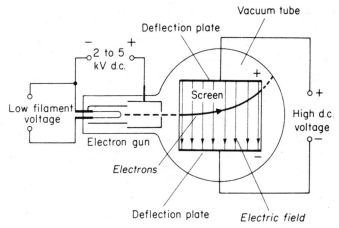

Fig. 44.11. Thermionic electron tube (electric deflection)

celerated towards a cylindrical anode and emerge as a thin flat beam from a horizontal slit at the end. Their path is rendered visible when they cross a vertical fluorescent screen.

At the top and bottom of the screen are two horizontal metal plates with connections to the outside of the tube. If a potential difference is applied to these plates, the electric field set up between them deflects the electron beam towards the positive plate. This suggests that the electrons carry a negative charge.

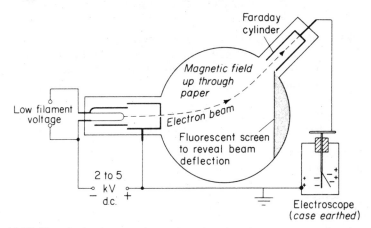

Fig. 44.12. Thermionic electron tube to show that electrons carry a negative charge

Fig. 44.12 shows the thermionic equivalent of the Perrin tube (page 512). Its electron gun is similar to that in the tube just described except that its anode has a small hole instead of a slit. This provides a thin beam of electrons which can be deflected by a magnetic field at right angles to their path so that they bend round and enter a collecting cylinder connected to a gold-leaf electroscope. The charge collected on the electroscope can be shown to be negative in the way described for the original Perrin tube (page 512).

Maltese cross tubes with an electron gun similar to that of the Perrin tube are also available.

The cathode ray oscilloscope

The cathode ray oscilloscope is an instrument used for studying the current and voltage waveforms in various electric circuits. In this connection it is very useful for checking laboratory electric equipment and television and radio receivers.

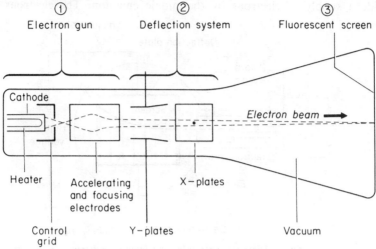

Fig. 44.13. Main features of the cathode ray tube

The chief feature of an oscilloscope is the cathode ray tube, which is a vacuum tube containing three main components (Fig. 14.13). These are:

(1) The **electron gun** which consists of an electron-emitting hot cathode, followed by a series of cylindrical and annular (ring-like) anodes at high positive potentials with respect to the cathode. These not only accelerate the electrons but also focus them into a fine beam. A variable, low, negative voltage can be applied to the control grid in order to vary the number of electrons passing through it. This alters the *brightness* of the spot on the screen.

(2) The **deflection system** which can deflect the electron beam either vertically or horizontally. This comprises two pairs of plates; a horizontal pair called the Y-plates and a vertical pair called the X-plates. These plates are parallel in their transverse directions but open out at an angle lengthwise in order to allow a bigger angular deflection of the beam.

When a potential difference is applied across either pair of plates, the electric field set up between them will deflect the beam in the appropriate direction.

(3) A **fluorescent screen** at the end of the tube on to which the electron beam is focused to form a bright spot which will trace out a pattern on the screen according to the voltage variations (signals) applied to the X- and Y-plates.

To illustrate the action of the oscilloscope we shall now describe some of its more simple applications.

To study a.c. waveforms, using a cathode ray oscilloscope

In addition to the tube, a cathode ray oscilloscope contains a **sweep generator** or time-base circuit. When this is switched on it applies a potential difference to the X-plates which builds up uniformly with time to a maximum and then repeats the process at regular intervals. The result is that the spot moves horizontally across the screen with steady velocity, returns to zero instantaneously (fly-back) and repeats the cycle continuously. Thus at low sweep frequencies the trace appears as a moving spot but, owing to persistence of vision it becomes a continuous line at higher frequencies.

If a signal in the form of a simple a.c. voltage is applied across the Y-plates, the spot will oscillate up and down in time with the voltage. The resultant trace on the screen will be a combination of the Y-plate a.c. signal and the X-plate sweep. In effect, the moving spot plots on the screen a luminous graph of a.c. voltage variation with time (see Fig. 44.14).

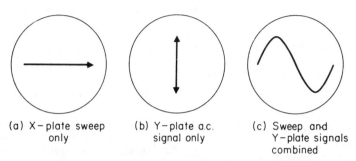

    (a) X−plate sweep   (b) Y−plate a.c.   (c) Sweep and
        only   signal only   Y−plate signals
                                                    combined

Fig. 44.14. How a simple waveform is produced on an oscillograph screen

Obviously, a single complete oscillation or wave will be observed when the sweep frequency is equal to the a.c. signal frequency. Otherwise two or more complete waves will be seen when the signal frequency is twice or some integral (whole-number) multiple of the sweep frequency.

Using a cathode ray oscilloscope the complex waveforms from an amplifier fed by a microphone or record player may be examined.

QUESTIONS: 44

1. What is an electron and how does its mass compare with that of a hydrogen atom?

2. The arrow in Fig. 44.15 represents a stream of electrons moving in the plane of the paper

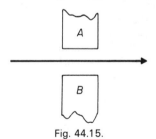

Fig. 44.15.

from left to right. How would the stream be deflected, relative to the paper, if:

(a) *A* is a N pole and *B* is a S pole;

(b) *A* is positively charged, *B* is negatively charged? (*J.M.B.*)

3. With the aid of a large clear diagram describe the essential features of a simple form of cathode ray tube.

State FOUR properties of cathode rays. Describe how two of these properties may be demonstrated, using modified tubes if necessary. (*S.*)

4. Draw a labelled diagram to show the components of a cathode ray tube *with a hot cathode*. Show the electrical connections required to enable the tube to produce a beam of cathode rays.

Name *three* important properties of cathode rays. State briefly how these properties may be demonstrated.

A charge of 1.8×10^{-3} C passes through a cathode ray tube per second when the voltage between the anode and cathode is 4000 V. Calculate the energy carried by the cathode ray beam in 8.0 s. What becomes of this energy? (*C.*)

5. (*a*) State **two** properties of an electron.

(*b*) Sketch a diagram of apparatus for *producing* a continuous stream of electrons. On your diagram indicate what power supplies you would use and how they would be connected to the apparatus. How may the path of the electron stream through the apparatus be made visible?

(*c*) Describe experiments, one in each case, to show how a narrow beam of electrons behaves when it passes through (i) a uniform electric field and (ii) a uniform magnetic field.

In each case include
(i) a sketch of the apparatus,
(ii) a circuit diagram,
(ii) an account of what observations you would expect to make and the conclusions which could be drawn from them.

(*J.M.B.*)

6. Make a labelled diagram of a sectioned side view of a cathode-ray oscilloscope tube showing the essential parts.

What is the function of the time-base circuit?

Explain how a C.R.O. is used to examine the variation with time of:

(*a*) the p.d. between the terminals of an a.c. generator, and

(*b*) the current supplied by the generator through a given circuit.

If the frequency of the generator fluctuates slightly, what effect will this have on the trace seen on the screen?   (*O.C.*)

7. Make clear the terms *frequency* and *wavelength* with reference to a sound wave.

A tuning fork is vibrated near an instrument, e.g., a microphone connected to an oscilloscope, which depicts the wave-form of the sound. Show clearly by means of labelled diagrams, the changes observed in the waveform when:

(*a*) the fork is replaced by one of twice the frequency;

(*b*) the fork emits a louder note; indicate clearly the important features of these changes.

(*C.*)

8. Describe the essential features of a cathode ray tube using thermionic emission. What is the effect on the beam of particles:

(*a*) of having a hotter filament;

(*b*) of increasing the anode voltage?

A stream of electrons proceeds horizontally from left to right in the plane of the paper. A magnetic field is then applied perpendicularly into the paper by placing a N pole in front of and a S pole behind the paper. Show, on a diagram, the path of the rays and explain how you arrive at the result. What would an electric field in the same direction as the magnetic field do to the rays?

(*O.C.*)

9. Draw a labelled diagram of a *hot-cathode X-ray tube* together with a simplified circuit diagram to indicate the kind of electric power supplies required to work it. Mention two practical uses of X-rays.

State the factors which govern:

(*a*) the *wavelength* of the X-rays obtained;

(*b*) the *intensity* of the radiation.

10. Electrons are accelerated from rest on to a target in a hot-cathode vacuum tube by a p.d. of 25 V. Find:

(*a*) the velocity of the electrons on reaching the target;

(*b*) the energy conveyed to the target per second if the electron beam current is 1.5 mA. (Mass of an electron $= 9.0 \times 10^{-31}$ kg; charge of an electron $= 1.6 \times 10^{-19}$ C.)

11. A beam of high-velocity electrons in a vacuum tube is directed towards a piece of tungsten embedded in a block of copper. Give an account of the energy changes which take place on impact. How and why should the apparatus be suitably shielded?

Why is it necessary to use:

(*a*) a vacuum tube;

(*b*) a copper block? How could the electrons be:

(*c*) produced;

(*d*) accelerated?

Draw a labelled diagram of such a tube to illustrate the essential features of your answer.

45. Electrons in solids, and semiconductor devices

'Electronics' is the name given to the study of electron movement in gases, vacuo and semiconductors. Semiconductor devices, mainly transistors, are what we shall be dealing with in this chapter.

Semiconductor devices of various kinds are used in a very wide range of electrical equipment: radios and televisions, tape recorders and alarm systems as well as washing machines and a host of other domestic equipment. Computers based on this technology are used not only in the home and office (in cash registers, word processors and so on), but also in heavy industry and manufacturing processes. Here computers using only a few watts of power operate low current relays which switch on and off electrical machines using many hundreds of watts.

Whole books have been written on the subject of semiconductor devices, so in this chapter we have tried only to explain what the various components *do* and not *how* they work. The study of so-called *solid-state physics* belongs to more advanced work.

Resistors in circuits

In any electronic circuit, there are many components. Inside a transistor radio, or on the circuit board of a computer, there are lots of small, striped cylinders. These are *resistors*. They are there to help other components to operate properly. They help to

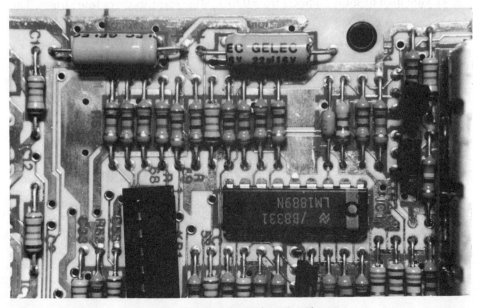

Fig. 45.1. The striped cylinders are resistors

control the voltage at different points in a circuit, and thus to control the current through the circuit. Low value resistors may be measured in *ohms* (Ω), but larger units are used for higher values. 1 kilohm (1 kΩ) means a resistance of 1000 Ω, 1 megohm (1 MΩ) means 1000 kΩ or one million ohms.

Fig. 45.1 shows some resistors on a circuit board inside a computer. Resistors may be the same size, but with very different values, perhaps ranging from 10 Ω to 10 MΩ. *Carbon composition* resistors (Fig. 45.2) are made from a compressed mixture of carbon grains in a ceramic (clay) base. Carbon does conduct electricity, though it is not as good a conductor as metals. To make lower value resistors (like the 10 Ω value mentioned), the proportion of carbon in the mixture is increased.

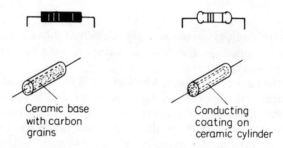

Ceramic base with carbon grains

Conducting coating on ceramic cylinder

Fig. 45.2. Carbon or a metal film conducts in the resistor

Carbon film and *metal film* resistors are made differently. The hot vapour of carbon or metal is allowed to settle and condense on a cold ceramic cylinder. For a higher value resistor, a thinner condensed layer is allowed to form. In both kinds of resistor, the 'package' is completed by fixing leads to the resistor, and encasing it in a ceramic surround. Finally, the coloured identifying stripes are added.

The resistor colour code

For a resistor of, for example 470 kΩ, it would be rather difficult to write the value on the small resistor itself; 470 000 Ω would not fit easily. Instead, the colours listed in Fig. 45.3 are used to show numbers from 0 to 9. Most resistors carry four stripes. The

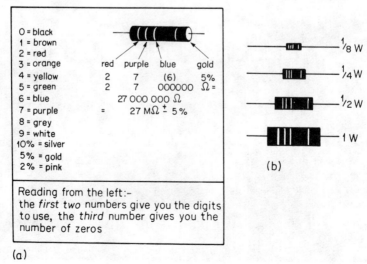

```
0 = black
1 = brown
2 = red
3 = orange        red   purple  blue    gold
4 = yellow         2      7      (6)      5%
5 = green          2      7    000000   Ω =
6 = blue              27 000 000 Ω
7 = purple         =     27 MΩ ± 5%
8 = grey
9 = white
10% = silver
5% = gold
2% = pink
```

Reading from the left:-
the *first two* numbers give you the digits to use, the *third* number gives you the number of zeros

(a)

(b)

$\frac{1}{8}$ W

$\frac{1}{4}$ W

$\frac{1}{2}$ W

1 W

Fig. 45.3. Stripes or bands on resistors

pink, silver or gold band tells you two things. First, to start counting at the other end! Second, it gives the resistor's *tolerance* or limits of accuracy. In the example in Fig. 45.3, the gold fourth band shows that the resistance value is within 5% of its stated value.

When resistors are chosen for a circuit, they must be able to handle the power which is to be dissipated in them. If a resistor of value 100 Ω is to carry a current of 40 mA, then the power dissipated is given by

$$P = I^2 \times R$$
$$= (0.04 \text{ A})^2 \times 100 \text{ Ω} = 0.16 \text{ W}$$

Resistors are available in different sizes to withstand different amounts of power. The carbon resistors in Fig. 45.3 (*b*) are roughly full size. In this example the smallest resistor would not do the job. It is designed to dissipate only 1/8 of a watt (0.125 W). The next largest resistor, or one of the others, should be chosen for the circuit.

The voltage or potential divider

In many electronic circuits, two resistors are used to fix the value of the voltage at a point in the circuit. The value of this voltage or potential depends on the relative values of the two resistors chosen.

Fig. 45.4 shows two examples. In Fig. 45.4 (*a*) the 'output voltage' of the divider chain is 2.5 V. The two resistors have the same value, 10 kΩ each. So the power supply voltage is split into two equal halves. Here the power supply negative connection (or rail) continues to the next stage of the circuit, so that the voltage output of the divider is the potential difference between the mid-point of the resistors and this negative power supply rail.

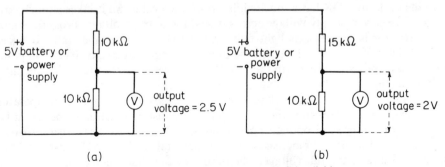

Fig. 45.4. Altering the output voltage of the divider

Note: The voltage supply terminals of the various circuits described, from this page onwards in this chapter, are mostly marked + V and 0 V instead of + V and − V. This has been done to emphasise that the negative terminal is to be regarded as a zero level of voltage or potential. The actual supply itself may be a battery or a low voltage supply unit.

In Fig. 45.4 (*b*) the output is less—it is 2.0 V. Here the resistor ratio is 10 kΩ:15 kΩ, 10:15 or 2:3. The 5 V supply is split into two parts in this ratio. Thus the 'output' from the divider is (2/5) × 5 V or 2 V. In both these cases fixed resistors are used. If a potentiometer (a variable resistor—see page 460) replaces one of them, the output voltage can be varied.

Light- and temperature-sensitive resistors

Another method of voltage variation is to use sensors in place of one of the resistors. A *light-dependent resistor* (LDR) is sensitive to light. Fig. 45.5 shows one type. The LDR contains a semiconducting material, and when it is exposed to light its resistance changes. In the dark it has a high resistance value, perhaps hundreds of kΩ, but when light shines on it the energy of the light releases more electrons for conduction and the resistance value of the component falls. In Fig. 45.5 (b) the LDR is in series with a fixed 10 kΩ resistor to form a voltage divider chain.

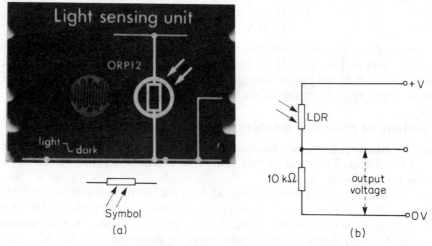

Fig. 45.5. The LDR varies its resistance with light intensity

In the dark the LDR has a much higher resistance value. So it takes a much larger share of the power supply voltage across it, and the output voltage across the smaller 10 kΩ resistor is low. But as light falls on the LDR, its resistance falls. It takes a smaller share of the full voltage, so the output voltage across the 10 kΩ component increases. If the fixed resistor and the LDR are interchanged in the chain, the output voltage of the circuit would fall as the light brightened.

Fig. 45.6 shows a similar chain using a *thermistor*, which is a temperature-dependent resistor. This time the extra energy to assist conduction comes from heat energy warming the semiconductor material inside the case. As the temperature of the thermistor in Fig. 45.6 (b) increases, its resistance falls. The 10 kΩ resistor takes an increased share of the full voltage. The output of the chain increases.

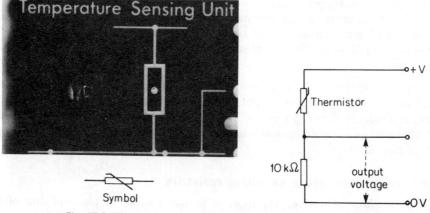

Fig. 45.6. The thermistor resistance decreases with temperature

The semiconductor diode

We have already mentioned semiconductors and the solid-state diode on page 408.

The materials usually used in their manufacture are either silicon or germanium which have been alloyed or doped with small quantities of a chosen impurity. According to the impurity chosen, they are classed as either p-type or n-type, where the letters p and n stand for positive and negative respectively. A junction diode consists of a single crystal of the material, one half being p-type and the other n-type.

Current will flow easily from the p-type or anode to the n-type or cathode but in the reverse direction only with difficulty. In these two cases the diode is said to be on forward or reverse bias respectively.

In external appearance, most semiconductor diodes look like very small resistors. The cathode end may have a coloured band, usually red, to indicate that the current will flow out of the diode at that end. Sometimes the symbol for a diode is printed on the outside, facing in the correct direction. See also Fig. 45.7 (*a*).

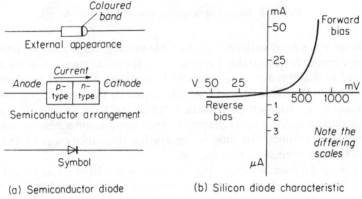

(a) Semiconductor diode   (b) Silicon diode characteristic

Fig. 45.7. Semiconductor diode

An explanation of the action of the diode is outside our scope, but this information can be found in more advanced textbooks or else in books devoted entirely to solid-state devices. The diode has been included here by reason of its importance in electronic circuits, particularly those in computers.

To plot the characteristic curve for a semiconductor diode

Diodes usually available for study in this experiment are likely to have maximum current outputs of 10 to 500 mA, rarely more than 1 A. Reverse bias voltages of up to 25 V or 100 V, or even more, may be permissible without harming the diodes. However, if the maker's data sheet is available for the particular diode used, it would be wise to consult it and not exceed the safe maximum output current and reverse bias voltage, lest the diode becomes overheated and destroyed.

For the forward bias characteristic, the circuit shown in Fig. 45.8 (*a*) is used. Here the e.m.f. is supplied by a potential divider or potentiometer. This consists of a wire-wound resistor connected directly across a cell or other low-voltage d.c. supply. A sliding contact, C, on the resistor, enables the voltage to be varied from zero to the maximum obtainable from the supply.

The meters used should be of the multimeter type shown on page 447 for preference, so that the widely differing voltage and current ranges required for forward and reverse bias conditions are readily available.

Fig. 45.8 (*b*) shows the circuit used for studying the reverse bias characteristic. In this part of the experiment, the low-voltage potential divider has been replaced by a variable h.t. (high tension) unit with an output of, say, up to 300 V.

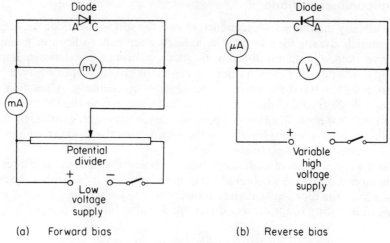

Fig. 45.8. Diode characteristic investigation circuits

The voltmeter has been switched to a suitable voltage range and the ammeter to a microampere range. Note also that the position of the microammeter in the circuit has been changed so that it is directly in series with the diode. This has been done in order to reduce errors in the diode current readings caused by the fact that, on reverse bias, the diode resistance is comparable with, or much greater than, that of the voltmeter. A similar case regarding the correct placing of the meters arose when we were dealing with the ammeter–voltmeter method for measuring resistance on page 453. It might be well worthwhile re-reading that section.

Here is a piece of advice before starting the experiment. In order to avoid damage to the diode through overload, the supply voltage should be switched off before changing the meter ranges or altering the circuit connections in any way. Finally, before switching on again, the supply voltage should be set to its lowest value.

In both parts of this experiment, current and voltage readings are tabulated in the usual way and a current–voltage graph is plotted. Fig. 45.7 (*b*) shows the general shape of the curve obtained for a silicon diode. Note the differing current and voltage scales for the forward and reverse conditions.

Half-wave rectification

If a source of alternating current is connected in series with a semiconductor diode the output from the circuit will flow in one direction only in a series of pulses (Fig. 45.9).

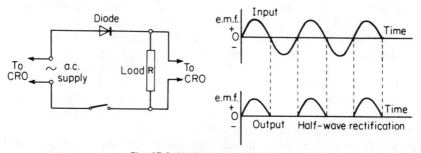

Fig. 45.9. Half-wave rectification

This is called *half-wave rectification*. The variation in the input and output voltages with time may be seen by connecting the input and output terminals, in turn, to a cathode ray oscilloscope (CRO). If a double-beam oscilloscope is available, the two voltage traces may be viewed and compared simultaneously. The load, shown as R in these diagrams, represents any circuit which is being supplied from the output.

Full-wave rectification

If four diodes are used, connected in what is described as *bridge* connection, then the whole of the input a.c. cycle may be rectified. The result is a unidirectional pulsating output at twice the frequency of the input.

The circuit may best be remembered if drawn as shown in Fig. 45.10. Note that the

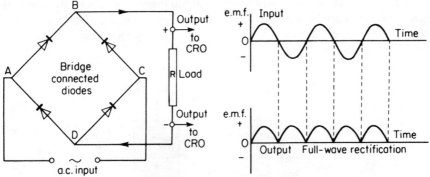

Fig. 45.10. Full-wave rectification

diodes are all pointing round the sides of a square towards B and away from D. If the current direction is traced through the diodes, as A and C become alternately positive and negative from the a.c. input, then the output current will always flow out of B, through the load and back to D.

Power pack. Smoothing

Half-wave and full-wave rectified alternating current is very useful for some purposes, e.g. charging secondary cells. However, it cannot be used to replace a direct current source in radio or similar electrical equipment, as it gives rise to intolerable hum. This hum may be reduced considerably by connecting a smoothing capacitor in parallel with the output.

The circuit of a simple power pack circuit is shown in Fig. 45.11. As the output

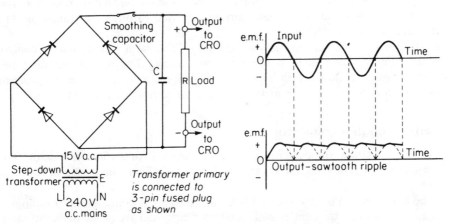

Fig. 45.11. Simple low voltage power pack circuit

voltage from the rectifiers rises to its maximum it charges up the *reservoir* or *smoothing* capacitor to its peak value. As the output voltage falls towards zero and then builds up to its next peak value, the capacitor discharges into the circuit thus preventing the final output from reaching zero.

The final output voltage has a slight ripple sometimes described as a *sawtooth* output. Even better smoothing can be obtained by the addition of filter circuits, but for details of these the reader is referred to more advanced texts.

Transistors

A junction diode is formed by *two* semiconductor pieces, one of which is p-type and the other n-type. A *junction transistor* is a 'sandwich' of *three*. By using three connections, one current going into the device can be used to control the size of another current. There are two possible 'sandwiches': *npn* and *pnp*. The npn type is the one used in this book.

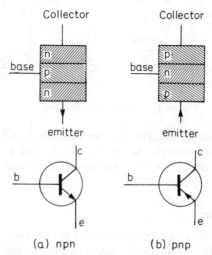

Fig. 45.12. A junction transistor is 3 semiconductor 'slices'

The two transistor symbols shown in Fig. 45.12 are similar, but the arrow is drawn differently. The connection to the central slice is the *base*. The arrowed connection is the *emitter*. The arrow points *in* on a pnp transistor, and *out* on an npn transistor. The arrow on the emitter shows the main current direction through the transistor. The other connection is the *collector*.

Transistors vary in current-carrying capacity. The larger ones, which control more power, may have a *heat sink* to remove heat energy dissipated within the transistor and prevent its destruction by overheating. Some have air-convection cooling vanes while others have a thick metal base to remove heat by conduction.

Current through a transistor

Fig. 45.13 shows how the currents are usually labelled for an npn transistor. The emitter current, here leaving the transistor, is made up from the base and collector currents.

I_b (the base current) is very low and is usually measured in microamps (millionths of an amp), I_e and I_c (the emitter and a collector currents) are in the milliamp range (thousandths of an amp). To connect a single transistor properly into a circuit, the different leads must be identified correctly. On different cases, the base, emitter and collector are not laid out in the same way. On the cases shown, the *base* is the middle terminal. Both diagrams in Fig. 45.13 (*b*) show the bottom view of the case.

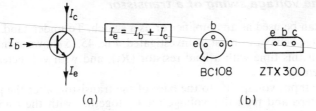

Fig. 45.13. The three terminals on an npn transistor

Showing the current gain of a transistor

The *amplification* or *gain* of the transistor tells you how much bigger the collector current is than the base current.

$$\text{gain} = \frac{I_c}{I_b}$$

The circuit shown in Fig. 45.14(*a*) can be used to show the operation of an npn transistor. It is used in the *common–emitter* mode. A transistor has three terminals. To use two for input, and two for output, one terminal must be used twice. In a common–emitter circuit, the input is connected between base and *emitter*. The output is measured between collector and *emitter*. The common–emitter circuit is one of the easiest to set up and use.

Here a potentiometer or voltage divider V_R is used to vary the base voltage, and to change the base current I_b, through R_1. This is only a few microamps, measured by a sensitive meter at A_1. The current I_c, through the collector is measured by a milliammeter at A_2. A graph of collector current against base current is called the *transfer characteristic* of the transistor. The base current is varied in small steps of, say, 10 μA and the corresponding values of the collector current I_c are read on A_2.

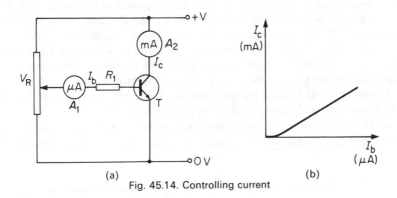

Fig. 45.14. Controlling current

The base resistor R_1 protects the transistor from overload. Fig. 45.14(*b*) shows the form of the resulting graph. After the initial curve it is almost straight. Notice that the axes of the graph use different scales. One scale is about 100 times bigger than the other. The slope of this graph is the *gain* of the transistor. For general-purpose npn transistors, the value is about 100 (somewhere between 50 and 200).

This gain is not constant, even for a particular transistor. This does not mean that a transistor actually makes the base current 100 times bigger! What is happening is that the small base current is controlling the much larger current through the emitter and the collector. So if the input (base) current changes, the output (collector) current changes about 100 times as much.

Showing the voltage swing of a transistor

A transistor can be used as an amplifier or as a switch. To understand the second of these, its voltage behaviour must be investigated. Fig. 45.15 shows a similar circuit to Fig. 45.14, but this time with a load resistor (R_2), and with voltmeters to measure what is happening.

To vary the input voltage V_1 to the base of the transistor, alter the potentiometer V_R in small steps and read this voltage on V_1 together with the transistor output voltage V_2. The graph in Fig. 45.15 (*b*) shows the changes. It is an important graph. The result is the basis for the use of transistors in many digital circuits. As the input voltage V_1 increases from zero, the output voltage V_2 stays constant at almost the full supply voltage until the transistor suddenly switches 'on' when the input voltage is about 0.7 V. This voltage is a characteristic of the semiconductor used, in this case silicon. The suddenness of this change makes the transistor suitable for use as a switch.

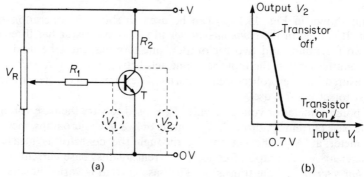

(a)   (b)

Fig. 45.15. The output voltage changes very quickly

With a small base current (Fig. 45.15), the amplified collector current is small. There is little voltage across the load resistor R_2. As the base voltage is increased above about 0.7 V, the base current increases. So the amplified collector current increases many times more. This large current through R_2 means there is a large voltage across it, and only a small voltage across the transistor. This sudden drop in the output voltage of the transistor is its most important feature.

The transistor seems to alter its resistance in the circuit, from high to low. This is where its name originally came from. It is a *transfer resistor*.

Fig. 45.16 shows the use of transistor as a switch. Although it can be operated mechanically with a simple two-way switch, it is often useful to use a sensor instead—

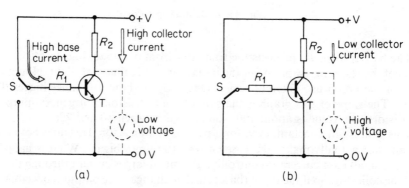

(a)   (b)

Fig. 45.16. The two states of the transistor switch

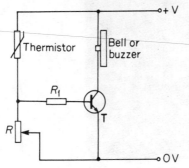

Fig. 45.17. When the temperature rises, the buzzer in this fire alarm circuit sounds

as, for example, in an automatic fire alarm. Here a buzzer or bell will be sounded if the temperature (sensed by the thermistor) becomes too high. See Fig. 45.17.

When the thermistor is cold, its resistance is high. It takes a large share of the power supply voltage and the potential difference across R is low, so the transistor is *off*. But as the temperature rises the thermistor resistance falls. The voltage across R rises. As it crosses 0.7 V, the transistor switches *on*.

Use of a reed switch and transistor in a burglar alarm

If the LDR in Fig. 45.5 of the thermistor in Fig. 45.17 is replaced by a reed switch we should have the basics of an efficient burglar alarm (see Fig. 45.18). A *reed switch* consists of two springy reeds made of magnetic material with contacts which are normally closed. If a magnet is brought near to them, the reeds become similarly magnetised and spring apart by magnetic repulsion. This opens the contacts and breaks the circuit.

If a reed switch is concealed in the frame of a door and is adjacent to a magnet in the door itself, the contacts will open when the door is closed. But if an intruder forces open the door, the magnet moves away and the contacts will close. A sudden flow of collector current will pass through the coil of the relay, which operates a bell circuit with a local voltage supply. Compare this with Fig. 45.16.

Uses of transistors in circuits

In amplifier circuits in radios, record players and similar devices, the 'old-fashioned' thermionic diode and triode have been superseded by semiconductor diodes and transistors. These components are much smaller and lighter. They operate from low voltages of only a few volts, compared with tens or hundreds of volts for thermionic devices. For this reason, among others, semiconductor circuits are safer than their predecessors. They are also more economical. A thermionic device, as its name suggests, operates using heat energy to emit electrons from the cathode. Semiconductors operate cold, and use much less energy than thermionic circuits.

The transistor has two basic uses. Both can be explained in terms of the voltage graph in Fig. 45.15 (*b*). Because the graph line is very steep, the transistor is an *amplifier*. A small change in the input voltage produces a large change in the output voltage. A small a.c. input signal gives a similar, larger, inverted output signal, if the rest of the circuit is properly designed. This is a use of transistor in an *analogue* circuit, where gradual voltage changes occur.

The fact that the voltage change on the graph is so sudden makes the transistor useful as a *switch*. There are two basic states: low input/high output (off) and high input/low output (on). This is a *digital* switch, like a light switch (on/off), compared with a dimmer or variable resistor (an *analogue* control). This switch can be operated

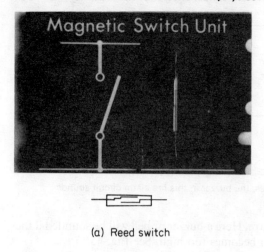

(a) Reed switch

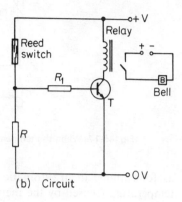

(b) Circuit

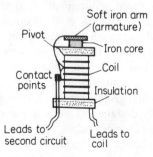

(c) Relay

Fig. 45.18. An alarm circuit. If the magnet is moved away, the bell sounds

by sensors as shown in Figs. 45.17 and 45.18. It can also be used as the basis for many logic circuits of the kind used in calculators or computers. In both analogue and digital applications, several (perhaps very many) transistors are often fitted into one microchip package.

Logic gates

Transistors are often used in digital circuits as *logic gates*. Here the output is either a high voltage or a low voltage, depending on the voltage values at the inputs to the circuit. There are five basic types of logic gate.

The NOT gate or inverter

A single transistor can be used as a NOT gate. This is the simplest type of logic gate. Many such gates, linked together, are used in digital electronic devices like clocks, computers and counters. The gate output voltage is *NOT* the same as the input voltage. Fig. 45.19 shows one version of such a gate, made using a single transistor.

When the input voltage value is low, there is no current into the base. The collector current is low, so the voltage across the load resistor is low. The output voltage between the transistor collector and emitter is high.

The reverse is true when a high voltage is connected to the input and a high current flows into the transistor base. This current is amplified by the transistor. The collector current is high, so the voltage across the load resistor is high. The output voltage *V* across the transistor is low.

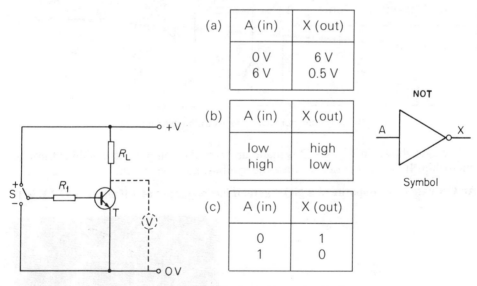

Fig. 45.19. A NOT gate and its truth tables

Several versions of a *truth table* for the gate can be worked out as shown in Fig. 45.19:

(a) gives actual voltage values for inputs of 0 V and 6 V, but the gate may operate at a different supply voltage. This is the least useful version of the truth table.
(b) uses the terms 'high' and 'low', without giving actual voltage values.
(c) uses '1' and '0' instead of 'high' and 'low'. These are easier to write, and show the binary connection more clearly.

The diagram also shows the logic symbol for a NOT gate. No components are shown; there are different ways of making gates, from diodes, transistors or microchips. No power supply is shown on gate symbols, but it must always be there. Only the 'box' to show the type of gate, and the inputs and outputs, are drawn.

The output state (0 or 1) of a NOT gate is opposite to the input (1 or 0).

This NOT gate is often called an *inverter*. It can be used to invert a series of electronic pulses. Fig. 45.20 shows this: the output pulses high when the input pulses low, and vice versa.

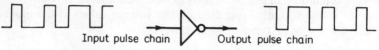

Input pulse chain   Output pulse chain

Fig. 45.20. Inverting a series of digital pulses using a NOT gate

Two inputs to the NOR gate

With two inputs to a gate (see Fig. 45.21), the truth table is more complicated. There are *four* ways of connecting high and low voltages to the two inputs. According to the two switch positions, the inputs at A and B can be: both 0, 0 and 1, 1 and 0, or both 1. You can see this in the truth table for the *NOR* gate shown in Fig. 45.21.

If both inputs are *high* (1), the output is still *low* (0). This is a *NOR* gate.

The NOR gate output is only 1 if neither input A *NOR* input B is 1.

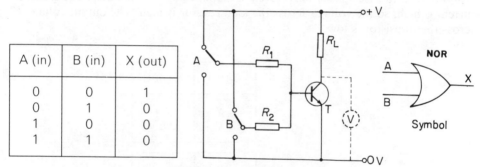

A (in)	B (in)	X (out)
0	0	1
0	1	0
1	0	0
1	1	0

Fig. 45.21. A transistor NOR gate, its truth table and symbol

The gate shown in Fig. 45.22 is made up from the components contained in one microchip. It is much easier to draw the symbol than the circuit.

An OR gate is the opposite of a NOR gate: it has an output of 1 if either A *OR* B is 1.

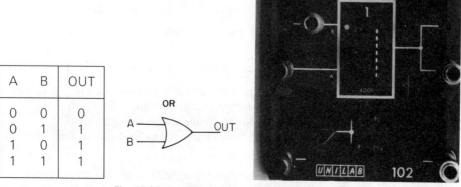

A	B	OUT
0	0	0
0	1	1
1	0	1
1	1	1

Fig. 45.22. An OR gate, its truth table and symbol

Two more logic gates (Fig. 45.23) are useful in processing inputs. The AND gate does just as the name implies.

The output of an AND gate is 1 only if inputs A *AND* B are 1.

The *NAND* gate is the inverse of the AND.

The NAND gate output is 1 unless both its inputs are 1.

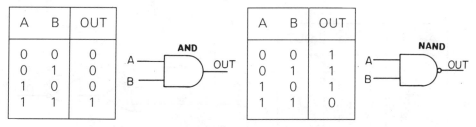

A	B	OUT
0	0	0
0	1	0
1	0	0
1	1	1

A	B	OUT
0	0	1
0	1	1
1	0	1
1	1	0

Fig. 45.23. The AND and NAND gates, their truth tables and symbols

Combinational logic

Often quite a large number of gates are used to solve a particular problem. When used together they make up a *combinational logic circuit*. You can use any of the fives gates to solve such a logic problem. In many circuits the NAND gate is used as the basic 'building block'. All the other gates mentioned can be made up from NAND gates. Fig. 45.24 shows the simplest of these, the NOT gate or inverter.

Here the two NAND gate inputs are linked together to make one input. If this input is 1 the output is 0 and vice versa.

Fig. 45.24. A NAND gate as an inverter

Fig. 45.25 (*a*) shows two NOT gates and a NAND gate, linked. NAND gates could be used as the two inverters or NOTs. A full truth table for the system is also shown. This table shows how to work out what this system does. Two middle points, X and Y, are labelled as well as the inputs and outputs.

The A and B columns are as usual for a 2-input gate. The X column is the opposite of A, since there is a NOT gate between them. The Y column is the opposite of B. The voltage signals at X and Y are fed into a NAND gate. The resulting OUT column shows that these three gates make up an OR gate.

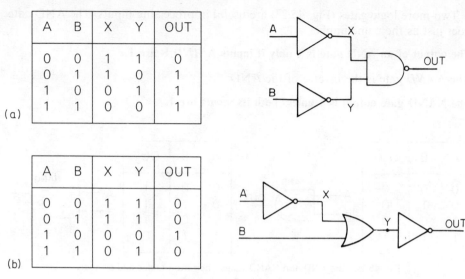

A	B	X	Y	OUT
0	0	1	1	0
0	1	1	0	1
1	0	0	1	1
1	1	0	0	1

(a)

A	B	X	Y	OUT
0	0	1	1	0
0	1	1	1	0
1	0	0	0	1
1	1	0	1	0

(b)

Fig. 45.25. More complex logic gate circuits

In the circuit in Fig. 45.25 (*b*), three different gates are shown in combination. The final truth table is found in the same way as for (*a*), by identifying the intermediate points in the layout. In this case there are two such points. The resulting logic state at each is found by considering the inputs to a gate and the truth table of the relevant gate. So X is simply NOT A—the X values are the opposite of the A values. Then Y is the result of linking X and B into an OR gate. The value of Y will be 1 if X OR B is 1. Finally the OUTPUT is the result Y into a NOT gate. The result is not one of the five standard gates. This truth table can only be achieved by combining several standard gates.

Bistable circuits

Logic gates can be used to solve logic problems. But with one extra feature—*feedback*—they can do much more. The word 'feedback' is a good one to describe this type of connection. The output of a circuit is fed back to the input.

Fig. 45.26 shows one example of this with two single transistor NOT gates or inverters. The output of gate 1 feeds into gate 2, and the output of gate 2 is fed right back into gate 1. On the circuit diagram, two voltmeters are connected to read the output voltages of the two gates.

To see how this type of circuit works, consider what happens if the transistor in

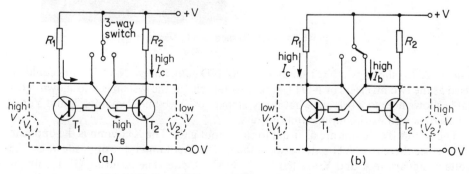

Fig. 45.26. Two NOT gates used to set up a bistable circuit

gate 1 is *off*. Its input voltage is low, so there is little base or collector current. Its output voltage will be high, since it is a NOT gate. The high voltage is fed into gate 2. This high voltage provides a high base current, and thus an amplified collector current for gate 2. Transistor 2 is *on*.

The low output voltage of transistor T_2 is fed back into gate 1, which is just what is needed to keep transistor T_1 *off*. This is a *stable* state—it will stay like this.

But now think what happens if the input to gate 1 is deliberately made high by linking it briefly with the high voltage line. The base current into transistor T_1 will be amplified, and it will turn *on*. The gate will have a low output voltage. So the input current to transistor T_2 will be low, as will its collector current.

Since gate 2 is a NOT gate too, its output voltage will be high. This is linked back to the input of gate 1. So the link from transistor T_1 to the high voltage line can be disconnected (Fig. 45.27). This input is already being kept high by the output of gate 2. The second state, with transistor T_1 *on* and transistor T_2 *off* is also stable.

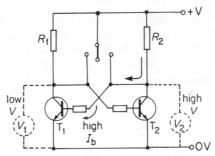

Fig. 45.27. Making and breaking the link to the input of gate 1

This type of circuit, with two linked transistors, is called a *multivibrator*. This may seem an odd name, but a later example will show that a circuit of this type can keep on changing from one state to the other and back again. This circuit does not keep changing. It has two stable states and is called a *bistable* multivibrator. This is usually shortened to *bistable*.

A bistable is a simple memory circuit. It 'remembers' the last change imposed on the circuit. Fig. 45.28 shows a way of setting this up with two microchip NOT gates or

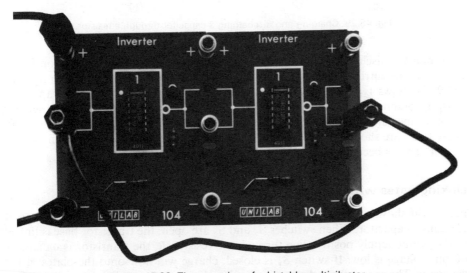

Fig. 45.28. The operation of a bistable multivibrator

inverters. They use LEDs (light-emitting diodes) to show whether the gate output is high (LED on) or low (LED off). As the flying lead is moved from one input to the other, the LEDs change and stay changed when the lead is disconnected.

The astable multivibrator

A multivibrator circuit containing resistors is a basic memory element. But the two transistor switches in such a circuit can also be linked by capacitors. Their behaviour will be different because the current through a capacitor is not constant. As the plates of the capacitor become charged, the potential difference between them increases. This voltage opposes further flow of charge, so the current decreases. The result of this is that the circuit switches continually from one state to the other, and is thus referred to as an *astable*.

Showing the current decay of a capacitor

Fig. 45.29 shows a voltage divider chain which includes a capacitor and a resistor. When the switch is in position A the battery is connected across the combination of the capacitor C and resistor R_1. To plot the current against time, move this switch, then note the value of the current on the centre-zero μA meter every 10 seconds until the value becomes too small to read.

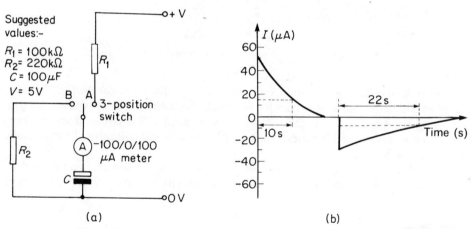

Fig. 45.29. Charging and discharging a capacitor through a resistor

To plot the discharge graph for the capacitor, move the switch to position B and again read the current value every 10 s. The capacitor is discharging through R_2. Fig. 45.29 (b) shows typical charge and discharge graphs for the suggested capacitor/resistor combinations. The *time constant* of the combination is roughly the time taken for the current to fall to 1/3 of its maximum (or to reach 2/3 if it is increasing). These times are shown on the graph. Notice that the discharge time constant here is greater. This is because R_2 is a larger value resistor than R_1.

Linking gates with capacitors

Fig. 45.30 shows a single transistor NOT gate or inverter. The input chain to it contains a capacitor. When switches S_1 and S_2 are open, the transistor base is linked to the power supply positive line through the resistor R, the transistor is *on* and its output voltage is low. If switch S_1 is closed, charge will flow onto the plates of the capacitor. The energy from the power supply is being stored on these two plates. The

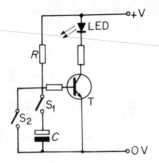

Fig. 45.30. The capacitor charges when switch S₁ is closed

input voltage to the transistor will increase as the charge stored on the capacitor and the voltage across it increase.

The capacitor will charge up fully, and the voltage across it will increase. When this voltage is sufficiently high, the transistor will switch *on* again as current flows into the base. The time T taken for this will depend on the value of the resistor R_1 and the capacitor C. For the 220 μF capacitor and a 100 kΩ resistor

$$T = R \times C$$
$$= 220 \times 10^{-6} \text{ F} \times 100 \times 10^3 \text{ } \Omega = 0.22 \text{ s}$$

This is not exactly the time it takes for the transistor to switch *off* again, but it does give an indication of the time scale of the switching operation. The system can be 'reset' by closing S_2 so that the capacitor discharges again.

With two cross-connected switches, each transistor can switch on another transistor as it is itself switched off. Fig. 45.31 shows such an arrangement, an *astable* multivibrator.

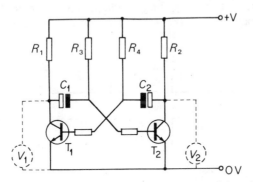

Fig. 45.31. The collector of each transistor is linked to the base of the other via a capacitor

The multivibrator does have two output states, but they are not stable. The times spent in the two states depend on the time constants of the two resistor–capacitor chains $(R_3 \times C_1)$ and $(R_4 \times C_2)$. The smaller these time constants, the quicker the states will change, and the higher the frequency of the pulses produced.

Fig. 45.32 shows a practical version of the astable circuit. The load resistors have been replaced by filament bulbs. With the component values shown, the time constant of the R/C pairs is (100 kΩ × 22 μF), or 2.2 s.

These various transistor circuits, logic gates, and bistable and astable multi-vibrators can be used together. Fig. 45.33 shows one example in 'block' form—the boxes represent the different electronic systems. The two resistors shown represent R_3 and R_4 from Fig. 45.31. They can be varied to alter the timing of the pulses from the slow astable. The whole system turns on and off three coloured LEDs in sequence.

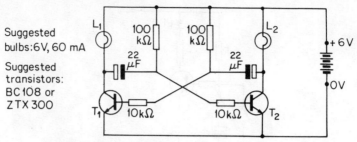

Fig. 45.32. Each bulb flashes on and off roughly every 5 s

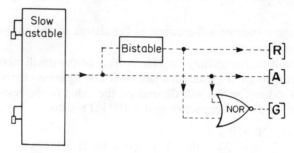

Fig. 45.33. Red, red + amber, green, amber, and back to red again

These glow red, amber and green, and this circuit gives a traffic-light sequence for the three colours.

The astable multivibrator operates the amber light. This keeps flashing on and off, as the middle row of Fig. 45.34 shows. The red light changes at half this frequency. This is operated by the bistable, being switched on and off by falling pulses at the output of the astable unit. Finally, the green light is on when neither red NOR amber is on. So it can be operated by linking these two outputs, via a NOR gate, to the green LED.

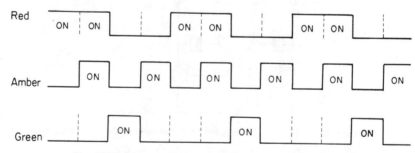

Fig. 45.34. The on-off sequence of the three LEDs in Fig. 45.33

This circuit will only work well if the logic gates can provide enough current to operate the indicators. With light-emitting diodes, this is possible. With higher powered lamps, electromagnetic relays (Fig. 45.18) could be used.

QUESTIONS: 45

1. Draw the circuit symbols for both npn and pnp transistors. Label the three connections on each of them.

2. A *reed switch* and a *relay* are both magnetic

devices. Which would you expect to see as part of the input of an electronic circuit, and which as part of the output?

Explain briefly how each uses magnetism to carry out its switching operation.

3. You are provided with a 6 V battery and three 10 kΩ resistors. Design three circuits, using some or all of these components, to provide output voltages of (a) 2 V; (b) 3 V; (c) 4 V.

4. Fig. 45.35 shows a circuit which can be used to find the *voltage characteristic* of a transistor. Sketch the resulting graph of output voltage against input voltage. Why is the transistor's use as a switch a consequence of the shape of this graph?

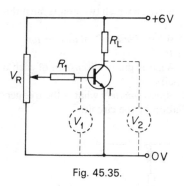

Fig. 45.35.

5. When the thermistor shown in the circuit in Fig. 45.36 is slightly warmed, what will be the effect on (a) the reading on ammeter A; (b) the reading on voltmeter V; (c) the power output from the battery?

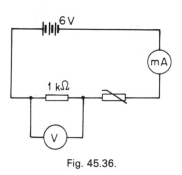

Fig. 45.36.

6. The truth table for a 2-input logic gate shows a '1' output only when both inputs have '0' values. What is the name of this logic gate? Draw out the truth table in full. Show the symbol for the logic gate.

7. Fig. 45.37 shows two transistor circuits, each operated by a resistor chain including a reed switch. Explain in each case what you would expect to happen when a strong magnet is brought near to the reed switch.

Note: The magnetically operated reed switches are of two types:

(a) Contacts normally *closed*.
(b) Contacts normally *open*.

This circuit uses type (b). The burglar alarm in Fig. 45.18 uses type (a).

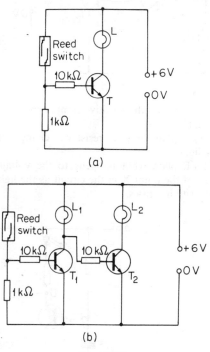

Fig. 45.37.

8. Fig. 45.38 shows a logic circuit containing two gates. Copy the truth table, filling in the gaps in the intermediate column and the output column. Which single gate could replace this circuit arrangement?

A	B	X	Y	OUT
0	0	*	*	*
0	1	*	*	*
1	0	*	*	*
1	1	*	*	*

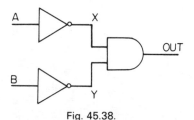

Fig. 45.38.

9. The circuit in Fig. 45.39 changes its operation as the light intensity alters.

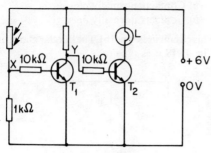

Fig. 45.39.

(a) Which is the sensitive component in the circuit?

(b) Explain how its resistance changes in daylight.

(c) Explain what happens to the voltage level at the point X in the circuit as the light intensity increases.

(d) Explain what happens to the voltage level at the point Y in the circuit as the light intensity increases.

(e) Under what conditions will the bulb come on?

(f) Explain briefly how this could be achieved using only one transistor in the circuit.

10. Fig. 45.40 shows two transistor switches, each with its input linked to the positive supply line via a resistor, and to the negative line via a capacitor. Another link is needed to convert this circuit into a multivibrator.

(a) Between which two points must this link be made?

(b) *What kind of multivibrator is then formed?*

(c) What is the *time constant* of each of the R/C pairs?

(d) *Approximately* what is the frequency of oscillation of the circuit?

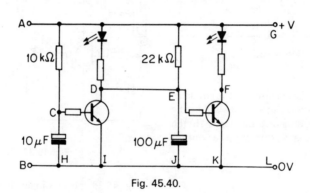

Fig. 45.40.

46. Atomic structure, the quantum theory and radioactivity

The discovery of the electron towards the end of the nineteenth century was the starting-point of new avenues of research in science which were to give physicists an insight into the structure and nature of the atoms of matter. Historical analysis of the various ideas which gradually evolved to explain the atom and the experiments carried out to test them presents a very complex picture, and any attempt to deal with all of them would only lead to confusion. So in this chapter we shall deal with the development of ideas on atomic structure and describe some of the more important experiments which have suggested or confirmed them.

The Rutherford–Bohr atom

On page 367 we mentioned the present-day notion of an atom as consisting of a central nucleus surrounded by electrons at various energy levels. Basically, this is a mathematical theory and, with our present knowledge, we cannot draw a simple picture or make a satisfactory model of it.

It will, however, be helpful to consider an early model of atomic structure first suggested by Ernest Rutherford and Niels Bohr. They visualized an atom as being constructed like a miniature solar system in which the planets are electrons and the sun is a small heavy nucleus. There are more than a hundred elements, some of which have been made artificially. The lightest, hydrogen, has a single orbital (planetary) electron and the heaviest natural element is uranium which has 92 electrons. They are all listed in a table at the end of chapter 47.

Rutherford–Bohr models of the four lightest atoms are shown in Fig. 46.1. The hydrogen atom is the simplest of all; its nucleus consists of a proton which is a particle charged with the smallest unit of positive electricity which has yet been

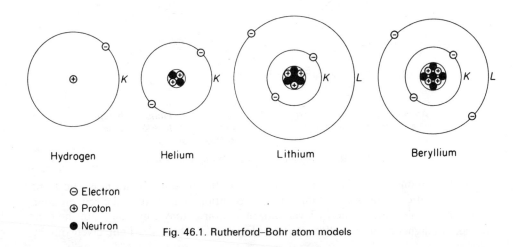

Hydrogen   Helium   Lithium   Beryllium

⊖ Electron
⊕ Proton
● Neutron

Fig. 46.1. Rutherford–Bohr atom models

isolated. This is equal in amount to the negative charge of the orbital electron, so that the two together form an electrically neutral combination. Bohr supposed that the electric force of attraction between the electron and the nucleus kept the electron in its orbit in the same way that gravitational attraction holds the planets in their orbits round the sun.

The next heaviest atom is that of the inert gas helium, which has a nucleus containing two protons together with two electrically neutral particles called neutrons; and there are two orbital electrons. Next comes a soft white metal, lithium, in which three electrons are revolving round a nucleus containing three protons and four neutrons. Following lithium is another metal, beryllium, with four orbital electrons, and so on until we come to the short lived hahnium, with a hundred and five.

Electron shells

The orbital electrons revolve continuously round the nucleus at very high speeds in orbits which alter their direction in space so that groups of electrons trace out *shells* rather than paths confined to one plane only. According to the size of the atom, it can have up to seven distinct shells of different radii which are designated by the letters K, L, M, N, O, P, and Q. There is a limit to the number of electrons which can occupy any given shell. For example, the innermost shell (K shell) cannot hold more than two; the second is complete when it has eight; the third can hold up to eighteen electrons, and so on.

The size of an atom taken as a whole is the volume enclosed by its outermost electron shell. The nucleus, in which the greater part of the mass is concentrated, is a mere speck whose diameter is only about $\frac{1}{100\,000}$ to $\frac{1}{10\,000}$ of that of the whole atom. The electrons are of almost negligible mass, since each one has a mass only $\frac{1}{1836}$ of that of a proton (hydrogen nucleus). It is obvious, therefore, that an atom contains far more empty space than solid matter.

How an atom gives out light

At ordinary temperatures the radii of the various electron shells are fixed and the electrons in them possess a fixed amount of energy. Under these conditions the atom is said to be in its lowest energy state or *ground state*. If an atom is given some extra energy, for example, when a substance is heated or if the atom is struck by a fast-moving ion in an electric discharge tube, then one or more of the electrons may

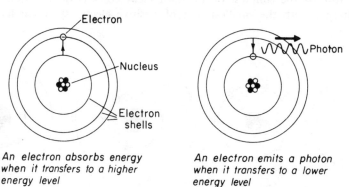

An electron absorbs energy when it transfers to a higher energy level

An electron emits a photon when it transfers to a lower energy level

Fig. 46.2. Mechanism of radiation from an atom

transfer from one energy level into a higher one. In this state the atom is said to be *excited*, but it does not stay for long in this condition. The disturbed electrons soon transfer back into lower energy level and, in so doing, they emit energy in the form of electromagnetic pulses called *photons* (Fig. 46.2). Thus, the light given out by any kind of lamp consists of billions of tiny photons pouring out from the excited atoms as they return to their lower energy states.

Such is the mechanism by which atoms give out not only visible light but also ultraviolet radiation, X-rays and infrared radiation. The wavelength of the radiation emitted depends simply on the two particular energy levels between which an electron transfer occurs.

Radiation and the quantum theory

Evidence for the existence of atomic nuclei came from Ernest Rutherford, and the explanation of the process of radiation from Niels Bohr; hence the term "Rutherford–Bohr atom".

Bohr's theory, as outlined above, is an application of the quantum theory of energy devised by the German physicist, Max Planck, who developed it for the purpose of explaining certain experimental observations in connection with the radiation of energy from "black bodies". Planck came to the conclusion that energy does not flow continuously from hot bodies but comes off in small packets called *quanta* or *photons*.

Planck's quantum theory has had far-reaching success in the realm of physics; it has established the principle that energy in all forms is discrete, i.e., it occurs only in individual units. In other words, energy, like matter, is atomic in nature. The reader may remember that another application of the quantum theory was mentioned on page 184, where it was explained that heat is conducted through certain substances by means of mechanical vibratory packets called *phonons* (not to be confused with photons, which are electromagnetic vibrations).

The photoelectric effect

In the year 1888, Wilhelm Hallwachs at Dresden discovered that if ultraviolet light is shone on to a clean zinc plate connected to a negatively charged gold-leaf electroscope, the electroscope slowly loses its charge. The reason for this was not understood at the time but we now realize that electrons were emitted from the zinc plate. These are called *photoelectrons* and the phenomenon is referred to as the photoelectric effect.

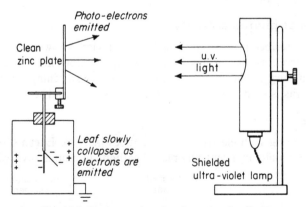

Fig. 46.3. Demonstrating the photoelectric effect

Hallwachs's experiment may be carried out as shown in Fig. 46.3. It is essential to clean the surface of the zinc thoroughly by means of emery paper. Having done this the plate is fixed to the cap of a gold-leaf electroscope and then given a negative charge by the usual method (page 371). When the ultraviolet lamp is switched on, the leaf divergence slowly decreases.

If the experiment is repeated with a positive charge on the electroscope, no loss of charge is observed. Any photoelectrons which may be emitted are immediately attracted back to the zinc.

Explanation of the photoelectric effect by the quantum theory

Following Hallwachs's experiment it was found that other metals behaved in the same way with ultraviolet light, and in addition, some showed the effect with visible light. The metal rubidium, for example, responds to red light.

Various wavelengths were used on each metal tested and it was shown that photoelectrons were emitted only if the wavelength of the light used was below a certain critical value. Speaking in terms of frequency, it meant that the frequency of the light had to be above a certain critical value called the *threshold value* for the particular metal concerned.

Moreover, even the weakest illumination worked, provided its frequency was equal to or above the threshold value, but below this even the most intense illumination failed to have any effect.

This was puzzling until finally, in 1905, an explanation was offered by Albert Einstein on the basis of Planck's quantum theory.

Planck had shown that the quantity of energy conveyed by a photon of radiation was directly proportional to its frequency. Thus, ultraviolet photons with their higher frequency have more energy than the photons of visible light.

Einstein suggested that, depending on the metal concerned, an electron requires a certain minimum quantity of energy to release it from the metal and *it must receive this energy in a single quantum, or lump so to speak. It will not accept several smaller quanta instead.* It will, of course, accept quanta larger than the critical value, in which case the balance of energy left over simply serves to give the photoelectron kinetic energy when it is ejected.

Einstein put this idea into the form of an equation:

Energy of incident photon = energy required to extract electron from metal + maximum kinetic energy of ejected electron.

Subsequently, Einstein's equation was verified experimentally for several different metals by Robert Millikan in America. It is one of many important applications of the quantum theory of radiation.

Applications of photoelectricity

The effect just described is put to practical use in devices known as *photocells*. These are of many different types and have numerous applications in science and industry, for example, in burglar alarms, automatic devices for switching on lights at dusk, television cameras, sound reproduction from film tracks, and so on.

Positive rays

While experimenting with electric discharge tubes in 1886, Eugen Goldstein found that if holes were made in a centrally placed cathode luminous rays, which he called

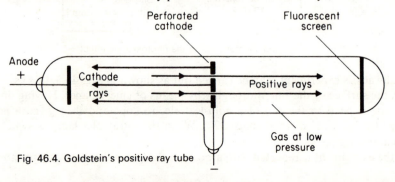

Fig. 46.4. Goldstein's positive ray tube

canal rays, were seen to pass through them. Obviously something was proceeding down the tube in a direction opposite to that of the cathode rays (Fig. 46.4).

Some years later, Wilhelm Wien showed that the rays could be deflected by electric and magnetic fields in the same manner as cathode rays but in the opposite direction. This and other tests indicated that they might consist of positively charged particles. Wein also measured the ratio of charge to mass $\frac{e}{m}$ for these canal rays (now called *positive rays*) by a method similar to that used by Sir J. J. Thomson to find $\frac{e}{m_e}$ for the electron (page 512). He found that the mass of the positive particles varied according to the kind of gas inside the tube. Even more important, the masses of the particles proved to be very nearly equal to those of the gas atoms in the tube. It therefore seemed highly probable that positive rays consisted of gas atoms from which electrons had been torn away.

About the same time that the properties of cathode rays and positive rays were being investigated, a new phenomenon, *radioactivity*, was discovered which was destined to provide still more information about the nature of matter.

Radioactivity

In the previous chapter we mentioned the fluorescence or emission of light which occurs when cathode rays fall on certain minerals or strike the walls of a discharge tube. Fluorescence is always associated with X-ray tubes, and this prompted Henri Becquerel, Professor of Physics at Paris, to investigate the possibility that X-rays might be associated with other forms of fluorescence.

Fluorescence can be produced in a number of different chemicals simply by exposing them to sunlight. Becquerel's method was to place some crystals on top of a photographic plate which had been well wrapped in black paper and then to place the whole lot in sunlight so that the crystals fluoresced. Afterwards he developed the plate to see if penetrating radiation had been given out.

Becquerel tested a number of fluorescent materials in this way with negative results, but eventually he did get a darkening of the plate when using a uranium salt. More important still, the experiment worked even when the uranium had not been rendered fluorescent by exposure to sunlight. Clearly, fluorescence was not a necessary condition for the emission of penetrating rays by this substance.

The Curies discover new radioactive substances

Two scientists in Paris became very interested in Becquerel's discovery. They were Marie Curie and her husband Pierre, who later became Professor of Physics at the Sorbonne.

They found that the rays from uranium caused ionization of air molecules and saw in this a means of measuring the intensity of the radiation. A simplified version of their apparatus is shown in Fig. 46.5. It consisted of two insulated metal plates

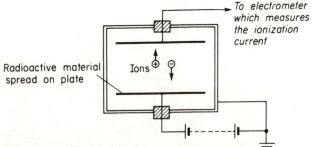

Fig. 46.5. How the Curies measured radioactivity by an ionization chamber

inside an earthed metal box or ionization chamber. The lower plate was raised to a high potential by an electric battery.

If some radioactive substance was spread on the lower plate the rays it gave off caused ionization of the air inside the chamber. The positive and negative ions so formed were then driven in opposite directions by the electric field between the two plates. Thus, a tiny ionization current flowed, and this was measured by an instrument called an *electrometer* connected to the upper plate. The magnitude of the ionization current was used as a measure of the radioactivity of the sample spread on the plate.

Using a radiation detector of this kind, the Curies tried various other chemicals and found that substances containing thorium also gave out ionizing radiation. But their most important discovery was the extreme activity of the ore of uranium, *pitchblende*.

On learning of this, the Austrian Government made them a gift of a ton of pitchblende residues from the uranium refineries in Bohemia, and after many weeks of arduous toil the Curies managed to extract from it a small quantity of a hitherto unknown radioactive element. This they called *polonium* in honour of Marie Curie's native country, Poland. Continuing their investigations, they subsequently isolated another new element which was more active still, and to this they gave the name *radium*.

The next step was to experiment with these new substances to find out more about the nature of the ionizing radiations which were given off.

Nature of the rays from radioactive substances

Both Becquerel and the Curies noticed that part of the radiation could be deviated by a magnetic field in exactly the same way as cathode rays, while the remainder carried a positive charge. Before long, P. Villard had found a third component which bore every resemblance to X-rays. In their experiments all these workers used both photographic and ionization methods to detect and measure the intensity of the different kinds of radiation.

In 1899 the study of radioactivity was taken up by Ernest Rutherford, who was at that time Professor of Physics at McGill University in Canada. Earlier, Rutherford had studied as a research student at Cambridge under Sir J. J. Thomson.

For convenience, Rutherford called the three types of radiation **alpha** (α), **beta** (β), and **gamma** (γ) rays respectively and set to work to investigate the properties of α-rays in particular. He placed a little radium at the bottom of a small lead box and subjected the rays that emerged from it to the action of a very strong magnetic field

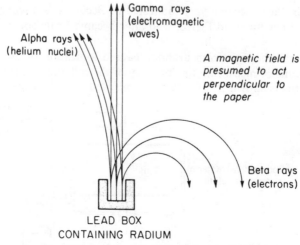

Fig. 46.6. An adaption of Marie Curie's diagram

at right angles to their direction. For this purpose he used a far stronger electromagnet than had hitherto been available and was able to show that the α-rays were deflected in a direction opposite to that of β-rays. This showed that the α-rays carried a positive charge. γ-rays are not affected by a magnetic field.

We cannot discuss all the experiments from which we have gained our present knowledge of the radiation, but it will be useful at this stage to summarize its chief properties. Marie Curie summed up the results of several experiments by a diagram similar to that shown in Fig. 46.6. *However, it must be emphasized that this is a composite diagram; it is not possible to study all three types of rays in a single experiment.*

Alpha rays are helium nuclei, i.e., helium atoms which have lost their two orbital electrons, and hence they have a net positive charge. From any particular radioactive substance, they are all ejected with approximately the same velocity. They have a range of several centimetres in air, but most are stopped by a very thin sheet of aluminium foil or by ordinary thicknesses of paper.

Beta rays are streams of high-energy electrons similar to cathode rays. They are emitted with variable velocities, approaching that of the velocity of light (3×10^8 m/s), and the more energetic ones are able to penetrate several millimetres thickness of aluminium.

Gamma rays consist of electromagnetic radiation and occupy a band among X-rays which are the shortest known wavelengths (see Fig. 26.19). The highest energy γ-rays are very penetrating and approach complete absorption (or attenuation) only after traversing a good many centimetres of lead.

The essential difference between γ-rays and X-rays is that γ-rays originate from energy changes in the nuclei of atoms while X-rays come from energy changes associated with the electron structure of atoms.

Crookes's spinthariscope, scintillations

Sir William Crookes found that when α-particles struck a screen coated with zinc sulphide a spark or scintillation was created at the point of impact. To demonstrate this effect he designed a simple instrument called a *spinthariscope* (Fig. 46.7). It

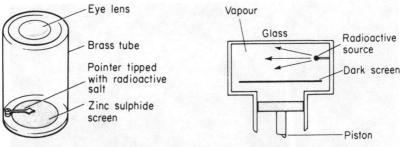

Fig. 46.7. Crookes's spinthariscope   Fig. 46.8. The cloud chamber

consists of a short brass tube having a zinc sulphide screen at one end and a magnifying glass at the other. Just above the surface of the screen is placed a watch-hand with a quantity of radioactive salt on its tip, and the scintillations can be viewed through the eye lens. Subsequently this method proved very useful for the purpose of counting α-particles.

The cloud chamber

If air is cooled sufficiently for the vapour present to reach saturation, it is possible to cool it still further without getting condensation. Under these conditions the vapour is said to be *supersaturated*. This, however, will occur only if the air is entirely free from dust or salt particles which act as nuclei on which the vapour can condense to form cloud droplets.

C. T. R. Wilson discovered that gaseous ions can also act as condensation nuclei and realized that this effect could be used to show the paths of ionizing radiations through air. Fig. 46.8 shows the principle of the cloud chamber designed by Wilson for this purpose.

A radioactive source emits particles into an air space saturated with water or alcohol vapour inside a vessel with a glass window. As the particles speed through the air they collide with the air molecules with such force that electrons are knocked off, leaving a trail of positive and negative ions. If the air space is now suddenly expanded by moving the piston, cooling occurs and vapour condenses out on ions, thus revealing the paths of the particles.

Cloud chamber tracks

The appearance of the cloud tracks depends on the particles concerned and can be used as a means of identification (Fig. 46.9).

Fig. 46.9. Appearance of cloud chamber tracks

The comparatively massive α-particles pursue straight paths, pulling electrons off atoms as they go and creating up to 10 000 ion-pairs per centimetre of their path. The resultant cloud tracks are straight and thick (Fig. 46.10).

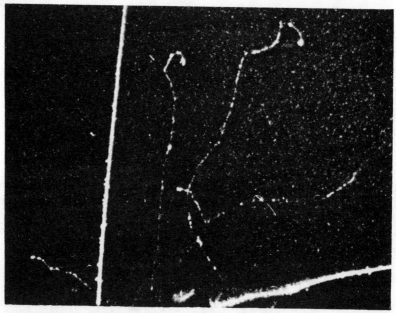

Fig. 46.10. α and β tracks showing the difference in ionizing power in the particles

By contrast, the very light β-particles suffer frequent repulsions from the electrons of atoms near which they pass and make ionizing collisions far less frequently. They make only a few hundreds of ion-pairs per centimetre of their path and consequently they display thin irregular cloud tracks.

γ-rays do not produce cloud tracks along their own paths. A gamma photon may, however, interact with an atom in its path and give up either a part or the whole of its energy in ejecting an electron from it. The electrons then behave like β-particles and produce irregular cloud tracks of their own which branch out from the direction of the gamma beam.

Nowadays, high-energy particles are mostly studied with the aid of *bubble chambers* which reveal the passage of particles by a trail of bubbles in liquid hydrogen or pentane (Fig. 47.12). But, as we shall see in the next chapter, the cloud chamber proved of great value in early research on particles and their interactions with matter. It not only revealed the tracks of otherwise invisible particles but also enabled an estimate of their energies to be made from the lengths of the tracks. In the case of electrically charged particles, the application of a magnetic field caused the tracks to curve and the direction of curvature was an indication of the sign of the charge carried. (See also under Cloud chamber studies on page 566 and Fig. 47.3.)

Methods of measuring the activity of radioactive substances

Since the time of the Curies many new and improved methods for measuring activity have been devised. A number of them involve some form of ionization chamber. In the following pages we shall describe two methods commonly used in elementary work.

The pulse (Wulf) electroscope

Fig. 46.11 shows the construction of a pulse electroscope and its use to detect the ionization of air by the radiation from an active source.

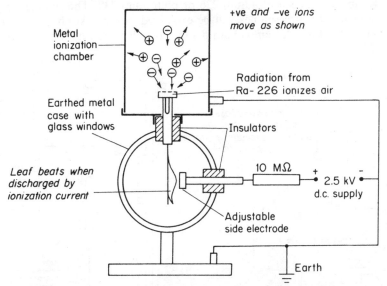

Fig. 46.11. Detecting ionization of air by a radioactive source

In some respects the pulse electroscope resembles a gold-leaf electroscope. The construction of the leaf system varies with different makes but the principle is the same for all. The leaf is charged by being attracted and making contact with a side electrode kept at a high potential. The leaf is then repelled, and its return to zero is assisted by some light spring device. Some have a fine quartz loop for this purpose:

in others the leaf takes the form of an aluminium flag on a taut phosphor-bronze suspension.

To demonstrate ionization of air by a radioactive source

A suitable source, e.g., radium-226, is picked up with forceps (never with fingers) and inserted into the top electrode of a Wulf electroscope. An ionization chamber is placed in position and connections made through a high value safety resistor to a 2.5 kV power pack as shown in Fig. 46.11.

Radiation from the source ionizes the air, and under the action of the electric field between the central electrode and the walls of the chamber, positive ions move towards the chamber walls and negative ions to the central electrode.

As the central electrode collects ions of opposite sign it eventually becomes discharged and is once more attracted to the side electrode. It then returns to zero and the discharging action continues as before. The leaf, therefore, pulses or beats at a rate which depends on the value of the ionization current. The ionization current, in turn, is governed by the activity of the source and the p.d. between the central electrode and the walls of the ionization chamber.

Function of the 10 MΩ *limiting resistor*

It is important to note that a high resistor of 10 MΩ or more is always placed between the high-voltage supply and the electroscope. This serves the double purpose of protecting the user from shocks and the leaf from damage. 10 MΩ may sound rather high, but it must be remembered that this is small compared with the resistance of the ionized air in the chamber and consequently it has little effect on the value of the ionization current.

To investigate the range of alpha-particles in air

Fig. 46.12 illustrates two different ways of using a pulse electroscope to measure the limited range of α-particles in air according to the type of apparatus available.

Suitable alpha sources are radium-226 or plutonium-239. The significance of the numbers attached to these names will be explained in the next chapter. Actually a sealed source of radium-226 emits both β and γ as well as α-radiation, but only the α-

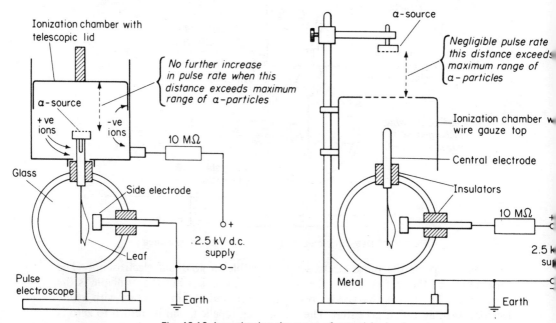

Fig. 46.12. Investigating the range of α-particles in air

particles produce a measurable effect in the ionization chambers we shall use. **All active sources must always be handled with forceps and never pointed towards oneself or anyone else, and when out of use, returned to the lead container provided.**

Method (1)

The source is fixed to the central electrode, and an ionization chamber with a telescopic lid is placed in position over it. An e.h.t. unit is connected as shown, adjusted to 2.5 kV, and switched on. The air becomes ionized by the α-particles and the leaf begins to pulse for reasons already explained in the previous experiment.

Now, if we start with the sliding lid very close to the source, the α-particles can travel only a very short distance before losing their ionizing power. The total number of ions produced will therefore be small and the pulse rate slow.

The lid is raised in 5 mm steps and each time the pulse rate is measured by counting and timing with the aid of a stop-clock. The shortest distance for which the pulse rate reaches its maximum value will be approximately equal to the range of the α-particles in air. The best way of obtaining the range is to plot a graph of pulse rate against distance between source and lid.

Method (2)

In this method, an ionization chamber with a wire gauze top is used and the source is positioned above it. The chamber is connected through its support to the negative terminal of the e.h.t. unit and earthed, while the side electrode is raised to 2.5 kV through the limiting resistor. Starting with the source close to the gauze, the pulse rate will be high since the path of the α-particles extends well into the chamber. As the source is raised by successive small distances the pulse rate decreases, and effectively drops to zero when the distance from source to gauze is approximately equal to the range of the α-particles.

This experiment can also be used to show that α-particles are completely absorbed by a piece of paper or very thin aluminium foil. The source is positioned just above the gauze and it is noticed that the pulses cease when the paper or foil is inserted between source and gauze.

The Geiger–Müller tube

This is a special form of ionization chamber which is operated at 400 V or more according to make (Fig. 46.13).

The type commonly used in elementary work consists of an aluminium tube which acts as the negative electrode while a wire down the centre forms the positive elec-

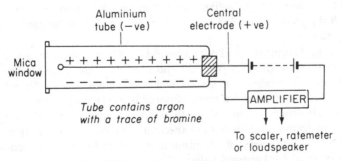

Fig. 46.13. The Geiger–Müller counter

trode. The gas inside the tube consists of argon at low pressure with an added trace of bromine.

A thin mica window at the end permits the entry of active particles or gamma photons. If one of these enters the tube it causes ionization of the gas inside.

The main advantage of a Geiger–Müller tube over the ordinary air chamber is that, by the time the electrons from the ionization process reach the central electrode, they are moving so fast under the high potential gradient that they create an avalanche of extra ions by collision. This process of *gas amplification* as it is called increases the sensitivity of the tube, enabling it to record the entry of β-particles and γ-photons which, on their own, produce far fewer ion-pairs per centimetre of their path than do α-particles.

The current pulses from the tube are amplified and used to operate either a *dekatron counter* or *scaler* or else a *ratemeter*. The former counts and records the pulses on special neon tubes while the latter measures the rate of arrival of pulses on a microammeter calibrated in *counts per second* (or *per minute*). Sometimes a small loudspeaker is incorporated which audibly indicates the pulses by a series of clicks.

Both the wavering of the ratemeter needle and the irregularity of the clicks from the loudspeaker are a clear demonstration of the random nature of the disintegration or decay of radioactive substances.

To study the absorption of beta-particles by aluminium

Beta-particles vary considerably in energy, so that while most of them are easily absorbed, the most energetic ones have a very long range in air. We, therefore, find it more convenient to investigate their absorption in some denser medium such as aluminium, rather than in air (Fig. 46.14).

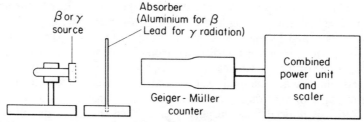

Fig. 46.14. Investigating the absorption of β-particles and γ-rays in aluminium or lead

For reasons explained in the previous sections, a Geiger–Müller counter is the best detector to use and we shall employ a strontium-90 source which gives very penetrating β-particles.

The Geiger–Müller tube is connected to a combined power unit and scaler and the voltage adjusted to the recommended value for the particular tube used. After the scaler has been switched on and allowed to warm up it is noticed that a random count is recorded even when no obvious source is present. This is called the *background count* and comes from active material in the earth and nearby surroundings together with the so-called *cosmic radiation* which penetrates the earth's atmosphere from outer space.

The background count is timed over a period of at least 2 minutes and after this the source is placed a short distance from the end window of the counter. The high count rate now obtained is measured as before, and at this stage it is worth noting that the count rate is scarcely affected if a piece of paper is inserted between tube and source.

Aluminium absorbers of increasing thickness are now inserted successively between tube and source and each time the count rate is recorded. It is found that several millimetres thickness of aluminium is required before the count rate approaches its original background value.

To show that beta-particles are deflected by a magnetic field

A beta source similar to that used in the previous experiment is set up with a brass tube in front of it to confine the β-particles to a fairly narrow beam. The beam enters a Geiger–Müller counter and the count rate is recorded (Fig. 46.15).

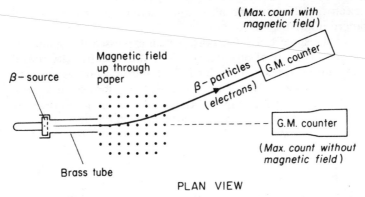

Fig. 46.15. β-particles may be deflected by a magnetic field

A strong magnet is now held vertically near the end of the brass tube so that a magnetic field is directed upwards and perpendicular to the path of the β-particles. Owing to deflection of the particles, the count rate drops. The deflected path of the particles can be determined by moving the tube to one side until the count rate increases to a maximum.

By applying Fleming's left-hand rule it can be shown that the β-particles are behaving in the same way as the cathode rays described on page 511, from which we may infer that they carry a negative charge.

To study the properties of gamma radiation

(1) *Absorption*

Using the apparatus shown in Fig. 46.14, together with a cobalt-60 source which provides high-energy γ-rays, it is found the thickness of aluminium which stopped β-particles has a negligible effect on γ-radiation, and that 2 or 3 cm thickness of lead is required before the count rate from high-energy γ-rays is reduced to a very low value.

(2) *Inverse square law for γ-rays in air*

The absorber is removed and count rates are measured when the source is respectively 20, 40, and 60 cm from the counter window.

Now, if γ-rays are a form of electromagnetic wave energy, and undergo very little absorption or attenuation in air we should expect them to obey an inverse square law as explained on page 306. This means that, after subtracting the background count rate from each of our readings, we should expect them to be in the ratio of $1 : \frac{1}{2^2} : \frac{1}{3^2}$. To a rough approximation this is found to be the case, but it must be borne in mind that: (*a*) our source is not a true point source, and (*b*) the distances measured ignore the distance of penetration of the rays into the counter, so that the percentage error will be large for small distances from the window.

(3) *Effect of magnetic field*

Even by applying the strongest magnetic field available it is found that no deviation of the γ-rays can be obtained. This indicates that γ-rays do not carry an electric charge as do α- and β-particles.

Emergence of ideas on atomic structure

By the end of the nineteenth century the notion that atoms were indivisible particles of matter was beginning to crumble. The study of cathode rays, positive rays and radioactivity had made it obvious that atoms contained particles of positive and negative electricity. The main problem now was to try to find out how these particles were arranged inside the atom.

Writing on the subject in 1902, Lord Kelvin expressed the opinion that an atom might consist of a sphere of positive electricity with negative electrons dotted about inside it. This idea was taken up by Sir J. J. Thomson, who was not happy about the electric forces involved in such an arrangement. He showed that, for stability, the electrons would have to be arranged in rings inside the atom. Furthermore, there is a limit to the number of electrons which can form a stable ring, after which there is a rearrangement and two rings are formed and so on. Now the Russian chemist Mendeleev had shown, many years previously, that if the chemical elements are written down in the order of their atomic weights they show a regularly recurring sequence in their chemical properties. It occurred to Sir J. J. Thomson that this might well be connected in some way with the number and arrangement of the electric charges inside the atoms. Here the matter rested for several years, during which time more knowledge accumulated to throw fresh light on the subject.

The next advance was made in 1911, when Lord Rutherford produced experimental evidence to show that the positive charge of an atom is concentrated in a small nucleus at its centre.

Evidence for the nuclear atom

In 1906, Rutherford had done some experiments at McGill University in which α-particles were passed through a thin sheet of mica. He noticed that they went easily through the mica without making holes in it as a bullet might. This led him to suspect that, quite possibly, the α-particles were passing right through the atoms themselves rather than pushing atoms out of the way.

Rutherford also noticed something else which, to his brilliant mind, was even more significant. Some of the particles were deflected or scattered out of their straight-line paths as they went through the mica, and he thought it highly probable that this was caused by electric repulsion between the positively charged part of the mica atoms and the positive α-particles.

Shortly afterwards Rutherford left Canada and became Professor at Manchester where, with the assistance of Hans Geiger and Ernest Marsden, he carried out a long series of experiments on the scattering of α-particles by thin metal films.

Geiger and Marsden's experiments

In order to explain the scattering Rutherford began by making calculations to see if the angle of deflection of the α-particles could be accounted for on the basis of Sir J. J. Thomson's theory that the positive charge on the atom was evenly distributed through it. He found, however, that some of the measured angles of deflection were far too large to be explained in this way.

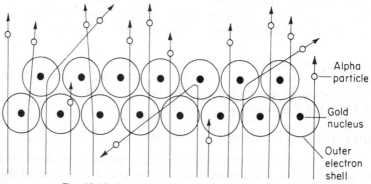

Fig. 46.16. Scattering of α-particles by gold nuclei

He next made a new set of calculations on the assumption that all the positive charge of the atom is concentrated in a tiny nucleus at the centre. The idea behind this approach to the problem was that the force of repulsion and consequent deflection of an α-particle would depend on how close it came near a nucleus (Fig. 46.16). Using Coulomb's *inverse square law*, which states that the force between two point electric charges is inversely proportional to the square of their distance apart, Rutherford worked out a formula giving the number of α-particles which ought to be scattered in any particular direction. The task of verifying this formula was undertaken in 1911 by Geiger and Marsden, who used the method shown in Fig. 46.17.

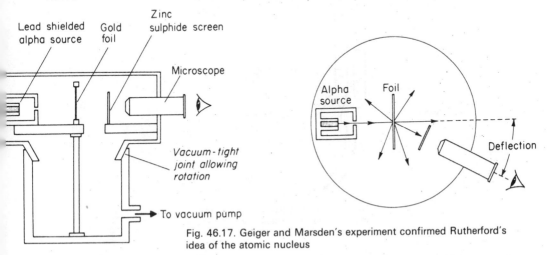

Fig. 46.17. Geiger and Marsden's experiment confirmed Rutherford's idea of the atomic nucleus

A radioactive source contained in a small lead box sent out a fine beam of α-particles through a hole in a lead screen and on to a very thin sheet of gold foil placed perpendicular to their direction of motion. The scattered particles produced scintillations on a glass screen coated with zinc sulphide and attached to a microscope which could be swung round through any angle. The whole arrangement was enclosed in an evacuated box. Geiger and Marsden spent many hours looking through the microscope, patiently counting the number of scintillations for a wider range of angles, and finally were able to show that their readings were in agreement with Rutherford's formula. Here was experimental evidence for the truth of Rutherford's assumption that the positive charge of an atom is concentrated in a small nucleus at its centre.

Rutherford and Royds's experiment

If the light from an electric discharge tube is examined through a prism by an instrument called a *spectrometer* the spectrum seen is not continuous, as in the case of white light, but consists of bright lines or bands. The same applies to the light from a bunsen flame coloured by the introduction of a small quantity of a chemical salt. Each element has its own particular spectrum, and spectrum analysis is an exceedingly sensitive test for the presence of a particular element.

This was the method used by Ernest Rutherford and one of his students named Thomas Royds to show that α-particles were helium nuclei. As a source of particles they used *radon*, the heavy radioactive gas which emanates from radium (page 563). This was introduced into the thin-walled glass inner tube of the apparatus shown in Fig. 46.18. α-particles from the radon passed through the thin glass and were stopped by the thick-walled outer tube, which had previously been evacuated. After about two days a sufficient number of α-particles had collected in the outer tube to enable a test to be carried out. Also by this time the α-particles had acquired electrons and

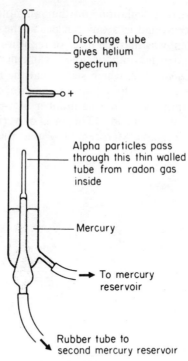

Discharge tube
gives helium
spectrum

Alpha particles pass
through this thin walled
tube from radon gas
inside

Mercury

To mercury
reservoir

Rubber tube to
second mercury reservoir

Fig. 46.18 Rutherford and Royds showed that α-particles were helium nuclei

become helium atoms. Mercury was then allowed to enter the outer tube at the bottom so that it pushed the helium gas up into a capillary discharge tube at the top. On passing an electric discharge through the gas the characteristic spectrum of helium appeared.

Atomic number and the periodic table

Experiments carried out by Wien and Thomson showed that the particles contained in positive rays had the same mass as the atoms from which they were derived. However, they did not always have the same charge. This arose from the fact that sometimes the atoms had more than one electron stripped away from them.

By 1914 Rutherford had satisfied himself that hydrogen, of all the gases, was consistent in its behaviour. The positive ray particles in a hydrogen discharge tube always gave the same value for the specific charge $\dfrac{e}{m_p}$, and always carried a single positive charge of the same magnitude as that of an electron. It therefore seemed feasible that the lightest atom, hydrogen, might consist of a single positive charge (proton) with a single orbital electron. This conclusion was confirmed by the observation that $\dfrac{e}{m_p}$ for gaseous hydrogen ions was equal to $\dfrac{e}{m_p}$ for hydrogen ions in electrolysis.

In what we have said above m_p refers to the mass of a proton or hydrogen nucleus in the same way that m_e refers to the mass of an electron (see page 513).

It seemed reasonable to suppose that the atoms of other substances were constructed in the same way except that their nuclei were made up of different numbers of protons with equal numbers of orbital electrons. But a difficulty arose straight away. Rutherford had shown that an α-particle or helium nucleus had a charge equal to that of two protons, while its mass was four times as much. Before proceeding further it was clearly necessary to measure the positive charge on a number of other nuclei and to see if this had any connection with their masses.

Now in the case of atoms heavier than hydrogen and helium it was not possible to strip off all their electrons in a discharge tube and so ascertain the number of protons in the nucleus. The problem had to be tackled from a different angle, and this was done in 1913 by Henry G. J. Moseley.

A full description of the theory and details of Moseley's experiments would take us far beyond the scope of this book, but they have to be mentioned here in view of the important conclusion which was drawn from them. Briefly, Moseley measured the wavelengths of the X-rays from tubes having anticathodes made of different elements and found that they varied in a regular pattern according to the position of the element in the periodic table. He then applied the equations relating to Bohr's theory of X-ray emission and obtained conclusive proof that **the atomic number of an element (i.e., its position in the periodic table) is equal to the number of protons in the nucleus.** Hence, also, the atomic number must be equal to the number of orbital electrons.

When Mendeleev first introduced the periodic system of classifying the elements he placed them in the order of their atomic masses. Moseley's work made it clear that an element's correct place in the periodic table was determined by the number of protons in its nucleus and not by its atomic mass. A list of the elements in order of their atomic numbers is given at the end of chapter 47.

The problem of nuclear mass—Isotopes

Moseley's contributions to knowledge of the nuclear charge was a big step forward, but there still remained the difficulty of accounting for nuclear mass. The problem had first presented itself about a hundred years previously. Following the introduction of Dalton's simple atomic theory at the beginning of the nineteenth century, chemists began to measure the atomic masses of the various elements, using the mass of the hydrogen atom as a unit. As the results gradually accumulated it was noticed that most of the atomic masses were very close to whole numbers. William Prout suggested in a paper written in 1816 that this might be explained by supposing that the atoms of the various elements were built up out of hydrogen atoms; in other words, hydrogen was the fundamental building brick of matter. But as time went on and improved methods gave more accurate values for the atomic masses, Prout's simple hypothesis had to be abandoned.

Some of the elements had atomic masses which were far from being whole numbers. Chlorine, as it occurs naturally, has an atomic mass of 35.5. One could not imagine that its atom contained half a hydrogen atom!

The first clue to the solution of the problem of fractional atomic masses came with the study of radioactive substances. By 1910, Frederick Soddy, who had worked with Rutherford in Canada, found that there were certain radioactive elements with identical chemical properties but different atomic masses. This meant that they had to be placed in the same position in the periodic table, and for this reason Soddy called them *isotopes*, a word derived from the Greek and meaning "occupying the same place".

It was soon to be shown that radioactive elements were not the only ones to possess isotopes. In 1913, Sir J. J. Thomson measured the value of $\dfrac{e}{m}$ for the positive ions of neon gas in a discharge tube and showed that there were two kinds of neon with masses 20 and 22 respectively. Now the atomic mass of ordinary neon is about 20.2; it was therefore inferred that ordinary neon is a mixture of these two isotopes in such proportions as to give an average mass of 20.2.

The final solution to the problem had to wait until 1932, when James Chadwick discovered the *neutron*. This is an uncharged particle with almost the same mass as a proton. It was then realized that the extra mass of nuclei is made up of neutrons. Thus, neon-22 simply contains two more neutrons than neon-20. Both nuclei, however, contain 10 protons each, which fix their position in the same place in the periodic table and give them identical chemical properties. Similarly, it was found that chlorine, with its atomic mass of 35.5, consists of two isotopes of masses 35 and 37 respectively.

Sir J. J. Thomson's discharge-tube method of separating isotopes was developed and improved by Francis W. Aston, who, in 1919, designed an instrument called a *mass spectrograph* which enabled him to weigh atoms very accurately. Since then several hundred nuclides have been discovered. The word *nuclide* refers to any atomic species, and hence is a term covering all the isotopes of the individual elements.

Hydrogen has three isotopes (Fig. 46.19). In addition to ordinary hydrogen there is

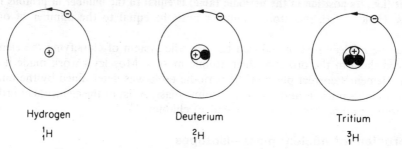

| Hydrogen | Deuterium | Tritium |
| $^1_1$H | $^2_1$H | $^3_1$H |

Fig. 46.19. Rutherford–Bohr models of three hydrogen isotopes

deuterium (heavy hydrogen), which has one neutron in its nucleus, and *tritium*, which has two. All three of the hydrogen isotopes contain one proton in their nuclei, and each has a single orbital electron.

We explained on page 558 that a hydrogen nucleus is called a *proton*. The deuterium nucleus is called a *deuteron*. Both protons and deuterons are used as high-energy missiles in particle accelerators which are used for the purpose of smashing atomic nuclei (see chapter 47).

Mass number, atomic number and nuclide symbols

The total number of protons and neutrons in a nucleus is called its mass number, and is denoted by A.

The atomic number is defined as the number of protons in the nucleus and is denoted by Z.

Thus if the number of neutrons is denoted by N we have

$$A = Z + N$$

It follows from this and what we have said in the previous section that

Isotopes of an element are atoms which have the same atomic number but different mass number.

If the reader has studied chemistry he will be aware that atoms of elements are represented by symbols. We have already met some of them in this book. Physicists use the same chemical symbols to represent the various nuclides, but with the addition of superscripts and subscripts giving their mass numbers and atomic numbers respectively.

Thus, the two isotopes of neon mentioned earlier are represented by $^{20}_{10}$Ne and $^{22}_{10}$Ne; the isotopes of hydrogen by $^1_1$H, $^2_1$H, and $^3_1$H; the chlorine isotopes by $^{35}_{17}$Cl and $^{37}_{17}$Cl. We shall meet others later on.

When discussing isotopes and we merely wish to distinguish one from another we often adopt a shorthand description by writing the name of the element followed by its mass number only, e.g., carbon-12 and carbon-14. From the list on page 562 we see that the atomic number of carbon is 6, so the full description of these two isotopes would be $^{12}_{6}C$, and $^{14}_{6}C$. From these symbols we are able to say that carbon–12 has 6 neutrons in its nucleus and carbon–14 has 8.

Atomic mass unit and binding energy

The mass number of an element, defined above, must not be confused with its *atomic mass*.

Since 1961 it has been agreed to use one-twelfth of the mass of the most commonly occurring isotope of carbon ($^{12}_{6}C$) as a unit for the measurement of atomic mass. This unit is called the **unified atomic mass unit** and is denoted by the symbol, u.

Protons and neutrons taken singly have approximately the same mass, but when combined in an atomic nucleus their total mass is always less than the sum of their individual masses. The difference is called the *mass defect* and represents the mass of the binding energy which holds the particles in the nucleus together. We shall have more to say about this in due course.

Symbols for protons, neutrons and electrons in nuclear equations

Let us summarize what we have learned so far.

An electron or β-particle possesses the fundamental unit of electric charge $(-e)$.

A proton or hydrogen nucleus has the fundamental unit of electric charge $(+e)$.

The atomic number (Z) of a nuclide is equal to the number of protons it contains and hence also represents both the number of fundamental charges $(+e)$ and the number of electron charges $(-e)$ in the neutral atom.

A proton has mass number 1 and charge $(+e)$ and so we represent it symbolically as $^{1}_{1}H$. It is also commonly represented as $^{1}_{1}p$.

A neutron has mass number 1 and charge (0) and is thus represented by $^{1}_{0}n$.

An electron or β-particle has a negligible mass and a charge $(-e)$. Hence, following the pattern used above it is symbolized in nuclear processes as $_{-1}^{0}e$.

Examples of the use of these symbols will be given in the next chapter.

QUESTIONS: 46

1. Write a brief account of the structure of atoms. Your account should make clear that you understand the terms neutron, proton, electron, and nucleus.

A certain atom of boron has a mass number of 11 and an atomic number of 5. What can you deduce concerning its structure?

Explain the meaning of the terms, positive and negative ions.

Briefly explain two methods of distinguishing between a stream of alpha particles and a stream of beta particles. (*W.*)

2. (*a*) What is meant by the *specific charge* of a particle?

(*b*) The specific charge of an electron is 1.76×10^{11} C/kg and the specific charge of a proton is 9.6×10^{7} C/kg.

(i) Why is it reasonable to assume that an electron and a proton carry the same quantity of charge?

(ii) Assuming that both particles carry the same quantity of charge, calculate the mass of a proton compared with the mass of an electron. (*J.M.B.*)

3. Name and describe three types of radiation emitted by radioactive substances, two of which are affected by a magnetic field. Draw two separate diagrams with labels to indicate:

(*a*) name of radiation;
(*b*) direction of magnetic field;
(*c*) effect on path of radiation.

4. A suitable detector is held very close to a radioactive source which is known to emit one form of radiation only. The count rate is observed to decrease considerably when either a thin sheet of cardboard is placed between detector and source, or when the detector is moved a few inches away. Name,

with a reason, the type of radiation which is probably emitted. Refer briefly to ONE other confirmatory test that you could apply.

(S.)

5. (a) Explain what is meant by "radioactive decay".

(b) You are provided with three radioactive sources each emitting a different kind of radiation. Describe an experiment which could be used to identify the radiation from each source.

(c) Name TWO other properties of each radiation, besides the one used to identify the radiation in the experiment you have described. (J.M.B.)

6. Describe any simple experiment to show that α-particles have a limited range in air.

7. Draw a diagram to show the essential features of a Geiger–Müller tube and briefly explain how it functions when an ionizing particle or photon enters it.

If you were provided with a Geiger–Müller tube and its associated counting equipment together with three radioactive sources which emit alpha, beta, and gamma radiation respectively, state what other simple equipment you would require and how you would use it to identify the three sources.

8. A radium source was placed some distance away from a Geiger–Müller tube connected to a scaler and the count rate measured over a period of 1 min with a good stop-watch was found to be 707 counts per minute. Two further readings over minute periods gave 690 and 715 counts per minute respectively. Are these unequal rates to be expected or do you suspect some fault in the counting equipment? Give your reasons.

The radium source was returned to its lead container and the count rate was again measured over three separate minute periods. The results obtained were 13, 8, and 10 counts per minute respectively. State two possible sources of this activity and give the name applied to it.

9. Give a labelled diagram of any one form of *cloud chamber* and explain its action. When an ionizing particle enters the chamber, what information can you obtain about the particle:

(a) from the appearance of its track;

(b) from the length of the track.

10. Define the terms *mass number* and *atomic number* in connection with the nucleus of an atom. The following symbols represent nuclides of copper and nickel respectively:

$$^{58}_{29}\text{Cu}, \quad ^{60}_{28}\text{Ni}.$$

What is the significance of the *superscript* and *subscript* numbers, and how many *neutrons* does each nuclide contain?

11. Carbon–14 is an isotope of carbon. Explain the meaning of this statement.

(O.C.)

12. What are *isotopes* of an element?

Tin (Sn) has twenty-five isotopes of which the lightest is represented by the symbol $^{108}_{50}\text{Sn}$. Knowing that all possible isotopes exist, write down the symbol for the heaviest tin isotope.

13. Distinguish between the *mass number* and the *atomic mass* of an element. What is meant by the term *isotopes of an element*? What do the isotopes of a particular element have in common and how do they differ from one another? Illustrate your answer by the aid of diagrams with reference to the three isotopes of hydrogen.

14. Write down the names of the particles represented by the following symbols and explain the meaning of the superscript and subscript numbers attached:

$$^{1}_{1}\text{p}, \quad ^{1}_{0}\text{n}, \quad ^{0}_{-1}\text{e}, \quad \alpha$$

Give alternative names and symbols for TWO of them.

15. If a proton is considered to have a mass m, what is the mass of:

(a) a neutron, and

(b) an electron? (J.M.B.)

16. What do you understand by the term *photoelectron*? Describe a simple demonstration of the photoelectric effect and mention one practical application of it.

17. State TWO ways in which free electrons may be emitted from a surface. (J.M.B.)

47. Splitting the nucleus

By the year 1903, Rutherford and Soddy had formed the opinion that radioactivity is the result of a spontaneous decay or disintegration of an atom during which it shoots out an α- or a β-particle. Simultaneously the atom changes into another element, which is itself radioactive, and this, in turn, disintegrates to become something else and so on. This sequence of changes is known as a *transformation series*. The radium and polonium discovered by the Curies are simply two links in a transformation series which begins with uranium-238 and ends with a stable isotope of lead. For example, radium emits α-particles and turns into a heavy gas called radon, which after emitting β-particles and α-particles becomes polonium. The polonium then emits a further α-particle and becomes a stable isotope of lead.

Half-life

It is not yet understood what causes a particular atom to disintegrate at a particular moment. The activity is entirely random and it makes no difference whether a radioactive element is used in its pure state or in chemical combination with something else. Likewise, heating or cooling have no effect on the rate of decay. Experiments show, however, that every radioactive element has a definite rate of decay which may be conveniently represented by its *half-life* period.

The half-life period of a substance is defined as the time taken for half the atoms in any given sample of the substance to decay.

Radium itself has a half-life of 1620 years. This means that if we start with 1 g of radium, then 0.5 g of it will have disintegrated in 1620 years. After another 1620 years half of what remains will have disintegrated, leaving 0.25 g and so on. Half-lives vary considerably from one element to another. That of radon, for example, is just under 4 days. Other short-lived radioisotopes have half-lives ranging from several hours down to less than a second. Proceeding in the other direction, we find that uranium-238, which is the naturally occurring parent of the whole series which includes radium, has the enormous half-life of 4.5×10^9 years.

Fig. 47.1 shows a typical decay curve for a radioisotope obtained by plotting the readings of a ratemeter which is used to measure the activity over a period of about three half-lives. From such a graph the half-life may be found as follows.

If the count-rate is n at some time t_1 and has fallen to $\frac{n}{2}$ at time t_2, then the half-life is $(t_2 - t_1)$.

Similarly, if the count-rate has fallen to $\frac{n}{4}$ at time t_3 the half-life is $(t_3 - t_2)$.

Laws of radioactive decay

Following a careful analysis of the various decay products in a transformation series, Rutherford and Soddy discovered two laws relating to radioactivity. They are:

(1) **When an element disintegrates by the emission of an α-particle it turns into an element with chemical properties similar to those of an element two places earlier in the periodic table.**

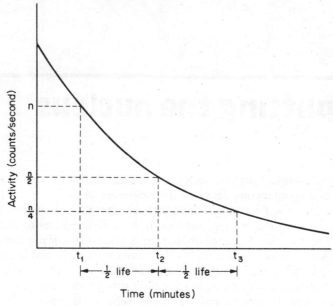

Fig. 47.1. Decay curve for a radioactive substance

(2) **When an element disintegrates by the emission of a β-particle it turns into an element with properties similar to those of an element one place later in the periodic table.**

Rutherford and Soddy's laws were, of course, based entirely on experimental observation, and at the time they were published no theoretical explanation was forthcoming. Since then, however, the discovery of the neutron has made it possible to explain the laws.

Alpha decay

We have seen that an α-particle consists of two protons and two neutrons. Hence, when an atom hurls out an α-particle its nucleus loses two units of positive charge, i.e., its atomic number decreases by 2, and consequently it moves two places further back in the periodic table. Radium, for example, has a mass number of 226 and an atomic number of 88 and is denoted by $^{226}_{88}$Ra. On shooting out an α-particle it loses 4 mass units and turns into radon, with a mass number of 222 and atomic number 86 ($^{222}_{86}$Rn).

In general, if any parent nuclide X of mass number A and atomic number Z emits an α-particle to form a daughter nuclide Y, we may express the process in symbols as follows:

α-decay

$$^{A}_{Z}X \longrightarrow {}^{A-4}_{Z-2}Y + {}^{4}_{2}He$$

| Parent nuclide | Daughter nuclide | α-particle (helium nucleus) |

Beta decay

The type of disintegration covered by the second law is not so easy to explain, and is involved with forces and energy exchanges inside the nucleus which are not yet fully understood. It appears that the spontaneous production of a β-particle (electron) from the nucleus is concerned with the simultaneous creation of other sub-atomic particles called *antineutrinos*, a discussion of which is outside our scope.

However, the end result is that an electron is created at the moment of ejection and, in the process, a neutron turns into a proton. Consequently, the mass number of the nucleus stays the same while its positive charge goes up by one. Chemically, this converts it into an element one place further on in the periodic table. An example of this type of decay is provided by the radioisotope of sodium ($^{24}_{11}$Na), which emits β-particles and turns into magnesium ($^{24}_{12}$Mg). (See page 573).

If, for our present purpose, we ignore the *antineutrino*, then if a parent nuclide X emits a β-particle ($_{-1}^{0}$e) to form a daughter nuclide Y the process may be represented as follows:

β-decay

$$^{A}_{Z}X \longrightarrow ^{A}_{z+1}Y + ^{0}_{-1}e$$

| Parent nuclide | Daughter nuclide | β-particle (electron) |

How Rutherford first split the atom

Following the success of their experiments on the scattering of α-particles by metal foils which gave undeniable proof of the existence of atomic nuclei, Rutherford and his colleagues began to try the effect of firing α-particles into various gases.

Marsden set up an evacuated tube containing an alpha source and a zinc sulphide screen and noticed the scintillations produced when the α-particles struck the screen. He then increased the distance of the screen from the source until it was too far away for the α-particles to reach it. Then, on introducing a little hydrogen into the tube, the scintillations reappeared. Since these could not be caused by α-particles, Marsden came to the conclusion that they were produced by the impact of hydrogen nuclei (protons) which had been struck by α-particles and projected forward with sufficiently high velocity to reach the screen. Calculations based on the ordinary laws of conservation of momentum and energy confirmed this conclusion. It was as though a ball of mass 4 units (α-particle) moving at, say, 16 000 km/s had hit a ball of mass 1 unit (proton) and projected it forward at about 24 000 km/s, being itself slowed up to about 8000 km/s.

Rutherford became very interested in these simple experiments and began to make further investigations of his own. Eventually he made the discovery that fast α-particles could be used not only to project a nucleus forwards but also to break it into two pieces.

Fig. 47.2 shows the apparatus he used. It consists of a metal tube containing an adjustable alpha source. At one end there is a window of silver foil, and two tubes are provided for the purpose of introducing various gases. Any particles which pass through the silver foil fall on a zinc sulphide screen and their scintillations can be observed through a microscope.

Rutherford first of all tried oxygen in the tube, but with no result, and in any case he did not expect *projection* of the oxygen nuclei, since they were far heavier than α-particles. But when nitrogen was allowed to enter the tube scintillations appeared on the screen. Rutherford made some tests and found that the particles which struck the screen were protons or hydrogen nuclei. Now the exceedingly small amount of hydrogen known to be present in the tube as an impurity was quite insufficient to account for the large number of scintillations observed. Rutherford came to the conclusion that the protons which caused the scintillations had been knocked out of the nitrogen nuclei by the fast α-particles and, in the process, the nitrogen nuclei were transmuted into oxygen nuclei. It was an exciting discovery. Here, for the first time,

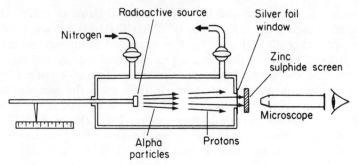

Fig. 47.2. Rutherford showed that nitrogen nuclei could be disintegrated by α-particle bombardment

the alchemist's dream of turning one substance into another had been realized. True, Rutherford had not turned lead into gold; he had only changed nitrogen into oxygen. But it was a vital link in the progress of research into the secrets of the atom, which has enabled man to transmute one substance into another, accompanied by the production of something of greater value than much gold, namely, energy.

The nuclear reaction which occurs in Rutherford's nitrogen experiment is represented as follows:

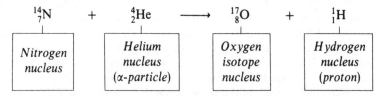

This may be interpreted thus. A nitrogen nucleus containing 7 protons and 7 neutrons is struck by an α-particle (helium nucleus) consisting of 2 protons and 2 neutrons, forming an unstable collective mass of 9 protons and 9 neutrons. This ejects a single high-energy proton and becomes transmuted into an isotope of oxygen of mass number 17. The shorthand notation used to represent this reaction is

$$^{14}_{7}N(\alpha,p)^{17}_{8}O$$

Following the nitrogen experiment, Rutherford and his colleague Chadwick succeeded in disintegrating more than a dozen other elements by alpha bombardment and obtaining fast protons. Calculations made from the range of these protons showed that they had energies far greater than those of the α-particles. *This was very strong evidence that the alphas were actually triggering off nuclear disintegrations and thereby releasing nuclear energy.* It was not a case of simple collision, whereby the α-particle conveyed part of its own energy to the proton, as had happened when Marsden first projected protons by alpha bombardment.

Cloud chamber studies

Supporting evidence regarding the nature of the artificial disintegrations was obtained in 1925 by Prof. P. M. S. Blackett, who became President of the Royal Society in 1965.

Blackett shot α-particles into a cloud chamber containing various gases and was able to study ordinary collision processes as well as disintegrations. Fig. 46.9 illustrates the common straight-line α-particle tracks which are observed. Occasionally, however, a cloud track displays a forked end (Fig. 47.3). This results from one of the rare occasions when an α-particle approaches so close to the nucleus of a gas atom that it is deflected out of its path. The spur track is caused by the recoiling nucleus. Thus, when helium was bombarded, the two tracks were approximately at right

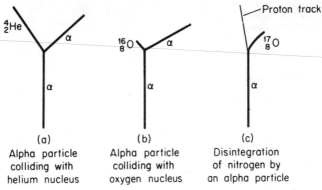

(a)
Alpha particle
colliding with
helium nucleus

(b)
Alpha particle
colliding with
oxygen nucleus

(c)
Disintegration
of nitrogen by
an alpha particle

Fig. 47.3. Some interesting cloud chamber tracks

angles and equal in length, as might be expected from the collision of two bodies of equal mass (an α-particle is a helium nucleus). But in the case of oxygen the spur track is much shorter owing to the greater mass of the oxygen nucleus compared with that of the α-particle.

Blackett's most striking achievement was the evidence he obtained to confirm the disintegration of nitrogen by alpha bombardment. After many trials he obtained a photograph which showed the path of the ejected proton on leaving the nitrogen nucleus (Fig. 47.3 (*c*)). The thin track going off to the left is that of the proton, and the thicker track that of the oxygen nucleus which results from the disintegration of the nitrogen nucleus.

Cockcroft and Walton's experiment

By the end of the 1920s the use of α-particles for bringing about nuclear disintegration had been studied intensively, and the need was felt for more powerful atom-smashing missiles.

Two research physicists at the Cavendish Laboratory, John Cockcroft and Ernest Walton, conceived the idea of accelerating protons in a powerful electric field and using these instead of α-particles to bombard atoms. Encouraged by Lord Rutherford, they set to work and, by 1932, had built the necessary apparatus. The general scheme of the method they used is shown in Fig. 47.4. The electric field for accelerating the protons was produced by three metal tubes set up in line vertically inside an evacuated glass tube about 2 m long. By means of a special voltage-quadrupling circuit consisting of diodes and capacitors fed from a step-up mains transformer, the upper tube was raised to a potential of 400 000 V, the middle one to 200 000 V, while the lower one was earthed. The original apparatus which is now in the Science Museum, London, is shown in Fig. 47.5.

One of Sir J. J. Thomson's hydrogen discharge tubes was used as a proton source. Protons from it were injected into the upper tube, where they were accelerated by the powerful electric field between the tubes and finally emerged from the bottom with a velocity of about 8000 km/s. Here they impinged on a plate made of lithium, where some managed to score direct hits on lithium nuclei, causing them to explode into two fragments. These fragments, which were afterwards proved to be helium nuclei, revealed their presence by causing scintillations on a zinc sulphide screen. Cockcroft and Walton were convinced that the scintillations could not have been caused by protons which had merely bounced off the lithium, since these had insufficient energy to give them the necessary range and the scintillations were characteristic of α-particles rather than protons.

Subsequently the disintegration products were passed into a cloud chamber. The tracks of the two helium nuclei were clearly visible, thus confirming the conclusion drawn from the original experiment.

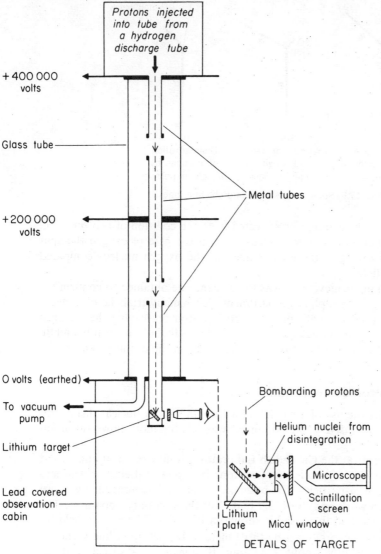

+400 000 volts

Glass tube

Metal tubes

+200 000 volts

0 volts (earthed)

To vacuum pump

Lithium target

Lead covered observation cabin

Protons injected into tube from a hydrogen discharge tube

Bombarding protons

Helium nuclei from disintegration

Microscope

Scintillation screen

Lithium plate

Mica window

DETAILS OF TARGET

Fig. 47.4. Cockroft and Walton's proton accelerator

Fig. 47.5. Cockroft and Wal original proton accelerator in the Science Museum, Lon (The metal stay-rods and c have been added by the Mus authorities for greater safety form no part of the original paratus)

As we have already explained on page 544, the lithium nucleus consists of three protons and four neutrons. When it is penetrated by a proton it splits up to form two helium nuclei, each containing two protons and two neutrons.

This nuclear reaction is represented as follows:

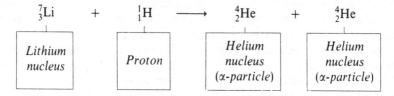

$$^{7}_{3}\text{Li} \quad + \quad ^{1}_{1}\text{H} \quad \longrightarrow \quad ^{4}_{2}\text{He} \quad + \quad ^{4}_{2}\text{He}$$

| Lithium nucleus | Proton | Helium nucleus (α-particle) | Helium nucleus (α-particle) |

Equivalence of mass and energy

We have already referred on page 82 to Einstein's theory of the equivalence of mass and energy. In 1905 he had shown that the relationship between mass and energy is given by the equation,

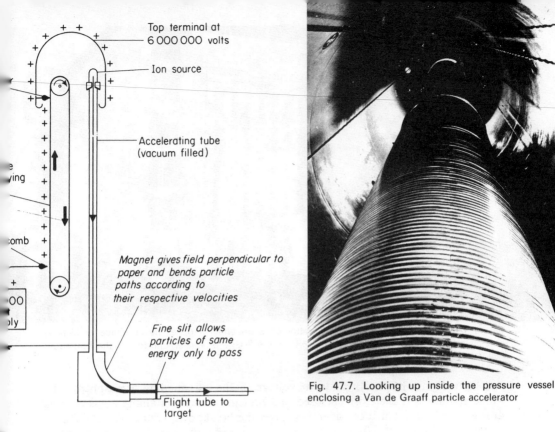

Top terminal at
6 000 000 volts

Ion source

Accelerating tube
(vacuum filled)

Magnet gives field perpendicular to
paper and bends particle
paths according to
their respective velocities

Fine slit allows
particles of same
energy only to pass

Flight tube to
target

Fig. 47.7. Looking up inside the pressure vessel
enclosing a Van de Graaff particle accelerator

47.6. Van de Graaff particle accelerator (insulating support
pressure vessel not shown)

$$E = mc^2$$

where, in appropriate units,
$E = $ energy;
$m = $ mass;
$c = $ velocity of light.

Apart from its importance as the first completely man-controlled splitting of the atom, Cockcroft and Walton's experiment also provided the first piece of experimental evidence for the truth of Einstein's equation.

When a nucleus disintegrates the sum of the masses of the fragments produced is always slightly less than the mass of the original nucleus. This loss in mass appears as energy of the fragments. Cockcroft and Walton calculated the total energy of the two flying helium particles and showed that it was related to the loss in mass when the lithium nucleus disintegrated, exactly in accordance with Einstein's equation.

The Van de Graaff particle accelerator

About the same time that Cockcroft and Walton were building their proton accelerator at Cambridge, a physicist in America, named Van de Graaff, was developing an accelerator of a different kind. This one has an electrostatic generator which employs point action and the principle that the charge on a hollow conductor resides on the outside.

Fig. 47.6 shows how it works. A transformer–rectifier circuit provides an initial high potential to a *spraycomb* consisting of a series of sharp points adjacent to a long moving belt made of special insulating paper. Electric charge is sprayed off the comb and onto the belt by point action (page 376) and is carried up inside a large hollow terminal at the top of an insulating column. Here it is removed from the belt by another spraycomb connected to the inside of the terminal. Thence the charge passes to the outside surface of the terminal and cumulatively builds up a very high poten-

Fig. 47.8. A modified 70 MeV Cockcroft–Walton generator which supplies the ion source for the linear accelerator which injects protons into NIMROD, the 7 GeV proton synchrotron at the Rutherford Laboratory, Chilton Berkshire

tial, limited only by the breakdown voltage of the surrounding atmosphere. The whole apparatus is therefore enclosed in a pressure vessel containing a gas such as nitrogen or freon, which considerably raises the breakdown voltage (Fig. 47.7).

Potentials up to 10 million volts can be obtained with Van de Graaff machines and are used for producing beams of high-energy ions down the accelerating tube shown in the diagram. The ion source used is a discharge tube containing hydrogen, deuterium or helium, depending on whether protons, deuterons or α-particles are required. At the bottom of the accelerating tube the particle beam passes through a magnetic field at right angles to its direction. This bends the beam by different amounts according to the particle velocities. The beam then impinges on a narrow

Fig. 47.9. Proton linear accelerator with vacuum tank cover removed showing drift tubes. This accelerator was in use at the Rutherford Laboratory, Chilton until the middle 60s. At present, accelerators like this are used for injecting protons into synchrotrons

slit which allows only particles of uniform energy to enter the flight tube on their way to the target.

The linear accelerator

Another type of accelerator is shown in Fig. 47.9. It is called a linear accelerator and consists of a series of co-axial tubes connected alternately as shown in Fig. 47.10 and given an alternating potential difference from a very high frequency a.c. source.

A narrow beam of particles is injected into the tubes from a suitable ion source and the frequency of the a.c. voltage is adjusted so that it reverses direction while the

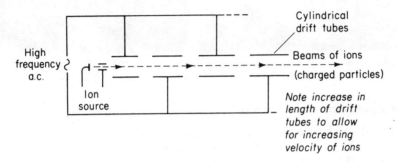

Fig. 47.10. Action of the linear accelerator

particles are traversing a tube. They are thus given an impulse and accelerated across the gap between tubes. Inside a tube they coast along with uniform velocity, since there is no electric field inside a hollow charged tube. By this method particles can be speeded up to very high energies. Its main advantage is that insulation problems are minimized. The particles are accelerated by a series of small potential differences instead of one or two large ones as in the Van de Graaff and Cockcroft and Walton methods respectively.

The synchrotron

At present, research in particle physics is mostly carried out by the use of *proton synchrotrons*. In these, protons are accelerated by an alternating electric field and guided by a powerful magnetic field at right angles to it. By this means the particles are accelerated along a circular path from which they may be extracted by *kicking magnets* for use in various experiments. The picture on **page 507** gives a view inside the tunnel containing the super proton synchrotron at CERN (European Organization for Nuclear Research). This tunnel has a diameter of 2.2 km and in it protons are accelerated to energies of 400 GeV.

The unit of energy used in particle physics is the electronvolt (eV), which is defined as the energy acquired by an electron (or a proton) in moving through a potential difference of 1 volt. Other units are:

$$1 \text{ keV} = 1 \text{ thousand } (10^3) \text{ electronvolts}$$
$$1 \text{ MeV} = 1 \text{ million } (10^6) \text{ electronvolts}$$
$$1 \text{ GeV} = 1 \text{ thousand million } (10^9) \text{ electronvolts}$$

With the very high-energy particles obtained from these machines, atomic nuclei in selected targets can be disintegrated into fragments and particles, the nature of which are investigated by the *bubble chambers* mentioned earlier on page 551. See Fig. 47.11 and 47.12.

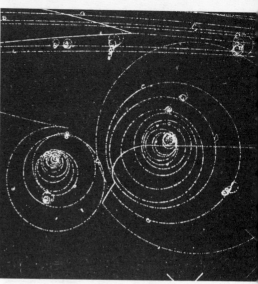

Fig. 47.11. Hydrogen bubble chamber at the Rutherford Laboratory, Chilton Berkshire (for some particle tracks see Fig. 47.12)

Fig. 47.12. Bubbles showing passage of particles bubble chamber at CERN

The discovery of the neutron

The neutron, which we have already described as one of the nucleons or constituent particles of which the nuclei of atoms are composed, was discovered by James Chadwick in 1933. The story behind it really started in 1930, when two German physicists, W. Bothe and H. Becker, found that when beryllium and certain other light elements were bombarded with α-particles some very penetrating radiation was produced which could easily pass through a good many centimetres of lead. Irene Curie (daughter of the famous Marie) and her husband Jean Joliot also experimented with the new radiation and found that it caused protons of very high energy to be knocked out of compounds containing hydrogen. They formed the opinion that this radiation was simply gamma radiation of unusually high energy. However, when they made calculations to measure the energies of the supposed gamma photons the results did not agree with the laws of conservation of momentum and energy.

The problem was finally solved by Chadwick, who pointed out that all these difficulties disappeared if the radiation was regarded as being composed of *uncharged particles* instead of gamma photons. The apparatus he used is shown in Fig. 47.13. A plate made of the metal beryllium was bombarded by α-particles and the new radiation from it allowed to impinge on a plate of paraffin wax. Protons were ejected from the paraffin and detected by means of an ionization chamber. Chadwick measured the energies of the protons and was then able to show that if the process was treated as a case of simple collisions between protons and uncharged particles of the same mass, then all the calculations agreed with the laws of momentum and energy. This newly discovered uncharged particle was given the name of *neutron*.

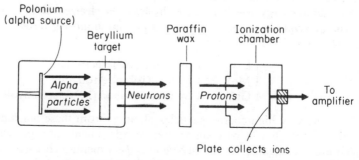

Fig. 47.13. Chadwick discovers neutrons in 1933

Disintegration by neutrons

The discovery of the neutron placed a very important atom-splitting missile at the disposal of physicists. Apart from α-particles, only two controllable particles had been available for the bombardment of nuclei, namely, the proton and deuteron (heavy hydrogen nucleus). The positive charge on both of these particles enables them to be speeded up to high energies by electric fields, but at the same time their charge was also a disadvantage. It meant that they were strongly repelled away from nuclei, and consequently very few were able to penetrate nuclei. Owing to the fact that it has no electric charge, the neutron does not possess this disadvantage.

In the early 1930s it was found that the majority of the elements could be successfully disintegrated by neutron bombardment, and many interesting and useful products were obtained. Take the case of lithium-6 for example. If this is struck by a neutron it produces the radioisotope of hydrogen called tritium ($^3_1$H), which is a beta emitter with a half-life of 12.26 years.

In nuclear reactions a neutron is represented by the symbol $^1_0$n, signifying that it is a particle of mass number one and zero electric charge. We can therefore represent the lithium–neutron reaction thus:

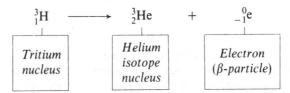

$$^6_3\text{Li} + {}^1_0\text{n} \longrightarrow {}^3_1\text{H} + {}^4_2\text{He}$$

| Lithium nucleus | Neutron | Tritium nucleus | Helium nucleus |

The subsequent decay of the tritium is given by

$$^3_1\text{H} \longrightarrow {}^3_2\text{He} + {}^{\ 0}_{-1}\text{e}$$

| Tritium nucleus | Helium isotope nucleus | Electron (β-particle) |

Another interesting case is the action of a neutron on magnesium, which is typical of a number of reactions in which the end-product is identical with the original. On capturing a neutron the magnesium is transmuted into sodium-24, accompanied by the ejection of a proton. The sodium nucleus formed is a radioactive isotope with a half-life of about 15 hours, which eventually decays back into magnesium with the emission of a β-particle. These reactions are represented as follows:

Irradiation of magnesium by neutrons

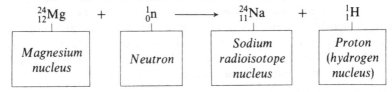

$$^{24}_{12}\text{Mg} + {}^1_0\text{n} \longrightarrow {}^{24}_{11}\text{Na} + {}^1_1\text{H}$$

| Magnesium nucleus | Neutron | Sodium radioisotope nucleus | Proton (hydrogen nucleus) |

Beta decay of the sodium isotope

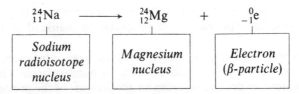

$$^{24}_{11}\text{Na} \longrightarrow {}^{24}_{12}\text{Mg} + {}^{\ 0}_{-1}\text{e}$$

| Sodium radioisotope nucleus | Magnesium nucleus | Electron (β-particle) |

Sodium-24 decays into the harmless magnesium-24 so that minute quantities of it are suitable as a tracer on blood circulation studies. The use of radioisotopes in medicine is discussed more fully later.

Nuclear fission

The examples of neutron capture just described are concerned with the disintegration of nuclei into two parts of *unequal* size. In 1939, Hahn and Strassman investigated the action of neutrons on uranium-235 and found that it was split into two roughly·*equal* pieces, one of which proved to be barium and the other krypton. Otto Frisch coined the expression *nuclear fission* to describe such cases as this. The importance of fission is that it is accompanied by the release of about ten times as much energy as in the case of ordinary disintegration, where only a small piece is broken off the nucleus. The uranium fission reaction was first used successfully to produce heat energy on a large scale by the Italian physicist, Enrico Fermi, at the University of Chicago in 1942.

In chapter 7 we have already described the fission process which goes on inside a commercial nuclear reactor used to produce heat for the production of electric energy. The reactions which go on inside these reactors are, of course, highly complex, but the one responsible for the bulk of the heat production is:

$$^{235}_{92}U + ^{1}_{0}n \rightarrow ^{236}_{92}U \longrightarrow \underbrace{^{144}_{56}Ba + ^{90}_{36}Kr} + 2\,^{1}_{0}n + energy$$

or other fission products

Nuclear weapons

When describing the Magnox nuclear reactor we explained, on page 81, what is meant by a chain reaction and how it occurs when a mixture of uranium-235 and graphite moderator reaches a critical size. We also explained how the speed of this reaction is safely controlled by the use of boron rods which may be raised or lowered into the core.

In the case of pure uranium-235 alone, a chain reaction is possible without the need of a moderator, provided the uranium exceeds a critical size. The first atomic bomb, used in August 1945, contained two pieces of uranium-235, each of which was not large enough to sustain a chain reaction on its own. When these were forced into contact with one another, they formed a single lump which exceeded the critical mass. As soon as this occurred a chain reaction started and proceeded at such a rate that a large proportion of the nuclei disintegrated almost instantaneously with tremendous destructive violence. Moreover, this was accompanied by the release of radiation and radioactive by-products which wrought incalculable biological damage.

Since then, bombs and missiles have been made containing plutonium, another fissile material which is a by-product from nuclear reactors. Furthermore, a hydrogen bomb based on *nuclear fusion* was tested in 1952 and shown to be of much greater destructive power than the uranium bomb. Nuclear fusion is mentioned under *thermonuclear energy* on page 82.

Artificial isotopes

Many of the end products resulting from the capture of neutrons by various elements are radioactive and have proved to be beneficial to mankind in the spheres of medicine and industry.

Radioisotopes are manufactured mainly by irradiating substances with neutrons in a nuclear reactor, but they can also be made by bombardment with high-energy particles (e.g., protons or deuterons) from an accelerator. In the previous chapter we mentioned the use of X-rays for the treatment of cancer. It is now possible to employ suitable artificial isotopes for the same purpose. One of the most important of these is radiocobalt ($^{60}_{27}$Co). Ordinary cobalt has a nucleus containing 27 protons and 32 neutrons ($^{59}_{27}$Co). If placed in a reactor where there is an abundance of neutrons a large proportion of the cobalt atoms capture an extra neutron and turn into cobalt-60. This decays, with the emission of β-particles together with very high-energy γ-radiation equivalent to that obtainable from 2.5 million volt X-rays. When properly shielded by lead the rays from cobalt-60 may be brought under control and employed instead of the more elaborate X-ray equipment for cancer therapy (Fig. 47.14). These cobalt units or *gammatrons* are now being used to alleviate human suffering in hospitals all over the world.

Certain other radioisotopes are taken internally, where they are selectively absorbed by certain organs, and so concentrate the radiation where it is most required. Radioiodine, for example, is absorbed mostly by the thyroid gland.

Radioisotopes are also used as *tracers*. Small quantities of low-activity substances are administered by injection to patients, and their passage through the body and location in diseased tissue may be ascertained by means of the Geiger–Müller counter described in the previous chapter. Originally designed by Geiger for the purpose of counting ionizing particles, this device has come to be one of the standard methods for detecting radioactivity (Fig. 46.13).

Radioisotopes in industry

Tracer techniques employing radioisotopes have found increasing application in industry. They are used to investigate such things as the flow of liquids in chemical plants and for the study of wear in machinery. In the latter case a small quantity of a radioisotope is incorporated in metal parts, e.g., bearings, piston rings, etc., and the rate of wear may then be calculated by measuring the activity of the abraded material carried away in the lubricating oil.

Elsewhere, the ability of substances to absorb γ-radiation has been adapted to give automatic control of the thickness of paper, plastic and metal sheeting as it goes through the production plant.

Gamma radiography. Autoradiography

Cobalt-60 and other gamma emitters are used as an alternative to X-ray apparatus to produce *radiographs* for the examination of forgings and the welded seams of boilers and other pressure vessels. The main advantage here is that of easier portability and the fact that a high-voltage electricity supply is not required. Some examples of radiography using radioactive substances are shown in Fig. 47.15.

In *autoradiography*, a thin specimen in which some of the atoms have been replaced by radioactive isotopes, is placed in contact with a photographic film and takes its own photograph by the emission of ionizing radiation. An example of this technique, widely used in biological studies, in seen in Fig. 47.15 (*d*).

Biological hazards and safety precautions

The need for extreme care when dealing with radioactive substances cannot be too strongly emphasized. People soon learn to keep away from fire because the pain is immediate and obvious, but the great danger where radioactivity is concerned is that one can receive a severe dose of radiation without being aware of it. A number of small doses received over a long period build up cumulatively in the system, and may eventually lead to leukaemia or cancer later in life. Radiation was a contributory factor to the deaths of both Marie Curie and Enrico Fermi.

Strong doses of radiation will lead to burning of the skin and body tissues similar to that caused by fire, but a more serious hazard comes from the extreme penetrating

Right: Patient being treated with radiation from a Gammatron

Below Section through the radiation head:

(1) Housing, cast iron
(2) Protection, lead
(3) Source, cobalt-60
(4) Shutter, tungsten
(5) Light source
(6) Reflecting mirror
(7) Diaphragm, tungsten
(8) Diaphragm-control
(9) Exit port

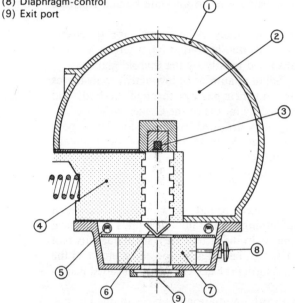

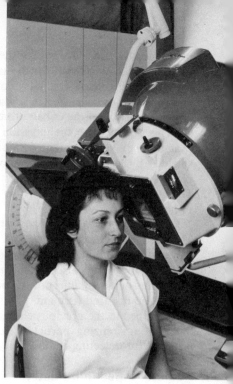

Fig. 47.14. *Gamma-ray therapy with cobal*

Below This Theratron 80 therapy unit at Churchill Ho Oxford contains a cobalt source supplied by the Is Production Unit, Harwell. In this model the source is tected by depleted uranium. The radiographer is positi the source over the patient

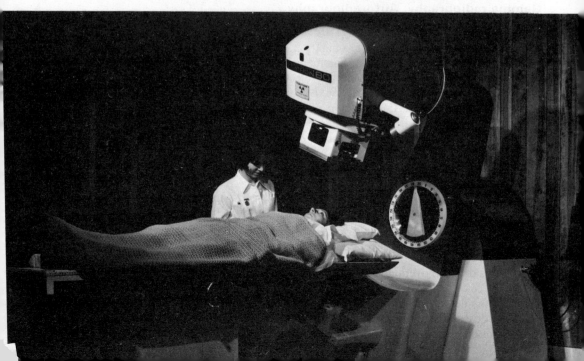

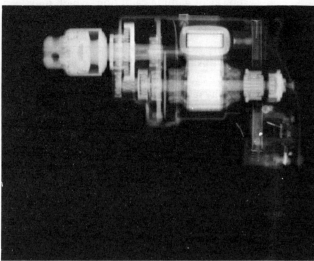

Fig. 47.15. (a). (**Left**) A radioisotope being used for radiographic examination of a complete circumferential weld with one exposure

Fig. 47.15. (b). (**Above**) This radiograph of an electric drill was taken with an iridium-192 gamma source placed about 1 metre away

7.15. (c). (**Below**) Document dating at the British m. A distributed source of low energy beta-radiation ed on one side of the paper and a photographic film other. Dating is achieved from the character of the mark. This, like a trademark, varies from one manufac- o another and is changed from time to time. The graph shows a beta-radiograph of a page from 's "Golden Legend", circa 1484

Fig. 47.15. (d). (**Below**) An example of autoradiography, by Dr. L. G. Lajtha, Churchill Hospital, Oxford. These human blood cells have been "labelled" with tritium, a low energy beta emitter. The black spots show where tritium labelled thymidine is incorporated in the chromosomes. (A labelled compound is one in which one or more of the atoms in a proportion of the molecules has been replaced by an active isotope.) Magnified approx. 7 000 diameters

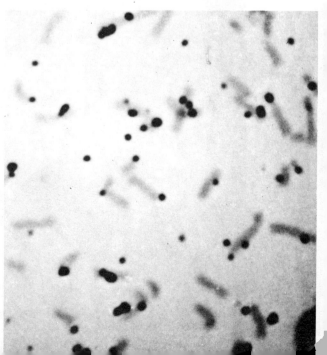

power of gamma radiation, which may lead to damage deep in the body tissues and particularly to the cell nuclei. An individual emission may be absorbed and affect a chromosome. As a result, subsequent cells may be abnormal, and the genetic effects arising may be passed on to future generations. On a more pleasant note we may mention the controlled effect of radiation on seeds. Such agricultural research has led to the production of short-stalked wheat, produced by genetic change, to beat the ravages of storms before harvest.

One of the earliest commercial applications of radium compounds was in the manufacture of luminous paint for the numbers on watch and other instrument dials. The paint consists of a minute quantity of a radium salt mixed with zinc sulphide, which glows in the dark as a result of scintillations from α-particles. Incidentally, the individual scintillations may be seen if the luminous numbers on a watch dial are viewed through a microscope in a dark room. The girls who painted the dials had a habit of applying the brush to their lips in order to put a fine point on

Fig. 47.16. *Safeguarding health*. A radiation monitor in use in the changeoom at Dounreay Experimental Reactor Establishment. Hands are checked for contamination by inserting them in special openings, one of which is visible in the front of the cabinet

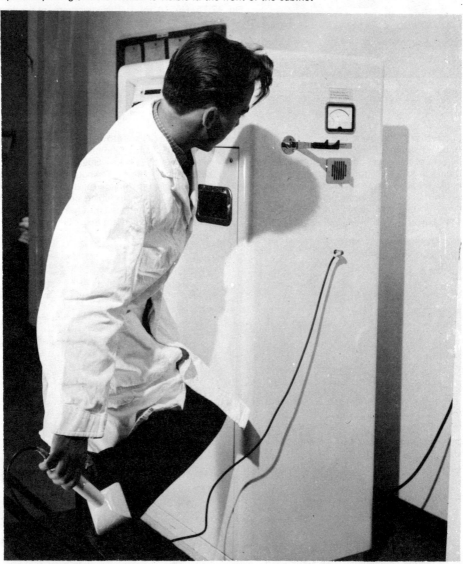

it. Unfortunately the danger was not realized in those early days and, over a long period, some of the workers absorbed enough radium to produce fatal illness many years later.

Dangers of this sort do not, of course, attend the use of radioisotopes under proper medical control. The isotopes used have half-lives measured in days or weeks only, and hence cease to act after their proper function has been completed. Also these isotopes either decay into harmless substances or else are eliminated from the system. Radium on the other hand, remains in bones in the body. It has a half-life of 1620 years, and consequently undergoes negligible decay over a normal lifetime. The same objection applies to many other substances, particularly the radiostrontium and radiocaesium from nuclear bomb tests, which can ultimately find its way into the food we eat.

Nowadays the strictest safety precautions are taken in nuclear research laboratories and nuclear power stations so that the workers there receive no more than the permitted maximum level of radiation (Fig. 47.16). No one, of course, can avoid receiving the normal *background radiation* which originates from radioactive compounds in the earth's crust and from particles and rays entering the earth's atmosphere.

Radioisotopes are handled by mechanical tongs operated by remote-control equipment from behind thick walls made of lead, concrete or other suitable material which absorbs the dangerous radiations. For transport, thick-walled lead containers are employed, and in laboratories radioactive samples are carefully shielded in lead castles built of lead bricks. This form of protection works both ways; in some experiments a lead castle is used to keep the background radiation out!

QUESTIONS: 47

1. A radioactive nucleus is denoted by the symbol $^{288}_{92}Y$.

Write down how you would expect the nucleus to be composed and state the composition of the nucleus at the end of each of the following stages of disintegration, illustrating by means of a symbol similar to the example given:

(i) the emission of an alpha particle;
(ii) the *further* emission of a beta particle;
(iii) the *further* emission of some gamma radiation. (L.)

2. (a) Explain what is meant by (i) atomic mass number and (ii) atomic number.

(b) The isotope $^{238}_{92}U$ decays by alpha-emission to an isotope of thorium (Th). Compare the $^{238}_{92}U$ and thorium nuclei, explaining the changes which have occurred in the uranium nucleus.

(ii) The thorium nucleus decays by beta-emission to an isotope of protactinium (Pa).

Compare the thorium and Pa nuclei, accounting for the changes you describe. (J.M.B.)

3. In 168 seconds, the activity of thoron falls to one-eighth of its original value. What is its half-life? Explain this term. (S.)

4. (a) Explain the terms *atomic number*, *atomic mass number*, and *isotope*.

(b) State *one* medical and *one* non-medical use of radioactive tracers.

(c) Name and describe the three principal types of emission which may accompany radioactive disintegration.

(d) The half-life of uranium X_1 is 24 days. Calculate the mass remaining unchanged of 0.64 g of the substance after:

(i) 24 days;
(ii) 48 days;
(iii) 72 days;
(iv) 96 days;
(v) 120 days.

Plot your answers on a graph and hence determine:

(vi) the mass remaining unchanged after 84 days;
(vii) after how many days there will be exactly 0.25 g unchanged. (A.E.B.)

5. Outline ONE way of detecting α-particles and describe, as far as you can, how it works.

The counting rate recorded by a detector fixed in front of an α-particle emitter is 256 per second. This figure is an average rate worked out from a count lasting several minutes. What is the average counting rate twenty days later, for the same arrangement, if the half-life of the emitter is five days?

If the number of α-particles were recorded

at this time for 1 second precisely, would you expect to find the number you have just calculated? Give a reason for your answer. (O.C.)

6. A radioactive source emits α- and β-particles. Describe the experiments you would carry out to distinguish between these radiations by investigating their ability:

(a) to penetrate matter;

(b) to produce ionisation. Explain why a particle which is able to produce high ionisation will not also possess high penetrability in the same medium.

Radium ($^{226}_{88}$Ra) disintegrates by the emission of an α-particle to form radon. What is the atomic number and the atomic mass of radon? Write the equation for this reaction and suggest the source of the energy of the α-particle which is emitted. (C.)

7. When a suitable counter was placed near a radioactive source of β-particles, the following rates of emission were obtained at the times shown:

Time (minutes)	0	5	10	15	20
Count rate	295	158	86	47	25

Plot a suitable graph to enable you to find the half-life of the source. State its value clearly and explain your method of calculation.

Describe briefly, with a clear diagram, how you could use a strong bar magnet with the above apparatus to show that the particles have a negative charge.

What would be the effect of placing:

(a) a sheet of thin paper;

(b) a thick sheet of aluminium between the source and counter? What differences would you expect in these cases if the source were replaced by an α emitter? (S.)

8. What effect does a transverse magnetic field have on narrow beams of:

(i) γ-rays;

(ii) β-rays;

(iii) α-rays from a radioactive source, as they pass through an evacuated box? What conclusions can be drawn from observations of these effects?

An atomic nucleus A is composed of Z protons and N neutrons. What will be the composition of nucleus B left when A emits an α-particle and of nucleus C left when B emits a β-particle? What can we say about the masses of A, B, and C? (O.C.)

9. With the aid of a diagram, explain the construction and action of a Van de Graaff electrostatic generator. For what purpose are these generators used in nuclear research?

10. Draw a fully labelled diagram to show how you would demonstrate the deflection of β radiation in a magnetic field. Indicate clearly the relative directions of field and deflection. (S.)

11. What are *artificial isotopes*? Mention two methods for making them. Give ONE example in each case of their application in:

(a) medicine;

(b) industry.

12. The radioisotope of barium, $^{139}_{56}$Ba, emits a β-particle and is transformed into the stable isotope of lanthanum (La). Express the transformation in symbolic form.

13. What is meant by *radioactive decay* and the *half-life* of a radioactive element?

What changes take place in the atomic structure of an element X with atomic number 88 and mass number 226 if it emits an alpha particle? (W.A.E.C.)

14. Describe the essential features of the nuclear atom model, making clear the meaning of the terms *proton, neutron, atomic number,* and *mass number.*

Radioactive sodium $^{24}_{11}$Na decays by β-particle emission. Account for the fact that the new atom produced from an atom of radioactive sodium has almost the same mass but different chemical properties.

A radioactive source emits β-particles with different speeds. How would you demonstrate the effect on the β-particles of *either* a magnetic field *or* different thicknesses of aluminium? How would you show, from your observations, that the particles have different speeds? (C.)

15. An early theory about atoms considered them to be very tiny hard spheres which were unchangeable. Write brief notes on *two* discoveries or *two* experiments which caused this theory to be modified.

State what information you can deduce about the structure of an atom of carbon-14 (mass number = 14), given that its atomic number is 6.

Explain, with special reference to the words printed in *italics,* the meaning of the following statement: "carbon-12 and carbon-14 are *isotopes,* but carbon-14 *disintegrates* by emitting a *beta particle.*" Deduce the atomic structure of the atom formed when an atom of carbon-14 disintegrates. (C.)

16. State **two** facts about nuclear fission. (W.A.E.C.)

List of elements

Atomic number Z	Symbol	Name	Atomic number Z	Symbol	Name
1	H	Hydrogen	54	Xe	Xenon
2	He	Helium	55	Cs	Caesium
3	Li	Lithium	56	Ba	Barium
4	Be	Beryllium	57	La	Lanthanum
5	B	Boron	58	Ce	Cerium
6	C	Carbon	59	Pr	Praseodymium
7	N	Nitrogen	60	Nd	Neodymium
8	O	Oxygen	61	Pm	Promethium
9	F	Fluorine	62	Sm	Samarium
10	Ne	Neon	63	Eu	Europium
11	Na	Sodium	64	Gd	Gadolinium
12	Mg	Magnesium	65	Tb	Terbium
13	Al	Aluminium	66	Dy	Dysprosium
14	Si	Silicon	67	Ho	Holmium
15	P	Phosphorus	68	Er	Erbium
16	S	Sulphur	69	Tm	Thulium
17	Cl	Chlorine	70	Yb	Ytterbium
18	A	Argon	71	Lu	Lutetium
19	K	Potassium	72	Hf	Hafnium
20	Ca	Calcium	73	Ta	Tantalum
21	Sc	Scandium	74	W	Tungsten
22	Ti	Titanium	75	Re	Rhenium
23	V	Vanadium	76	Os	Osmium
24	Cr	Chromium	77	Ir	Iridium
25	Mn	Manganese	78	Pt	Platinum
26	Fe	Iron	79	Au	Gold
27	Co	Cobalt	80	Hg	Mercury
28	Ni	Nickel	81	Tl	Thallium
29	Cu	Copper	82	Pb	Lead
30	Zn	Zinc	83	Bi	Bismuth
31	Ga	Gallium	84	Po	Polonium
32	Ge	Germanium	85	At	*Astatine*
33	As	Arsenic	86	Rn	Radon
34	Se	Selenium	87	Fr	*Francium*
35	Br	Bromine	88	Ra	Radium
36	Kr	Krypton	89	Ac	Actinium
37	Rb	Rubidium	90	Th	Thorium
38	Sr	Strontium	91	Pa	Protactinium
39	Y	Yttrium	92	U	Uranium
40	Zr	Zirconium	93	Np	*Neptunium*
41	Nb	Niobium	94	Pu	*Plutonium*
42	Mo	Molybdenum	95	Am	*Americium*
43	Tc	*Technetium*	96	Cm	*Curium*
44	Ru	Ruthenium	97	Bk	*Berkelium*
45	Rh	Rhodium	98	Cf	*Californium*
46	Pd	Palladium	99	E	*Einsteinium*
47	Ag	Silver	100	Fm	*Fermium*
48	Cd	Cadmium	101	Mv	*Mendelevium*
49	In	Indium	102	No	*Nobelium*
50	Sn	Tin	103	Lw	*Lawrencium*
51	Sb	Antimony	104	Ku	*Kurchatovium*
52	Te	Tellurium	105	Ha	*Hahnium*
53	I	Iodine			

The elements printed in italic do not occur naturally, but have been produced artificially.

Definitions and laws

(with page numbers, in the order in which they appear in the text)

Answers to problems

QUESTIONS : 1 (page 12)

1. 12.27 cm; 3.63 cm
2. 3.47 mm; 17.74 mm
4. 3.231 mm

QUESTIONS : 3 (page 33)

1. (a) 38.7 km/h; (b) 10.7 m/s
2. (i) 1.33 m/s^2; (ii) 225 m
3. (a) 8.5 m/s; (b) 51 m
4. 80 m
5. (a) 24 m, 30 m; (b) 8.6 m/s
6. (i) 40 m/s; (ii) 162 m
7. (a) 5 cm/s^2; (b) 0.05 m/s^2
8. 1.25 m/s^2
9. 0.14 m/s^2; 139 m
10. 10 m
11. 32.4 km/h
12. (i) 0.5 m/s^2; (ii) 12 775 m; (iii) 19.7 m/s
14. (b) 80 m; 8 s
15. 125 m
16. 0.4 m/s^2
18. 10 m/s; 18 m/s; 4 m/s^2
19. 50 m/s; 5 s
20. (a) 3 s; (b) 6.6 s
21. (a) 25 m/s; (b) 31.25 m; 2500 m/s^2

QUESTIONS : 4 (page 51)

1. (a) 50 m/s^2; (b) 1000 m/s; (c) 10 000 m
2. 60 kg m/s; 60 kg m/s
3. 500 N; 10 m
4. 1.25 m/s^2
5. (i) 2.5 m/s^2; (ii) 1.25 N; (iii) 5 kg m/s
6. (a) 0.6 m; (b) 72 N
7. 0.53 m/s
8. 3 m/s; 4.5 m/s
10. 1.5 m/s^2; They either stop or else rebound with velocities such that their total momentum is zero
11. 0.0032 N
12. 1000 m/s
13. 87.5 kg; (a) 70 kg; (b) 35 kg

15. 22.5 mm; 117.5 mm
16. 50 g

QUESTIONS : 5 (page 61)

3. 8.7 N
4. 11.5 N at 145° to each 7 N force
5. 13.6 N at 25°; 6.3 N at 65°
6. 7.9 N at 19° south of west
7. BC, 5.2 N; AC, 3.9 N
8. Forward, 140 N; lift, 50 N
9. (a) 560 N; (b) 840 N
10. 5 N at 38° west of north; 25 m/s^2
11. String, 1.82 N; wire, 5.32 N
12. Spring, 50 N; horizontal, 58 N
13. 19° west of north; 32 min; 1 km

QUESTIONS : 6 (page 73)

2. 52 g
3. 320 N
5. 2 cm
6. 30 N; 60 N
7. (a) 29 cm; (b) 43.5 cm
9. 18.6 cm
12. 200 N

QUESTIONS : 7 (page 85)

2. 360 W
3. 1000 J; (a) 500 J; (b) 1000 J
4. (i) 7.2 MJ; (ii) 7.2 MJ; (iii) power is doubled
8. (a) 2.5 m/s^2; (b) 10 m/s; (c) 20 m; (d) 400 J
9. (a) 70 J; (b) 9.5 m/s, 45 J
10. (i) 25 000 kg m/s; (ii) 313 kJ; (iii) 3130 N
11. 10 m/s; (a) 20 m/s; (b) 14 m/s
12. 32 000 N (or 32 kN)
14. (a) 690 N; (b) 9.6 kW

15. 640 J; work independent of angle θ between planks and horizontal; effort, 400 $\sin \theta$ N; k.e. 640 J; velocity 5.7 m/s
16. 80 kW: 1600 kg/s
17. 10.7 kW; (a) 1800 N; (b) 24 kW
18. 3.1 m; 76 J
19. (a) 7 t; (b) 27 kJ; (c) 6.8 kJ

QUESTIONS : 8 (page 98)

1. (a) 3; (b) 2.4
2. (a) 20; (b) 80%
3. (a) 3.6; (b) 80 W
5. 1.3 kW
6. 4 pulleys in lower block; (i) 115 N; (ii) 3.9; (iii) 78%; (c) 62.5 W
7. 75%; 3333 J
9. 300 J
10. (a) 450 N; (b) 3; (c) 1.67; (d) 1125 J
11. (b) 15 kW
12. 2400; 216 kN
13. 70 N; 943; 6601 N
14. 0.18; 77%
15. (a) 8; (b) 6.8
16. (b) (i) 10 J; (ii) 4 W; (c) (i) 10 m/s; (ii) 100 J; (iii) 500 N

QUESTIONS : 9 (page 104)

1. (a) 1.8 g/cm^3; (b) 1800 kg/m^3
2. 605 kg
3. (i) 7.5 N; (ii) 0.75 kg; (iii) 12 500 kg/m^3
4. 0.385 m^3
5. (i) 1300 kg/m^3; (ii) 6.41 cm^3
6. 53.2 kg
7. 13.6 g/cm^3
8. 0.84 g/cm^3

QUESTIONS : 10 (page 117)

2. 400 N/m^2 (or Pa); 7.2 N
3. 12 120 N
5. 2×10^8 Pa
7. 1500 N/m^2 (or Pa)
8. 1200 N/m^2 (or Pa); 13.6 cm
9. (a) 2.45 N; (b) 4.08 N; (c) 1.63 N
12. (i) 5000 Pa; (ii) 5000 Pa; 125 000 Pa
13. 52 kg
14. More; 4.95 N
15. 1632 m
17. 102 kPa

QUESTIONS : 11 (page 126)

4. 0.8; 17.2 cm
5. 1000 kg/m^3

QUESTIONS : 12 (page 135)

1. (a) 1.8; (b) 0.15 N
2. 0.492 N; 7g/cm^3
3. 0.024 m
4. 8000 kg/m^3; 0.418 kg
5. 0.196 N
6. 0.1024

7. (i) $\dfrac{\text{Volume of water}}{\text{Volume of mercury}}$

   $= \dfrac{\text{density of mercury}}{\text{density of water}}$; (ii) equal

8. 9 N; 62.5 N
9. 0.8
10. 700 kg/m^3 (or 0.7 g/cm^3)
11. 0.024 m^3; 750 kg/m^3
12. (i) 0.75; (ii) 400 N
13. (a) 3.7 N; (b) 10 N; (c) 1 kg; (d) 0.001 m^3

QUESTIONS : 14 (page 157)

1. 35 °C
5. 51.5 °C

QUESTIONS : 15 (page 168)

1. 0.000 02/K
3. 765 °C
4. − 102 °C
5. (a) increases; (b) increases; (c) unchanged; (d) decreases
6. 38 cm
7. 0.018
8. 250 K
11. Too large; 1.1 mm

QUESTIONS : 16 (page 182)

1. 15 cm^3
2. 750 mmHg
3. 750 mmHg
4. 75 cm
6. 5.33 cm^3
7. 19.25 litre (or 0.019 25 m^3)
9. 1247 cm^3
10. 46 °C
11. 89.5 °C
12. 162 °C
13. 75 cm^3
14. Pressure doubles in each part
15. 142 000 N/m^2
16. (b) reduced by one quarter
17. (c) (iii) $p = \dfrac{1}{3} \dfrac{Nmc^2}{V}$;

    $pV = $ constant

QUESTIONS : 18 (page 202)

1. (*a*) 690 J; (*b*) 120 J/kg K; (*c*) 1.56×10^8 J
2. (*a*) 123 J; (*b*) 123 J; 1.89 K
3. 4.1×10^3 J/kg K
5. 22.2 K
6. 50 °C
7. 7.5 min
8. 2250 J/kg K
9. 22 °C
11. 189 J/kg K; too high.

QUESTIONS : 19 (page 211)

1. 2.2×10^6 J
3. 6 082 400 J
4. 16.3 J/s
5. 62 400 J; (*a*) 4 min 40 s; (*b*) 30 min
6. 105 000 J/kg; 36 800 J
7. 30 370 J
8. (*a*) 504 000 J; (*b*) 1 695 000 J; (*c*) 8 min 24 s
9. 7.05 g
13. 198 kJ
16. 120 g
17. 15.8 g
20. (*a*) 8400 J; (*b*) 25 g; (*c*) 4.4 °C

QUESTIONS : 20 (page 219)

4. 758 mmHg; 749 mmHg
6. 51.5 °C
8. (*a*) 2 308 J/g; (*b*) 100 J/s. *Hint:* write down two equations containing the unknowns *l* (specific latent heat of steam in J/g) and *h* (rate of loss of heat in J/s). Then solve for *l* and *h*.

QUESTIONS : 21 (page 234)

6. 4.5 m
7. Length of mirror 90 cm; height of base above floor 84 cm

QUESTIONS : 22 (page 247)

1. 333 mm from mirror
2. 30 cm from mirror on same side as object; magnification 2; real
3. 1.2 cm tall; 24 cm in front of mirror
4. Virtual; erect; 2.5 cm tall; 6.7 cm behind mirror
7. 0.4 cm tall; 12 cm behind mirror
8. 6 cm behind mirror
10. 7.5 cm in front of mirror; 1.25 cm tall

QUESTIONS : 23 (page 259)

2. 27°; 1.5; 28°
3. 33°
4 (i) 35° 16′; (ii) 40° 31′; (iii) $\frac{9}{8}$
5. 1.5
7. 37° 18′
8. 1.73
9. 12 cm below the surface; appears larger
10. 7.5 cm
12. 27° 55′
13. 1.33 cm below surface
14. Angle of emergence 48° 36′
15. 1.6
16. (*a*) 70°; (*b*) 41°

QUESTIONS : 24 (page 275)

1. Virtual; erect; 15 cm from lens on same side as object; *m* = 2.5
2. (*a*) 20 cm from lens on same side as object; (*b*) 20 cm
3. 3 cm tall; 87 cm from lens on same side as object
4. Object distance 15 cm; image distance 30 cm
5. 25 cm
6. (*a*) 13.3 cm; (*b*) 6.7 cm from lens
8. 40 cm; 40 cm; reduce screen distance to approx 12 m
9. 13.3 cm from lens on same side as object; 0.67 cm tall; virtual; erect
11. (i) 20 D; (iii) sum of focal lengths

QUESTIONS : 26 (page 308)

4. 6 m/s
5. (*a*) 2 m; (*b*) 2×10^8 m/s
6. (*b*) 660 Hz
7. 3 cm; 21 cm/s; 7 Hz; angle of refraction 28° 7′
8. 5.5 Hz
10. 176 m
11. (i) 1.1; (ii) 33° 22′
13. 0.2 m; 330 m/s
14. 3.6×10^{-4} m; 4.6×10^{-7} m
16. 9.8×10^{-5} cm

QUESTIONS : 27 (page 320)

3. 330 m/s
4. 320 m/s; (ii) 1280 m
5. 333 m/s; 4.45 m/s
7. 320 m/s; 1280 m
8. 1100 Hz
9. 333 m/s; 100 m
11. 400 m/s

QUESTIONS : 28 (page 326)

1. 512 Hz
2. 16 rev/s; 26 holes
3. 1200 rev/min
5. 340 Hz; 1 beat per second

QUESTIONS : 29 (page 337)

2. Reduce length by a factor of $\frac{2}{3}$; increase tension by a factor of $2\frac{1}{4}$
3. 33.3 cm
4. (i) doubled; (ii) doubled; (iii) halved
5. (i) 0.5 m; (ii) 0.35 m/s; (iii) 0.52 mm
6. 47 m/s
8. 480 Hz
9. 680 Hz
12. 320 Hz

QUESTIONS : 34 (page 403)

5. (i) $\dfrac{\text{Storage capacity of P}}{\text{Storage capacity of Q}} = \dfrac{4}{1}$;

(ii) $\dfrac{\text{internal resistance of P}}{\text{internal resistance of Q}} = \dfrac{1}{2}$

QUESTIONS : 35 (page 417)

1. 6 V each
2. $\frac{1}{3}$; $\frac{1}{2}$; $\frac{2}{3}$; 1; $1\frac{1}{2}$; 2; 3 Ω
3. (i) 4 A; (ii) 5 A
4. 0.25 A; 0.5 V; 0.75 V
5. $\frac{1}{3}$ A
6. (a) 1.8 A; (b) 3.0 A
7. 1.14 V; (Internal resistance of cell = 0.6 Ω)
8. (i) 0.8 Ω; (ii) 5 Ω
9. 1.8 V; 0.5 Ω
10. (a) (i) 2 E volts; (ii) 2 R ohms;

(iii) $\dfrac{E}{R+1}$ amperes; (iv) $\dfrac{2E}{R+1}$ volts

(b) (i) E volts; (ii) $\dfrac{R}{2}$ ohms;

(iii) $\dfrac{2E}{R+4}$ amperes; (iv) $\dfrac{4E}{R+4}$ volts

(c) 2 volts; 4 ohms
11. (a) 1.5 V; (b) 4 Ω
12. 0.5 Ω
13. 4.5 V
14. $E = 1.5$ V; $r = 1$ Ω; p.d. = 1.0 V
15. (a) 5 Ω; (b) 2.5 Ω
16. 0.3 A
17. 0.5 Ω; 6 V
18. 0.04 A; 150 Ω; 12 A; 0.5 Ω

QUESTIONS : 36 (page 427)

4. 10 Ω; 0.143 A

QUESTIONS : 38 (page 441)

5. 50%

QUESTIONS : 39 (page 449)

1. 0.08 V
2. 20.2 cm of wire as shunt
3. 2.22 Ω
4. (a) 0.026 3 Ω shunt; (b) 999.5 Ω in series
5. 25 Ω
6. (i) 0.075 V; (ii) 0.025 Ω shunt; (iii) 995 Ω
7. 0.4 Ω; assume main circuit current is unaltered
9. 9900 Ω
10. (i) 400 Ω; (ii) 0.1 Ω

QUESTIONS : 40 (page 461)

2. (i) 0.2 A; (ii) 1.1 V
5. In parallel; (i) bright; (ii) 0.3 A; (iii) 0.6 A
In series; (i) dim; (ii) 0.15 A; (iii) 0.15 A
6. (i) 1.06×10^{-6} Ω m; (ii) 212 V
7. 11.3 m

QUESTIONS : 41 (page 470)

2. (a) 4.0 A; (b) 3.0 Ω
3. 4 h
4. 86 400 J; 2 A
5. (a) 28.8 Ω; (b) 35p
6. (a) 750 W, 3 A; 1250 W, 5 A
(b) 750 W, 83.3 Ω; 1250 W, 50 Ω; at 200 V, 1280 W
7. 1 : 4
8. 11 693 s; 49p
10. (i) 6.25 A; (ii) 5.4×10^7 J; (iii) 75p
11. 470 s
12. £5.26
13. (a) 3.96 Ω (b) 15.8 V
14. 39%
15. 19.2 Ω; (i) 12 A; (ii) 115 W
17. (i) 5 A; (ii) 13 A
18. 4:1

QUESTIONS : 42 (page 481)

1. (i) 7.2×10^4 J; (ii) 1350 J/kg K; (iii) 0.5 A; (iv) 480 Ω; (v) 40 m
2. 21.6 °C/min; 4170 J/kg K
3. (i) 337.5 kJ; (ii) 302.4 kJ; 2.25 MJ/kg

5. 1920 s
6. (i) 672 s; (ii) 2.25 MJ/kg
7. 2.09 MJ/kg
8. 2.32 MJ/kg
9. 2.24 MJ/kg
10. 21 kJ/kg
11. 280 kJ/kg
13. (i) 2500 J/kg K (low); (ii) 6250 J/kg K (high); (iii) cannot be determined; (iv) 270 K; (v) 390 K; (vi) cannot be determined
14. 1.5×10^3 J/kg K; 15 K

QUESTIONS : 43 (page 503)

8. (a) 189 kW; (b) 0.01 kW
9. (a) 91%; (b) 8p
10. (i) 12 W; (ii) 11.25 W; (iii) 93.75%
12. 3.3 A
13. 5.6 A
14. (b) $\dfrac{V^2}{R}$ (in J/s); (i) 5 A; (ii) 400 W; (iii) 80 V
15. 133 turns; (iv) 0.08 A

QUESTIONS : 44 (page 519)

4. 57.6 J
10. (a) 2.98×10^6 m/s; (b) 37.5×10^{-3} J/s

QUESTIONS : 45 (page 540)

6. AND gate
8. NOR gate
10. (a) F and C;
(b) 100 ms, 2.2 s;
(c) about 0.4 Hz

QUESTIONS : 46 (page 561)

2. $\dfrac{\text{mass of proton}}{\text{mass of electron}} = 1833$
12. $^{132}_{50}$Sn

QUESTIONS : 47 (page 579)

3. 56 s
4. (i) 0.32; (ii) 0.16; (iv) 0.04; (v) 0.02; (vi) 0.056; (vii) 32.5
5. 16 counts/s
6. Atomic number 86; mass number 222

Index

The index is compiled mainly in subjects: i.e. Atomic matters, Electricity, Magnetism, Static electricity etc., but wherever possible individual cross-references have been added to assist the reader. If a subject is not found listed in its own right, reference should be made to the branch of physics where it would be discussed.